Acute Kidney Injury

This book has been made possible by the generous support of

Contributions to Nephrology

Vol. 156

Series Editor

Claudio Ronco, Vicenza

Acute Kidney Injury

Volume Editors

Claudio Ronco, Vicenza
Rinaldo Bellomo, Melbourne, Vic.
John A. Kellum, Pittsburgh, Pa.

63 figures, 2 in color, and 40 tables, 2007

Basel · Freiburg · Paris · London · New York ·
Bangalore · Bangkok · Singapore · Tokyo · Sydney

Contributions to Nephrology

(Founded 1975 by Geoffrey M. Berlyne)

....

Claudio Ronco
Department of Nephrology
St. Bortolo Hospital
I-36100 Vicenza (Italy)

Rinaldo Bellomo
Department of Intensive Care
Austin Hospital
Melbourne, Vic. 3084 (Australia)

John A. Kellum
The CRISMA Laboratory
Department of Critical Care Medicine
University of Pittsburgh, 3550, Terrace Street
Pittsburgh, PA 15261 (USA)

Library of Congress Cataloging-in-Publication Data

Acute kidney injury / volume editor[s], Claudio Ronco, Rinaldo Bellomo, John A. Kellum.
     p. ; cm. – (Contributions to nephrology, ISSN 0302-5144 ; v. 156)
   Includes bibliographical references and index.
   ISBN-13: 978-3-8055-8271-1 (hard cover : alk. paper)
   1. Acute renal failure. 2. Kidney–Wounds and injuries. 3. Intensive care nephrology. I. Ronco, C. (Claudio), 1951- II. Bellomo, R. (Rinaldo), 1956- III. Kellum, John A. IV. Series.
     [DNLM: 1. Kidney Failure, Acute. 2. Critical Care. 3. Kidney–injuries. 4. Kidney Failure, Acute–therapy. 5. Renal Replacement Therapy. 6. Sepsis–complications. W1 CO778UN v.156 2007 / WJ 342 A1885 2007]
   RC918.R4A327 2007
   616.6′14–dc22
                                                2007012324

Bibliographic Indices. This publication is listed in bibliographic services, including Current Contents® and Index Medicus.

Disclaimer. The statements, options and data contained in this publication are solely those of the individual authors and contributors and not of the publisher and the editor(s). The appearance of advertisements in the book is not a warranty, endorsement, or approval of the products or services advertised or of their effectiveness, quality or safety. The publisher and the editor(s) disclaim responsibility for any injury to persons or property resulting from any ideas, methods, instructions or products referred to in the content or advertisements.

Drug Dosage. The authors and the publisher have exerted every effort to ensure that drug selection and dosage set forth in this text are in accord with current recommendations and practice at the time of publication. However, in view of ongoing research, changes in government regulations, and the constant flow of information relating to drug therapy and drug reactions, the reader is urged to check the package insert for each drug for any change in indications and dosage and for added warnings and precautions. This is particularly important when the recommended agent is a new and/or infrequently employed drug.

All rights reserved. No part of this publication may be translated into other languages, reproduced or utilized in any form or by any means electronic or mechanical, including photocopying, recording, microcopying, or by any information storage and retrieval system, without permission in writing from the publisher.

© Copyright 2007 by S. Karger AG, P.O. Box, CH–4009 Basel (Switzerland)
www.karger.com
Printed in Switzerland on acid-free paper by Reinhardt Druck, Basel
ISSN 0302–5144
ISBN 978–3–8055–8271–1

Contents

XI Preface
Ronco, C. (Vicenza); Bellomo, R. (Melbourne, Vic.); Kellum, J.A. (Pittsburgh, Pa.)

Critical Care Nephrology Issues

1 Pre-Renal Azotemia: A Flawed Paradigm in Critically Ill Septic Patients?
Bellomo, R.; Bagshaw, S.; Langenberg, C. (Melbourne, Vic.); Ronco, C. (Vicenza)

10 The Concept of Acute Kidney Injury and the RIFLE Criteria
Kellum, J.A. (Pittsburgh, Pa.); Bellomo, R. (Melbourne, Vic.); Ronco, C. (Vicenza)

17 The Liver and the Kidney: Mutual Clearance or Mixed Intoxication
Arroyo, V. (Barcelona)

24 Critical Care Nephrology: A Multidisciplinary Approach
Vincent, J.-L. (Brussels)

Epidemiology and Pathogenesis of AKI, Sepsis and MOF

32 Incidence, Classification, and Outcomes of Acute Kidney Injury
Hoste, E.A.J. (Ghent/Pittsburgh, Pa.); Kellum, J.A. (Pittsburgh, Pa.)

39 Pathophysiology of Acute Kidney Injury: Roles of Potential Inhibitors of Inflammation
Bonventre, J.V. (Boston, Mass.)

47 Sepsis and Multiple Organ Failure
Pinsky, M.R. (Pittsburgh, Pa.)

64 Classification, Incidence, and Outcomes of Sepsis and Multiple Organ Failure
Vincent, J.-L.; Taccone, F.; Schmit, X. (Brussels)

Evaluation of Illness Severity

75 Genetic Polymorphisms in Sepsis- and Cardiopulmonary Bypass-Associated Acute Kidney Injury
Haase-Fielitz, A.; Haase, M. (Melbourne, Vic./Berlin);
Bellomo, R. (Melbourne, Vic.); Dragun, D. (Berlin)

92 Predictive Capacity of Severity Scoring Systems in the ICU
Schusterschitz, N.; Joannidis, M. (Innsbruck)

101 Determining the Degree of Immunodysregulation in Sepsis
Cavaillon, J.-M.; Adib-Conquy, M. (Paris)

Metabolism, Electrolytes and Acid-Base Disorders

112 Nutritional Management in Acute Illness and Acute Kidney Insufficiency
Leverve, X.M. (Grenoble); Cano, N.J.M. (Marseille)

119 Fundamentals of Oxygen Delivery
Yassin, J.; Singer, M. (London)

133 Principals of Hemodynamic Monitoring
Polanco, P.M.; Pinsky, M.R. (Pittsburgh, Pa.)

158 Acid-Base Disorders and Strong Ion Gap
Kellum, J.A. (Pittsburgh, Pa.)

167 Fluid Resuscitation and the Septic Kidney: The Evidence
Licari, E.; Calzavacca, P. (Melbourne, Vic.); Ronco, C. (Vicenza);
Bellomo, R. (Melbourne, Vic.)

Nursing Issues in Critical Care Nephrology

178 Factors Affecting Circuit Patency and Filter 'Life'
Baldwin, I. (Melbourne, Vic.)

185 Starting up a Continuous Renal Replacement Therapy Program on ICU
De Becker, W. (Leuven)

191 Is There a Need for a Nurse Emergency Team for Continuous Renal Replacement Therapy?
Baldwin, I. (Melbourne, Vic.)

197 Information Technology for CRRT and Dose Delivery Calculator
Ricci, Z. (Rome/Vicenza); Ronco, C. (Vicenza)

Early Diagnosis and Prevention of AKI

203 Emerging Biomarkers of Acute Kidney Injury
Devarajan, P. (Cincinnati, Ohio)

213 Diagnosis of Acute Kidney Injury: From Classic Parameters to New Biomarkers
Bonventre, J.V. (Boston, Mass.)

220 Endotoxin and Cytokine Detection Systems as Biomarkers for Sepsis-Induced Renal Injury
Opal, S.M. (Providence, R.I.)

227 Quantifying Dynamic Kidney Processes Utilizing Multi-Photon Microscopy
Molitoris, B.A.; Sandoval, R.M. (Indianapolis, Ind.)

236 Diuretics in the Management of Acute Kidney Injury: A Multinational Survey
Bagshaw, S.M. (Melbourne, Vic./Edmonton, Alta.); Delaney, A. (Sydney/St. Leonards); Jones, D. (Melbourne, Vic.); Ronco, C. (Vicenza); Bellomo, R. (Melbourne, Vic.)

250 Stem Cells in Acute Kidney Injury
Bussolati, B.; Camussi, G. (Turin)

Practice Patterns for RRT in the ICU

259 Anticoagulation Options for Patients with Heparin-Induced Thrombocytopenia Requiring Renal Support in the Intensive Care Unit
Davenport, A. (London)

267 Nutritional Support during Renal Replacement Therapy
Chioléro, R.; Berger, M.M. (Lausanne)

275 Vascular Access for HD and CRRT
Schetz, M. (Leuven)

287 Dialysate and Replacement Fluid Composition for CRRT
Aucella, F.; Di Paolo, S.; Gesualdo, L. (Foggia)

297 Results from International Questionnaires
Ricci, Z.; Picardo, S. (Rome); Ronco, C. (Vicenza)

Which Treatment for AKI in the ICU

304 Intermittent Hemodialysis for Renal Replacement Therapy in Intensive Care: New Evidence for Old Truths
Van Biesen, W.; Veys, N.; Vanholder, R. (Ghent)

309 Continuous Renal Replacement in Critical Illness
Ronco, C.; Cruz, D. (Vicenza); Bellomo, R. (Melbourne, Vic.)

320 Sustained Low-Efficiency Dialysis
Tolwani, A.J.; Wheeler, T.S.; Wille, K.M. (Birmingham, Ala.)

325 The Role of the International Society of Nephrology/Renal Disaster Relief Task Force in the Rescue of Renal Disaster Victims
Vanholder, R.; Van Biesen, W.; Lameire, N. (Ghent); Sever, M.S. (Istanbul)

Extracorporeal Treatment for Specific Indications

333 Renal Replacement Therapy for the Patient with Acute Traumatic Brain injury and Severe Acute Kidney Injury
Davenport, A. (London)

340 Cardiopulmonary Bypass-Associated Acute Kidney Injury: A Pigment Nephropathy?
Haase, M.; Haase-Fielitz, A. (Melbourne, Vic./Berlin); Bagshaw, S.M. (Melbourne, Vic.); Ronco, C. (Vicenza); Bellomo, R. (Melbourne, Vic.)

354 CRRT Technology and Logistics: Is There a Role for a Medical Emergency Team in CRRT?
Honoré, P.M. (Ottignies-Louvain-la-Neuve); Joannes-Boyau, O. (Pessac); Gressens, B. (Ottignies-Louvain-la-Neuve)

365 Continuous Hemodiafiltration with Cytokine-Adsorbing Hemofilter in the Treatment of Severe Sepsis and Septic Shock
Hirasawa, H.; Oda, S. (Chiba); Matsuda, K. (Yamanashi)

371 Blood and Plasma Treatments: High-Volume Hemofiltration – A Global View
Honoré, P.M. (Ottignies-Louvain-la-Neuve); Joannes-Boyau, O. (Pessac); Gressens, B. (Ottignies-Louvain-la-Neuve)

387 Blood and Plasma Treatments: The Rationale of High-Volume Hemofiltration
Honoré, P.M. (Ottignies-Louvain-la-Neuve); Joannes-Boyau, O. (Pessac); Gressens, B. (Ottignies-Louvain-la-Neuve)

396 Liver Support Systems
    Santoro, A.; Mancini, E.; Ferramosca, E.; Faenza, S. (Bologna)

405 Coupled Plasma Filtration Adsorption
    Formica, M.; Inguaggiato, P.; Bainotti, S. (Cuneo); Wratten, M.L. (Mirandola)

411 Albumin Dialysis and Plasma Filtration Adsorption Dialysis System
    Nalesso, F.; Brendolan, A.; Crepaldi, C.; Cruz, D.; de Cal, M. (Vicenza); Bellomo, R. (Melbourne, Vic.); Ronco, C. (Vicenza)

419 Renal Assist Device and Treatment of Sepsis-Induced Acute Kidney Injury in Intensive Care Units
    Issa, N.; Messer, J.; Paganini, E.P. (Cleveland, Ohio)

428 Renal Replacement Therapy in Neonates with Congenital Heart Disease
    Morelli, S.; Ricci, Z.; Di Chiara, L.; Stazi, G.V.; Polito, A.; Vitale, V.; Giorni, C.; Iacoella, C.; Picardo, S. (Rome)

New Trials and Meta-Analyses

434 The DOse REsponse Multicentre International Collaborative Initiative (DO-RE-MI)
    Monti, G. (Milan); Herrera, M. (Malaga); Kindgen-Milles, D. (Düsseldorf); Marinho, A. (Porto); Cruz, D. (Vicenza); Mariano, F. (Turin); Gigliola, G. (Cuneo); Moretti, E. (Bergamo); Alessandri, E. (Rome); Robert, R. (Poitier); Ronco, C. (Vicenza)

444 Clinical Effects of Polymyxin B-Immobilized Fiber Column in Septic Patients
    Cruz, D.N. (Vicenza/Quezon City); Bellomo, R. (Melbourne, Vic.); Ronco, C. (Vicenza)

452 Author Index

454 Subject Index

Preface

Four million people will die this year of a condition whose pathophysiology we do not understand and for which no effective treatment exists. Millions more will sustain complications and prolonged hospitalizations. Acute kidney injury (AKI) is complex syndrome for which treatment is lacking and understanding is limited. Defined as an abrupt change in serum creatinine and/or urine output and classified according the RIFLE (Risk, Injury, Failure, Loss and End-stage kidney disease) criteria, AKI is associated with a more than twofold increase in the risk of death in hospital – even after controlling for other conditions. When severe enough to require renal replacement therapy, AKI results in hospital mortality rates of approximately 60%. Yet, even this severe form of AKI is surprisingly common – nearly 6% of all patients admitted to the ICU.

Moreover, as many as two thirds of patients admitted to the ICU have some evidence of AKI and virtually all may be at risk of this condition. Sepsis, shock, advanced age and exposure to nephrotoxins lead the list of risk factors and many patients have more than one. As such, it is absolutely clear that in order for patients to receive optimal care, the treating physician needs a detailed working knowledge of multiple aspects of care so that appropriate multidisciplinary assistance is sought at the right time and new techniques of organ support are applied in a safe, timely and effective way. In this volume we have combined the contributions of experts in various fields to tackle some of the fundamental and complex aspects of AKI from pathophysiology to epidemiology to diagnosis and treatment; from emerging biomarkers to genetic polymorphisms. We have also included contributions which focus on the many complications of AKI and comorbid conditions encountered in patients with

AKI. From abnormalities in oxygen delivery, hemodynamics and acid-base balance to multi-organ failure, leading experts cover the fundamentals as well as the latest developments.

Because the immune response to infection is central in determining organ injury, this volume also focuses on the role of immune dysregulation in determining renal and lung injury, on the role of immune mediators in inducing dysregulation of the immune response, and on the role of genetics in determining such a response. The roles of nutritional support, metabolic management and fluid resuscitation in modulating the immune response and influencing patient outcomes are also considered. As extracorporeal therapies are being increasingly used in the care of these complex patients, we focus on important technical aspects of such therapies, including vascular access, anticoagulation, and fluid composition as well as the logistics of starting continuous renal replacement therapy programs and keeping them running. As the choice of treatment modality remains controversial, we also discuss different approaches to renal support from intermittent dialysis to continuous therapies and hybrid techniques. Finally, we conclude with a description of advanced extracorporeal techniques of organ support and discuss their role in the management of sepsis and AKI in the context of an overall strategy of multi-organ failure management.

The overall aim of this volume of the *Contributions to Nephrology* series is to provide the medical community involved in the care of critically ill patients with AKI with a practical and up-to-date summary of current knowledge and technology as well as a fundamental understanding of pathogenesis and likely future developments in this field. Just as importantly, this volume serves to challenge and reexamine the fundamental underlying assumptions we hold with regard to critical illness in general and AKI in particular. By reexamining age-old paradigms such as 'pre-renal azotemia' and looking to redefine the concepts of acute renal disease, we can expect to stumble and fall but in the end find ourselves in a new and better place. Critical care nephrology is an interdisciplinary field and it is through the continued work of basic, clinical, and translational researchers from numerous disciplines, and clinicians who will always drive the field forward, that together we will realize our ambition to improve the standard of care for patients with AKI, worldwide.

Claudio Ronco, Vicenza
Rinaldo Bellomo, Melbourne, Vic.
John A. Kellum, Pittsburgh, Pa.

Pre-Renal Azotemia: A Flawed Paradigm in Critically Ill Septic Patients?

Rinaldo Bellomo[a], Sean Bagshaw[a], Christoph Langenberg[a], Claudio Ronco[b]

[a]Department of Intensive Care, Austin Hospital, Melbourne, Vic., Australia;
[b]Department of Nephrology, St. Bortolo Hospital, Vicenza, Italy

Abstract

The term pre-renal azotemia (or on occasion 'pre-renal renal failure') is frequently used in textbooks and in the literature to indicate an acute syndrome characterized by the presence of an increase in the blood concentration of nitrogen waste products (urea and creatinine). This syndrome is assumed to be due to loss of glomerular filtration rate but is not considered to be associated with histopathological renal injury. Thus, the term is used to differentiate 'functional' from 'structural' acute kidney injury (AKI) where structural renal injury is taken to indicate the presence of so-called acute tubular necrosis (ATN). This paradigm is well entrenched in nephrology and medicine. However, growing evidence from experimental animal models, systematic analysis of the human and experimental literature shows that this paradigm is not sustained by sufficient evidence when applied to the syndrome of septic AKI, especially in critically ill patients. In such patients, several assumptions associated with the 'pre-renal azotemia paradigm' are violated. In particular, there is no evidence that ATN is the histopathological substrate of septic AKI, there is no evidence that urine tests can discriminate 'functional' from 'structural' AKI, there is no evidence that any proposed differentiation leads or should lead to different treatments, and there is no evidence that relevant experimentation can resolve these uncertainties. Given that septic AKI of critical illness now accounts for close to 50% of cases of severe AKI in developed countries, these observations call into question the validity and usefulness of the 'pre-renal azotemia paradigm' in AKI in general.

Copyright © 2007 S. Karger AG, Basel

Introduction

Acute renal failure or, to use the more recent term, acute kidney injury (AKI) is a common syndrome in hospitals [1] and intensive care units [2]. It is associated with a mortality rate which varies according to definition. However, if the RIFLE consensus definition is applied [3, 4], the hospital mortality of severe injury (RIFLE category F) approximates 50% [1]. Although, it has been argued that patients die with AKI rather than of AKI and that AKI is simply an expression of illness severity, consistent and strong evidence supports the notion that AKI has an independent impact on outcome, even after all other possible variables determining outcome have been corrected for [5]. Finally, because of the frequent need for extracorporeal supportive therapies, the treatment of established AKI is complex, labor intensive and costly [6]. Clearly, therefore, AKI is a major issue in acute medicine and increasing our understanding of its pathogenesis is vital. As the assessment of renal function in man is only indirect and limited by the inability to obtain tissue, to reliably and continuously measure renal blood flow and the need to derive indirect information on glomerular filtration rate and/or tubular cell status through urine analysis, animal experimentation has been used to advance our understanding of AKI. Unfortunately, the animal models that have been used to study AKI have been predominantly based on ischemia (occlusion of the renal artery for a given period of time) or the administration of a selected group of toxins instead of sepsis where renal blood flow may be increased [3, 7, 8]. These ischemic/toxic models have limited relevance to clinical reality where renal artery occlusion is an uncommon cause of AKI and the administration of toxins is a similarly less common trigger of renal injury, especially if severe [2]. In the clinical environment, sepsis is now the number one trigger of AKI [2] especially in critically ill patients who now make up the vast majority of cases of severe AKI [9]. Yet realistic models of septic AKI, which fully simulate the clinical situation, have not yet been developed. Most studies have been short-term in design and have not reported any information on systemic hemodynamics [7]. Importantly, many such studies have shown clear evidence of a hypodynamic state thus confounding septic AKI with AKI secondary to cardiogenic shock and renal ischemia [7].

Despite all the serious limitations in our understanding of AKI in general and septic AKI in particular, a variety of paradigms have emerged over the last 30 years to explain AKI, its pathogenesis and its histopathology. They have also been progressively incorporated into textbooks of medicine, critical care and nephrology [10–13]. These paradigms provide explanations and descriptions of the underlying functional and structural changes associated with AKI and offer a diagnostic and prognostic map to guide clinicians in the interpretation of their findings. They have progressively become 'standard teaching' and, finally, have ossified into dogma. In this article, we will review some of the problems associated with one such paradigm (the 'pre-renal azotemia' paradigm) and argue that such a paradigm suffers from major flaws when applied to critically ill septic patients. We will then further argue that, given that critically ill septic patients make up close to 50% of all patients with severe AKI, a paradigm that cannot apply to such patients cannot be said to reasonably and adequately apply to AKI in general. Finally, we will argue that a paradigm with such serious flaws is in need of much rethinking and reviewing.

The Meaning of 'Pre-Renal'

Any physician in the world is or should be familiar with the term 'pre-renal'. The term is widely used in textbooks and the daily practice of clinical medicine to indicate that the events that might be affecting kidney function are occurring 'outside' the kidney itself and do not include obstruction of the urinary excretory pathways (so-called 'post-renal' conditions) [11–13]. However, the use of the term 'pre-renal' often becomes less rigorous as it is also applied to describe two different but overlapping processes. The first process is the presence of a trigger outside the kidney which causes AKI. Thus, for example, a patient with myocardial infarction and low cardiac output syndrome and rising serum creatinine with oliguria is said to have 'pre-renal' renal failure. Similarly, a patient with severe infectious gastroenteritis and major fluid losses through diarrhea who has an acute increase in serum creatinine is also said to have 'pre-renal' renal failure. The term 'pre-renal' is used here to emphasize the fact that no 'parenchymal' disease (e.g. glomerulonephritis or interstitial nephropathy) is responsible for the changes in renal function. The second process describes a condition where the azotemia (increased urea and creatinine concentrations) seen in a given patient in the clinical scenario given above is also taken not only to involve no intrinsic parenchymal disease (see above) but also to be 'functional' and thus relatively rapidly reversible. Here the functional aspect is emphasized to differentiate this pathophysiological state from one where the pre-renal injury (low cardiac output or severe volume depletion as described above) has caused 'structural' injury or so-called acute tubular necrosis (ATN; fig. 1). This conceptual framework assumes that, of course, ATN cannot arise de novo but must by necessity be secondary to sustained or uncorrected functional injury which has progressed to the point of cell necrosis (hence the term ATN). We believe that, this paradigm, although simple and attractive, should still trigger several logical and pertinent questions before acceptance:
- How does one know when 'functional' AKI (pre-renal azotemia) becomes 'structural' AKI (ATN)?
- How much structural injury does one need to be able to say that the kidney has transitioned from 'pre-renal azotemia' to ATN?

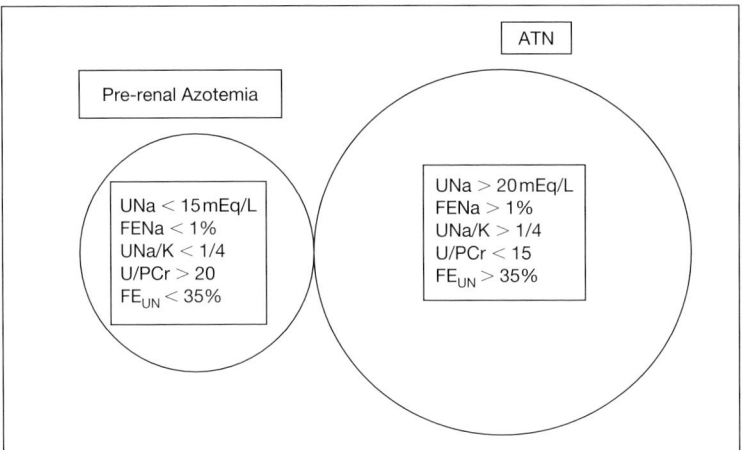

Fig. 1. Convention view of pre-renal azotemia and acute tubular necrosis (ATN). UNa = Urinary sodium; FENa = fractional excretion of sodium; UNa/K = urinary sodium/potassium ratio; U/PCr = urinary to plasma creatinine ratio; FE_{UN} = fractional excretion of urea.

- What are the treatment implications of being able to accurately classify AKI into these two types?
- What experimental data do we have that in the most common condition leading to AKI (severe sepsis) these pathophysiological states exist?

We will try and use the available evidence and reasoning to address these questions.

How Do We Know when Pre-Renal Azotemia Becomes ATN?

Logically, in order to know when a clinical entity or syndrome (pre-renal azotemia) becomes a histopathological entity (ATN), one first needs to have widely accepted definitions of each. Unfortunately this is currently not the case. Pre-renal azotemia does not come with 'consensus criteria' (whether sufficiently based on evidence or not). Thus it is impossible to know/test/prove that a given clinician's pre-renal azotemia is not another clinician's ATN. This lack of data and consensus on diagnostic criteria is a huge problem as it makes it impossible to know the epidemiology, natural history, inter-rater reliability and robustness of this paradigm. Further, if we cannot say with some degree of reproducibility whether a patient does or does not have pre-renal azotemia, how can we say the he or she does or does not have ATN? Even more problematically, if we say (as the textbook and literature consistently do) that pre-renal azotemia can in several cases progress to ATN, how can we decide when the line between these two conditions that we cannot define has actually been crossed?

The protagonists of the pre-renal azotemia/ATN paradigm will say that the answer is simple: either a biopsy will provide the diagnosis or the clinical course will. Yet both responses are flawed. First, we have absolutely no renal biopsy series from critically ill patients with clinically suspected ATN let alone 'pre-renal azotemia'. Thus, we do not actually know that ATN is the histopathological substrate of non-resolving AKI, which, presumably, most clinicians would classify as 'ATN' and/or, conversely, that it is not the histological substrate of rapidly resolving AKI, which presumably, most clinicians would classify as 'pre-renal azotemia'. Second, early postmortem studies of patients in the intensive care unit (ICU) with sepsis and AKI sufficiently severe to cause death show that, in >90% cases, the histopathology of the kidney is normal (Hotchkiss CCM). If people who die of sepsis with AKI have normal renal histopathology, how can people who are alive with it be expected to have ATN? Given that sepsis accounts for close to 50% of cases of severe AKI in the ICU, this is a huge problem. Pre-renal azotemia would then be separated from a diagnosis (ATN) that does not exist in the largest group of patients supposed to have it. If we have no evidence of ATN in sepsis, how can we know that septic patients with sustained or non-resolving AKI have ATN as the histopathological substrate while patients with less sustained more rapidly resolving AKI do not? Again, it is worthy of note that no specific meaning is allocated to words like 'sustained' or 'non-resolving' with values between 48 and 72 h being most commonly used but not universally so [10, 14–18]. Finally, if there is an assumed link between the duration of progressive AKI/azotemia and ATN, how long should this be? Why 72 h? Why 48? Why 24? Why not 56? The problems with these nosological entities are obvious.

How Much Structural Injury Is Enough to Change from 'Pre-Renal Azotemia' into 'ATN'?

The protagonists of the pre-renal azotemia/ATN paradigm will see the previous arguments as insufficient to affect their view of AKI, its course and histology. They will say that ATN is a patchy disease that will be easily missed by a histopathological analysis not specifically searching for its presence and that it can be easily missed by using renal biopsy methods to diagnose it. In other words, one can never 'refute' its presence to believers. Even accepting this scientifically untenable position, one is still left asking the question, how much necrosis of tubular cells is needed to diagnose ATN? Is one necrotic cell per low-powered microscopy field enough? Or does one need 10? How does the 'necrotic cell count' correlate with the duration of AKI and its ability to resolve in less than 48 h in response to appropriate therapy? Does a patient with 100 necrotic cells per low-powered field who improves with appropriate treatment have pre-renal azotemia or ATN? Could he/she have both simultaneously? What does 'improvement in response to therapy' mean? A return to pre-illness creatinine levels? A halt in the progression of azotemia? All of these issues remain unresolved but cannot be ignored.

What Are the Treatment Implications of Pre-Renal Azotemia Compared with ATN?

Many of the issues raised above may be easily seen as 'academic' rather than clinical. They appear to deal with scientific rigor rather than the messy but tangible world of clinical medicine. As such they miss the point. In the clinical world boundaries are imperfect and recognized as such by physicians. In the clinical world, paradigms, no matter how flawed, still work reasonably well in guiding clinicians to the right therapeutic measure. In the clinical world, flawed paradigms still work a bit like an old map that does not accurately direct the driver to the right place the way a sophisticated electronic navigation system might, but is always better than no map at all. In other words, in clinical medicine a 'flawed paradigm' is better than no paradigm and always much better than academic skepticism. The problem with such views is that they are not sustained by sufficient evidence of therapeutic implications. Would a clinician stop resuscitating a volume-depleted patient with gastroenteritis if he/she thought that the patient had established ATN instead of pre-renal azotemia? We hope not. Would a cardiologist not seek immediate revascularization, adequate inotropic support, pacing if necessary and/or intra-aortic balloon counterpulsation to treat a patients with myocardial infarction and a rapidly rising creatinine if he or she thought the patient had ATN instead of pre-renal azotemia? We hope not. Would a critical care physician and/or nephrologist alone or together not optimize intravascular volume, cardiac output, oxygenation and blood pressure while administering appropriate antibiotics and seeking to identify the focus of infection in a patient with septic AKI if they thought it was due to ATN instead of pre-renal azotemia? We hope not.

We cannot see any evidence or reason why the paradigm which separates so-called pre-renal azotemia from so-called ATN can, at this stage, have any useful therapeutic implications.

What Experimental Data Do We Have that ATN Occurs in Severe Sepsis?

The protagonists of the pre-renal azotemia/ATN paradigm might be forced to acknowledge that the evidence for the presence of ATN in human septic AKI simply does not exist and ATN or its absence can only be inferred. This is because the factual data are clear. However, they can still argue that, even in the absence of a tissue diagnosis, the presence or absence of ATN can be confirmed or refuted by means of urinalysis. This approach introduces a new paradigm, which is also widely published in textbooks and the literature [11–18], namely that the measurement of analytes in urine (urea, creatinine, sodium, potassium) and/or the calculations of derived variables (fractional excretion of sodium, fractional excretion of urea, various analyte ratios and osmolarity) can be used to accurately infer preserved tubular function (which equate to pre-renal azotemia) or lost tubular function (which equates to ATN) [14–18].

This argument is time-honored, widely accepted and is supported by several studies [14–18]. However, the problem with this paradigm is that it would ideally require histopathological confirmation that there is or there is not ATN for a given set of urinary findings. Unfortunately no such data exist in humans to confirm the validity of the urinalysis-based approach. Surely the presence of tubular cell casts proves that there is ATN. Unfortunately, it does not. The cells in these casts are often viable (accordingly they do not necessarily indicate 'necrosis') [19]. Surely, evidence of clearly abnormal urinary biochemistry proves that tubular function is lost and that tubular damage (ATN) has taken place. Unfortunately, it does not. Animal experiments show that loss of cell polarity and relocation of ATPase activity are likely responsible for such changes, not tubular cell necrosis [20]. Surely, abnormal urinary biochemistry and microscopy can at least be used to distinguish between sustained (see above for problems with this concept) from rapidly resolving AKI. This may be true in ward patients (even though the data are of limited strength) but is certainly not true of septic critically ill patients. For such patients, often on vasopressor drugs, receiving large amounts of fluid and often diuretics as well, the data have been systematically reviewed and found lacking [10].

Even in experimental septic models where the timing of injury is known, the evidence is simply absent that urinary biochemistry or microscopy can identify the type, course and outcome of AKI [14].

If we cannot use data derived from urine analysis to identify the 'type' of AKI (pre-renal azotemia vs. ATN or sustained vs. rapidly resolving) in septic patients who make up close to 50% of all patients, how can the accuracy of such tests be trusted overall?

Conclusions

In this article, we have argued that the paradigm of pre-renal azotemia is seriously flawed when applied to septic AKI. Given that close to 50% of cases of severe AKI are associated with severe sepsis, it appears likely that this paradigm may be flawed in a more general sense. Until our understanding of the histopathology, clinical course, response to therapy and pathogenesis of septic AKI has increased sufficiently, we believe it more intellectually honest to talk about the clinical syndrome of AKI without making any assumptions about the structural substrate or our ability to predict or reverse its course. If we believe that, within the syndrome of septic AKI, specific subgroups exist, which can be usefully identified by clinical tests, epidemiologically separated, shown to carry a different prognosis, require different treatments and have a different histopathology, then we should develop prospective working definitions, diagnostic criteria and outcome measures and test them in appropriately designed epidemiological and interventional studies. Until such time, we consider that the flaws of the current paradigm are too great to go unchallenged.

References

1. Uchino S, Bellomo R, Goldsmith D, Bates S, Ronco C: An assessment of the RIFLE criteria for acute renal failure in hospitalized patients. Crit Care Med 2006;34:1913–1917.
2. Uchino S, Kellum J, Bellomo R, et al: Acute renal failure in critically ill patients – a multinational, multicenter study. JAMA 2005;294:813–818.
3. Bellomo R, Ronco C, Kellum JA, Mehta RL, Palevsky P; Acute Dialysis Quality Initiative workgroup: Acute renal failure – definition, outcome measures, animal models, fluid therapy and information technology needs: the Second International Consensus Conference of the Acute Dialysis Quality Initiative (ADQI) Group. Crit Care 2004;8:R204–R212.
4. Bellomo R: Defining, quantifying and classifying acute renal failure. Crit Care Clin 2005;21: 223–237.
5. Metnitz PG, Krenn CG, Steltzer H, et al: Effect of acute renal failure requiring renal replacement therapy on outcome in critically ill patients. Crit Care Med 2002;30:2051–2058.
6. Bellomo R, Ronco C: Continuous hemofiltration in the intensive care unit. Crit Care 2000;4: 339–345.
7. Langenberg C, Bellomo R, May C, Wan L, Egi M, Morgera S: Renal blood flow in sepsis. Crit Care 2005;9:R363–R374.
8. Langenberg C, Wan L, Egi M, May CN, Bellomo R: Renal blood flow in experimental septic acute renal failure. Kidney Int 2006;69:1996–2002.

9   Cole L, Bellomo R, Silvester W, Reeves J: A prospective study of the epidemiology and outcome of severe acute renal failure patients treated in a 'closed' ICU model. Am J Resp Crit Care Med 2000;162:191–196.
10 Bagshaw SM, Langenberg C, Bellomo R: Urinary biochemistry and microscopy in septic acute renal failure: a systematic review. Am J Kidney Dis 2006;48:695–705.
11 Kasper DL, Braunwald E, Fauci AS, et al. (eds): Harrison's Principles of Internal Medicine, ed 15. New York, McGraw-Hill, 2001.
12 Davison A, Cameron JS, Grunfeld JP, Ponticelli C (eds): Oxford Textbook of Clinical Nephrology, ed 3. Oxford, Oxford University Press, 2005.
13 Fink MP, Abraham E, Vincent J-L, Kochanek PM (eds): Textbook of Critical Care, ed 5. Philadelphia, Elsevier, 2005.
14 Langenberg C, Wan L, Bagshaw SM, Moritoki E, May CN, Bellomo R: Urinary biochemistry in experimental septic acute renal failure. Nephrol Dial Transplant 2006;21:3389–3397.
15 Carvounis CP, Nisar S, Guro-Razuman S: Significance of the fractional excretion of urea in the differential diagnosis of acute renal failure. Kidney Int 2002;62:2223–2229.
16 Espinel CH: The FENa test. Use in the differential diagnosis of acute renal failure. JAMA 1976;236:579–581.
17 Perlmutter M, Grossman SL, Rothenberg S, Dobkin G: Urine/serum urea nitrogen ratio: simple test of renal function in acute azotemia and oliguria. JAMA 1959;170:1533–1537.
18 Miller TR, Anderson RJ, Linas SL, Henrich WL, Berns AS, Gabow PA, Schrier RW: Urinary diagnostic indices in acute renal failure: a prospective study. Ann Intern Med 1978;89:47–50.
19 Graber M, Lane B, Lamina R, Pastoriza-Munoz E: Bubble cells: renal tubular cells in the urinary sediment with characteristics of viability. J Am Soc Nephrol 1991;1:999–1004.
20 Kwon O, Nelson WJ, Sibley R, et al: Backleak, tight junctions and cell-cell adhesion in post-ischemic injury in the renal allograft. J Clin Invest 1998;101:2054–2064.

Prof. Rinaldo Bellomo
Department of Intensive Care, Austin Hospital
Melbourne, Vic. 3084 (Australia)
Tel. +61 3 9496 5992, Fax +61 3 9496 3932, E-Mail rinaldo.bellomo@austin.org.au

The Concept of Acute Kidney Injury and the RIFLE Criteria

John A. Kellum[a], *Rinaldo Bellomo*[b], *Claudio Ronco*[c]

[a]Department of Critical Care Medicine, University of Pittsburgh Medical Center, Pittsburgh, Pa., USA; [b]Department of Intensive Care and Department of Medicine, Austin Hospital and University of Melbourne, Melbourne, Vic., Australia; [c]Department of Nephrology, Ospedale San Bortolo, Vicenza, Italy

Abstract

Over last half century, the concept of acute renal failure has evolved and with it our estimates of the incidence, prevalence and mortality. Indeed, until very recently no standard definition of acute renal failure was available, and this lack of a common language created confusion and made comparisons all but impossible. In response to the need for a common definition and classification of acute renal failure, the Acute Dialysis Quality Initiative group of experts developed and published a set of consensus criteria for defining and classifying acute renal failure. These criteria which make up acronym 'RIFLE' classify renal dysfunction according to the degree of impairment present: risk (R), injury (I), and failure (F), sustained loss (L) and end-stage kidney disease (E). However, as these criteria were developed, a new concept immerged. Renal dysfunction was no longer only considered significant when it reached the stage of failure, but a spectrum from early risk to long-term failure was recognized and codified. Subsequent studies have validated these criteria in various populations and have shown that relatively mild dysfunction is associated with adverse outcomes. The term acute kidney injury has subsequently been proposed to distinguish this new concept from the older terminology of failure.

Copyright © 2007 S. Karger AG, Basel

Acute Kidney Injury: What's in a Name?

Would a syndrome of any other name still be as deadly? In 1993 the late Roger Bone penned the following words:

'Too often, the way we describe a disorder influences, and often limits, the way we think about that disorder' [1].

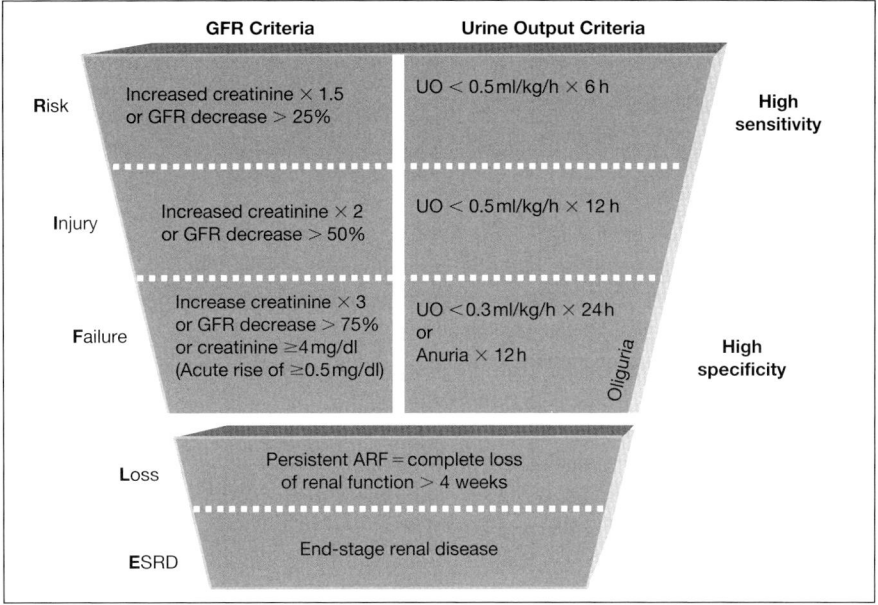

Fig. 1. The RIFLE criteria for AKI (used with permission). ARF = Acute renal failure; GFR = glomerular filtration rate; UO = urine output.

Although he was talking about sepsis, not renal failure, the similarities are striking. By focusing on 'failure' we might conclude that earlier phases are less important. By using the term 'acute tubular necrosis' we might assume that ischemia is a predominant pathophysiological mechanism. The entire concept of 'pre-renal azotemia' may have, in its description, influenced our thinking about a disorder which comprises not discrete entities but a dynamic spectrum from an early reversible condition to an established disease. Furthermore, given that acute renal failure (ARF) has been reported to affect from 1 to 25% of intensive care unit (ICU) patients and has lead to mortality rates from 15 to 60% [2–5], the existing terminology has not influenced our thinking uniformly.

Over the last few years the case for a consensus definition and a classification system has repeatedly been made [6, 7]. In these articles we have argued that the major aim of such a system would be to bring one of the major intensive care syndromes (acute kidney injury, AKI) to a standard of definition and a level of classification similar to that achieved by two other common ICU syndromes (sepsis and acute respiratory distress syndrome, ARDS). Following such advocacy and through the persistent work of the Acute Dialysis Quality Initiative (ADQI) group, such a system was developed through a broad consensus of experts [8]. The characteristics of this system are summarized in figure 1.

Since its publication, the RIFLE classification system has received much attention with more than 70,000 electronic hits for its publication site and more than 70 citations in 2 years. It has also spawned several investigations of its predictive ability, internal validity, robustness and clinical relevance in a variety of settings. This review will focus on some of the key findings from these investigations and their broader implications for the concept of AKI.

Validation Studies Using RIFLE

In one of the earliest studies to evaluate RIFLE, Abosaif et al. [9] sought to evaluate its sensitivity and specificity in patients in the ICU. These investigators studied 247 patients admitted to the ICU with a serum creatinine of $>150\,\mu mol/l$. This approach identified patients with renal dysfunction on the first day of admission but not those who developed ARF while in the ICU. Thus, its findings are limited in scope. The investigators found that the ICU mortality was greatest among patients classified as RIFLE class 'F' (failure) with a 74.5% mortality compared to 50% among those classified as class 'I' (injury), and 38.3% in those classified as RIFLE class 'R' (risk). In a significantly larger single-center multi-ICU study, Hoste et al. [10] evaluated RIFLE as an epidemiological and predictive tool in 5,383 critically ill patients. They found that AKI occurred in a staggering 67% of the patients with 12% achieving a maximum class of 'R', 27% a maximum 'I' class, and 28% a maximum 'F' class. Of the 1,510 patients who reached 'R', 56% progressed to either 'I' or 'F'. Patients with a maximum score of 'R' had a mortality rate of 8.8%, compared to 11.4% for 'I' and 26.3% for 'F'. On the other hand, patients who had no evidence of renal dysfunction had a mortality rate of 5.5%. Furthermore, maximum RIFLE class 'I' (hazard ratio of 1.4) and maximum RIFLE class 'F' (hazard ratio of 2.7) were independent predictors of hospital mortality after controlling for other variables known to predict outcome in critically ill patients. These findings are important as they involve a large cohort of heterogeneous critically ill patients. They suggest that the incidence of AKI is much greater than previously appreciated or reported. Even when only patients in RIFLE class 'F' are considered, the incidence of 28% is striking and makes this syndrome, at least in academic ICUs, more common than ALI or ARDS. The RIFLE classification also made it possible for these investigators to describe the progression of renal dysfunction over time. Of note, more than 50% of the patients progressed to a more severe form of renal impairment (RIFLE class 'I' or 'F') each carrying an independent increase in the risk of death.

A further assessment of the validity of the RIFLE classification has been performed in a heterogeneous population of hospitalized patients. Uchino et al. [11]

focused on the predictive ability of the RIFLE classification in a cohort of 20,126 patients admitted to a teaching hospital for >24 h over a 3-year period. The authors used the electronic laboratory database and the MDRD equation to classify patients into the 3 main RIFLE classes and followed them to hospital discharge or death to assess outcome. They separately analyzed patients who were admitted once and patients who were readmitted. For both groups the findings were similar. Close to 10% of patients achieved a maximum RIFLE class of 'R', close to 5% achieved a maximum of 'I' and close to 3.5% achieved a maximum of 'F'. There was an almost linear increase in hospital mortality with increasing RIFLE class with patients at 'R' having more than 3 times the mortality rate of patients without AKI. Patients with 'I' had close to twice the mortality of 'R' patients and patients with 'F' had 10 times the mortality rate of hospitalized patients without AKI. The investigators performed multivariate logistic regression analysis to test whether the RIFLE classification was an independent predictor of hospital mortality. They found that class 'R' carried an odds ratio of hospital mortality of 2.5, class 'I' of 5.4, and class 'F' of 10.1. These observations are particularly striking when compared to other important predictors of outcome such as admission to the ICU (odds ratio of 2.9), mechanical ventilation (odds ratio of 4.8), or admission to the hematology unit (odds ratio of 3.1). They suggest that the development of renal injury has greater prognostic implications that the need for ICU or mechanical ventilation and that the development of RIFLE class 'F' (renal failure) is one of the most powerful identifiable outcome predictors for hospitalized patients. The findings of this study are important in the assessment of the RIFLE system, not only because they link it to outcome but also because they validate its predictive ability in yet another large population of patients. It is important to note, however, that hospital mortality may not be the ideal outcome measure for both syndrome classification systems such as RIFLE or future interventional trials in such patients. A consensus view is that 60-day mortality may be more appropriate [8].

RIFLE in Specific Diagnostic Groups

Recently, the RIFLE criteria were evaluated in a cohort of 813 consecutive patients undergoing cardiac surgery in a university hospital in Finland [12]. These investigators found that 10.9% of patients were classified as 'R', and 3.5% as having 'I', while 5% developed 'F' criteria. Compared to a control population of patients who had no evidence of AKI (mortality 0.9%), mortality for the three groups R, I and F were 8, 21.4 and 32.5%, respectively. Furthermore, the AUC for the ROC for mortality at 90 days showed good discrimination for the RIFLE classification with a value of 0.824. Finally, multivariate logistic regression analysis identified the RIFLE classification as an independent risk factor for 90-day mortality. These observations are important because they suggest clinical usefulness and good discrimination in a large cohort of patients and because they are prospective in design. The RIFLE system has also very recently been tested in a unique population of patients requiring extracorporeal membrane oxygenation for post-cardiotomy cardiogenic shock [13]. These patients are uncommon but almost always develop evidence of renal impairment. In a study of 46 such patients, Lin et al. [13] found that there was a progressive increase in mortality with an increase in RIFLE level, and found that such a classification calibrated well when tested using the Hosmer-Lemeshow goodness-of-fit test and that it discriminated well with AUC for the ROC curve of 0.86. However, this population was small and the confidence intervals wide. In another restricted population of patients admitted to the ICU with major burns [14], the RIFLE system also performed well as it did in a recent cohort of patients with bone marrow transplantation [15]. The RIFLE system has also been adapted by the Acute Kidney Injury Network, a group involving all major international societies of critical care medicine and nephrology in an effort to improve the care of patients with AKI.

Conceptual Developments and Implications

The concept of AKI, as defined by RIFLE, creates a new paradigm. Rather than focusing exclusively on patients with ARF or on those who receive renal replacement therapy, the strong association of AKI with hospital mortality demands that we change the way we think about this disorder. In the study by Hoste et al. [10], only 14% of patients reaching RIFLE 'F' received renal replacement therapy, yet these patients experienced a hospital mortality more than 5 times that of the same ICU population without AKI. Is renal support underutilized or delayed? Are there other supportive measures that should be employed for these patients? Sustained AKI leads to profound alterations in fluid, electrolyte, acid-base and hormonal regulation. AKI results in abnormalities in the CNS, immune system and coagulation system. Many patients with AKI already have multisystem organ failure. What is the incremental influence of AKI on remote organ function and how does it affect outcome? A recent study by Levy et al. [16] examined outcomes of over 1,000 patients enrolled in the control arms of two large sepsis trials. Early improvement (within 24 h) in cardiovascular ($p = 0.0010$), renal ($p < 0.0001$), or respiratory ($p = 0.0469$) function was significantly related to survival. This study suggests that outcomes for patients with severe sepsis in the ICU are closely related to early resolution of AKI. While rapid resolution of AKI may simply be a marker of a good prognosis, it may also indicate a window of therapeutic opportunity to improve outcome in such patients.

Conclusion

AKI, as defined by the RIFLE criteria, is now recognized as a important ICU syndrome along side other syndromes used in ICU patients for the purpose of epidemiology and trial execution such as the ALI/ARDS consensus criteria [17] and the consensus definitions for SIRS/sepsis/severe sepsis and septic shock [18]. The introduction of the RIFLE system into the clinical arena represents a useful step in the field of critical care nephrology.

References

1. Bone RC: Why new definitions of sepsis and organ failure are needed. Am J Med 1993;95: 348–350.
2. Silvester W, Bellomo R, Cole L: Epidemiology, management, and outcome of severe acute renal failure of critical illness in Australia. Crit Care Med 2001;29:1910–1915.
3. Schaefer JH, Jochimsen F, Keller F, et al: Outcome prediction of acute renal failure in medical intensive care. Intensive Care Med 1991;17:19–24.
4. Liano F, Pascual J; Madrid Acute Renal Failure Study Cluster: Epidemiology of acute renal failure: a prospective, multicenter, community-based study. Kidney Int 1996;50:811–818.
5. Brivet FG, Kleinknecht DJ, Philippe L, et al: Acute renal failure in intensive care units – causes, outcome, and prognostic factors of hospital mortality: a prospective, multicenter study. Crit Care Med 1996;24:192–198.
6. Bellomo R, Kellum J, Ronco C: Acute renal failure: time for consensus. Intensive Care Med 2001;27:1685–1688.
7. Kellum JA, Mehta RL, Ronco C: Acute Dialysis Quality Initiative (ADQI); in Ronco C, Bellomo R, La Greca G (eds): Blood Purification in Intensive Care. Contrib Nephrol. Basel, Karger, 2001, vol 132, pp 258–265.
8. Bellomo R, Ronco C, Kellum JA, Mehta RL, Palevsky P; Acute Dialysis Quality Initiative workgroup: Acute renal failure – definition, outcome measures, animal models, fluid therapy and information technology needs: the Second International Consensus Conference of the Acute Dialysis Quality Initiative (ADQI) Group. Crit Care 2004;8:R204–R212.
9. Abosaif NY, Tolba YA, Heap M, et al: The outcome of acute renal failure in the intensive care unit according to RIFLE: model applicability, sensitivity, and predictability. Am J Kidney Dis 2005;46: 1038–1048.
10. Hoste EAJ, Clermont G, Kersten A, et al: RIFLE criteria for acute kidney injury are associated with hospital mortality in critically ill patients: a cohort analysis. Crit Care 2006;10:R73–R83.
11. Uchino S, Bellomo R, Goldsmith D, et al: An assessment of the RIFLE criteria for acute renal failure in hospitalized patients. Crit Care Med 2006;34:1913–1917.
12. Kuitunen A, Vento A, Suojaranta-Ylinen R, et al: Acute renal failure after cardiac surgery: evaluation of the RIFLE classification. Ann Thorac Surg 2006;81:542–546.
13. Lin C-Y, Chen Y-C, Tsai F-C, et al: RIFLE classification is predictive of short-term prognosis in critically ill patients with acute renal failure supported by extra-corporeal membrane oxygenation. Nephrol Dial Transplant 2006;21:2867–2873.

14 Lopes JA, Jorge S, Neves FC, et al: An assessment of the RIFLE criteria for acute renal failure in severely burned patients. Nephrol Dial Transplant 2007;22:285.
15 Lopes JA, Jorge S, Silva S, et al: An assessment of the RIFLE criteria for acute renal failure following myeloablative autologous and allogenic haematopoietic cell transplantation. Bone Marrow Transplant 2006;38:395.
16 Levy MM, Macias WL, Vincent JL, Russell JA, Silva E, Trzaskoma B, Williams MD: Early changes in organ function predict eventual survival in severe sepsis. Crit Care Med 2005;33: 2194–2201.
17 Bernard G, Artigas A, Briggham KL, et al: The American-European Consensus Conference on ARDS. Definitions mechanisms, relevant outcomes, and clinical trial coordination. Am J Respir Crit Care Med 1994;149:818–824.
18 American College of Chest Physicians/Society of Critical Care Medicine Consensus Committee: Definitions for sepsis and organ failure and guidelines for the use of innovative therapies in sepsis. Chest 1992;101:1658–1662.

John A. Kellum, MD
The CRISMA Laboratory
Department of Critical Care Medicine
University of Pittsburgh, 3550 Terrace Street
Pittsburgh, PA 15261 (USA)
Tel. +1 412 647 6966, Fax +1 412 647 3791, E-Mail kellumja@ccm.upmc.edu

The Liver and the Kidney: Mutual Clearance or Mixed Intoxication

Vicente Arroyo

Liver Unit, Institute of Digestive and Metabolic Diseases, Hospital Clinic, University of Barcelona, Barcelona, Spain

Abstract

Hepatorenal syndrome is a frequent complication in patients with cirrhosis, ascites and advanced liver failure. Its annual incidence in patients with ascites has been estimated at 8%. Hepatorenal syndrome is a functional renal failure due to low renal perfusion. Renal histology is normal or shows lesions that do not justify the decrease in glomerular filtration rate. The traditional concept is that hepatorenal syndrome is due to deterioration in circulatory function secondary to an intense vasodilation in the splanchnic circulation (peripheral arterial vasodilation hypothesis). Over the last decade, however, several features have suggested a much more complex pathogenesis. In this article new concepts on the pathogenesis of hepatorenal syndrome are reported, the current options for prophylaxis are shown, and the most applicable treatments are described.

Copyright © 2007 S. Karger AG, Basel

Hepatorenal Syndrome: Concept and Clinical Types

Hepatorenal syndrome (HRS) is a frequent complication in patients with cirrhosis, ascites and advanced liver failure. Its annual incidence in patients with ascites has been estimated at 8% [1]. HRS is a functional renal failure due to low renal perfusion. Renal histology is normal or shows lesions that do not justify the decrease in glomerular filtration rate (GFR). The traditional concept is that HRS is due to deterioration in circulatory function secondary to an intense vasodilation in the splanchnic circulation (peripheral arterial vasodilation hypothesis). Over the last decade, however, several features suggest a much more complex pathogenesis.

Type-1 HRS is characterized by a severe and rapidly progressive renal failure, which has been defined as a doubling of serum creatinine reaching a level of

>2.5 mg/dl in less than 2 weeks. Although type-1 HRS may develop spontaneously, it frequently occurs in close relationship with a precipitating factor, such as severe bacterial infection, mainly spontaneous bacterial peritonitis (SBP), gastrointestinal hemorrhage, major surgical procedure or acute hepatitis superimposed to cirrhosis. The association of HRS and SBP has been carefully investigated [2–4]. Type-1 HRS develops in approximately 25% of patients with SBP despite a rapid resolution of the infection with non-nephrotoxic antibiotics. Besides renal failure, patients with type-1 HRS associated with SBP show signs and symptoms of a rapid and severe deterioration in liver function (jaundice, coagulopathy and hepatic encephalopathy) and circulatory function (arterial hypotension, very high plasma levels of renin and norepinephrine) [2–5]. It is interesting to note that in contrast to SBP, sepsis related to other types of infection in patients with cirrhosis induces type-1 HRS only when there is a lack of response to antibiotics [6]. In most patients with sepsis unrelated to SBP responding to antibiotics, renal impairment, which is also a frequent event, is reversible. Without treatment, type-1 HRS is the complication of cirrhosis with the poorest prognosis with a median survival time after the onset of renal failure of only 2 weeks [1].

Type-2 HRS is characterized by a moderate (serum creatinine of <2.5 mg/dl) and slowly progressive renal failure. Patients with type-2 HRS show signs of liver failure and arterial hypotension but to a lesser degree than patients with type-1 HRS. The dominant clinical feature is severe ascites with poor or no response to diuretics (a condition known as refractory ascites). Patients with type-2 HRS are predisposed to develop type-1 HRS following SBP or other precipitating events [2–4]. The median survival of patients with type-2 HRS (6 months) is worse than that of patients with non-azotemic cirrhosis with ascites.

Pathogenesis of Type-2 HRS

Portal hypertension in cirrhosis is associated with arterial vasodilation in the splanchnic circulation due to the local release of nitric oxide and other vasodilatory substances. The peripheral arterial vasodilation hypothesis proposes that renal dysfunction and type-2 HRS in cirrhosis is related to this feature (fig. 1). Type-2 HRS represents the extreme expression in the progression of splanchnic arterial vasodilation. Homeostatic stimulation of the renin-angiotensin system, the sympathetic nervous system and antidiuretic hormone is very strong leading to intense renal vasoconstriction and a marked decrease in renal perfusion and GFR.

Most hemodynamic investigations in cirrhosis have been performed in non-azotemic patients with and without ascites, and the peripheral arterial

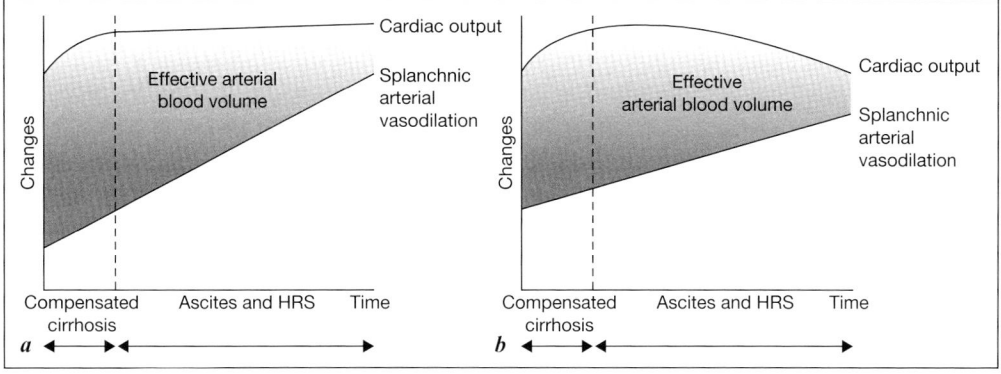

Fig. 1. Peripheral vasodilation hypothesis (***a***) and modified peripheral vasodilation hypothesis (***b***). According to this latter hypothesis, impairment in arterial blood volume in cirrhosis could be the consequence of a progression of splanchnic arterial vasodilation and of a decrease in cardiac output.

vasodilation hypothesis was based on these studies. It assumed that type-2 HRS develops when there is a hyperdynamic circulation and increased cardiac output. However, recent studies assessing cardiovascular function in patients with type-2 HRS show that cardiac output is significantly reduced compared to patients without HRS, suggesting that circulatory dysfunction associated with HRS is due not only to arterial vasodilation but also to a decrease in cardiac function (fig. 1) [7].

Type-1 HRS Associated with SBP: A Special Form of Multi-Organ Failure

Type-1 and type-2 HRS have two features in common. They occur in patients with cirrhosis and ascites, and renal failure is an important component of both syndromes. However, they show important differences. Type-2 HRS develops imperceptibly in patients with cirrhosis and ascites who are otherwise in a stable clinical condition. Circulatory function, although severely deteriorated, remains steady or progresses slowly during months as it occurs with the renal failure. Patients have advanced cirrhosis but the degree of liver failure is also stable. Hepatic encephalopathy is infrequent. The main clinical problem of patients with type-2 HRS is refractory ascites. In contrast, type-1 HRS is an extremely unstable condition. It frequently develops in the setting of an important clinical event that acts as a precipitating factor. On the other hand, there is

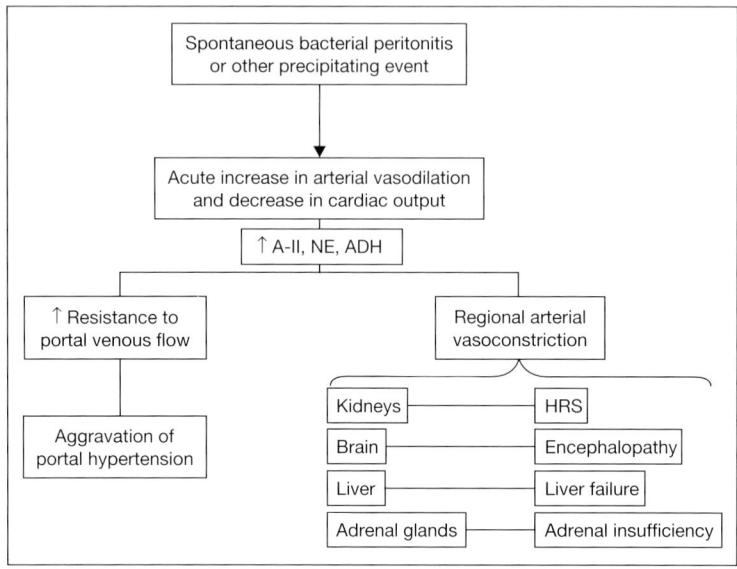

Fig. 2. Hepatorenal syndrome as part of multi-organ failure. A-II = Angiotensin II; NE = norepinephrine; ADH = antidiuretic hormone; HRS = hepatorenal syndrome.

a rapid deterioration in circulatory and renal function within days after the onset of the syndrome leading to severe arterial hypotension and acute renal failure with intense oliguria. Finally, there is also rapid deterioration in hepatic function, with an increase in jaundice and encephalopathy.

Recent studies in patients with SBP have presented data indicating that type-1 HRS represents a special form of acute multi-organ failure related to the rapid deterioration in circulatory function (fig. 2). The syndrome develops when there is a significant decrease in arterial pressure and a marked stimulation of the renin-angiotensin and sympathetic nervous systems in the absence of changes in systemic vascular resistance, which is consistent with an increase in arterial vasodilation obscured by the vascular effect of these vasoconstrictor systems. There is also an acute decrease in the cardiac output that contributes to effective arterial hypovolemia [5]. The mechanism of this impairment in cardiac function is complex. There is cirrhotic cardiomyopathy that decreases the cardiac response to stress conditions. On the other hand, in patients developing type-1 HRS associated with SBP there is a decrease in cardiopulmonary pressures suggesting a decrease in cardiac preload. Finally, despite the stimulation of the sympathetic nervous system there is no increase in heart rate indicating an impaired cardiac chronotropic function.

In addition to renal vasoconstriction, patients with type-1 HRS associated with SBP develop vasoconstriction in the intrahepatic circulation, with a marked reduction in hepatic blood flow and an increase in portal pressure [5]. The acute deterioration of hepatic function and hepatic encephalopathy may be related to this feature. Cerebral vascular resistance is increased in patients with decompensated cirrhosis. A reduction in cerebral blood flow could, therefore, play a contributory role in hepatic encephalopathy.

Prevention of Type-1 HRS Associated with SBP

Two randomized controlled studies in large series of patients have shown that type-1 HRS associated with SBP can be prevented either by selective intestinal decontamination in patients at high risk of developing SBP and HRS [8] or by circulatory support with intravenous albumin at SBP diagnosis [9].

The first study was performed in cirrhotic patients with a high risk of developing SBP and type-1 HRS [8]. Primary prophylaxis of SBP, using long-term oral norfloxacin, was given to patients with low protein ascites of <15 g/l and serum bilirubin of ≥ 3 mg/dl or serum creatinine of ≥ 1.2 mg/dl. Norfloxacin administration was associated with a significant decrease in the 1-year probability of developing SBP (7 vs. 61%) and type-1 HRS (28 vs. 41%) and with a significant increase in the 3-month and 1-year probabilities of survival (94 vs. 62 and 60 vs. 48%, respectively).

In the second study [9], the administration of albumin (1.5 g/kg i.v. at infection diagnosis and 1 g/kg i.v. 48 h later) to patients with cirrhosis and SBP markedly reduced the incidence of circulatory dysfunction and type-1 HRS (10% in patients receiving albumin vs. 33% in the control group). The hospital mortality rate (10 vs. 29%) and 3-month mortality rate (22 vs. 41%) were lower in patients receiving albumin. Albumin administration to cirrhotic patients with SBP induces not only an expansion of the plasma volume but also an increase in systemic vascular resistance. The efficacy of albumin in the prevention of type-1 HRS could, therefore, be related to both an increase in cardiac preload and cardiac output and a vasoconstrictor effect of albumin in the arterial circulation related to an attenuation of endothelial dysfunction [10].

Treatment of Type-1 HRS Associated with SBP

Several therapeutic measures can be used in patients developing type-1 HRS associated with SBP. The most effective is liver transplantation. However, the applicability of this procedure is low. The most applicable treatment consists of the administration of plasma volume expansion with albumin and vasoconstrictors. The insertion of a transjugular intrahepatic portosystemic shunt is another possibility that can be used either alone or after reversal of HRS with vasoconstrictors plus albumin. Finally, extracorporeal albumin dialysis can be used in these patients. Each of these treatments should be considered after the resolution of infection, since HRS may reverse following effective antibiotic treatment in a significant number of patients.

Liver Transplantation

Liver transplantation is the treatment of choice for patients with type-1 HRS [11]. Immediately after transplantation a further impairment in GFR may be observed and many patients require hemodialysis (35% of patients with HRS as compared with 5% of patients without HRS). After this initial impairment in renal function, GFR starts to improve. This moderate renal failure persists during follow-up.

The main problem of liver transplantation in type-1 HRS is its applicability. Due to their extremely short survival, most patients die before transplantation. The introduction of the MELD score, which includes serum creatinine, bilirubin and the international normalized ratio for listing, has partially solved the problem as patients with HRS are generally allocated the first places on the waiting list. Treatment of HRS with vasoconstrictors and albumin (see below) increases survival in a significant proportion of patients and therefore the number of patients reaching living transplantation, and decreases early morbidity and mortality after transplantation and prolongs long-term survival.

Vasoconstrictors and Albumin

The intravenous administration of vasoconstrictor agents (terlipressin or noradrenaline) and intravenous albumin over 1–2 weeks is an effective treatment for type-1 HRS [12–14]. The rate of positive response, as defined by a decrease in serum creatinine to <1.5 mg/dl, is reported in several pilot studies to be approximately 60%. Remarkably, type-1 HRS does not recur after discontinuation of treatment in most patients.

Reversal of type-1 HRS when terlipressin is given alone (25%) [13] is lower than that observed in studies in which vasoconstrictors are associated with intravenous albumin, suggesting that albumin administration is an important component in the pharmacological treatment of type-1 HRS. Two randomized controlled trials comparing terlipressin plus albumin versus terlipressin alone have recently been reported in abstract form. Their results confirm that terlipressin plus albumin is an effective therapy in patients with type-1 HRS and that reversal of HRS improves survival. However, the effectiveness of the treatment reported by these trials (40% rate of reversal of HRS) is lower than those reported in the pilot studies.

References

1. Gines A, Escorsell A, Gines P, et al: Incidence, predictive factors, and prognosis of the hepatorenal-syndrome in cirrhosis with ascites. Gastroenterology 1993;105:229–236.
2. Navasa M, Follo A, Filella X, et al: Tumor necrosis factor and interleukin-6 in spontaneous bacterial peritonitis in cirrhosis: relationship with the development of renal impairment and mortality. Hepatology 1998;27:1227–1232.
3. Follo A, Llovet JM, Navasa M, et al: Renal impairment after spontaneous bacterial peritonitis in cirrhosis – incidence, clinical course, predictive factors and prognosis. Hepatology 1994;20:1495–1501.
4. Toledo C, Salmeron JM, Rimola A, et al: Spontaneous bacterial peritonitis in cirrhosis – predictive factors of infection resolution and survival in patients treated with cefotaxime. Hepatology 1993;17:251–257.
5. Ruiz-del-Arbol L, Urman J, Fernandez J, et al: Systemic, renal, and hepatic hemodynamic derangement in cirrhotic patients with spontaneous bacterial peritonitis. Hepatology 2003;38:1210–1218.
6. Terra C, Guevara M, Torre A, et al: Renal failure in patients with cirrhosis and sepsis unrelated to spontaneous bacterial peritonitis: value of MELD score. Gastroenterology 2005;129:1944–1953.
7. Ruiz-del-Arbol L, Monescillo A, Arocena C, et al: Circulatory function and hepatorenal syndrome in cirrhosis. Hepatology 2005;42:439–447.
8. Navasa M, Fernandez J, Montoliu S, et al: Randomized, double-blind, placebo, controlled trial evaluating norfloxacin in the primary prophylaxis of spontaneous bacterial peritonitis in cirrhotics with renal impairment, hyponatremia or severe liver failure (abstract). J Hepatol 2006;44:S51.
9. Sort P, Navasa M, Arroyo V, et al: Effect of intravenous albumin on renal impairment and mortality in patients with cirrhosis and spontaneous bacterial peritonitis. N Engl J Med 1999;341:403–409.
10. Fernandez J, Monteagudo J, Bargallo X, et al: A randomized unblinded pilot study comparing albumin versus hydroxyethyl starch in spontaneous bacterial peritonitis. Hepatology 2005;42:627–634.
11. Gonwa TA, Morris CA, Goldstein RM, Husberg BS, Klintmalm GB: Long-term survival and renal function following liver transplantation in patients with and without hepatorenal syndrome – experience in 300 patients. Transplantation 1991;51:428–430.
12. Uriz J, Gines P, Cardenas A, et al: Terlipressin plus albumin infusion: an effective and safe therapy of hepatorenal syndrome. J Hepatol 2000;33:43–48.
13. Ortega R, Gines P, Uriz J, et al: Terlipressin therapy with and without albumin for patients with hepatorenal syndrome: results of a prospective, nonrandomized study. Hepatology 2002;36:941–948.
14. Duvoux C, Zanditenas D, Hezode C, et al: Effects of noradrenalin and albumin in patients with type I hepatorenal syndrome: a pilot study. Hepatology 2002;36:374–380.

V. Arroyo, MD
Liver Unit, Hospital Clínic
Villarroel 170
ES–08036 Barcelona (Spain)
Tel. +34 93 2275400, Fax +34 93 4515522, E-Mail varroyo@clinic.ub.es

Critical Care Nephrology: A Multidisciplinary Approach

Jean-Louis Vincent

Department of Intensive Care, Erasme Hospital, Free University of Brussels, Brussels, Belgium

Abstract

Background/Aims: Acute renal failure is a common complication in critically ill patients, affecting some 25% of intensive care unit (ICU) admissions, and is associated with high mortality rates of around 40–50%. Acute renal failure in the ICU frequently occurs as part of multiple organ failure (MOF). **Methods:** We reviewed the pertinent medical literature related to the occurrence of acute renal failure in the ICU and its association with other organ failures. We also reviewed the literature related to different patient management strategies, notably the differences between 'closed' and 'open' ICU formats. **Results:** The increasingly common association of acute renal failure with other organ failures, in the context of a more generalized MOF, has important implications on patient care, moving management away from the realm of nephrologists and towards a more multidisciplinary approach. Closed ICU formats with intensivist-led care, supported by specialist consultation, have been shown to be associated with improved ICU outcomes. **Conclusion:** ICU patients with acute renal failure should be managed using a multidisciplinary team approach led by an intensivist. Good collaboration and communication between intensivists and renal and other specialists is essential to insure the best possible care for ICU patients with renal disease.

Copyright © 2007 S. Karger AG, Basel

Introduction: Acute Renal Failure and Multiple Organ Failure

Acute renal failure is a common complication in critically ill patients, affecting some 25% of intensive care unit (ICU) admissions, and is associated with high mortality rates of around 40–50% [1]. While isolated renal failure may occur, acute renal failure in the ICU frequently occurs as part of multiple organ failure (MOF), and is commonly associated with a septic etiology [1–3]. Indeed, patients with severe sepsis and septic shock are at an increased risk of developing acute renal failure and have higher mortality rates than patients without infection [1, 2, 4].

The degree of organ dysfunction in patients with MOF can be assessed using various organ dysfunction scoring systems, one of the most widely used being the sequential organ failure assessment (SOFA) score [5]. The SOFA score (table 1) considers the function of 6 organ systems – cardiovascular, respiratory, neurological, hepatic, renal, and coagulation. Data from the SOFA database showed that 69% of patients with acute renal failure developed MOF [1]. In addition, while 23% of patients with organ failure had isolated renal failure on admission, renal failure was often associated with other organ failures – cardiovascular and hepatic failure (each in 74% of the patients), coagulation (in 72% of the patients), neurologic (67%), and respiratory (57%) failures [6].

It is interesting to note that the mortality from renal failure has not changed markedly over the years [7] (fig. 1) despite the development and introduction of new techniques. There are several reasons for this including the lack of effectiveness of new therapies; for example, even though we have good reason to believe hemofiltration is superior to intermittent dialysis, there are no real data to support this [8]. However, perhaps more importantly, the demographics of critically ill patients with renal failure have changed so that, although therapy and support may have improved, we are increasingly treating older, sicker patients, with multiple comorbidities, who survive longer and develop renal failure as part of a more complex MOF picture with its associated higher mortality rates [1, 9, 10].

Critical Care Nephrology – The Bigger Picture?

Several studies have investigated the importance of acute renal failure in the ICU in terms of its association with other organ failures. In patients undergoing cardiac surgery, a moderate (20%) increase in plasma creatinine shortly after surgery occurred in 15.6% of the patients and was associated with other organ failures in 79% of the patients [11]. The mortality rate in patients with a postoperative 20% increase in creatinine was 12% in patients with other organ dysfunctions, but 0% if no other organ dysfunctions were present, demonstrating the important role of other organ failures on outcomes in this group of patients. In patients who developed contrast-mediated renal failure, Levy et al. [12] reported that the mortality rate was 34% compared to 7% in patients who did not develop renal failure. After adjusting for differences in comorbidity, renal failure was associated with an odds ratio of dying of 5.5. Patients who died after developing renal failure had complicated clinical courses with sepsis, bleeding, delirium, and respiratory failure. Van Biesen et al. [13], showed that patients with sepsis who later developed acute renal failure needed higher

Table 1. The sequential organ failure score (SOFA) [5]

Organ system	SOFA score				
	0	1	2	3	4
Respiration					
PaO$_2$/FiO$_2$, mm Hg	>400	≤400	≤300	≤200	≤100
				with respiratory support	
Coagulation					
Platelets ×10^3/mm^3	>150	≤150	≤100	≤50	≤20
Liver					
Bilirubin, mg/dl	<1.2	1.2–1.9	2.0–5.9	6.0–11.9	>12.0
(μmol/l)	(<20)	(20–32)	(33–101)	(102–204)	(>204)
Cardiovascular					
Hypotension	No hypotension	MAP <70 mm Hg	dopamine ≤5 or dobutamine (any dose)[a]	dopamine >5 or epinephrine ≤0.1 or norepinephrine ≤0.1[a]	dopamine >15 or epinephrine >0.1 or norepinephrine >0.1[a]
Central nervous system					
Glasgow coma score	15	13–14	10–12	6–9	<6
Renal					
Creatinine, mg/dl	<1.2	1.2–1.9	2.0–3.4	3.5–4.9	>5.0
(μmol/l)	(<110)	(110–170)	(171–299)	(300–440)	(>440)
or urine output				or <500 ml/day	or <200 ml/day

[a] Adrenergic agents administered for at least 1 h (doses given are in μg/kg/min).

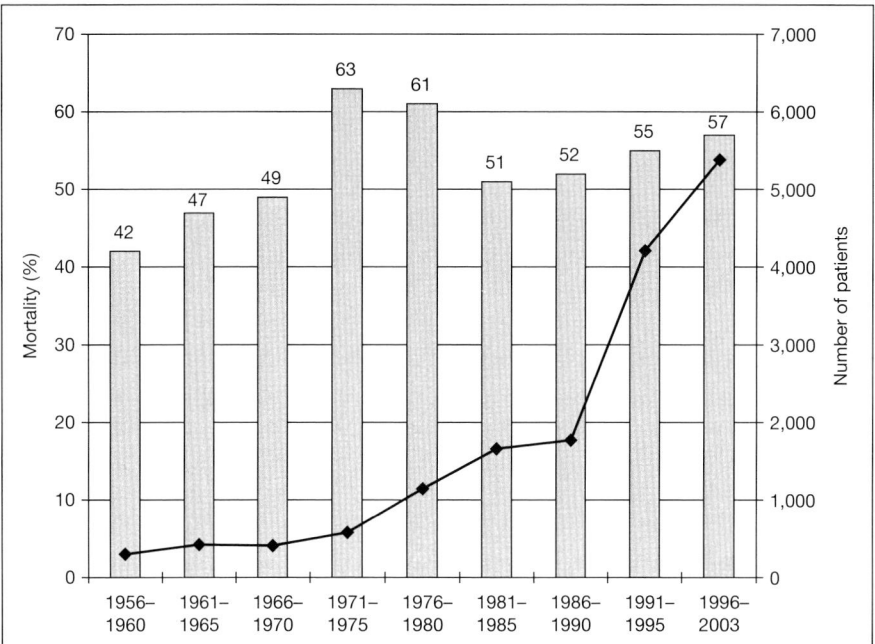

Fig. 1. Changes in mortality in patients with acute renal failure over the past 48 years. Line represents numbers of patients included in the studies during each time period. From Ympa et al. [7] with permission.

inspired oxygen fractions (FiO_2), even before serum creatinine levels increased, than patients who did not develop acute renal failure. Using data from the SOFA database, de Mendonça et al. [1] noted that the number of associated organ failures on admission in patients with acute renal failure was an independent predictor of mortality from acute renal failure (OR 1.24, 95% CI 1.03–1.5, $p = 0.02$). Acute renal failure with cardiovascular failure was associated with the highest mortality rate.

Studies have also suggested that renal failure may play a proactive role in the development and maintenance of MOF [14]. Experimental data indicate systemic effects of acute renal failure via various mediators [15, 16], and clinical studies have reported that critically ill patients with acute renal failure have increased oxidative stress compared to other critically ill patients and patients with end-stage renal disease [17].

In the context of a more generalized MOF, the increasingly common association of acute renal failure with other organ failures has important implications on patient care, moving management away from the realm of the nephrologists and towards a more multidisciplinary approach.

The Role of the General Intensivist in Critical Care Nephrology

Intensivists must be trained and ready to take care of a vast range of acute problems of various etiologies and involving any, or many, organ system(s). As part of their training, therefore, today's intensivist must learn how to diagnose, monitor and manage acute renal failure, acute respiratory failure, circulatory shock, and any number of other acute and life-threatening conditions. Intensivists can perhaps be considered as 'multidisciplinary specialists'.

Importantly, studies have shown that intensivists can really improve patient outcomes. One way in which the importance of the intensivist has been demonstrated is in comparing so-called 'open' ICUs (those in which patients are cared for by their admitting physician with limited intensivist input) with 'closed' ICUs (in which patient care is transferred to an intensivist); often these studies are cohort studies assessing outcomes before and after conversion of an open system to the now more common closed ICU. Several studies have reported improved resource utilization [18–20] and improved outcomes [19, 21, 22] for patients managed in a closed compared to an open system. High intensity and 24-hour intensivist coverage have also been associated with improved patient outcomes [23–25]. In the 'closed' ICU system of Victoria, Australia, Cole et al. [26] reported that patients with acute renal failure were managed by intensivists, with outcomes comparing favorably to those predicted by illness severity scores, supporting the 'closed' model of care for patients with acute renal failure.

The general intensivist is, therefore, a key element to providing effective ICU care for all critically ill patients, including those with renal failure, and other systems of care should be avoided. The nephrologist, therefore, should not be responsible for ICU patients with acute renal failure (any more than the pneumologist should care for patients with acute respiratory failure or the cardiologist for patients with cardiac failure), unless they have received specialist training in intensive care medicine. This does not, of course, exclude the possibility that many intensivists will have a favorite aspect of intensive care medicine, some preferring respiratory problems, others neurological cases, and others renal diseases, and their chosen field of expertise will likely result in research in this area.

In addition, there is no advantage to having multiple small 'specialist' ICUs; for example, a renal ICU plus a respiratory ICU plus a trauma ICU, etc. The problems faced by critically ill patients are basically similar, regardless of the underlying etiology, and no benefit can be served by spreading resources and staff among multiple units. Even the separation between surgical and medical ICUs is historical and should be avoided. We need to promote the system of multidisciplinary departments of intensive care. In our Department of Intensive Care at Erasme Hospital, Brussels, Belgium, we have 36 beds, with close to

20 clinical doctors trained in intensive care medicine (and a number of research fellows), supported by a team of dedicated nurses, physiotherapists, and other health care personnel.

Collaboration with Nephrologists and Other Specialists

Having promoted above the importance of the closed ICU and the role of the intensivist in the management of critically ill patients with acute renal failure, this does not mean there is no place for the nephrologist in the management of these patients. Indeed, collaboration may be needed with other specialists also depending on the severity of other associated organ dysfunctions. For many patients, optimum care will involve collaboration with several specialties; nevertheless, overall responsibility must lie with the intensivist.

Collaboration with a nephrologist, or other specialist, can be important in several situations:

Advice and Opinion in Complex Cases

Despite the broad training of modern intensivists, there are cases for which a specialist consultation is always welcome – however diligent and enthusiastic, one cannot cover all aspects of every disease state! But such requests must be reserved for complex cases: a nephrologist should not be called for every oliguric patient, just as a neurologist will not be called for every comatose patient or an endocrinologist for every raised blood glucose level! Excessive consultation may be a useless consumption of resources and may even have untoward effects if it causes delay in treatment while waiting for the specialist opinion (although, in patients with acute renal failure there is usually sufficient time to discuss the various options). In addition, if specialist consultation is requested for many patients, the role of the intensivist will be undermined, and the system becomes increasingly similar to the aforementioned open format again. Importantly very complex cases with several associated organ failures may require consultation from several different specialists and good communication is essential to decide on optimal approaches to treatment; here, the coordinating role of the intensivist can be paramount in integrating and combining specialist opinion.

Need for New Techniques

We are in an era of rapid developments in medical therapies and technology, and it is not physically or mentally possible for every physician to keep abreast of all the latest techniques in all the specialties. Hence, input from a specialist may be welcome in a patient needing treatment with a new and unfamiliar agent or technique.

Scientific Discussions, Research, and Training

Importantly, specialists should be invited to seminars and other scientific discussions within the intensive care department whenever the topic includes aspects relevant to their specific field. Similarly, members of other specialties should be included in clinical research when it involves specialized aspects of their particular field. In addition, specialists from all departments, who are most likely to be able to provide the latest specialist data and results in their field, must be involved in the training and supervision of junior intensivists and students.

Conclusion

Critically ill patients with renal failure often have multiple ongoing acute processes and require a multidisciplinary approach to management. This is ideally performed in a multidisciplinary ICU with patients cared for by trained intensivists available 24 h/day, 7 days/week. This does not mean there is no place for the nephrologist in the care of such patients, but overall responsibility must lie with the intensivist. Any consultant from any specialty should be welcome on the ICU at any time; the patient must be at the center of attention, and visits from other specialists must be welcomed if they can improve patient care. In collaborating with other specialties it is important that each party acknowledges the respective competences of the other, and that there be open and honest discussion of all issues associated with patient management. Critical care nephrology is a multidisciplinary field, and only a multidisciplinary team approach to patient management, under the ultimate coordination and control of a trained intensivist, will provide the best possible care for ICU patients with renal disease.

References

1. de Mendonca A, Vincent JL, Suter PM, et al: Acute renal failure in the ICU: risk factors and outcome evaluated by the SOFA score. Intensive Care Med 2000;26:915–921.
2. Brivet FG, Kleinknecht DJ, Loirat P, et al: Acute renal failure in intensive care units – causes, outcome, and prognostic factors of hospital mortality: a prospective, multicenter study. French Study Group on Acute Renal Failure. Crit Care Med 1996;24:192–198.
3. Uchino S, Kellum JA, Bellomo R, et al: Acute renal failure in critically ill patients: a multinational, multicenter study. JAMA 2005;294:813–818.
4. Neveu H, Kleinknecht D, Brivet F, et al: Prognostic factors in acute renal failure due to sepsis: results of a prospective multicenter study. Nephrol Dial Transplant 1996;11:293–299.
5. Vincent JL, Moreno R, Takala J, et al: The SOFA (Sepsis-related Organ Failure Assessment) score to describe organ dysfunction/failure. Intensive Care Med 1996;22:707–710.
6. Vincent JL, de Mendonça A, Cantraine F, et al: Use of the SOFA score to assess the incidence of organ dysfunction/failure in intensive care units: results of a multicentric, prospective study. Crit Care Med 1998;26:1793–1800.

7 Ympa YP, Sakr Y, Reinhart K, et al: Has mortality from acute renal failure decreased? A systematic review of the literature. Am J Med 2005;118:827–832.
8 Vinsonneau C, Camus C, Combes A, et al: Continuous venovenous haemodiafiltration versus intermittent haemodialysis for acute renal failure in patients with multiple-organ dysfunction syndrome: a multicentre randomised trial. Lancet 2006;368:379–385.
9 Cappi SB, Sakr Y, Vincent JL: Daily evaluation of organ function during renal replacement therapy in intensive care unit patients with acute renal failure. J Crit Care 2006;21:179–183.
10 Mehta RL, Pascual MT, Soroko S, et al: Spectrum of acute renal failure in the intensive care unit: the PICARD experience. Kidney Int 2004;66:1613–1621.
11 Ryckwaert F, Boccara G, Frappier JM, et al: Incidence, risk factors, and prognosis of a moderate increase in plasma creatinine early after cardiac surgery. Crit Care Med 2002;30:1495–1498.
12 Levy EM, Viscoli CM, Horwitz RI: The effect of acute renal failure on mortality. A cohort analysis. JAMA 1996;275:1489–1494.
13 Van Biesen W, Yegenaga I, Vanholder R, et al: Relationship between fluid status and its management on acute renal failure (ARF) in intensive care unit (ICU) patients with sepsis: a prospective analysis. J Nephrol 2005;18:54–60.
14 Van Biesen W, Lameire N, Vanholder R, et al: Relation between acute kidney injury and multiple-organ failure: the chicken and the egg question. Crit Care Med 2007;35:316–317.
15 Rabb H, Wang Z, Nemoto T, et al: Acute renal failure leads to dysregulation of lung salt and water channels. Kidney Int 2003;63:600–606.
16 Kelly KJ: Distant effects of experimental renal ischemia/reperfusion injury. J Am Soc Nephrol 2003;14:1549–1558.
17 Himmelfarb J, McMonagle E, Freedman S, et al: Oxidative stress is increased in critically ill patients with acute renal failure. J Am Soc Nephrol 2004;15:2449–2456.
18 Multz AS, Chalfin DB, Samson IM, et al: A 'closed' medical intensive care unit (MICU) improves resource utilization when compared with an 'open' MICU. Am J Respir Crit Care Med 1998;157:1468–1473.
19 Ghorra S, Reinert SE, Cioffi W, et al: Analysis of the effect of conversion from open to closed surgical intensive care unit. Ann Surg 1999;229:163–171.
20 Hanson CW, Deutschman CS, Anderson HL, et al: Effects of an organized critical care service on outcomes and resource utilization: a cohort study. Crit Care Med 1999;27:270–274.
21 Baldock G, Foley P, Brett S: The impact of organisational change on outcome in an intensive care unit in the United Kingdom. Intensive Care Med 2001;27:865–872.
22 Topeli A, Laghi F, Tobin MJ: Effect of closed unit policy and appointing an intensivist in a developing country. Crit Care Med 2005;33:299–306.
23 Pronovost PJ, Angus DC, Dorman T, et al: Physician staffing patterns and clinical outcomes in critically ill patients: a systematic review. JAMA 2002;288:2151–2162.
24 Blunt MC, Burchett KR: Out of hours consultant cover and case-mix-adjusted mortality in intensive care. Lancet 2000;356:735–736.
25 Higgins TL, McGee WT, Steingrub JS, et al: Early indicators of prolonged intensive care unit stay: impact of illness severity, physician staffing, and pre-intensive care unit length of stay. Crit Care Med 2003;31:45–51.
26 Cole L, Bellomo R, Silvester W, et al: A prospective, multicenter study of the epidemiology, management, and outcome of severe acute renal failure in a 'closed' ICU system. Am J Respir Crit Care Med 2000;162:191–196.

Dr. J.-L. Vincent
Erasme University Hospital
Route de Lennik 808
BE–1070 Brussels (Belgium)
Tel. +32 2 555 3380, Fax +32 2 555 4555, E-Mail jlvincen@ulb.ac.be

Incidence, Classification, and Outcomes of Acute Kidney Injury

Eric A.J. Hoste[a,b], John A. Kellum[b]

[a]Intensive Care Unit, Ghent University Hospital, Ghent, Belgium; [b]The Clinical Research, Investigation, and Systems Modeling of Acute Illness (CRISMA) Laboratory, Department of Critical Care Medicine, University of Pittsburgh, School of Medicine, Pittsburgh, Pa., USA

Abstract

Background: Traditionally the epidemiology of acute renal failure was assessed in patients requiring renal replacement therapy. Recent data emphasized the importance of less severe impairment of kidney function, hence the terminology acute kidney injury (AKI) was introduced. **Methods:** In this paper we present a review of current published data on the epidemiology of AKI. **Results:** The RIFLE classification categorizes the whole severity range of AKI into 3 severity categories and 2 outcome classes. AKI is associated with increased costs and worse outcomes. Increasing severity classes are associated with increasing morbidity and mortality. There is an increasing incidence of AKI, while mortality seems to decrease. **Conclusion:** Small changes in kidney function have an impact on outcomes and this knowledge has led to the introduction of the terminology AKI, encompassing both discrete and severe impairment of kidney function. The RIFLE classification describes the whole range of AKI and has been validated in multiple cohorts. As a consequence of increasing comorbidity, the incidence of AKI is increasing. The incidence of acute renal failure requiring renal replacement therapy even compares to that of acute lung injury, and up to two thirds of general ICU patients meet RIFLE criteria for AKI.

Copyright © 2007 S. Karger AG, Basel

The epidemiology of acute renal failure (ARF) has undergone considerable change over the years. This is not only caused by a change in patient characteristics, but probably even more importantly, by a change in definition of the disease. It was Homer W. Smith [1] who introduced the term 'acute renal failure' in a chapter on Acute Renal Failure Related to Traumatic Injuries, in his textbook *The Kidney: Structure and Function in Health and Disease* (1951). Since then ARF has become an established term in medical parlance. Confusingly, the term ARF covers many different meanings. Workgroup 1 of the second conference of the Acute Dialysis Quality Initiative (ADQI) found in 2002 that over 30 definitions of ARF were used in medical literature, ranging from a 25% increase of serum creatinine to the need for renal replacement therapy (RRT) (http://www.ccm.upmc.edu/adqi/ADQI2/ADQI2g1.pdf) [2]. Different definitions of acute kidney injury (AKI) describe different cohorts. This is nicely illustrated in a study on the occurrence of AKI in 9,210 hospitalized patients in whom 2 or more serum creatinine levels were assessed [3]. The incidence of AKI ranged from 1 to 44%.

Initially most emphasis in ARF research was on patients with severe impairment of kidney function, e.g. defined by the need for RRT. More recently, several authors have demonstrated that small changes in serum creatinine are independently associated with increased morbidity and mortality. Levy et al. [4] demonstrated that a 25% increase of serum creatinine following radiocontrast administration was associated with a 5-fold increased risk for in-hospital mortality. Lassnigg et al. [5] demonstrated similar findings in cardiac surgery patients. Finally, Chertow et al. [3] demonstrated that a $\geq$0.3 mg/dl increase of serum creatinine was associated with greater cost, morbidity and mortality in hospitalized patients. The emphasis of interest is therefore more and more shifting to less severe derangements of kidney function consistent with several calls toward definitions which identify this patient population [6, 7]. Hence the terminology acute kidney injury (AKI) was introduced. In order to meet the need for a uniform definition which included different severity grades of AKI the RIFLE classification was developed by ADQI [7]. This classification system (table 1) has now been validated in numerous settings [8].

Epidemiology of AKI and ARF

ARF
ARF severe enough to require RRT occurs in approximately 5% of general ICU patients [9]. This proportion may vary according to the ICU cohort described, e.g. cardiac surgery versus medical ICU but appears to be surprisingly uniform around the world. Over a period of almost 20 years the incidence of ARF treated with RRT has more than doubled. Feest et al. [10] and Waikar et al. [11] found in a cohort from the late 1980s that the incidence of ARF patients treated with RRT was less than 50 patients per million population. Recent data from the beginning of this century report an incidence rate of even 270 patients per million population [11]. In comparison, the incidence of acute lung injury was estimated at 112–320 patients per million population [12]. ARF requiring RRT has therefore a comparable incidence to that of acute lung injury. The

Table 1. RIFLE classification

	Glomerular filtration rate criteria	Urine output criteria
Risk	serum creatinine × 1.5	UO < 0.5 ml/kg/h × 6 h
Injury	serum creatinine × 2	UO < 0.5 ml/kg/h × 12 h
Failure	serum creatinine × 3	UO < 0.3 ml/kg/h × 24 h
	or	or
	serum creatinine ≥4 mg/dl with an acute rise >0.5 mg/dl	anuria × 12 h
Loss	persistent ARF = complete loss of kidney function >4 weeks	
ESKD	ESKD >3 months	

For conversion of creatinine expressed in conventional units to SI units, multiply by 88.4. RIFLE class is determined based on the worst of either glomerular filtration criteria or urine output (UO) criteria. Glomerular filtration criteria are calculated as an increase of serum creatinine above the baseline serum creatinine level. Acute renal dysfunction should be both abrupt (within 1–7 days) and sustained (more than 24 h). When the baseline serum creatinine is not known and patients are without a history of chronic renal insufficiency, it is recommended to calculate a baseline serum creatinine using the modification of diet in renal disease (MDRD) equation for assessment of renal function, assuming a glomerular filtration rate of 75 ml/min/1.73 m^2. When baseline serum creatinine is elevated, an abrupt rise of at least 0.5 mg/dl to more than 4 mg/dl is all that is required to achieve 'failure'.

---

reason for the increasing incidence of ARF can probably be explained by the change in baseline characteristics of patients. Patients nowadays are older, have more comorbid disease, and are more severely ill at the start of RRT [13, 14].

From ARF to AKI

The incidence of less severe AKI has also increased over time. Perhaps, the best evidence of this comes from the studies of Hou et al. [15] and Nash et al. [16]. These authors evaluated the occurrence of AKI in a single hospital in 1979 and 1996, and demonstrated that the proportion of patients with AKI had increased from 4.9 to 7.2% of all hospitalized patients. Two large multicenter, longitudinal databases on AKI in the United States also showed that the incidence has increased. Both studies used administrative databases, and scored occurrence of AKI on the basis of reporting of International Classification of Diseases, Ninth Revision (ICD-9) codes for ARF which include cases we would now consider to have AKI [11, 17]. Waikar et al. [11] found that the incidence quadrupled from 610 to 2,880 patients per million population during the 15-year study period. Xue et al. [17] found that during the period from 1992 to 2001, there was a 11%

yearly increase of the diagnosis of AKI. A limitation to these studies may be that ICD coding does not currently define strict cutoffs for AKI. Consequently, sensitivity to detect AKI is low, or in other words, many patients with AKI are missed when epidemiology is based on administrative databases. Also, reporting of AKI may vary across hospitals and during different time periods.

AKI Defined by RIFLE Criteria

The RIFLE classification was used to evaluate the in-hospital epidemiology of AKI in a single center setting in Melbourne, Australia [18]. In a cohort of over 20,000 hospitalized patients, 18% developed AKI according to RIFLE criteria. The maximum severity of AKI was risk in 9.1% of patients, injury in 5.2%, and failure in 3.7% of patients. Our group evaluated the incidence of AKI in a cohort of 5,383 ICU patients admitted during a 1-period to a tertiary care hospital. In this cohort two thirds developed AKI according to the RIFLE classification, 12.4% of patients had a maximum RIFLE Risk, 26.7% had maximum RIFLE Injury and 28.1% had maximum RIFLE Failure [19].

Outcome of ARF and AKI

Length of Stay

Patients with AKI or ARF are amongst the most severely ill in the ICU. Therefore, it is not surprising that ICU patients with AKI or ARF have a longer length of stay in the ICU and in the hospital compared to ICU patients without these conditions. Patients with less severe forms of AKI have an increased length of in-hospital stay compared to patients without AKI, and there is a stepwise increase of length of stay depending on the severity of AKI according to RIFLE criteria (length of stay for patients without AKI 6 days, RIFLE Risk 8 days, RIFLE Injury 10 days, RIFLE Failure 16 days; $p < 0.01$) [19].

End-Stage Kidney Disease

Although the majority of survivors regain kidney function, some patients do not recover from ARF and evolve to end-stage kidney disease (ESKD) with permanent need for RRT. In the large multicenter BEST Kidney study 13.8% (95% confidence interval 11.2–16.3%) of survivors developed ESKD [9] (RIFLE 'E').

Mortality

The in-hospital mortality rate for general ICU patients with ARF who are treated with RRT is approximately 60% [9]. Mortality rate for ICU patients with ARF will depend upon the severity of AKI, and the case mix of the

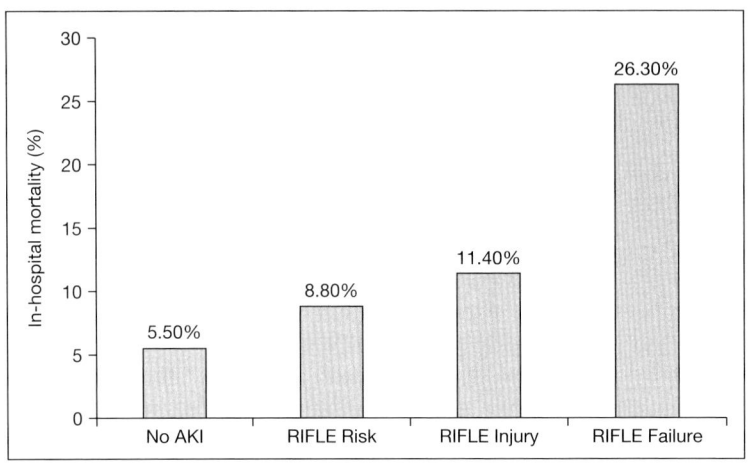

Fig. 1. In-hospital mortality for ICU patients without AKI and ICU patients with increasing RIFLE class [after 18].

observed cohort (fig. 1) [3, 5, 9, 18, 19]. Despite advances in treatment, the published mortality rates for ARF patients in individual studies remain more or less constant at 50% from 1956 on [20]. However, longitudinal collected data demonstrated an improvement of outcome [11, 17, 21].

Long-Term Outcome

Increasing severity of AKI is associated with increasing 1-mortality rate, and patients with AKI or ARF have a worse 1-year survival compared to non-ARF patients [22]. However, after hospital discharge survival curves for patients with different severity of AKI parallel each other. One-year mortality rate is comparable for patient groups with different severity of AKI [22].

Patients Die of AKI

Optimism from the past that dictated that renal function could be substituted by RRT, and therefore, patients died *with* ARF but not because *of* ARF, has been replaced by realism. Observed mortality in patients with ARF or AKI is significantly higher than predicted from underlying disease and AKI has been shown to be an independent predictor of mortality. This appears to be true for the entire spectrum of AKI, from patients with only minor severity of AKI to patients with ARF, and for ICU and non-ICU patients. Patients with 25% increase of serum creatinine after radiocontrast procedures had an in-hospital mortality of 34% compared to 7% for carefully matched patients without increase of serum creatinine (odds ratio 5.5) [4]. After this publication there were several others which confirmed the association of small absolute or relative increases of serum creatinine in hospitalized, non-ICU patients and in-hospital mortality [3]. In ICU patients minor severity of AKI is also associated with worse outcome, even after correction for covariates. This has been demonstrated in several studies that used the RIFLE classification for AKI, especially classes 'I' and 'F', and also in studies that evaluated small increases of serum creatinine [3, 5, 18, 19].

Conclusions

Small changes in kidney function have an impact on outcomes and this knowledge has led to the introduction of the terminology AKI, encompassing both discrete and severe impairment of kidney function. The RIFLE classification describes the whole range of AKI and has been validated in multiple cohorts. As a consequence of increasing comorbidity, the incidence of AKI is increasing. The incidence of ARF requiring RRT even compares to that of acute lung injury, and up to two thirds of general ICU patients meet RIFLE criteria for AKI.

References

1 Smith HW: Acute renal failure related to traumatic injuries; in The Kidney: Structure and Function in Health and Disease. Cary, Oxford University Press, 1951.
2 Kellum JA, Levin N, Bouman C, Lameire N: Developing a consensus classification system for acute renal failure. Curr Opin Crit Care 2002;8:509–514.
3 Chertow GM, Burdick E, Honour M, Bonventre JV, Bates DW: Acute kidney injury, mortality, length of stay, and costs in hospitalized patients. J Am Soc Nephrol 2005;16:3365–3370.
4 Levy EM, Viscoli CM, Horwitz RI: The effect of acute renal failure on mortality. A cohort analysis. JAMA 1996;275:1489–1494.
5 Lassnigg A, Schmidlin D, Mouhieddine M, Bachmann LM, Druml W, Bauer P, Hiesmayr M: Minimal changes of serum creatinine predict prognosis in patients after cardiothoracic surgery: a prospective cohort study. J Am Soc Nephrol 2004;15:1597–1605.
6 Bellomo R, Kellum J, Ronco C: Acute renal failure: time for consensus. Intensive Care Med 2001;27:1685–1688.
7 Bellomo R, Ronco C, Kellum JA, Mehta RL, Palevsky P; ADQI workgroup: Acute renal failure – definition, outcome measures, animal models, fluid therapy and information technology needs: the Second International Consensus Conference of the Acute Dialysis Quality Initiative (ADQI) Group. Crit Care 2004;8:R204–R212.
8 Hoste EA, Kellum JA: Acute kidney injury: epidemiology and diagnostic criteria. Curr Opin Crit Care 2006;12:531–537.
9 Uchino S, Kellum JA, Bellomo R, Doig GS, Morimatsu H, Morgera S, Schetz M, Tan I, Bouman C, Macedo E, et al: Acute renal failure in critically ill patients: a multinational, multicenter study. JAMA 2005;294:813–818.
10 Feest TG, Round A, Hamad S: Incidence of severe acute renal failure in adults: results of a community based study. BMJ 1993;306:481–483.
11 Waikar SS, Curhan GC, Wald R, McCarthy EP, Chertow GM: Declining mortality in patients with acute renal failure, 1988 to 2002. J Am Soc Nephrol 2006;17:1143–1150.

12 Goss CH, Brower RG, Hudson LD, Rubenfeld GD: Incidence of acute lung injury in the United States. Crit Care Med 2003;31:1607–1611.
13 Ostermann ME, Taube D, Morgan CJ, Evans TW: Acute renal failure following cardiopulmonary bypass: a changing picture. Intensive Care Med 2000;26:565–571.
14 McCarthy JT: Prognosis of patients with acute renal failure in the intensive care unit: a tale of two eras. Mayo Clin Proc 1996;71:117–126.
15 Hou S, Bushinsky D, Wish J, Cohen J, Harrington J: Hospital-acquired renal insufficiency: a prospective study. Am J Med 1983;74:243–248.
16 Nash K, Hafeez A, Hou S: Hospital-acquired renal insufficiency. Am J Kidney Dis 2002;39: 930–936.
17 Xue JL, Daniels F, Star RA, Kimmel PL, Eggers PW, Molitoris BA, Himmelfarb J, Collins AJ: Incidence and mortality of acute renal failure in Medicare beneficiaries, 1992 to 2001. J Am Soc Nephrol 2006;17:1135–1142.
18 Uchino S, Bellomo R, Goldsmith D, Bates S, Ronco C: An assessment of the RIFLE criteria for acute renal failure in hospitalized patients. Crit Care Med 2006;34:1913–1917.
19 Hoste EA, Clermont G, Kersten A, Venkataraman R, Angus DC, De Bacquer D, Kellum JA: RIFLE criteria for acute kidney injury are associated with hospital mortality in critically ill patients: a cohort analysis. Crit Care 2006;10:R73.
20 Ympa YP, Sakr Y, Reinhart K, Vincent JL: Has mortality from acute renal failure decreased? A systematic review of the literature. Am J Med 2005;118:827–832.
21 Desegher A, Reynvoet E, Blot S, De Waele J, Claus S, Hoste E: Outcome of patients treated with renal replacement therapy for acute kidney injury. Crit Care 2006;10:P296.
22 Bagshaw SM, Mortis G, Doig CJ, Godinez-Luna T, Fick GH, Laupland KB: One-year mortality in critically ill patients by severity of kidney dysfunction: a population-based assessment. Am J Kidney Dis 2006;48:402–409.

Eric A.J. Hoste
ICU, 2K12-C, Ghent University Hospital
De Pintelaan 185
BE–9000 Gent (Belgium)
Tel. +32 9 240 27 75, Fax +32 9 240 49 95, E-Mail Eric.Hoste@UGent.be

Pathophysiology of Acute Kidney Injury: Roles of Potential Inhibitors of Inflammation

Joseph V. Bonventre

Renal Division, Brigham and Women's Hospital and Department of Medicine, Harvard Stem Cell Institute, Harvard Medical School and Harvard-Massachusetts Institute of Technology, Division of Health Sciences and Technology, Boston, Mass., USA

Abstract

The pathogenesis of acute kidney injury (AKI) is complex and varies to some extent based on the particular cause. Inflammation contributes to this pathophysiology in a variety of contexts. Inflammation can result in reduction in local blood flow to the outer medulla with adverse consequences on tubule function and viability. Both the innate and adaptive immune responses are important contributors. With ischemia/reperfusion endothelial cells upregulate a number of adhesion molecules which have counterreceptors on leukocytes. A number of vasoactive mediators that are released with injury, such as nitric oxide, may also affect leukocyte-endothelial interactions. Tubule epithelial cells generate proinflammatory and chemotactic cytokines. We and others have found that injection of mesenchymal stem (stromal) cells is protective against renal injury as assessed by serum creatinine measured 24 h after ischemia. The mechanism of such protection may be through intrarenal paracrine effects to decrease inflammation or by systemic immune modulation. Resolvins (Rv) and protectins (PD) have been identified as two newly identified families of naturally occurring n–3 fatty acid docosahexaenoic acid metabolites. In collaboration with Serhan et al. we recently reported that, in response to bilateral ischemia/reperfusion injury, mouse kidneys produce D series resolvins (RvDs) and PD1 [J Immunol 2006;177:5902–5911]. Administration of RvDs or PD1 to mice prior to, or subsequent to, ischemia resulted in a reduction in functional and morphological kidney injury. Understanding how these anti-inflammatory processes are regulated may provide insight into how we might intervene to facilitate and enhance them so that we might prevent or mitigate the devastating consequences of AKI.

Copyright © 2007 S. Karger AG, Basel

Body homeostasis depends critically on the ability of the kidney to function normally. There are many potential causes of acute kidney injury (AKI), some of which are related to a mismatch between oxygen and nutrient delivery to the nephrons and energy demand of the nephrons. Other causes relate to direct toxic effects of substances on the epithelium. The kidney is particularly susceptible to toxic effects of many environmental substances or therapeutics since many of these compounds are increased in concentration as glomerular filtrate is reabsorbed from the tubule as the filtrate moves down the nephron. In many situations in humans acute injury is superimposed on chronic renal disease and AKI is increasingly being recognized as an important precipitant in progression to end-stage renal disease.

Whether the injury is related to oxygen deprivation or toxins there are many common features of the epithelial cell response. The processes of injury and repair to the kidney epithelium is depicted schematically in figure 1. Injury results in rapid loss of cytoskeletal integrity and cell polarity. There is shedding of the proximal tubule brush border, loss of polarity with mislocalization of adhesion molecules and other membrane proteins such as the $Na^+K^+ATPase$ and β-integrins [1], as well as apoptosis and necrosis [2]. With severe injury, viable and nonviable cells are desquamated leaving regions where the basement membrane remains as the only barrier between the filtrate and the peritubular interstitium. This allows for backleak of the filtrate, especially under circumstances where the pressure in the tubule is increased due to intratubular obstruction resulting from cellular debris in the lumen interacting with proteins such as fibronectin which enter the lumen [3]. This injury to the epithelium results in the generation of inflammatory and vasoactive mediators, which can act on the vasculature to worsen the vasoconstriction and inflammation. Thus inflammation contributes in a critical way to the pathophysiology of AKI [4]. In contrast to the heart or brain, the kidney can recover from an ischemic or toxic insult that results in cell death, although it is becoming increasingly recognized that there are longer-term detrimental effects of even brief periods of ischemia [5]. Surviving cells that remain adherent undergo repair with the potential to recover normal renal function. Whether there is a subpopulation of stem or progenitor cells is a matter of active study at this point in time [6]. When the kidney recovers from acute injury it relies on a sequence of events that include epithelial cell spreading and migration to cover the exposed areas of the basement membrane, cell dedifferentiation and proliferation to restore cell number, followed by differentiation which results in restoration of the functional integrity of the nephron [7]. The potential role of stem cells derived from the bone marrow to directly replace cells lost by the injury has been addressed in a number of publications. We and others have concluded that the bone marrow does not contribute directly to the replacement of cells but bone marrow-derived cells may have paracrine effects that may facilitate repair, potentially by reducing inflammation [6]. In this brief review I will focus on the role of

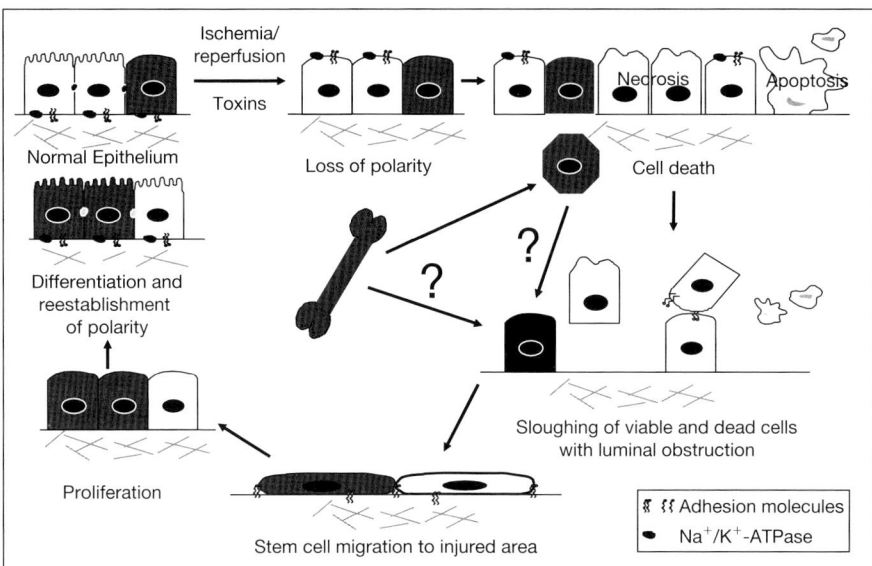

Fig. 1. Injury and repair to the epithelial cells of the kidney with ischemia/reperfusion. With injury to the kidney an early response is loss of the polarity of the epithelial cells with mislocation of adhesion molecules and Na^+/K^+-ATPase and other proteins. In addition there is a loss of the brush border. With increasing injury, there is cell death by either necrosis or apoptosis. Some of the necrotic debris is then released into the lumen, where it interacts with luminal proteins and can ultimately result in obstruction. In addition, because of the mislocation of adhesion molecules viable epithelial cells lift off the basement membrane and are found in the urine. The kidney can respond to the injury by initiating a repair process if there are sufficient nutrients and sufficient oxygen delivery and the basement membrane integrity has not been altered irreparably. Viable epithelial cells migrate and cover denuded areas of the basement membrane. The source of these cells appears to be from the kidney itself and not from the bone marrow. Bone marrow cells may contribute to the interstitial cellular infiltrate and may produce factors that may modulate inflammation and facilitate repair. Cells replacing the epithelium may derive from differentiated epithelial cells or from a subpopulation of progenitor cells in the tubule or in the interstitium. The cells which populate the recovering epithelium express proteins that are not normally expressed in an adult mature epithelial cell. The cells then undergo division and replace lost cells. Ultimately, the cells go on to differentiate and reestablish the normal polarity of the epithelium. In this figure the cells depicted as dark cells are those that might represent a progenitor pool. The existence and potential role of a subpopulation of stem/progenitor cells in this process of repair is controversial.

inflammation and the potential contribution of factors that can mitigate the inflammation such as bone marrow-derived stem cells via paracrine mechanisms, and naturally occurring anti-inflammatory compounds: resolvins and protectins.

Inflammation

The pathogenesis of ischemic acute renal failure has been attributed to abnormal regulation of local blood flow following the initial ischemic episode. Persistent preglomerular vasoconstriction may be a contributing factor; however, inflammation contributes in an important way to the reduction in local blood flow to regions of the cortex and the outer medulla with adverse consequences on tubule function and viability.

The Innate Immune Response

Both the innate and adaptive immune responses are important contributors to the pathobiology of ischemic injury. The innate component is responsible for the early response to infection or injury and is foreign antigen independent. Toll-like receptors (TLR) which are important for the detection of exogenous microbial products [8] and development of antigen-dependent adaptive immunity [9] also recognize host material released during injury [10]. The role of TLRs was evaluated using an ischemia/reperfusion (I/R) model in TLR2−/− and +/+ mice [11]. Significantly fewer granulocytes were present in the interstitium of the kidney 1 day post-I/R in the TLR2−/− mice and fewer macrophages were present 1–5 days after I/R. Kidney homogenate cytokines KC, MCP-1, interleukin-1β (IL-1β), and IL-6 were also significantly lower in the TLR2−/− animals as compared to the TLR+/+ mice. Hence the absence of TLR2 clearly had an anti-inflammatory effect on the response to I/R. This anti-inflammatory effect was associated with a functional protection as measured by serum creatinine at 1 day post-I/R and blood urea nitrogen and tubular injury score 1 and 5 days post-I/R.

Leukocyte-Endothelial Interactions

With I/R endothelial cells upregulate integrins, selectins, and members of the immunoglobulin superfamily, including intercellular adhesion molecule-1 (ICAM-1) and vascular cell adhesion molecule. A number of vasoactive compounds may also affect leukocyte-endothelial interactions. Vasodilators, such as nitric oxide, also can have effects to decrease inflammation. NO inhibits adhesion of neutrophils to endothelial cells stimulated by TNF-α, which would also be protective [12]. It has been known for quite some time now that there is less flow to the outer medulla in the postischemic kidney [13]. In addition, as endothelial cells are injured with resulting cell swelling and increased expression of cell adhesion molecules, leukocytes are also activated. Enhanced leukocyte-endothelial interactions can result in cell-cell adhesion, which can physically impede blood flow [14]. Furthermore, these interactions will additionally activate both leukocytes and endothelial cells and contribute to the generation of local factors that promote vasoconstriction especially in the presence of other vasoactive mediators, resulting in compromised local blood flow and impaired tubule cell metabolism [15]. Due to the anatomical relationships of vessels and tubules in the outer medulla these leukocyte-endothelial interactions likely impact the outer medulla to a greater extent than the cortex.

In an early study to evaluate the significance of endothelial-leukocyte interactions and inflammation to the pathobiology of ischemic injury we administered anti-ICAM-1 antibodies and found that when administered prior to or 2 h following renal I/R they protected the kidney from injury [16]. We further confirmed these results in finding that kidneys of ICAM-1 knockout mice also are protected [17]. We proposed that this upregulation of ICAM-1 was related to the upregulation of the proinflammatory cytokines TNF-α and IL-1 which we measured to be increased by I/R.

Later phases of AKI are characterized by infiltration of macrophages and T lymphocytes which predominate over neutrophils. ROS are generated during reperfusion and as a result of the inflammatory response then play a major role in cell injury. ROS are generated by activated infiltrating leukocytes and by epithelial cells. ROS are directly toxic to tubular epithelial cells, with ROS generating systems mimicking the effects of ischemic injury [18].

Tubule Contribution to Inflammatory Injury

Both the S3 segment of the proximal tubule and the medullary thick ascending limb are located in the outer stripe of the outer medulla. This region of the kidney is marginally oxygenated under normal conditions and after an ischemic insult, oxygenation is further compromised because the return in blood flow is delayed. Both segments of the nephron contribute to the inflammatory response in AKI [19]. The tubule epithelial cells are known to generate proinflammatory and chemotactic cytokines such as TNF-α, MCP-1, IL-8, IL-6, IL-1β, and TGF-β, MCP-1, IL-8, RANTES and ENA-78 [20]. Proximal tubular epithelia may respond to T lymphocyte activity through activation of receptors for T cell ligands that are expressed on the proximal tubule cell [21]. When CD40 is ligated in response to interaction with CD154, CD40 ligation stimulates MCP-1 and IL-8 production, TRAF6 recruitment, and MAPK activation [21]. CD40 also induces RANTES production by human renal tubular epithelia, an effect which is amplified by production of IL-4 and IL-13 by Th2 cells, a subpopulation of T cells [22]. B7-1 and B7-2 can be induced on proximal tubule epithelial cells in vivo and in vitro. After B7-1 and B7-2 induction, proximal tubule epithelial cells costimulate CD28 on T lymphocytes resulting in cytokine production [23].

Paracrine Effects of Bone Marrow-Derived Stem Cells

There is a potential role for interstitial bone marrow-derived cells in the production of protective paracrine factors that may facilitate repair of the epithelium. We and others have found that injection of mesenchymal stem (stromal) cells (MSCs) is protective against renal injury as assessed by serum creatinine measured 24 h after ischemia [24]. Other groups have also found that MSCs protect against ischemic renal injury by a differentiation-independent mechanism [25]. In the Adriamycin nephropathy model, injection of the side population is also protective in the absence of tubular integration [26]. The mechanism of such protection may be through intrarenal paracrine effects to decrease inflammation or by systemic immune modulation, since injected cells may be rapidly ingested by immune cells in the spleen, liver and lungs. It has been increasingly recognized that MSCs can modulate innate immunity by generating a large number of agents that modify this response [27].

Role of Resolvins and Protectins as Endogenous Modulators of the Inflammatory Response in Kidney Injury

Resolvins (Rv) and protectins (PD) are two newly identified families of naturally occurring n–3 fatty acid docosahexaenoic acid metabolites. These compounds were identified by Serhan et al. [28], and have been proposed to be important for the resolution of inflammation. In collaboration with the Serhan laboratory we recently reported that, in response to bilateral I/R injury, mouse kidneys produce D series resolvins (RvDs) and PD1 [29]. Administration of RvDs or PD1 to mice prior to and subsequent to the ischemia resulted in a reduction in functional and morphological kidney injury. In addition, initiation of RvD1 administration 10 min after reperfusion also resulted in protection of the kidney. Both RvDs and PD1 reduced the number of infiltrating leukocytes and inhibited TLR-mediated activation of macrophages. Interstitial fibrosis after I/R was reduced in mice treated with RvDs. Thus, production of resolvins and protectins may represent an endogenous mechanism of the kidney to control inflammation and may play an important role in resolution of AKI. Whether tissue injury is due, in part, to failure or inadequacy of this anti-inflammation response is not known at the present time.

Conclusions

Renal injury is a dynamic process that often exists in the context of multiple organ failure and involves hemodynamic alterations, inflammation and direct injury to the tubular epithelium followed by a repair process that restores epithelial differentiation and function. Inflammation plays a considerable role in the pathophysiology of AKI. Much emphasis has been placed on understanding mechanisms of inflammation that contribute to the pathophysiology. It is becoming increasingly recognized that there are endogenous mechanisms that the organism brings to bear to control the inflammation. Understanding how these anti-inflammatory processes are regulated may provide insight into how we might intervene to facilitate and enhance them so that we might prevent or mitigate the devastating consequences of AKI.

Acknowledgments

This work was supported by the National Institutes of Health (grants DK 39773, DK54741, DK72381).

References

1. Zuk A, Bonventre JV, Brown D, Matlin KS: Polarity, integrin and extracellular matrix dynamics in the post-ischemic rat kidney. Am J Physiol Cell Physiol 1998;275:C711–C731.
2. Thadhani R, Pascual M, Bonventre JV: Acute renal failure. N Engl J Med 1996;334:1448–1460.
3. Zuk A, Bonventre JV, Matlin KS: Expression of fibronectin splice variants in the postischemic rat kidney. Am J Physiol Renal Physiol 2001;280:F1037–F1053.
4. Bonventre JV, Zuk A: Ischemic acute renal failure: an inflammatory disease? Kidney Int 2004;66:480–485.
5. Park KM, Byun JY, Kramers C, Kim JI, Huang PL, Bonventre JV: Inducible nitric oxide synthase is an important contributor to prolonged protective effects of ischemic preconditioning in the mouse kidney. J Biol Chem 2003;278:27256–27266.
6. Humphreys BD, Duffield JD, Bonventre JV: Renal stem cells in recovery from acute kidney injury. Minerva Urol Nefrol 2006;58:13–21.
7. Bonventre JV: Dedifferentiation and proliferation of surviving epithelial cells in acute renal failure. J Am Soc Nephrol 2003;14(suppl 1):S55–61.
8. Aderem A, Ulevitch RJ: Toll-like receptors in the induction of the innate immune response. Nature 2000;406:782–787.
9. Kaisho T, Akira S: Toll-like receptor function and signaling. J Allergy Clin Immunol 2006;117:979–988.
10. Johnson GB, Brunn GJ, Platt JL: Activation of mammalian Toll-like receptors by endogenous agonists. Crit Rev Immunol 2003;23:15–44.
11. Leemans JC, Stokman G, Claessen N, Rouschop KM, Teske GJ, Kirschning CJ, Akira S, van der Poll T, Weening JJ, Florquin S: Renal-associated TLR2 mediates ischemia/reperfusion injury in the kidney. J Clin Invest 2005;115:2894–2903.
12. Linas S, Whittenburg D, Repine JE: Nitric oxide prevents neutrophil-mediated acute renal failure. Am J Physiol 1997;272:F48–F54.

13   Vetterlein F, Pethö A, Schmidt G: Distribution of capillary blood flow in rat kidney during postischemic renal failure. Am J Physiol 1986;251:H510–H519.
14   Bonventre JV, Weinberg JM: Recent advances in the pathophysiology of ischemic acute renal failure. J Am Soc Nephrol 2003;14:2199–2210.
15   Sheridan AM, Bonventre JV: Cell biology and molecular mechanisms of injury in ischemic acute renal failure. Curr Opin Nephrol Hypertens 2000;9:427–434.
16   Kelly KJ, Williams WW, Colvin RB, Bonventre JV: Antibody to intercellular adhesion molecule-1 protects the kidney against ischemic injury. Proc Natl Acad Sci USA 1994;91:812–816.
17   Kelly KJ, Williams WW, Colvin RB, Meehan SM, Springer TA, Gutierrez-Ramos JC, Bonventre JV: Intercellular adhesion molecule-1 deficient mice are protected against renal ischemia. J Clin Invest 1996;97:1056–1063.
18   Malis CD, Weber PC, Leaf A, Bonventre JV: Incorporation of marine lipids into mitochondrial membranes increases susceptibility to damage by calcium and reactive oxygen species: evidence for enhanced activation of phospholipase A_2 in mitochondria enriched with n–3 fatty acids. Proc Natl Acad Sci USA 1990;87:8845–8849.
19   Bonventre JV, Brezis M, Siegel N, Rosen S, Portilla D, Venkatachalam M: Acute renal failure. I. Relative importance of proximal vs. distal tubular injury. Am J Physiol 1998;275:F623–F632.
20   Bonventre JV, Zuk A: Ischemic acute renal failure: an inflammatory disease? Kidney Int 2004;66:480–485.
21   Li H, Nord EP: CD40 ligation stimulates MCP-1 and IL-8 production, TRAF6 recruitment, and MAPK activation in proximal tubule cells. Am J Physiol Renal Physiol 2002;282:F1020–F1033.
22   Deckers JG, De Haij S, van der Woude FJ, van der Kooij SW, Daha MR, van Kooten C: IL-4 and IL-13 augment cytokine- and CD40-induced RANTES production by human renal tubular epithelial cells in vitro. J Am Soc Nephrol 1998;9:1187–1193.
23   Niemann-Masanek U, Mueller A, Yard BA, Waldherr R, van der Woude FJ: B7-1 (CD80) and B7-2 (CD86) expression in human tubular epithelial cells in vivo and in vitro. Nephron 2002;92:542–556.
24   Duffield JS, Park KM, Hsiao LL, Kelley VR, Scadden DT, Ichimura T, Bonventre JV: Restoration of tubular epithelial cells during repair of the postischemic kidney occurs independently of bone marrow-derived stem cells. J Clin Invest 2005;115:1743–1755.
25   Togel F, Hu Z, Weiss K, Isaac J, Lange C, Westenfelder C: Administered mesenchymal stem cells protect against ischemic acute renal failure through differentiation-independent mechanisms. Am J Physiol Renal Physiol 2005;289:F31–F42.
26   Challen GA, Bertoncello I, Deane JA, Ricardo SD, Little MH: Kidney side population reveals multilineage potential and renal functional capacity but also cellular heterogeneity. J Am Soc Nephrol 2006;17:1896–1912.
27   Stagg J: Immune regulation by mesenchymal stem cells: two sides to the coin. Tissue Antigens 2007;69:1–9.
28   Serhan C, Hong S, Gronert K, Colgan SP, Devchand PR, Mirick G, Moussignac R: Resolvins: a family of bioactive products of omega-3 fatty acid transformation circuits initiated by aspirin treatment that counter proinflammation signals. J Exp Med 2002;196:1025–1037.
29   Duffield JS, Hong S, Vaidya VS, Lu Y, Fredman G, Serhan CN, Bonventre JV: Resolvin D series and protectin D1 mitigate acute kidney injury. J Immunol 2006;177:5902–5911.

Joseph V. Bonventre, MD, PhD
Renal Division, Brigham and Women's Hospital, Harvard Institutes of Medicine
4 Blackfan Circle
Boston, MA 02115 (USA)
Tel. +1 617 525 5960, Fax +1 617 525 5965, E-Mail joseph_bonventre@hms.harvard.edu

Sepsis and Multiple Organ Failure

Michael R. Pinsky

Department of Critical Care Medicine, Bioengineering and Anesthesiology, University of Pittsburgh, Pittsburgh, Pa., USA

Abstract

Background/Aims: Sepsis and multiple organ failure are complex processes that result from dysregulation of the immune response and its associated hematological, hemodynamic and metabolic disturbances. **Methods:** Review of the pathophysiological basis for sepsis and a review of the literature on its mechanisms of expression. **Results:** Sepsis is the host response to an injury, often infectious in origin, that creates both pro- and anti-inflammatory immune responses. The level and duration of this response roughly correlates with outcome. Subcellular injury characterized by increased oxidative stress defines the central mitochondrial component of this process. Treatments which minimize the amplification of this response are usually more effective at reducing tissue injury than are measures aimed at suppressing the inflammatory response. **Conclusions:** Sepsis is a complex process whose expression and treatment are just now being defined. Treatments that minimize the overall host response still represent the most effective strategies.

Copyright © 2007 S. Karger AG, Basel

Severe sepsis is characterized by an intravascular activation of the host's inflammatory pathways releasing potent inflammatory mediators into the circulation [1]. Often referred to as the systemic inflammatory response syndrome [2] it is a nonspecific inflammatory process to a wide variety of insults that is manifest by increased circulating levels of numerous active small molecules, each of which is often capable of inducing a generalized inflammatory response. Furthermore, if sustained, this system-wide inflammatory process may result in loss of the normal cardiovascular regulatory adaptations to stress and metabolic demand and nonischemic organ system dysfunction. Importantly, the distribution of organ system dysfunction and its severity appears to be related to the severity of the circulatory shock initially induced and its associated inflammatory insult, suggesting that the tissue hypoperfusion though important in determining tissue function is not the only process determining organ function in septic shock. Historical perspectives, recent clinical trials of immunomodulating agents and cellular and molecular data have created an exciting and productive story that only recently led to the first positive clinical outcome trial of pharmacotherapy for severe sepsis.

Systemic and Cellular Events of Inflammation

Inflammatory processes require immune cell response, immune effector cell recruitment into the local area and further activation and transcapillary migration if the inciting infectious stimulus is to be contained and the systemic inflammatory response is to be sustained. This innate immunity response involves initial cellular recognition of the stimulant. This initial recognition process involves a complex interaction of host-derived cofactors, such as lipopolysaccharide-binding protein, compliment activation and coating of foreign biological material (bacteria) and release of procoagulant materials. Inflammatory mediators bind to cell surface receptors inducing a transmembrane signal transduction and intracellular response via activation of several specific gene promoter proteins. Numerous unrelated exogenous stimulants, such as endotoxins (lipopolysaccharide or LPS) and exotoxins, and endogenous stimulants, such as activated compliment (C5a), Hageman's factor (XIIa), and products of generalized cell injury can induce immune competent cells (usually modeled as monocytes) to synthesize TNF-α and IL-1β. The closest thing we have to a 'universal receptor' is the cell surface molecule CD14. Many foreign substances, such as endotoxin, require initial binding with lipopolysaccharide-binding protein before they can bind to CD14. Thus, host-derived factors (usually proteins) are probably necessary for the host recognition of foreign materials as foreign and initiate the inflammatory response. However, no such intermediate binding appears necessary for stimulation of the inflammatory response via the intrinsic mediators, such as XIIa and C5a.

Several of these activation complexes nonspecifically bind to the host universal inflammation receptor, CD14, whose transmembrane signal transducing partners, toll-like receptors, activate an intracellular tyrosine kinase system that eventually activates the oxidant sensitive proinflammatory promoter, nuclear factor-kappa B (NF-κB) and other proinflammatory promoters. Simultaneously, expression of novel cell surface receptors, cell adhesion molecules and gene induction associated with new protein synthesis occur. Following this wave of intracellular metabolic activity, immune-competent cells and responsive parenchymal cells make novel protein species necessary to induce a localized inflammatory response essential to host survival in a host environment.

The release of immune active mediators stimulates primary target cell (immune cell) responses via similar mechanisms to those described by the primary

Table 1. Biological effects of various cytokines during acute infections

Cytokine	Biological effects
IL-1β	Induces COX-2, iNOS, and PG(E$_2$) expression Stimulates release of TNF, IL-6, chemokines and other adhesion molecules Stimulates myeloid progenitor cells thereby inducing neutrophilia as well as thrombocytosis Decreases response to erythropoietin and anemia
TNF	Activates macrophages, lymphocytes, neutrophils, eosinophils, fibroblasts, osteoclasts, chondrocytes, endothelial cells, and nerve cells Induces the expression of ICAM-1, ELAM, VACAM-1 Activates COX-2, phospholipase A$_2$, NOS, PAF, PGE$_2$, PGI$_2$, NO Activates complements Induces release of IL-1, IL-6, IL-8, MCP-1, IL-4, IL-10, IL-1ra, PDGF, IL-2, endothelin-1, PAI Downregulates thrombomodulin Increases the expression of surface adhesion molecules, C3bi receptors, L-selectin, superoxide production, and phagocytosis
IL-6	Regulates synthesis of ACTH in pituitary gland and neuronal growth factor Activates Janus kinase/signal transducer and mitogen-activated protein kinase cascades through binding of gp130 receptor molecules Induces thrombosis by enhancing the release of von Willebrand factor fragments and inhibiting its cleaving protease Induces gut barrier dysfunction
IL-10	Downregulates TNF and other proinflammatory cytokines Causes T cell anergy, suppresses T cell and Th1 proliferation Induced T and B cell apoptosis Causes defects in antigen presentation and induces macrophage 'paralysis'

COX-2 = Cyclooxygenase 2; iNOS = inducible nitric oxide synthase; IL-1ra = IL-1 receptor antagonist.

activation sequence, except now more complex and interacting mediator systems coexist, including activation of the contact system (compliment), fibrinolytic system (thrombin and activate protein C) and paired cytokine stimuli (TNF-α and IL-1β, or IL-8 and MIP-1). Secondary parenchymal cell response via these soluble mediators and formed cell interactions then occur via paracrine activation.

Cytokines are presumably present to modulate cellular response and metabolism on a local or paracrine level. A listing of some of biological actions of the early phase cytokines (TNF, IL-1, IL-6, and IL-10) are listed in table 1. With few exceptions, notably pre-pro-IL-1, cytokines have to be synthesized de novo in response to a specific external stimulus and do not exist in a dormant state inside the cell. However, once these cytokines are synthesized and secreted they rapidly gain access to the bloodstream. If the initiating stimulus is great or the host response heightened, the systemic delivery of cytokines can induce a systemic immune cell response. It is still not clear to what extent this systemic response is adaptive or maladaptive to the host. On the one hand, fever and the induction and release of acute phase proteins by the liver improve survival in bacterial infections, and malaise may be a very adaptive symptom in limiting the host activity and excess consumption of limited energy resources. However, the generalized inflammatory response may not confer a survival advantage for the host. Clearly, on a local level, a violent inflammatory response is highly effective at containing, killing and removing foreign biological material (alive or dead) and the lack of a vigorous local inflammatory response may impair survival. This was best exemplified by the recent documentation that patients with a mild TNF-α response, as characterized by a specific genetic genotype of a mild TNF-α responder, had a greater likelihood of dying of meningococcal meningitis than patients who had the vigorous TNF-α response genotype.

Still, persistent activation, rather than massive pulse activation, appears to be detrimental for survival. Pinsky et al. [3] measured the circulating levels of TNF-α, IL-1β, and IL-2 and the immunomodulating cytokines IL-6 and interferon-γ. Although levels of IL-1, IL-2 and interferon-γ did not correlate with any aspect of the generalized inflammatory response, severity of illness or mortality, sustained elevations of the proinflammatory cytokine TNF-α and the immunomodulating cytokine IL-6 did. Thus, rather than the maximal serum levels of any specific cytokines, those patients who subsequently develop multiple organ dysfunction and die display a persistent elevation of TNF-α and IL-6 in their blood [3–5].

Importantly, both proinflammatory cytokines, such as TNF-α, IL-1, IL-6 and IL-8, and anti-inflammatory species, such as IL-1 receptor antagonist, IL-10 and the soluble TNF-α receptors I and II (sTNFrI and sTNFrII, respectively), coexist in the circulation in patients with established sepsis and presumably within the tissues [6–8]. Thus, sepsis may be more accurately described as a dysregulation of the innate immunity rather than merely the overexpression of either proinflammatory substances. This combined pro- and anti-inflammatory mediator interaction may explain the observed failures of all the major anti-inflammatory drug trials in the treatment of septic shock [9]. Accordingly, we proposed 'malignant intravascular inflammation' to describe the systemic process of severe sepsis [10]. This paradoxical expression in the blood of proinflammatory mediators and anti-inflammatory species creates an internal milieu that in sustained sepsis induces impaired host immunity. Experimentally, this altered immune response state resembles 'endotoxin tolerance'. Cavaillon [11] termed this blunted immune response 'inflammatory-stimuli-induced anergy', because it is universally seen in severe stress states and can be induced by prior exposure to low levels of one of many proinflammatory stimuli, not only endotoxin. Hence, its original name 'endotoxin tolerance' is misleading, but will be used here to describe this nonspecific response. Inflammatory-stimuli-induced anergy can be induced by anti-inflammatory cytokines, such as TGF-β, IL-10, IL-4 and somewhat by IL-1β, but not TNF-α, IL-6 or IL-8. Furthermore, it is associated with altered intracellular metabolism of the important regulatory protein NF-κB. Leukocyte dysfunction also occurs in sepsis and in endotoxin-tolerant states and may be a central determinant of outcome [12–15].

Presumably, inflammatory stimuli-induced anergy minimizes the inflammatory response, preventing a chain reaction system-wide activation of the inflammatory processes. However, it also limits the subsequent ability to mount an appropriate inflammatory defense to infection. Since endotoxin tolerance usually requires a few hours to develop in isolated cells and 8–10 h in intact animals, its expression parallels that of clinical sepsis. Thus, we have previously hypothesized that endotoxin tolerance carries most if not all of the intracellular qualities of fully developed severe sepsis. However, the intracellular mechanisms by which inflammatory stimuli-induced anergy are poorly understood, but reflect in large part intracellular change regarding the activation of the inflammatory pathways.

Intracellular Inflammatory Response: Pro- or Anti-Inflammatory Response

NF-κB is an oxidant-inducible promoter protein of the proinflammatory response of immune effector cells [16]. Although other intracellular proinflammatory promoters are also present, none have the breadth of gene activation or complex feedback control mechanisms described for NF-κB. NF-κB-inducible proteins include TNF-α, IL-1β and IL-8 plus the proinflammatory enzyme-inducible nitric oxide synthase and cyclooxygenase 2 [11]. When viewed from a purely regulatory perspective NF-κB is an excellent target for modulating the cellular inflammatory response. It is not surprising, therefore, that several intracellular mechanisms exist that modulate NF-κB activity.

What is endotoxin tolerance and why can it be used to study the molecular mechanisms of sepsis? Exposure to small amounts of endotoxin induces an endotoxin-tolerant state in both cell culture in 4 h and animal models in about 8 h [17]. In this state, subsequent exposure to a previously lethal dose of endotoxin does not induce the fatal proinflammatory state. The endotoxin-tolerant state lasts in decreasing strength for between 24 and 36 h, depending on the species and the initial dose of endotoxin. Interestingly, following the induction of an endotoxin-tolerant state the initial steps of proinflammatory signal transduction up to cleavage of IκB can still occur. However, the liberated NF-κB appears to be dysfunctional [16]. The reasons for this dysfunction are multiple and not fully defined. However, some specific interactions have been defined that speak to the importance of the pro- and anti-inflammatory interactions. Importantly, there is much similarity between endotoxin tolerance and human sepsis. First, sepsis rarely starts with a massive exposure to an overwhelming noxious stimulus. Usually, infection or inflammation build over hours. Thus, the host is exposed to low levels of proinflammatory species with a time course similar to classic endotoxin tolerance. Second, just as endotoxin tolerance minimizes the subsequent proinflammatory response to an additional proinflammatory stimulation, so too sepsis is characterized by a blunted immune effector cell response. By what mechanisms do proinflammatory stimuli induce an apparently anti-inflammatory response? The answer appears to relate to the complex mechanisms by which NF-κB is activated and binds to the proinflammatory promoter regions.

Intracellular NF-κB Activation

The activation of NF-κB is central to immune effector cell activation. Endotoxin can induce the initial steps of signal transduction up to NF-κB, but NF-κB activation is required for much of the subsequent intracellular signaling. The NF-κB family is composed of various members, p50 (NF-κB1), p52 (NF-κB2), p65 (RelA), RelB and c-Rel, which can form homo- and heterodimers [18]. The p65 subunit has the most variability, with a common variation being the RelA subunit substitution for p65. The phosphorylation of the IκB-α subunit of the NF-κB complex following intracellular oxidative stress frees the dimer to translocate into the nucleus [19]. LPS induces IκB-α phosphorylation through activation of the IK kinase [20]. The phosphorylated IκB-α is rapidly degraded by proteosomes. Processes that inhibit IκB-α phosphorylation, such as 4-hydroxynoneal, prevent NF-κB activation [21]. The p65 moiety has a DNA-binding domain that allows it to bind to numerous specific DNA sites throughout the genome, regulating gene transcription for most, if not all, of the proinflammatory species, including TNF-α [22], IL-1β, inducible nitric oxide synthase [23], lipoxygenase, and cyclooxygenase. IκB-α is a heat shock protein (HSP) and its increased synthesis also downregulates NF-κB activation by dissociating the p65-p50 heterodimer from its responsive elements on the genome and keeping it in an inactive form in the cytoplasm [24].

NF-κB DNA-binding activity can be downregulated by processes independent of IκB-α. The NF-κB dimer can exist in one of two forms: a p65-p50 dimer and a p50-p50 homodimer. The p65 subunit has DNA-binding activity, whereas the p50 subunit does not. Thus, activation of p65-p50 dimers results in markedly increased transcription of mRNA following binding, whereas p50-p50 activation only minimally increases transcription rates. NF-κB dysfunction reflects both an excess p50 homodimer production, which has impaired transcription activity [16], and excess synthesis of IκB-α [24]. The balance of NF-κB species is very sensitive to transcription rates, with ratios of NF-κB p50-p65 heterodimer to p50 homodimer of 1.8 ± 0.6 conferring activation of the inflammatory pathways, and a ratio of 0.8 ± 0.1 conferring lack of stimulation in response to LPS (i.e. endotoxin tolerance). The p50 subunit may inhibit TNF mRNA synthesis. Using knockout mice for the p50 subunit of NF-κB, Bohuslav et al. [25] demonstrated that endotoxin tolerance was not achieved from p50−/− mice, long-term TNF mRNA synthesis was not blocked in p50−/− macrophages (in contrast to wild-type macrophages), and ectopic overexpression of p50 reduced transcriptional activation of the TNF promoter. Finally, analysis of the four κB sites from the murine TNF promoter demonstrated that binding of p50 homodimers to a positively acting κB3 element was associated with endotoxin tolerance. These different κB-binding activities may result from specific differences in their sequences. κB2 and κB4 sites do not meet the requirements for p50 binding due to the lack of a GGG motif at the 5′-end and the κB1 site lacks a 3′-end CCC (present on the κB3 site), favorable for binding p50 homodimers. However, these p65-p50/p50-p50-related processes can only explain the downregulation of the overall inflammatory process. They cannot explain why both pro- and anti-inflammatory activation is sustained in severe sepsis, or why immune suppression, a common characteristic of sepsis, coexists with a heightened inflammatory state.

Control of NF-κB activation can occur at many levels. Excess intracellular IκB excess can pull active p50-p65 dimers off their promoter sites or prevent their release altogether. IκB synthesis is stimulated by NF-κB binding to the genome. Thus, in a negative feedback loop process, NF-κB activation stimulates its own inhibition. NF-κB dysfunction due to endotoxin tolerance also reflects excess synthesis of the inhibitor of NF-κB, IκB-α [24], presumably due to the absence of induction of IκB kinase. When endotoxin-tolerant cells are challenged with a second dose of LPS, cytosolic levels of IκB are not reduced, as they are with the initial challenge, and IκB remains in cytoplasm where it sequesters free NF-κB dimers [26]. Furthermore, the IκB-α promoter can be upregulated by NF-κB, thus providing a negative feedback loop for further NF-κB activation [27]. As sustained proinflammatory activation would be highly detrimental to an organism, having intrinsic mechanisms to downregulate this proinflammatory response seems prudent.

As another form of downregulatory control, NF-κB can also alter its own intrinsic gene induction ability. Specifically, NF-κB dimers exist primarily in one of two forms depending on their subunit composition. The specific molecular species usually referred to by the moniker as NF-κB is the p50-p65 heterodimer. As the name p50-p65 implies, it is comprised of the p65 subunit with its active DNA consensus domain-binding site and a smaller p50 subunit devoid of the active binding site. Importantly, the p65 subunit with its DNA consensus domain-binding site allows gene activation once bound to such promoter regions on chromosomes. The p50 monomer has no such activity [16]. Although both p50-p50 and p65-p65 homodimers can theoretically exist, very little p65-p65 has been detected. The primary homodimer is the p50-p50 species. This p50-p50 homodimer can account for more than half the total amount of intracellular NF-κB with immune suppressed cells. Thus, a second adaptive mechanism involves the balance of NF-κB p50-p65 to p50-p50 species. The ratio of p65-p50 to p50-p50 determines NF-κB-induced gene transcription rates. Ratios of NF-κB p50-p65 heterodimer to p50-p50 homodimer of 1.8 ± 0.6 or greater are associated with NF-κB activation inducing mRNA synthesis of these genes, while a ratio of 0.8 ± 0.1 or less confers a lack of mRNA transcription following cleavage of IκB [16].

Downregulation of NF-κB-related intercellular processes is an important aspect of the overall intracellular inflammatory response. Potentially, if anti-inflammatory pathways were activated by the same stimuli that active proinflammatory pathways, then an intrinsic mechanism would exist to autoregulate the inflammatory response on a cellular level. This system may be induced by much of the same stimuli and act through parallel intracellular processes. The cell does have several intrinsic anti-inflammatory processes, including antiproteases, melanoproteins and free radical scavengers. However, the HSP system is by far the most prevalent, in terms of the mass of protein and scope of its oversight.

HSP and the Stress Response

The HSP system is the oldest phylogenetic cellular defense mechanism identified. It has widespread and overarching basic roles in cellular defense against numerous stresses, such as fever, trauma, and inflammation [28]. Intracellular chaperone proteins of the heat shock family appear to be pivotal in the regulation of the cellular response to inflammatory signals and external injury. An initial step in the intracellular activation of the inflammatory pathway is the production of reactive oxygen species (ROS). Mitochondria are not only very sensitive to oxidative stress from ROS but are a primary intracellular site of free radical production. Thus, measuring mitochondrial membrane potential ($\Psi\Delta$) allows one to monitor the degree of intracellular oxidative stress over time [29]. Detection of the mitochondrial permeability transition event also provides an early indication of the initiation of cellular apoptosis. This process is typically defined as a collapse in the mitochondrial membrane electrochemical gradient, as measured by the change in the $\Psi\Delta$. Loss of mitochondrial

ΨΔ can be detected by a fluorescent cationic dye, 5,5′,6,6′-tetrachloro-1,1′,3,3′-tetraethyl-benzamidazolocarbocyanin iodide, commonly known as JC-1. Using this sensitive bioassay of mitochondrial membrane ΨΔ [30] Polla et al. [31] showed that HSP70 prevents this mitochondrial oxidative stress injury and blunts the inflammatory response. Numerous other studies have shown that HSP are a basic cellular defense mechanism against numerous stresses, such as fever, trauma, and inflammation [28, 31]. They also minimize nitric oxide, ROS and streptozotocin cytotoxicity [32, 33]. HSP70 is also an inducible protective agent in myocardium against ischemia, reperfusion injury and nitric oxide toxicity [34], but requires some initial trigger for its activation. Such induced HSP synthesis blunts the inflammatory response to endotoxin and TNF-α in vitro. Thus, the intracellular pathway mobilized upon heat shock may be important in modulating the intracellular inflammatory signal acting at the level of NF-κB and appears to have many of the immune modulating characteristics of endotoxin tolerance. Since heat shock is a primordial cellular defense mechanism, upregulates in a matter of minutes and rapidly confers an immune-depressed state, we consider its actions to reflect a relatively pure antioxidant, anti-inflammatory response. Although the time courses for production of endotoxin tolerance and heat shock are different, similar gene activation and inhibition occur with both processes.

The heat shock factors (HSFs), particularly HSF-1, are primarily responsible for inducing the transcription of HSP genes and can be considered the NF-κB equivalent for HSP synthesis. HSF exist preformed in the cytosol bound to HSP70. The presence of denatured protein (resulting from heat or ROS-induced protein damage) strips the HSP70 proteins from HSF as HSP70 binds to the damaged proteins. Trimerized HSF migrates into the nucleus to bind to numerous regions within the genome. HSF-1 DNA-binding sites possess multiple repeats of a 5′-G-A-A-3′ triplet, often in an inverted orientation and separated by at least two nucleotides (e.g. 5′-G-A-A- N-N- T-T-C-3′). According to Amin et al. [35], 'a functional heat shock regulatory element includes a minimum of three GAA segments although these segments do not have to be consecutive'. Importantly in natural heat shock response elements (HSEs) single nucleotide substitutions of the GAA core triplet are found [35]. Although HSF-1 and NF-κB are not structurally related, the DNA recognition elements for both factors can be somewhat similar. For example, the core NF-κB recognition element is comprised of the sequence 5′-G-G-G-R-N-W-T-T-C-C-3′ (where R = purine, N = any nucleotide and W = A or T) [36]. Interestingly, one of the NF-κB-responsive elements within the TNF-α gene contains the sequence 5′-G-G-G-A-A-A-G-C-C-C-3′ which contains only one mismatch with a partial HSE core element [36]. Likewise, the interferon-γ gene promoter contains an NF-κB response element with the sequence 5′-G-A-A-T-T-T-T-C-C-3′, which contains a perfectly spaced and matched pair of HSE GAA triplets (one in an inverted orientation) [36]. Even in natural HSEs, a third GAA element does not have to be perfectly matched in sequence or spacing to constitute a functional element. Thus, the possibility exists that some NF-κB sites may be recognized by HSF-1.

Putative recognition of specific NF-κB response elements by HSF-1 does not necessarily imply that transcription would ensue. Transcriptional activation by DNA-bound HSF-1 requires site-specific phosphorylation, which apparently is brought about by the precise coordination of multiple stress-activated kinases. Indeed, HSF-1 has been implicated in transcriptional repression of the IL-1β gene promoter [29]. The mechanism for this repression appears to require the binding of HSF-1 to an HSE-like sequence within the IL-1β promoter [29]. One of the phenotypes observed in HSF-1 knockout mice also supports a role for HSF-1 in transcriptional repression of proinflammatory genes. Following an endotoxic challenge (i.e. *Echerichia coli* LPS), HSF-1 knockout mice exhibited a potentiation of proinflammatory TNF-α production [37]. Since HSP induction is severely limited in the HSF-1 knockout mice [37], the effect of HSF-1 on TNF-α induction does not appear to require HSF-1-induced HSP, such as HSP70. Thus, while HSP70 stabilization of IκB-α has been postulated to participate in downregulation of NF-κB [38], other levels of control may operate under conditions of endotoxic shock. The IκB-α promoter region has an NF-κB recognition site, and NF-κB activation increases IκB-α synthesis in a negative-feedback loop fashion [39, 40]. Potentially, HSF-1 may inhibit NF-κB-induced transcription promotion via tethering of NF-κB to its responsive element in a fashion similar to glucocorticoids [41]. The anti-inflammatory action of glucocorticoid hormones is mediated by its binding its cognate receptor, the glucocorticoid receptor. Glucocorticoid repression of NF-κB-directed transcription is brought about, in part, by the association of the glucocorticoid receptor with DNA-bound NF-κB. Glucocorticoids inhibit the inflammatory process by binding to a glucocorticoid receptor. The glucocorticoid receptor then can attach to the DNA-bound NF-κB in a site remote from the recognition site preventing transcription. Potentially, HSF-1 inhibition of NF-κB activation involves a tethering mechanism. If so, then competitive binding studies may not demonstrate NF-κB displacement by HSF-1.

HSP confer a survival advantage to their host. Thermal pretreatment is associated with attenuated lung damage in a rat model of acute lung injury induced by intratracheal instillation of phospholipase A_2 [42]. Thermal pretreatment reduces mortality rate and sepsis-induced acute lung injury produced by cecal ligation and perforation [43]. The subsequent increased expression of a broad variety of HSP confers a nonspecific protection from not only subsequent oxidative stress but also minimizes the cellular response to proinflammatory stimuli. Survival in cold-blooded animals given an infectious inoculum is linearly related to body temperature. Finally, if specific HSP are depleted, then multiple intracellular signaling processes can be affected.

The Role of Mitochondria in Intracellular Inflammation

An initial step in the intracellular activation of the inflammatory pathway between mediator binding to the cell surface and inflammatory gene activation is the production of ROS and an associated oxidative stress on the mitochondria [44]. Mitochondria operate by generating a chemiosmotic gradient via the Krebs cycle inside their inner membrane necessary to drive ATP formation across this membrane. Essentially they are intracellular batteries whose charge or polarization level defines their ability to create ATP. Loss of internal membrane polarization induces cytochrome c release from the mitochondria into the cytosol. Importantly, cytochrome c activates the intrinsic caspase system to initiate apoptosis or programmed cell death. Thus, preventing mitochondrial depolarization would be an important function for HSP if their role were to aid in cell survival. In vivo measures of mitochondrial polarization are possible using redox state-sensitive substances that can be introduced into cells in vitro.

HSP70 prevents this mitochondrial oxidative stress and blunts the inflammatory response. HSP70 also minimizes nitric oxide, oxygen-free radical and streptozotocin cytotoxicity [32]. Consistent with the pluripotential effects of the heat shock response, HSP70 is also an inducible protective agent in myocardium against ischemia, reperfusion injury and nitric oxide toxicity [34]. Nitric oxide-induced HSP synthesis blunts the inflammatory response to endotoxin and TNF-α in vitro. Evidence that HSP70 may be active in human sepsis comes from the observation that higher HSP70 expression is seen in peripheral mononuclear cells in septic patients. Although not conclusive, these data strongly suggest that HSP, and HSP70 in particular, may be important in modulating the intracellular inflammatory signal acting at the level of NK-κB. Importantly, Wong et al. [45] identified a potential heat shock-responsive element in the IκB-α promoter that can be activated by HSF after heat shock. Moreover, the heat shock response can also modulate NF-κB inactivation by this increased IκB-α expression. Thus, the primary inhibitor of NF-κB activation, IκB-α, is itself an HSP. The interaction between the HSP system and the proinflammatory pathway, however, has yet to be defined.

Since mitochondrial dysfunction causes leakage of cytochrome c that activates the caspase system leading to programmed cell death, one of the outcomes from such activation is a shut-down of all extrinsic cellular activities, like active transport of ions or, in the case of muscle, contraction. Thus, an early expression of mitochondrial dysfunction is a nonspecific organ dysfunction, which if sustained, leads to organ failure. Finally, mitochondria are essential free radical generators. Their primary role is to create free radicals that drive the electron transport chain to create the chemiosmotic gradient needs to synthesize ATP for ADP and inorganic phosphate. Thus, with disruption of mitochondrial membrane integrity, intracellular free radical release often occurs increasing the intracellular oxidative stress. Although the linkage between mitochondrial dysfunction and apoptosis is strong and well described, the linkage between mitochondrial disruption and intracellular oxidative stress is not. It is not clear if free radical poisoning does occur in sepsis. It does occur with ischemic necrosis and induces a profound inflammatory response.

Sepsis Is Characterized by Excessive Pro- and Anti-Inflammatory Activity

Both pro- and anti-inflammatory processes are ongoing in the cell and both pro- and anti-inflammatory mediators are present in the bloodstream. Thus, the circulating and fixed immune effector cells also receive mixed messages. However, the phenotypic response that they make is more difficult to predict. The normal cellular inflammatory response is essential to survival. It localizes and eliminates foreign material, including microorganisms. Similarly, some degree of systemic inflammatory response is useful. Fever reduces microorganism growth, malaise causes the host to rest, and acute phase protein secretion minimizes oxidative injury. However, inflammation is also destructive. Local abscess formation and multiple system organ failure are its very real byproducts. However, sepsis also carries a strong anti-inflammatory response. The exact balance or interaction among these two processes with their multiple layers of feedback, activation and control is difficult, if not impossible, to assess clinically.

Prior studies documented that sepsis and all acute severe processes result in the expression in the systemic circulation of cytokines. However, it is difficult to assess the degree of inflammatory stimulation by measuring serum cytokine levels. Serum cytokine levels can change within minutes and may be very different in adjacent tissue compartments [46]. Although TNF and IL-6 serum levels are excellent markers of disease severity and a good positive predictor of the subsequent development of remote organ system dysfunction, measuring blood levels of cytokines does not aid in defining the pro- to anti-inflammatory balance in predicting response to therapy.

Attention has subsequently shifted to examination of the functional status of circulating immune effector cells. Since polymorphonuclear leukocyte (PMN) activation and localization represent the initial cellular host defense against infection, their tight control is essential to prevent widespread nonspecific injury. Subsequently, monocytes localize at the site of inflammation. Their activity appears to become the predominant process in both host defense and repair, especially during the second and third day onward in the course of acute illness. Thus, inhibition of monocyte immune responsiveness is a powerful mechanism to downregulate the inflammatory response. Anergy is a cardinal characteristic of severe illness and reflects macrophage inhibition of antigen processing. Importantly, antigen processing reflects a primary aspect of this cellular response. In this regard, the cell surface receptor family, HLA-DR, is responsible for antigen presentation to antigen processing cells. Immature monocytes cannot process antigen and have lower cell surface HLA-DR levels. Docke et al. [47] demonstrated that monocytes require HLA-DR levels $>20\%$ for normal cell-mediated immunity. Lower levels of HLA-DR expression confirm immune suppression. Consistent with the overall theme of increased anti-inflammatory responses in severe sepsis, these workers and others have found a profound decrease in HLA-DR expression on circulating monocytes from patients with sepsis.

Rosenbloom et al. [48] examined the relation between circulating cytokine levels and the expression of the strong β_2-integrin surface cell adhesion molecules on circulating immune effector cells. The activation state of circulating immunocytes can be indirectly assessed by measuring the intensity of display on their β_2-integrin surface cell adhesion molecules. They showed that all circulating immune effector cells, including PMNs, lymphocytes and monocytes are activated in critical illness. Furthermore, the level of circulating immune effector cell activation is proportional to mean circulating IL-6 levels [48]. Importantly, the degree of organ dysfunction but not the level of shock severity correlated with CD11b expression. CD11b is part of the cell integrin system essential for cell adhesion in the inflammatory response. Since the level of activation of circulating PMN, as measured by total PMN count and its display of immature forms, is used clinically as an indicator of the host response to systemic inflammation, assessment of PMN responsiveness should also be a good measure of the pro- and anti-inflammatory balance in severe sepsis. They reasoned that immune effector cell responsiveness to a known dose of a pro-inflammatory stimulant would define the host's immune responsiveness. In essence, they used the circulating immune effector cells as a bioassay. Prior studies have shown that PMNs can be both overactive [49] and dysfunctional [12], and that their CD11b display can be either decreased [50] or increased [51] in critically ill patients.

Rosenbloom et al. [52] demonstrated that the de novo display of CD11b on circulating PMNs and its subsequent change in expression of both total CD11b and its avid form, CBRM1/5 epitope, in response to in vitro stimulation to TNF-α characterized the in vivo state of PMN activation and responsiveness. Circulating PMNs of septic humans have a similar phenotype characterized by high CD11b and low L-selectin expression. This is the phenotype of acute activation of the inflammatory response. Thus, severe sepsis is associated with an increased de novo activation of circulating immune effector cells. Paradoxically, however, those same PMNs with a sustained inflammatory state are also impaired in their ability to upregulate CD11b further or to change surface CD11b to the avid state [53] in response to an ex vivo challenge by exposure to biologically significant levels of TNF-α. Furthermore, circulating PMN for subjects with severe sepsis have impaired phagocytosis, reduced oxygen burst capacity and diminished in vitro adhesiveness [54]. This desensitization to exogenous TNF-α was not due to a loss of TNF receptors because the cell surface TNF-α receptor density was not reduced in the cells of these septic patients. Importantly, the hyporesponsiveness observed was extended to all circulating PMN and monocytes in these critically ill patients.

Finally, when peripheral blood monocytes from septic subjects were analyzed for the NF-κB activity and in vitro responsiveness to LPS, Adib-Conquy et al. [17] observed an NF-κB pattern of response similar to that seen with endotoxin tolerance. The cause of the reduced nuclear translocation of NF-κB was not due to excess IκB but rather to an increase in the proportion of the inactive p50-p50 species relative to the active p65-p50 species. Interestingly, survivors had higher levels of NF-κB than did nonsurvivors, suggesting that although downregulation of inflammation is a normal aspect of sepsis, excessive inhibition of the process is associated with a poor prognosis. Clearly, this is an area of active investigation and the complex interactions amongst pro- and anti-inflammatory processes are just now being teased out.

Resuscitation and Reversal of Organ Injury

Given the above complex processes one may reasonably ask: how does any individual survive a severe inflammatory insult like massive bacteremia, pancreatitis or massive trauma? Clearly many patients do survive. Several aspects of the above processes lend themselves to survival. First, as described above, the inflammatory response is immediately downregulated by the anti-inflammatory heat shock response limiting immune effector cell activation and mitochondrial injury. Thus, if the inciting stimulus is removed, the system tends to right itself. Second, inflammation does not exist in isolation, but in concert with a deranged circulatory state, wherein vasomotor tone is reduced and the responsiveness of the cardiovascular system to increased sympathetic stimulation is diminished. Within this context, Rivers et al. [55] aggressively resuscitated patients presenting in septic shock to the emergency department of a large inner city hospital. They compared resuscitation to either a normal blood pressure and sensorium or further increasing O_2 delivery until indirect measures of tissue ischemia were resolved. They found that not only did the more aggressively resuscitated patients have a better outcome in terms of reduced mortality and shorter length of stay in the hospital, but they also had longer circulating levels of TNF and IL-6. These preliminary data suggest that aggressive goal-directed resuscitation of patients early in their septic event may reduce the overall inflammatory response.

Acknowledgment

This work was supported in part by the NIH grants HL67181, HL07820 and HL073198.

References

1 Schlag G, Redl H: Mediators of injury and inflammation. World J Surg 1996;20:406–410.
2 Bone RC: Toward a theory regarding the pathogenesis of the systemic inflammatory response syndrome: what we do and do not know about cytokine regulation. Crit Care Med 1996;24:163–172.
3 Pinsky MR, Vincent JL, Deviere J, Alegre M, Kahn R J, Dupont E: Serum cytokine levels in human septic shock. Relation to multiple-system organ failure and mortality. Chest 1993;103: 565–575.
4 Thijs LG, Hack CE: Time course of cytokine levels in sepsis. Intensive Care Med 1995;21(suppl 2):S258–S263.
5 Blackwell TS, Christman JW. Sepsis and cytokines: current status. Br J Anaesth 1996;77:110–117.
6 Goldie AS, Fearon KC, Ross JA, Barclay GR, Jackson RE, Grant IS, Ramsay G, Blyth AS, Howie JC: Natural cytokine antagonists and endogenous anti-endotoxin core antibodies in sepsis syndrome. The Sepsis Intervention Group. JAMA 1995;274:172–217.
7 Vanderpoll T, Malefyt RD, Coyle SM, Lowry SF: Anti-inflammatory cytokine responses during clinical sepsis and experimental endotoxemia – sequential measurements of plasma soluble interleukin (IL)-1 receptor type II, IL-10, and IL-13. J Infect Dis 1997;175:118–122.
8 Ertel W, Scholl FA, Trentz O: The role of anti-inflammatory mediators for the control of systemic inflammation following severe injury; in Faist E, Baue AE, Schildberg FW (eds): The Immune Consequences of Trauma, Shock, and Sepsis – Mechanisms and Therapeutic Approaches. Lengerich, Pabst Science Publishers, 1998, pp 453–470.
9 Abraham E: Why immunomodulatory therapies have not worked in sepsis. Intensive Care Med 1999;25:556–566.
10 Pinsky MR: Clinical studies on cytokines in sepsis: role of serum cytokines in the development of multiple-systems organ failure. Nephrol Dial Transplant 1994;9(suppl 4):94–98.
11 Cavaillon JM: The nonspecific nature of endotoxin tolerance. Trends Microbiol 1995;3:320–324.
12 McCall CE, Grosso-Wilmoth LM, LaRue K, Guzman RN, Cousart SL: Tolerance to endotoxin-induced expression of the interleukin-1 beta gene in blood neutrophils of humans with the sepsis syndrome. J Clin Invest 1993;91:853–861.
13 Wenisch C, Parschalk P, Hasenhundl M, et al: Polymorphonuclear leukocyte dysregulation in patients with gram-negative septicemia assessed by flow cytometry. Eur J Clin Invest 1995;25:418–424.
14 Vespasiano MC, Lewandoski JR, Zimmerman JJ: Longitudinal analysis of neutrophil superoxide anion generation in patients with septic shock. Crit Care Med 1993;21:666–672.
15 Sorrell TC, Sztelma K, May GL: Circulating polymorphonuclear leukocytes from patients with gram-negative bacteremia are not primed for enhanced production of leukotriene B4 or 5-hydroxyeicosatetraenoic acid. J Infect Dis 1994;169:1140–1151.
16 Ziegler-Heitbrock HWL, Wedel A, Schraut W, Strobel M, Wendelgass P, Sterndorf T, Bauerle PA, Haas JG, Riethmuller G: Tolerance to lipopolysaccharide involves mobilization of nuclear factor κB with predominance of p50 homodimers. J Biol Chem 1994;269:17001–17004.

17 Adib-Conquy M, Adrie C, Moine P, Ashnoune K, Fitting C, Pinsky MR, Dhainaut J-F, Cavaillon J-M: NF-κB expression in mononuclear cells of septic patients resembles that observed in LPS tolerance. Am J Respir Crit Care Med 2000;162:1877–1883.
18 Schleiffenbaum B, Fehr J: The tumor necrosis factor receptor and human neutrophil function. Deactivation and cross-deactivation of tumor necrosis factor-induced neutrophil responses by receptor down-regulation. J Clin Invest 1990;86:184–195.
19 Kamata H, Hirata H: Redox regulation of cell signaling. Cell Signal 1999;11:1–14.
20 Sanlioglu S, Williams CM, Samavati L, et al: Lipopolysaccharide induces Rac1-dependent reactive oxygen species formation and coordinates tumor necrosis factor – a secretion through IKK regulation of NF-κB. J Biol Chem 2001;276:30188–30198.
21 Page S, Fischer C, Baumgartner B, et al: 4-Hydroxynoneal prevents NF-κB activation and tumor necrosis factor expression by inhibiting IκB phosphorylation and subsequent proteolysis. J Biol Chem 1999;274:11611–11618.
22 Young S-H, Ye J, Frazer DG, et al: Molecular mechanism of tumor necrosis factor-α production in 1?3-β-glucan (zymosan)-activated macrophages. J Biol Chem 2001;276:20781–20787.
23 Taylor BS, de Vers ME, Ganster RW, et al: Multiple NF-κB enhancer elements regulate cytokine induction of human inducible nitric oxide synthase gene. J Biol Chem 1998;273:15148–15156.
24 Larue KEA, McCall CE: A liable transcriptional repressor modulates endotoxin tolerance. J Exp Med 1994;180:2269–2275
25 Bohuslav J, Kravchenko VV, Parry GCN, Erlich JH, Gerondakis S, Mackman N, Ulevitch RJ: Regulation of an essential innate immune response by the p50 subunit of NF-κB. J Clin Invest 1998;102:1645–1652.
26 Kohler NG, Joly A: The involvement of an LPS inducible IκB kinase in endotoxin tolerance. Biochem Biophys Res Commun 1997;232:602–607.
27 De Martin R, Vanhove B, Cheng Q, Hofer E, Csizmadia V, Winckler H, Bach FH: Cytokine-inducible expression in endothelial cells of an IκBα-like gene is regulated by NFκB. EMBO J 1993;12:2773–2779.
28 Jäättelä M, Wising D: Heat shock proteins protect cells from monocyte cytotoxicity: possible mechanism of self-protection. J Exp Med 1993;177:231–236.
29 Cahill CM, Waterman WR, Xie Y, et al: Transcriptional repression of the prointerleukin 1beta gene by heat shock factor 1. J Biol Chem 1996;271:24874–24879.
30 Cossarizza A, Baccarani-Contri M, Kalashnikova G, et al: A new method for the cytofluorimetric analysis of mitochondrial membrane potential using the J-aggregate forming lipophilic cation 5,5?,6,6'-tetrachloro-1,1',3,3?-tetraethylbenzimidazolyl-carbocyanine iodide (JC-1). Biochem Biophys Res Commun 1993;197:40–45.
31 Polla BS, Kantegwa D, François S, et al: Mitochondria are selective targets for the protective effects of heat shock against oxidative injury. Proc Natl Acad Sci USA 1996;93:6458–6463.
32 Bellmann K, Wenz A, Radons J, Burkart V, Kleemann R, Kolb H: Heat shock induces resistance in rat pancreatic islet cells against nitric oxide, oxygen radicals and streptozotocin toxicity in vitro. J Clin Invest 1995;95:2840–2845.
33 Gabai VL, Kabakov AE: Rise in heat-shock protein level confers tolerance to energy deprivation. FEBS Lett 1993;327:247–250.
34 Malyshev IY, Malugin AV, Golubeva LY, Zenina TA, Manukhina EB, Mikoyan VD, Vanin AF Nitric oxide donor induces HSP70 accumulation in the heart and in cultured cells. FEBS Lett 1996;391:21–23.
35 Amin J, Ananthan J, Voellmy R: Key features of heat shock regulatory elements. Mol Cell Biol 1988;8:3761–3769.
36 Chen FE, Ghosh G: Regulation of DNA binding by Rel/NF-κB transcription factors: structural views. Oncogene 1999;18:6845–6852.
37 Xiao ZX, Zuo XX, Davis AA, et al: HSF1 is required for extra-embryonic development, postnatal growth and protection during inflammatory responses in mice. EMBO J 1999;18:5943–5952.
38 Santoro MG: Heat shock factors and the control of the stress response. Biochem Pharmacol 2000;59:55–63.
39 van der Poll T, Calvano SE, Kumar A, et al: Endotoxin induces downregulation of tumor necrosis factor receptors on circulating monocytes and granulocytes in humans. Blood 1995;86:2754–2759.

40 Detmers PA, Powell DE, Walz A, et al: Differential effects of neutrophil-activating peptide 1/IL-8 and its homologues on leukocyte adhesion and phagocytosis. J Immunol 1991;147:4211–4217.
41 Steer JH, Kroeger KM, Abraham LJ, et al: Glucocorticoids suppress tumor necrosis factor-alpha expression by human monocytic THP-1 cells by suppressing transactivation through adjacent NF-κB and c-Jun-activating transcription factor-2 binding sites in the promoter. J Biol Chem 2000; 275:18432–18440.
42 Villar J, Edelson JD, Post M, Mullen BM, Slutsky AS: Induction of the heat stress proteins is associated with decreased mortality in an animal model of acute lung injury. Am Rev Respir Dis 1993;147:177–181
43 Villar J, Ribiero SP, Mullen BM, Kuliszewski M, Post M, Slutsky AS: Induction of the heat shock response reduces mortality rate and organ damage in a sepsis-induced acute lung injury model. Crit Care Med 1994;22:914–921.
44 Polla BS, Jacquier-Sarlin MR, Kantengwa S, Mariethoz E, Hennet T, Russo-Marie F, Cossarizza A: TNF-α alters mitochondrial membrane potential in L929 but not in TNFα-resistance L929.12 cells: relationship with the expression of stress proteins, annexin 1 and superoxide dismutase activity. Free Radic Res 1996;25:125–131.
45 Wong HR, Ryan M, Wispé JR: Stress response decreases N-FκB nuclear translocation and increases I-κBα expression in A549 cells. J Clin Invest 1997;99:2423–2428.
46 Boutten A, Dehoux MS, Seta N, Ostinelli J, Venembre P, Crestani B, Dombret MC, Durand G, Aubier M: Compartmentalized IL-8 and elastase release within the human lung in unilateral pneumonia. Am J Respir Crit Care Med 1996;153:336–342.
47 Docke WD, Syrbe U, Meinecke A, Platzer C, Makki A, Asadullah K, Klug C, Zuckermann H, Reinke P, Brunner H, von Baehr R, Volk HD: Improvement of monocyte function – a new therapeutic approach? in Reinhart K, Eyrich K, Sprung C (eds): Sepsis: Current Perspectives in Pathophysiology and Therapy. Berlin, Springer, 1994, pp 473–500.
48 Rosenbloom AJ, Pinsky MR, Bryant JL, Shin A, Tran T, Whiteside T: Leukocyte activation in the peripheral blood of patients with cirrhosis of the liver and SIRS. Correlation with serum interleukin-6 levels and organ dysfunction. JAMA 1995;274:58–65.
49 Trautinger F, Hammerle AF, Poschl G, Micksche M: Respiratory burst capability of polymorphonuclear neutrophils and TNF-alpha serum levels in relationship to the development of septic syndrome in critically ill patients. J Leukoc Biol 1991;49:449–454.
50 Nakae H, Endo S, Inada K, Takakuwa T, Kasai T: Changes in adhesion molecule levels in sepsis. Res Commun Mol Pathol Pharmacol 1996;91:329–338
51 Lin RY, Astiz ME, Saxon JC, Rackow EC: Altered leukocyte immunophenotypes in septic shock. Studies of HLA-DR, CD11b, CD14, and IL-2R expression. Chest 1993;104:847–853.
52 Rosenbloom AJ, Pinsky MR, Napolitano C, Nguyen T-S, Levann D, Pencosky N, Dorrance A, Ray BK, Whiteside T: Suppression of cytokine mediated β2-integrin activation on circulating neutrophils in critically ill patients. J Leukoc Biol 1999;66:83–89.
53 Diamond MS, Springer TA: A subpopulation of Mac-1 (CD11b/CD18) molecules mediates neutrophil adhesion to ICAM-1 and fibrinogen. J Cell Biol 1993;120:545–556.
54 Terregino CA, Lubkin C, Thom SR: Impaired neutrophil adherence as an early marker of systemic inflammatory response syndrome and severe sepsis. Ann Emerg Med 1997;29:400–403.
55 Rivers E, Nguyen B, Havstad S, Ressler J, Muzzin A, Knoblich B, Peterson E, Tomlanovich E: Early goal-directed therapy in the treatment of severe sepsis and septic shock. N Engl J Med 2001;345:1368–1377.

Michael R. Pinsky, MD
606 Scaife Hall
3550 Terrace Street
Pittsburgh, PA 15261 (USA)
Tel. +1 412 647 5387, Fax +1 412 647 8060, E-Mail pinskymr@upmc.edu

Classification, Incidence, and Outcomes of Sepsis and Multiple Organ Failure

Jean-Louis Vincent, Fabio Taccone, Xavier Schmit

Department of Intensive Care, Erasme Hospital, Free University of Brussels, Brussels, Belgium

Abstract

Background/Aims: Sepsis and multiple organ failure are common complications in intensive care unit (ICU) patients and are associated with considerable morbidity and mortality. **Methods:** We reviewed pertinent medical literature related to sepsis and multiple organ failure to determine strategies of classification, the current incidence, and the outcomes associated with these disease processes. **Results:** Sepsis affects some 40% of ICU admissions, severe sepsis occurs in about 30%, and septic shock in 15%. Recent consensus has improved the definition of sepsis and proposed a new classification system based on predisposing factors, infection, immune response, and organ dysfunction. We discuss the possible components of each of these four categories. **Conclusion:** Although there is some evidence that mortality rates may have decreased in recent years, the incidence of sepsis is increasing so that overall deaths from this disease are increasing. Improved diagnostic techniques and classification may help target therapies more rapidly and more appropriately.

Copyright © 2007 S. Karger AG, Basel

Sepsis is an increasingly common problem in the intensive care unit (ICU) and severe sepsis and septic shock are associated with high morbidity, mortality, and costs. Attempts have been made recently to improve definitions and characterization of sepsis and multiple organ failure (MOF). Large epidemiological observational studies conducted in recent years have provided important information about the scale of this problem and its impact, and suggest that it is becoming more common, likely due to aging populations, the prolonged survival of patients who would previously have died earlier, and increased use of immunosuppressive agents. Such data are valuable in determining appropriate resource allocation and in providing a baseline for ongoing assessment of the impact of the introduction of new therapeutic interventions and strategies.

Classification of Sepsis and MOF

One of the difficulties in the field of sepsis over the years has been in determining how best to define it. Unlike many other disease processes, where there are relatively clear signs and symptoms, a diagnosis of sepsis in a critically ill patient is often complicated; signs and symptoms are not specific, there is no simple marker, no magical imaging technique, and microbiological cultures are unreliable, being negative in some 40% of patients [1]. A Consensus Conference organized by the American College of Chest Physicians (ACCP) and the Society of Critical Care Medicine (SCCM) in 1991 [2] coined the term systemic inflammatory response syndrome (SIRS) and defined sepsis as SIRS occurring when infection was present. To meet the SIRS criteria, patients needed to satisfy at least two of the following: fever or hypothermia, tachycardia, tachypnea or hyperventilation, leukocytosis or leukopenia. Severe sepsis was defined as sepsis complicated by organ dysfunction, and septic shock as severe sepsis with persistent arterial hypotension. However, the SIRS criteria can be met by many ICU patients and are too nonspecific to be widely useful in the diagnosis of sepsis or in classifying patients with sepsis. A more recent sepsis definition conference, sponsored by the SCCM, the ACCP, the American Thoracic Society (ATS), the European Society of Intensive Care Medicine (ESICM), and the Surgical Infection Society (SIS), decided that the SIRS concept should be abandoned and replaced by an expanded list of signs and symptoms of sepsis that are more representative of the clinical response to infection [3].

Sepsis is a complex disease process and the population of patients who develop sepsis is highly heterogeneous. Sepsis can affect all patients of any age and sex, who may have multiple comorbidities or none at all. It can be caused by bacteria, viruses, or fungi, and can arise in any organ of the body. Patients with sepsis develop different degrees of immune response and their response varies over time. One way of characterizing patients with sepsis is to use mortality prediction or severity of illness scores, like the acute physiology and chronic health evaluation (APACHE) and simplified acute physiology score (SAPS) systems. Rather than an aggregate score, an organ dysfunction score, like the sequential organ failure assessment (SOFA) score, describes better the degree of dysfunction of each individual organ and can be calculated repeatedly. However, organ dysfunction and illness severity are only two small components of the intricate septic process, and a more complex and inclusive staging system is required. The Sepsis Consensus Conference suggested a way in which all the heterogeneous characteristics could be included into a staging system to enable patients with sepsis to be better grouped, thus facilitating research and perhaps enabling therapies to be more appropriately targeted. They called this system, PIRO, with the initials of the acronym standing for four key features of the sepsis response:

predisposing factors, infection, response, and organ dysfunction [3]. Broadly based on the TNM (tumor, nodes, metastases) staging system for cancers, it is envisaged that points could be allocated to each of these aspects, such that a patient with sepsis could, for example, be staged as $P_1I_2R_1O_0$ [4], depending on the features present for each of the four PIRO components.

Predisposing Factors

There are many factors that predispose a patient to developing sepsis, including age, sex, comorbidities, and genetic makeup. Older patients are more likely to develop severe sepsis than their younger counterparts [5], and more likely to have a worse outcome. Men seem to be at greater risk of developing sepsis and MOF [6–8] although the effect of sex on outcome from sepsis is less clear [7, 9], and may be related in part to hormonal status. Various genetic factors also influence the development, severity and outcome from sepsis [10–14], as do the presence of certain comorbidities or ongoing or recent immunosuppressive medication.

Infection

Sepsis is associated with an underlying infection, which needs to be characterized, typically by the source of infection and the incriminated organisms. These features can also be related to severity and outcome. For example, pneumonia is more likely to be associated with severe sepsis than urinary tract infection, certain microorganisms, e.g., *Pseudomonas* species, are associated with a greater risk of death [1], and organisms resistant to multiple antimicrobial agents will be less likely to respond to therapy and, therefore, may also be associated with worse outcomes.

Response

The host response in sepsis varies among patients and in the same patient over time and the severity or nature of the response may influence how a patient responds to particular therapies. Various factors will influence a patient's response to sepsis, as listed under predisposing factors, and techniques to assess and monitor the degree of host response are urgently needed. Markers of sepsis, including C-reactive protein and procalcitonin, are useful [15–17], but due to the complexity of the sepsis response a combination of markers is likely to be needed to fully assess a patient's immune status. Advances in proteomic and microarray techniques are likely to facilitate classification of the host response.

Organ Dysfunction

The presence of sepsis-associated organ dysfunction is an indication of the severity of the sepsis response, and is associated with outcome, the greater the

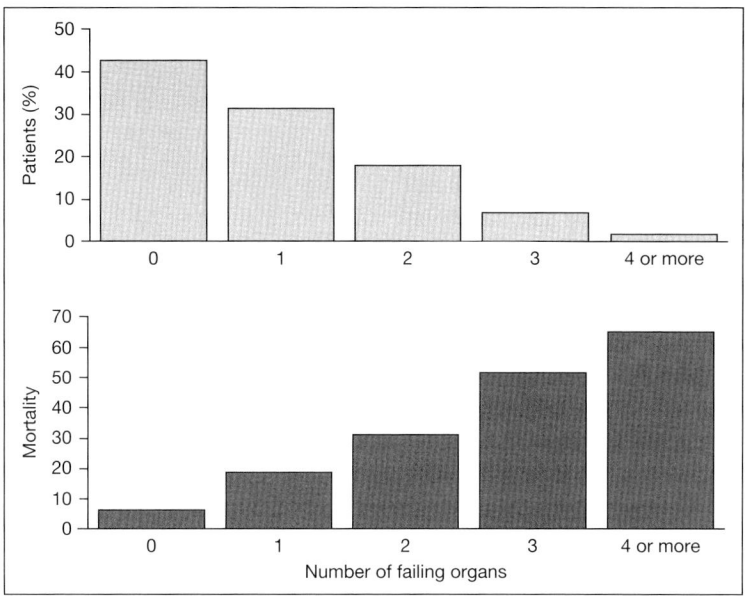

Fig. 1. Frequency of organ failure in patients in the SOAP study on admission to the ICU and corresponding ICU mortality [from 1 with permission].

number of failing organs, the higher the mortality (fig. 1) [1, 6, 18]. Organ dysfunction can be measured and monitored using various scoring systems, of which the most widely used is the SOFA score [18]. Any organ can be involved, but the cardiovascular, respiratory, and renal organs are most commonly affected [1, 6, 19]. The pattern of organ dysfunction in sepsis varies among patients and with time in the same patient and careful monitoring can help improve our understanding of the pathophysiology of sepsis-related organ dysfunction, as well as guiding therapy and predicting outcome.

Incidence of Sepsis and MOF

Many observational or epidemiological studies have been conducted in recent years assessing the incidence of sepsis, some in individual centers, others in multiple centers in one geographical area, and others crossing international boundaries [1, 6, 19–26]. Two key techniques are used to assess how many people are affected by sepsis: prevalence, i.e., the percentage of a population that is affected with sepsis at a given time, and incidence, i.e., the number of new episodes of sepsis that commence during a specified period of time in a specified population. The different techniques, different definitions used for sepsis and organ dysfunction, and different study populations make it difficult to compare study results (table 1).

Using discharge records from hospital in seven states in the United States, Angus et al. [20] identified 192,980 cases of severe sepsis in 1995, giving annual estimates of 751,000 cases (3.0 cases/1,000 population and 2.26 cases/100 hospital discharges), although these numbers may be an overestimate [27]. In 23 ICUs across New Zealand and Australia, Finfer et al. [24] calculated the incidence of severe sepsis to be 0.77/1,000 population. In the UK, the incidence of severe sepsis was estimated to be 0.51/1,000 population [19], and in Finland 0.38/1,000 population [25]. In the EpiSepsis study, which included 3,738 admissions to ICUs in France over a 2-week period, 14.6% of patients had severe sepsis. In the SOAP study, which studied 3,147 patients in 198 ICUs across Europe, 37% of the patients had sepsis at some point during their ICU stay, 30% of the patients had severe sepsis, and 15% had septic shock [1]; the frequency varied considerably in the different countries, from 10% in Switzerland to 63% in Portugal. This study also assessed the incidence of organ failure, as defined by a SOFA >2 for the organ in question [18], and reported that 71% of patients had at least one organ failure at some point during their ICU stay. Sepsis was present in 41% of episodes of organ failure.

Attempts have also been made to determine whether the numbers of patients with sepsis is changing over time [6, 28]. Using discharge data from 750 million hospitalizations in the United States, Martin et al. [6] assessed the occurrence of sepsis in 2000 to be 2.4/100,000 population, an increase of some 8.7% compared to the 0.83/100,000 population estimated for 1979. In an analysis of data from 100,554 ICU admissions across France over an 8-year period, Annane et al. [26] reported that the frequency of septic shock had increased from 7% in 1993 to 9.7% in 2000. In England, Wales and Northern Ireland data from 343,860 admissions to 172 adult, general critical care units between December 1995 and January 2005 showed an increase in the numbers of ICU patients admitted with severe sepsis, from 23.5% in 1996 to 28.7% in 2004 [28].

Outcome from Sepsis and MOF

Sepsis was associated with an ICU mortality rate of 27% in the SOAP study [1], ranging from 10% in Switzerland to 35% in Italy [1]. Severe sepsis was associated with ICU mortality rates of 32% and in patients with septic shock, the mortality rate rose to 54% [1]. In a multivariate analysis, the

Table 1. Results from recent large, multicenter epidemiological studies assessing the incidence of and outcome from sepsis

Authors	Year of data	Study population	Size of study population	Methodology	Diagnosis of sepsis used	Incidence/ prevalence/ frequency	Mortality
Angus et al. [20]	1995	Nonfederal acute hospitals in 7 US states	6,621,559 hospitalizations	Analysis of discharge records	Severe sepsis defined as documented infection and acute organ dysfunction using criteria from the International Classification of Diseases	3.0 cases of severe sepsis/1,000 population	Hospital mortality 28.6%
Martin et al. [6]	1979–2000	Nonfederal acute hospitals in USA	750 million hospitalizations	Analysis of discharge records	Severe sepsis defined as documented infection and acute organ dysfunction using criteria from the International Classification of Diseases	82.7 cases of severe sepsis/100,000 population in 1979 to 240.4/100,000 population in 2000	Hospital mortality 27.8% for 1979–1984; 17.9% for 1995–2000
Finfer et al. [24]	May 1, 1999 to July 31, 1999	23 ICUs in Australia and New Zealand	5,878 ICU admissions	Inception cohort study	Severe sepsis as defined by presence of SIRS criteria plus organ dysfunction	0.77 cases of severe sepsis/1,000 population; 11.8/100 ICU admissions	ICU: 26.5%; 28-day: 32.4%; hospital: 37.5%
Padkin et al. [19]	1995–2000	172 ICUs in England, Wales, Ireland	56,673 ICU admissions	Retrospective data analysis	Severe sepsis as defined in the PROWESS study [32] occurring in first 24 h of ICU admission	0.51 cases of severe sepsis/1,000 population	ICU: 35%; hospital: 47%

Table 1. (continued)

Authors	Year of data	Study population	Size of study population	Methodology	Diagnosis of sepsis used	Incidence/ prevalence/ frequency	Mortality
Guidet et al. [29]	1997–2001	35 ICUs in France	65,910 ICU admissions with stay >24 h	Review of prospectively collected data	Severe sepsis defined as infection plus at least one organ dysfunction	27.7%	ICU mortality 10.5% in patients with one organ failure, 42.7% in patients with at least two organ failures
Brun-Buisson et al. [22]	November 19, 2001 to December 2, 2001	206 ICUs in France	3,738 ICU admissions	Inception cohort study	Severe sepsis as defined by ACCP/SCCM consensus conference definitions [2]	Severe sepsis in 14.6% of patients	35% 30-day mortality
Silva et al. [23]	May 2001 to January 2002	5 ICUs in Brazil	1,383 ICU admissions	Observational cohort study	Sepsis, severe sepsis, and septic shock as defined by ACCP/SCCM consensus conference definitions [2]	Sepsis in 61.4%, severe sepsis in 35.6%, septic shock in 30%	33.9% for sepsis, 46.9% for severe sepsis and 52.2% for septic shock
Vincent et al. [1]	May 1–15, 2002	198 ICUs in 24 European countries	3,147 ICU admissions	Prospective cohort observational study	Sepsis, severe sepsis and septic shock as defined by ACCP/SCCM consensus conference definitions [2]	Sepsis in 37% of patients, severe sepsis in 30%, septic shock in 15%	ICU mortality in sepsis 27%, severe sepsis 32.2%, septic shock 52.1%

Zahorec et al. [33]	July to December 2002	12 ICUs in Slovak Republic	1,533 ICU admissions	Observational cohort study	Severe sepsis as defined in the PROWESS study [32]	Severe sepsis in 7.9% of ICU admissions	Hospital mortality 51.2%
Karlsson et al. [25]	November 1, 2004 to February 28, 2005	24 ICUs in Finland	4,500 ICU admissions	Prospective observational study	Severe sepsis as defined by ACCP/SCCM consensus conference definitions [2]	0.38 cases of severe sepsis/1,000 population	ICU mortality 15.5%, hospital mortality 28.3%

SAPS II score on admission, the cumulative fluid balance within the first 72 h of the onset of sepsis, age, SOFA score at the onset of sepsis, bloodstream infection, cirrhosis, *Pseudomonas* sp. infection, and being a medical admission were associated with an increased risk of death [1]. Other studies have reported similar findings, although again definitions differ among studies making a direct comparison difficult (table 1). In the UK, Padkin et al. [19] noted an ICU mortality rate of 35% in patients with severe sepsis; in the USA hospital mortality was reported as 28.6% [20]; in Australia and New Zealand the ICU mortality rate for patients with severe sepsis was 27.5% [24], and in Brazil mortality rates of 34.7, 47.3 and 52.2% were reported for sepsis, severe sepsis, and septic shock, respectively [23]. In patients with an ICU stay >24 h, Guidet et al. [29] noted that patients with at least two organ dysfunctions had longer ICU stays (20.4 vs. 11.6 days) and greater ICU mortality (42.7 vs. 5.5%) than patients with just one organ dysfunction. In the EpiSepsis study, 30-day mortality was 35% and chronic liver and heart failure, acute renal failure and shock, SAPS II at onset of severe sepsis, and 24-hour total SOFA scores were the independent factors most associated with an increased risk of death [22].

Interestingly, sepsis is not only associated with high mortality during the immediate ICU and hospital stay, but may increase the risk of death for up to 5 years after the septic episode even after adjusting for the presence of comorbidities [30]. Health-related quality of life may also be reduced in survivors of sepsis compared to other ICU survivors [31].

Conclusion

Sepsis affects some 40% of ICU patients at some point during their stay, and when severe is associated with mortality rates of around 30%. Although studies have suggested that mortality rates from sepsis may be decreasing [6, 28], as the incidence of sepsis continues to increase, the numbers of patients dying from sepsis are also still increasing [28]. The heterogeneous nature of the patients who develop sepsis, the range of infectious organisms and sources of infection, and the complexities of the immune response to severe sepsis make the classification of patients with sepsis a challenge, and yet the ability to do so is seen as increasingly important for clinical trial development and appropriate therapeutic targeting. With further development and validation, the PIRO system may provide a means by which this can be done, aiding the development of new therapeutic strategies and ultimately improving outcomes for patients with sepsis.

References

1 Vincent JL, Sakr Y, Sprung CL, et al: Sepsis in European intensive care units: results of the SOAP study. Crit Care Med 2006;34:344–353.
2 ACCP-SCCM Consensus Conference: Definitions of sepsis and multiple organ failure and guidelines for the use of innovative therapies in sepsis. Crit Care Med 1992;20:864–874.
3 Levy MM, Fink MP, Marshall JC, et al: 2001 SCCM/ESICM/ACCP/ATS/SIS International Sepsis Definitions Conference. Crit Care Med 2003;31:1250–1256.
4 Marshall JC, Vincent JL, Fink MP, et al: Measures, markers, and mediators: toward a staging system for clinical sepsis. A report of the Fifth Toronto Sepsis Roundtable, Toronto, Ontario, Canada, October 25–26, 2000. Crit Care Med 2003;31:1560–1567.
5 Martin GS, Mannino DM, Moss M: The effect of age on the development and outcome of adult sepsis. Crit Care Med 2006;34:15–21.
6 Martin GS, Mannino DM, Eaton S, et al: The epidemiology of sepsis in the United States from 1979 through 2000. N Engl J Med 2003;348:1546–1554.
7 Wichmann MW, Inthorn D, Andress HJ, et al: Incidence and mortality of severe sepsis in surgical intensive care patients: the influence of patient gender on disease process and outcome. Intensive Care Med 2000;26:167–172.
8 Frink M, Pape HC, van Griensven M, et al: Influence of sex and age on MODS and cytokines after multiple injuries. Shock 2007;27:151–156.
9 Eachempati SR, Hydo L, Barie PS: Gender-based differences in outcome in patients with sepsis. Arch Surg 1999;134:1342–1347.
10 Appoloni O, Dupont E, Andrien M, et al: Association of TNF2, a TNFα promoter polymorphism, with plasma TNFα levels and mortality in septic shock. Am J Med 2001;110:486–488.
11 Arnalich F, Lopez-Maderuelo D, Codoceo R, et al: Interleukin-1 receptor antagonist gene polymorphism and mortality in patients with severe sepsis. Clin Exp Immunol 2002;127:331–336.
12 Arbour NC, Lorenz E, Schutte BC, et al: TLR4 mutations are associated with endotoxin hyporesponsiveness in humans. Nat Genet 2000;25:187–191.
13 Lorenz E, Mira JP, Cornish KL, et al: A novel polymorphism in the toll-like receptor 2 gene and its potential association with staphylococcal infection. Infect Immun 2000;68:6398–6401.
14 Stassen NA, Leslie-Norfleet LA, Robertson AM, et al: Interferon-gamma gene polymorphisms and the development of sepsis in patients with trauma. Surgery 2002;132:289–292.
15 Claeys R, Vinken S, Spapen H, et al: Plasma procalcitonin and C-reactive protein in acute septic shock: clinical and biological correlates. Crit Care Med 2002;30:757–762.
16 Luzzani A, Polati E, Dorizzi R, et al: Comparison of procalcitonin and C-reactive protein as markers of sepsis. Crit Care Med 2003;31:1737–1741.
17 Christ-Crain M, Stolz D, Bingisser R, et al: Procalcitonin-guidance of antibiotic therapy in community-acquired pneumonia: a randomized trial. Am J Respir Crit Care Med 2006;174:84–93.
18 Vincent JL, de Mendonça A, Cantraine F, et al: Use of the SOFA score to assess the incidence of organ dysfunction/failure in intensive care units: results of a multicentric, prospective study. Crit Care Med 1998;26:1793–1800.
19 Padkin A, Goldfrad C, Brady AR, et al: Epidemiology of severe sepsis occurring in the first 24 hrs in intensive care units in England, Wales, and Northern Ireland. Crit Care Med 2003;31:2332–2338.
20 Angus DC, Linde-Zwirble WT, Lidicker J, et al: Epidemiology of severe sepsis in the United States: analysis of incidence, outcome, and associated costs of care. Crit Care Med 2001;29:1303–1310.
21 Alberti C, Brun-Buisson C, Burchardi H, et al: Epidemiology of sepsis and infection in ICU patients from an international multicentre cohort study. Intensive Care Med 2002;28:108–121.
22 Brun-Buisson C, Meshaka P, Pinton P, et al: EPISEPSIS: a reappraisal of the epidemiology and outcome of severe sepsis in French intensive care units. Intensive Care Med 2004;30:580–588.
23 Silva E, Pedro MA, Sogayar AC, et al: Brazilian Sepsis Epidemiological Study (BASES study). Crit Care 2004;8:R251–R260.
24 Finfer S, Bellomo R, Lipman J, et al: Adult-population incidence of severe sepsis in Australian and New Zealand intensive care units. Intensive Care Med 2004;30:589–596.

25 Karlsson S, Varpula M, Ruokonen E, et al: Incidence, treatment, and outcome of severe sepsis in ICU-treated adults in Finland: the Finnsepsis study. Intensive Care Med 2007;33:435–443.
26 Annane D, Aegerter P, Jars-Guincestre MC, et al: Current epidemiology of septic shock: the CUB-Rea Network. Am J Respir Crit Care Med 2003;168:165–172.
27 Wenzel RP, Edmond MB: Severe sepsis-national estimates. Crit Care Med 2001;29:1472–1474.
28 Harrison DA, Welch CA, Eddleston JM: The epidemiology of severe sepsis in England, Wales and Northern Ireland, 1996 to 2004: secondary analysis of a high quality clinical database, the ICNARC Case Mix Programme Database. Crit Care 2006;10:R42.
29 Guidet B, Aegerter P, Gauzit R, et al: Incidence and impact of organ dysfunctions associated with sepsis. Chest 2005;127:942–951.
30 Quartin AA, Schein RM, Kett DH, et al: Magnitude and duration of the effect of sepsis on survival. Department of Veterans Affairs Systemic Sepsis Cooperative Studies Group. JAMA 1997;277:1058–1063.
31 Heyland DK, Hopman W, Coo H, et al: Long-term health-related quality of life in survivors of sepsis. Short form 36: a valid and reliable measure of health-related quality of life. Crit Care Med 2000;28:3599–3605.
32 Bernard GR, Vincent JL, Laterre PF, et al: Efficacy and safety of recombinant human activated protein C for severe sepsis. N Engl J Med 2001;344:699–709.
33 Zahorec R, Firment J, Strakova J, et al: Epidemiology of severe sepsis in intensive care units in the Slovak Republic. Infection 2005;33:122–128.

Dr. J.-L. Vincent
Erasme University Hospital
Route de Lennik 808
BE–1070 Brussels (Belgium)
Tel. +32 2 555 3380, Fax +32 2 555 4555, E-Mail jlvincen@ulb.ac.be

Genetic Polymorphisms in Sepsis- and Cardiopulmonary Bypass-Associated Acute Kidney Injury

Anja Haase-Fielitz[a,b], Michael Haase[a,b], Rinaldo Bellomo[a], Duska Dragun[c]

[a]Department of Intensive Care, Austin Health, Melbourne, Vic., Australia;
[b]Department of Nephrology, Charité University Medicine, Berlin, and
[c]Department of Nephrology and Intensive Care and Center for Cardiovascular Research, Charité, University Berlin, Berlin, Germany

Abstract

Acute kidney injury (AKI) is a major medical problem in critical illness, and has a separate independent effect on the risk of death. Septic shock and cardiac surgery utilizing cardiopulmonary bypass are the two most common factors contributing to AKI. Clinical predictors and biochemical markers identified for the development of AKI can only explain a part of this individual risk. Another tool to predict the risk of AKI and to improve individualized patient care focuses on the identification of genetic risk factors which might be involved in the development of AKI. However, to date our knowledge on the importance of such genetic polymorphisms in influencing the susceptibility to and severity of AKI remains limited. There is evidence that several genetic polymorphisms accounting for sepsis- or cardiopulmonary bypass-associated AKI involve genes which participate in the control of inflammatory or vasomotor processes. In this article, we will review current knowledge concerning the role of genetic polymorphism in the pathogenesis of sepsis- and cardiopulmonary bypass-associated AKI and discuss possible areas for future developments and research in this field.

Copyright © 2007 S. Karger AG, Basel

Acute kidney injury (AKI), as shown by rapid oliguria or anuria and an increase in the blood concentration of kidney-dependent waste products, is a major medical problem occurring in 5% of all patients admitted to hospital and 30% of those admitted to an intensive care unit [1]. A recent observational study found that 4.2% of intensive care unit patients developed renal replacement therapy-dependent AKI [2] and that AKI, defined according to the RIFLE criteria [3], occurred in 5.2% of hospital patients.

Not only is AKI a common complication of critical illness but it also has a separate independent effect on the risk of death [4, 5].

In a multinational, multicenter study, septic shock was the most common contributing factor to AKI [2] accounting for around 47.5% of patients. It was followed in incidence by cardiac surgery utilizing cardiopulmonary bypass (CPB), which accounted for 23.2% of cases. Other rare causes comprise the use of nephrotoxic drugs, hepatorenal syndrome, rhabdomyolysis, rapid glomerulonephritis and obstructive uropathy.

Many clinical predictors and biochemical markers for the development of AKI have been identified. Though risk stratification based on these factors can only explain a part of the individual risk of developing AKI [6, 7].

Another tool to predict the risk of AKI and to improve individualized patient care focuses on the identification of genetic risk factors, which might be involved in the development of AKI. However, to date our knowledge about the importance of such genetic polymorphisms in influencing the susceptibility to and severity of AKI remains limited. In this article, we will review current knowledge concerning the role of genetic polymorphism in the pathogenesis of AKI and discuss possible areas of future developments and research in this field.

Single Nucleotide Polymorphisms

In order to understand the role of genetic factors in the pathogenesis of AKI, one first needs to become familiar with some fundamental aspects of genetics. One such fundamental aspect relates to the concept of genetic polymorphism. Perhaps the first definition of genetic polymorphism originated in 1940 from Ford [8] who described it as 'the occurrence together in the same habitat of two or more discontinuous forms, or "phases", of a species in such proportions that the rarest of them cannot be maintained merely by recurrent mutation'. Most frequently, however, genetic polymorphism is defined as a difference in DNA sequence among individuals, groups or populations, or as the presence of two genotypes in a population.

Single nucleotide polymorphisms (SNPs) are DNA sequence variations that occur when a single nucleotide in the genome sequence is altered. For a variation to be considered a SNP, at least 1% of the population must present with it. SNPs make up about 90% of all human genetic variation and occur every 100–300 bases along the 3-billion-base human genome. Many SNPs have no effect on cell function, whereas others can predispose people to disease or influence their response to a drug. Although more than 99% of human DNA sequences are the same across the population, variations in the remaining

DNA sequence can have a major impact on how humans respond to disease, environmental factors, such as toxins, drugs and other therapies. This makes SNPs of great value for biomedical research.

However, currently investigated SNPs may not necessarily be a causal or functional SNP, but rather, they might be in partial linkage disequilibrium with a causal SNP and thus, can contribute to conflicting reports.

One of the first studies involving genetic polymorphism was published in 1958 [9], and 50 years later almost 100,000 citations can be found using electronic search engines such as PubMed for 'genetic polymorphism'; a number which is growing daily.

During the last decade, tremendous effort has been made to establish an association or wherever possible a causality between genetic predisposition and AKI.

However, the number of genetic polymorphism studies essentially focusing on selected patient populations and specific clinical outcomes or organ dysfunction is still limited.

In the following, we will highlight the relationship of genetic polymorphism and acute kidney injury in the context of sepsis and cardiac surgery.

Sepsis-Associated Acute Kidney Injury

Epidemiology
AKI secondary to sepsis is a common diagnosis in the intensive care unit [10] and occurs in approximately 19% of patients with moderate sepsis, 23% with severe sepsis, and 51% with septic shock when blood cultures are positive [11]. The combination of AKI and sepsis shows a higher mortality rate than AKI alone and hence constitutes a serious medical problem [10, 12]. Despite our increasing ability to support vital organs and resuscitate patients, the incidence and mortality of patients with sepsis-associated AKI remains high [13].

Genetic Polymorphisms
Substantial progress has been made toward understanding the mechanisms whereby sepsis is associated with AKI. Although hemodynamic factors might play a role in the pathogenesis of sepsis-associated AKI, other mechanisms are likely to be involved, which are immunologic, toxic or inflammatory in nature [10, 14]. The initial event in the pathogenesis of sepsis-associated AKI is a systemic infection, which triggers a generalized humoral and cellular immunologic response to it. Much clinical and molecular biology research suggests that

pro- and anti-inflammatory cytokines, the generation of reactive oxygen species (ROS) and the formation of glomerular microthrombi cause generalized injury to glomerular endothelial cells [15–19]. In addition, there is evidence that human renal tubular cells die by apoptosis as well as necrosis in experimental models of sepsis [20].

Once sufficient evidence had accumulated for the possible pathophysiological role of septic mediators in sepsis-associated AKI, the significance of genetic polymorphism in patients with sepsis became a field of investigation [21–23]. There is already evidence for a positive association between genetic polymorphisms of immune response genes and sepsis in general [24]. However, the number of studies of genetic variability in sepsis focusing on AKI has remained limited. We included studies which fulfilled the following criteria: patients with sepsis-associated AKI and for whom the relationship or distribution of various genotypes to renal outcome was reported.

Below, we will discuss the evidence available at this stage and focus on the results of clinical studies focusing on genetic variants of tumor necrosis factor (TNF)-α, interleukin (IL)-6 and IL-10 genes and their role in sepsis-associated AKI (table 1).

Tumor Necrosis Factor

The role of TNF-α in endotoxin-related AKI has been studied in both animals and human studies [23, 33, 34]. TNF has also been demonstrated to play a major role in the pathogenesis of gram-negative septic shock mediating a broad spectrum of host responses to endotoxin. In the kidney, endotoxin stimulates the release of TNF-α from glomerular mesangial cells [35]. TNF-α is a cytokine that initiates the inflammatory cascade and induces the production of numerous additional mediators which are associated with the endothelial and tissue injury seen in septic multiple organ failure [17, 36, 37]. The TNF-α gene is located on the short arm of chromosome 6. Polymorphism within the 5′ flanking region of the TNF-α gene – a region which influences transcriptional activity – has been associated with high promoter activity and may therefore be of functional relevance [38–40].

In a retrospective study, genotyping blood from 92 newborns, Treszl et al. [26] found that the constellation of high TNF-α-producer and low IL-6-producer genetic variants (i.e., TNF-α/IL-6 AG/GC or AG/CC haplotype) was associated with AKI in neonates with very low birth weight and severe infection (table 1).

In an international multicentric prospective study of 213 patients with severe sepsis or septic shock, plasma levels of TNF-α were significantly higher in those who died in the intensive care unit compared to those who survived.

Table 1. Previous studies highlighting the association between a genetic polymorphism and sepsis-associated acute kidney injury

Gene	Study design	Patients	Definition of AKI	Definition of sepsis	Primary inclusion criteria (all patients)	Outcome	Reference
TNF-α/IL-6	Retrospective cohort study	n = 92 Neonates born with birth weight <1,500 g	ARF as serum creatinine >120 μmol/l after 2nd postnatal day or serum urea >9 mmol/l, diuresis <1.0 ml urine/kg body weight/h	Early onset sepsis (at least 2 of the categories) [25]	Sepsis	Constellation of (TNF-α/IL-6) AG/GC or AG/CC haplotypes were more often present in AKI (26 vs. 6%)	Treszl et al. [26]
TNF-α/TNF-αR	Prospective, observational, multicenter study	n = 213	AKI as part of the SOFA score [27]	Severe sepsis or septic shock [28]	Sepsis	No association between candidate TNF or TNF receptor polymorphism (SF1A, B) and renal function or mortality	Gordon et al. [29]
IL-10	Genetic association study	n = 550	AKI defined as days alive and free of organ dysfunction using the Brussels criteria [30]; renal support defined as HD, PD or any continuous renal replacement mode	Sepsis [28]	Sepsis	IL-10 haplotype (−592C/734G/3367G) CGG haplotype more frequently associated with AKI and with increased need for renal support compared with other haplotypes	Wattanathum et al. [31]

Table 1. (continued)

Gene	Study design	Patients	Definition of AKI	Definition of sepsis	Primary inclusion criteria (all patients)	Outcome	Reference
NADPH oxidase p22phox/ catalase	Prospective cohort study	n = 200	Incremental increase in serum creatinine by 0.5, 1.0 or 1.5 mg/dl (see [1])	Sepsis [28]	AKI	Incidence of sepsis not significantly different according to various NADPH oxidase or catalase genotype groups	Perianayagam et al. [32]

AKI = Acute kidney injury; HD = hemodialysis; IL = interleukin; PD = peritoneal dialysis; TNF-αR = tumor necrosis factor receptor; TNF-α = tumor necrosis factor.

However, there were no significant associations between the selected candidate TNF-α or TNF-α receptor polymorphisms, or their haplotypes, and susceptibility to sepsis, illness severity or outcome including the degree of AKI [29]. Thus, the role of TNF-α-related genetic polymorphism in the pathogenesis of sepsis-associated AKI remains uncertain.

Interleukin-10

IL-10 is an important component of the anti-inflammatory cytokine network in sepsis [41, 42]. The IL-10 gene is located on the long arm of chromosome 1 and its SNP is also located in the promoter region.

In a genetic association study of 550 patients with sepsis, the IL-10 gene CGG haplotype ($-592C/734G/3367G$) was associated with increased 28-day mortality in critically ill patients who had sepsis from pneumonia [31]. Acute kidney injury and the need for renal replacement therapy was evaluated as days alive and free of organ dysfunction (for a maximum of 28 days). Patients who carried one or two copies of the CGG haplotype were alive for longer and free of AKI including the need for renal support compared to patients carrying no copies of the CGG haplotype (table 1). Thus it appears that the IL-10 CGG haplotype might be protective from sepsis-associated AKI.

Pro- and Antioxidant Enzymes

NADPH oxidase is a membrane-associated enzyme that catalyses the production of superoxide and is highly expressed in neutrophils and endothelial cells. In contrast, catalase is an antioxidant enzyme which metabolizes hydrogen peroxide and thereby limits oxidative stress-mediated injury. In a recent study [32] the relationship of SNP in the coding region (C to T substitution at position $+242$) of the NADPH oxidase p22phox subunit gene to adverse clinical outcomes was evaluated in a cohort of 200 patients with established AKI of mixed cause and severity. Within this cohort the incidence of sepsis (average 45%) was not significantly different according to the various NADPH oxidase or catalase genotype groups.

Other studies have also investigated further inflammatory SNPs [43] and heat shock protein genetic polymorphisms in patients with AKI [44]. However, in those studies patients with sepsis as the underlying reason for AKI were only a subgroup, for which further data on a specific association between the genetic polymorphism and renal outcome was not provided. These studies were excluded from further review.

In summary, although the pathogenesis of sepsis-associated AKI appears to be multifactorial, most genetic association studies with renal outcome have focused on cytokine and pro- and antioxidant enzyme SNPs. Further studies specifically designed to investigate the relationship between sepsis-associated AKI and genetic polymorphisms seem desirable.

Cardiopulmonary Bypass-Associated Acute Kidney Injury

Epidemiology

AKI remains a serious complication for patients undergoing cardiac surgery with an incidence ranging between 5 and 50%, depending on the definition used [45–48]. AKI following cardiac surgery is associated with an increase in mortality, morbidity, and prolonged stay in hospital and intensive care, particularly for patients requiring renal replacement therapy [4, 49]. Indeed, the development of AKI requiring renal replacement therapy has been identified as the strongest independent risk factor for death with an odds ratio of 7.9 (95% CI 6–10) in a large retrospective study of cardiac surgical patients [4]. Even minor degrees of postoperative AKI are associated with increased mortality [47].

Genetic Polymorphisms

Prognostic risk stratification is used to predict AKI and identify patients who are at a greater risk of developing postoperative AKI. Several independent risk factors of AKI including age, pre-existing renal disease, diabetes and low cardiac output have been identified [7, 50, 51]. Preoperative renal risk stratification provides an opportunity to develop strategies for early diagnosis and intervention. There is the potential that existing clinical scores for renal risk stratification can be strengthened by taking into account variability in genetic polymorphisms predisposing to postoperative AKI.

Multiple causes of AKI following cardiac surgery have been proposed, including perioperative hemodynamic instability and impaired renal blood flow, ischemia-reperfusion injury, CPB-induced activation of inflammatory pathways and the generation of ROS [48, 52–55]. In addition, CPB causes hemolysis and free hemoglobin-induced AKI may be another contributor to CPB-associated AKI.

Ischemia-reperfusion injury and the generation of oxido-inflammatory stress represent two conventionally accepted major mechanisms in the pathogenesis of CPB-associated AKI.

Ischemia-reperfusion injury of the kidney frequently occurring during cardiac surgery may be an important factor contributing to postoperative AKI [56, 57]. Decreased tissue oxygen tension under such circumstances might promote mitochondrial generation of ROS [58] which, in turn, might cause AKI.

Such ischemia may be further aggravated by over-activation of the angiotensinogen/angiotensin II pathways or decreased endothelial nitric oxide synthase (eNOS) with resulting excessive renal vasoconstriction.

CPB has also been shown to stimulate neutrophils and to induce the generation of ROS and inflammatory mediators [59–62]. Increased levels of serum lipid peroxidation products and an intra- and postoperatively decreased total serum antioxidative capacity have also been found following CPB [52, 53]. Apolipoprotein-E (APO-E) is a lipoprotein involved in lipid metabolism, tissue repair and immune response, which might also affect endothelial repair after lipid peroxidation. The gene is located on chromosome 19q13.2 [63]. APO-E genetic polymorphisms have been linked to atherosclerosis [64–66] and neurocognitive dysfunction after cardiac surgery [67] suggesting a role in injury and repair for this protein during and after cardiac surgery. There is much evidence indicating that the generation of ROS may contribute to the initiation and maintenance of acute tubular necrosis [68]. Oxidative stress is considered to be an important cause of AKI in patients exposed to CPB [52–55]. CPB is proinflammatory and activates components of the nonspecific immune system. The inflammatory response to CPB generates cytokines (e.g. TNF-α and IL-6), both systemically and locally and in the kidney, that have major effects on the renal microcirculation and may lead to tubular injury [69–71].

ROS from ischemia-reperfusion injury and oxido-inflammatory response to CPB also contribute significantly to the deactivation of catecholamines via oxidation into their corresponding vaso-inactive degradation products and thereby to systemic vasodilatation and AKI [72, 73].

Reduced degradation of catecholamines by the enzyme, catechol-O-methyltransferase (COMT), may result in alternative metabolic pathways, which goes along with increased formation of chemically reactive intermediates and enhanced generation of ROS [74, 75].

As a consequence of the above observations, genetic polymorphism of several of the proteins involved in these postulated injurious pathways has been considered for investigation (table 2). We included studies for further review if the relation of a SNP to CPB-associated AKI was evaluated.

Interleukin-6

IL-6 is a pleiotropic cytokine with both pro- and anti-inflammatory properties. It is located on the long arm of chromosome 7. A polymorphism has been identified within the promoter region of the IL-6 gene at position -174 (G to C) [77] and -572 (G to C) [78].

In a prospective study [76] of 111 patients receiving coronary artery bypass surgery, the -174 G/C polymorphism of the IL-6 gene was determined and correlated with the postoperative plasma IL-6 levels and the development

Table 2. Previous studies highlighting the association between a genetic polymorphism and CPB-associated acute kidney injury

Gene	Study design	Patients	Definition of AKI	Type of cardiac surgery	Outcome	Reference
IL-6 (−174 G/C)	Prospective, observational, single center study	n = 111	Perioperative change in serum creatinine	Coronary bypass graft surgery	IL-6 (G homozygous) was associated with greater increase in perioperative creatinine	Gaudino et al. [76]
12 Polymorphisms in 7 candidate genes	Prospective, observational, single center study	n = 1,671	Perioperative relative change in serum creatinine	Coronary bypass graft surgery	IL-6 −572C, angiotensinogen 842C allele, APO-E 448C ε4, AGTR1 1166C, eNOS 894T and ACE deletion/insertion associated with AKI	Stafford-Smith et al. [48]
APO-E alleles	Prospective, observational, single center study	n = 564	Postoperative changes in serum creatinine	Coronary bypass graft surgery	ε4 allele was associated with a smaller increase in postoperative serum creatinine than ε2, ε3 allele	Chew et al. [79]

APO-E ε4 allele	Prospective, observational, single center study	n = 130	Peak in-hospital postoperative serum creatinine (also perioperative change in serum creatinine, change in serum creatinine clearance)	Coronary bypass graft surgery	Interaction between ε4 status (ε4/non-ε4; n = 26/106) and atheroma burden with a greater peak in-hospital postoperative serum creatinine for increases in ascending aorta atheroma load for non-ε4 patients	MacKensen et al. [80]

ACE = Angiotensin-converting enzyme; AGTR = angiotensin receptor1; AKI = acute kidney injury; APO-E = apolipoprotein E; eNOS = endothelial nitric oxide synthase; IL = interleukin.

of postoperative renal complications. The investigators found evidence that the IL-6 −174 G/C polymorphism modulated postoperative IL-6 levels and was associated with the degree of postoperative kidney and pulmonary dysfunction and with the duration of in-hospital stay after coronary surgery [76].

IL-6, Angiotensin and NO Pathway Interaction

Stafford-Smith et al. [48] assessed twelve candidate genes for polymorphism to test the hypotheses that selected gene variants might be associated with AKI. Two alleles (IL-6 −572C and angiotensinogen 842C) showed a strong association with AKI. The combination of these polymorphisms was associated with major postoperative AKI with an average peak serum creatinine level increase of 121%, which was four times greater than average for the overall study population. Additional vasoconstrictor gene polymorphisms were also identified to account for AKI after cardiac surgery (angiotensin, eNOS, and angiotensin receptor 1). Finally, genetic polymorphism for APO-E was also found to be associated with the development of AKI.

Apolipoprotein E

Chew et al. [79] further studied the impact of genetic polymorphism for APO-E and showed a reduced postoperative increase in serum creatinine and a lower peak serum creatinine after cardiac surgery in patients with the APO-E ε4 allele compared with the APO-E ε2 allele and APO-E ε3 allele.

The same working group evaluated the atheromatous burden of ascending, arch, and descending aorta and APO-E status in 130 coronary bypass patients. They found that an equivalent ascending aortic atheromatous burden is associated with a greater susceptibility to postoperative AKI among patients undergoing cardiac operation who lack the APO-E ε4 allele [80].

The above observations provide consistent and reliable evidence that genetic factors and, in particular SNP for IL-6, the angiotensin-generating pathways, a nitric oxide regulation enzyme and lipid pathways significantly influence the likelihood of a patient developing AKI after cardiac surgery, and do so in a clinically important way. These observations, however, do not tackle another likely important pathway in the regulation of the response to stress, the production, release and uptake of catecholamines and the generation of ROS: the COMT system.

Potential Role of COMT Polymorphism in
Sepsis- and CPB-Associated AKI

The activity of this crucial enzyme involved in the degradation of circulating catecholamines, COMT, is controlled by a common autosomal co-dominantly inherited SNP [81].

This SNP, located at chromosome 22 (22q11.21-q11.23; 22q11.21), results in a trimodal distribution of low, intermediate and high levels of COMT activity [82]. Reduced degradation of catecholamines by COMT, associated with the COMT LL genotype, may result in alternative metabolic pathways with increased formation of chemically reactive intermediates and enhanced generation of ROS [74, 75]. The prevention of ROS generation by catecholamines is closely related to the activity of COMT and thus, where a high activity prevents their conversion to semiquinones and quinones and therefore blocks the generation of ROS [83], a low activity is associated with increased generation of ROS by oxidative pathways. The biological and therapeutic efficacy of catecholamines to regulate vasomotor tone may be reduced by inactivation through ROS [72].

The formation of ROS by auto-oxidation of catecholamines on the one hand and the inactivation of catecholamines by ROS on the other hand may contribute to endothelial damage, microcirculatory dysfunction and vasodilatation [84, 85]. Thus, such SNPs may lead to systemic vasodilatation, may significantly aggravate CPB-induced vasoplegia and thereby participate in inducing or sustaining AKI. Unfortunately, this SNP has not yet been investigated in these patients or in patients with sepsis-associated AKI. Hopefully, future investigations will help elucidate the role of COMT-related SNP in the development of both sepsis-associated AKI and CPB-associated AKI.

Conclusion

There is limited but growing evidence for an important role of genetic polymorphism in the pathogenesis of AKI in sepsis and after CPB. This evidence shows that most genetic polymorphisms accounting for sepsis- or CPB-associated AKI appear to involve genes which participate in the control of inflammatory or vasomotor processes. Thus, it provides further indirect evidence of the importance of these pathways in the pathogenesis of these syndromes. Based on this observation, we raise the question whether individual variability in genetic polymorphism of the COMT enzyme (another powerful controller of vasomotor state) might not also predispose to sepsis- or CPB-associated AKI and suggest that studies targeting its genetic variability should be the next step in the investigation of how such variability affects the likelihood of developing sepsis- and CPB-associated AKI.

In addition, studies involving genetic polymorphism may further help us to understand the pathogenesis of sepsis- or CPB-associated AKI; discover potential markers of susceptibility, severity and clinical outcomes; identify markers for responders vs. non-responders in therapeutic trials, and recognize targets for therapeutic intervention.

The use of genetic epidemiology may stratify those who may benefit from preventive measures from those who will not, depending on the individual genotype.

Genotyping may prove one day to be a routine tool in an individualized patient care model, which is clinically useful and cost-effective.

References

1. Hou SH, Bushinsky DA, Wish JB, et al: Hospital-acquired renal insufficiency: a prospective study. Am J Med 1983;74:243–248.
2. Uchino S, Kellum JA, Bellomo R, et al; Beginning and Ending Supportive Therapy for the Kidney (BEST Kidney) Investigators: Acute renal failure in critically ill patients: a multinational, multicenter study. JAMA 2005;294:813–818.
3. Uchino S, Bellomo R, Goldsmith D, et al: An assessment of the RIFLE criteria for acute renal failure in hospitalized patients. Crit Care Med 2006;34:1913–1917.
4. Chertow GM, Levy EM, Hammermeister KE, et al: Independent association between acute renal failure and mortality following cardiac surgery. Am J Med 1998;104:343–348.
5. De Mendonca A, Vincent JL, Suter PM, et al: Acute renal failure in the ICU: risk factors and outcome evaluated by the SOFA score. Intensive Care Med 2000;26:915–921.
6. Zanardo G, Michielon P, Paccagnella A, et al: Acute renal failure in the patient undergoing cardiac operation. Prevalence, mortality rate, and main risk factors. J Thorac Cardiovasc Surg 1994;107:1489–1495.
7. Chertow GM, Lazarus JM, Christiansen CL, et al: Preoperative renal risk stratification. Circulation 1997;95:878–884.
8. Ford EB: Genetic polymorphism. Proc R Soc Lond B Biol Sci 1966;164:350–361.
9. Hsu TC, Klatt O: Mammalian chromosomes in vitro. IX. On genetic polymorphism in cell populations. J Natl Cancer Inst 1958;21:437–473.
10. Klenzak J, Himmelfarb J: Sepsis and the kidney. Crit Care Clin 2005;21:211–222.
11. Rangel-Frausto MS, Pittet D, Costigan M, et al: The natural history of the systemic inflammatory response syndrome (SIRS). A prospective study. JAMA 1995;273:117–123.
12. Schrier RW, Wang W: Acute renal failure and sepsis. N Engl J Med 2004;351:159–169.
13. Marshall JC: Inflammation, coagulopathy, and the pathogenesis of multiple organ dysfunction syndrome. Crit Care Med 2001;29(suppl):S99–S106.
14. Wan L, Bellomo R, Di Giantomasso D, et al: The pathogenesis of septic acute renal failure. Curr Opin Crit Care 2003;9:496–502.
15. Parrillo JE: Pathogenetic mechanisms of septic shock. N Engl J Med 1993;328:1471–1477.
16. Bone RC, Grodzin CJ, Balk RA: Sepsis: a new hypothesis for pathogenesis of the disease process. Chest 1997;112:235–243.
17. Pinsky MR, Vincent JL, Deviere J, et al: Serum cytokine levels in human septic shock. Relation to multiple-system organ failure and mortality. Chest 1993;103:565–575.
18. Hack CE, Aarden LA, Thijs LG: Role of cytokines in sepsis. Adv Immunol 1997;66:101–195.
19. Messmer UK, Briner VA, Pfeilschifter J: Basic fibroblast growth factor selectively enhances TNF-alpha-induced apoptotic cell death in glomerular endothelial cells: effects on apoptotic signaling pathways. J Am Soc Nephrol 2000;11:2199–2211.
20. Lieberthal W, Triaca V, Levine J: Mechanisms of death induced by cisplatin in proximal tubular epithelial cells: apoptosis vs. necrosis. Am J Physiol 1996;270:F700–F708.
21. Kohan DE: Role of endothelin and tumour necrosis factor in the renal response to sepsis. Nephrol Dial Transplant 1994;9(suppl 4):73–77.
22. Karnik AM, Bashir R, Khan FA, et al: Renal involvement in the systemic inflammatory reaction syndrome. Ren Fail 1998;20:103–116.

23 Knotek M, Rogachev B, Wang W, et al: Endotoxemic renal failure in mice: Role of tumor necrosis factor independent of inducible nitric oxide synthase. Kidney Int 2001;59:2243–2249.
24 Jaber BL, Pereira BJ, Bonventre JV, et al: Polymorphism of host response genes: implications in the pathogenesis and treatment of acute renal failure. Kidney Int 2005;67:14–33.
25 Dollner H, Vatten L, Linnebo I, et al: Inflammatory mediators in umbilical plasma from neonates who develop early-onset sepsis. Biol Neonate 2001;80:41–47.
26 Treszl A, Toth-Heyn P, Kocsis I, et al: Interleukin genetic variants and the risk of renal failure in infants with infection. Pediatr Nephrol 2002;17:713–717.
27 Vincent JL, de Mendonca A, Cantraine F, et al: Use of the SOFA score to assess the incidence of organ dysfunction/failure in intensive care units: results of a multicenter, prospective study. Working group on 'sepsis-related problems' of the European Society of Intensive Care Medicine. Crit Care Med 1998;26:1793–1800.
28 American College of Chest Physicians/Society of Critical Care Medicine Consensus Conference: Definitions for sepsis and organ failure and guidelines for the use of innovative therapies in sepsis. Crit Care Med 1992;20:864–874.
29 Gordon AC, Lagan AL, Aganna E, et al: TNF and TNFR polymorphisms in severe sepsis and septic shock: a prospective multicentre study. Genes Immun 2004;5:631–640.
30 Sibbald WJ, Vincent JL: Round table conference on clinical trials for the treatment of sepsis. Brussels, March 12–14, 1994. Intensive Care Med 1995;21:184–189.
31 Wattanathum A, Manocha S, Groshaus H, et al: Interleukin-10 haplotype associated with increased mortality in critically ill patients with sepsis from pneumonia but not in patients with extrapulmonary sepsis. Chest 2005;128:1690–1698.
32 Perianayagam MC, Liangos O, Kolyada AY, et al: NADPH oxidase p22phox and catalase gene variants are associated with biomarkers of oxidative stress and adverse outcomes in acute renal failure. J Am Soc Nephrol 2007;18:255–263.
33 Reinhart K, Menges T, Gardlund B, et al: Randomized, placebo-controlled trial of the anti-tumor necrosis factor antibody fragment afelimomab in hyperinflammatory response during severe sepsis: The RAMSES Study. Crit Care Med 2001;29:765–769.
34 Gallagher J, Fisher C, Sherman B, et al: A multicenter, open-label, prospective, randomized, dose-ranging pharmacokinetic study of the anti-TNF-alpha antibody afelimomab in patients with sepsis syndrome. Intensive Care Med 2001;27:1169–1178.
35 Baud L, Oudinet JP, Bens M, et al: Production of tumor necrosis factor by rat mesangial cells in response to bacterial lipopolysaccharide. Kidney Int 1989;35:1111–1118.
36 Pinsky MR: The critically ill patient. Kidney Int Suppl 1998;66:S3–S6.
37 Camussi G, Ronco C, Montrucchio G, et al: Role of soluble mediators in sepsis and renal failure. Kidney Int Suppl 1998;66:S38–S42.
38 Wilson AG, de Vries N, Pociot F, et al: An allelic polymorphism within the human tumor necrosis factor alpha promoter region is strongly associated with HLA A1, B8, and DR3 alleles. J Exp Med 1993;177:557–560.
39 Morse HR, Olomolaiye OO, Wood NA, et al: Induced heteroduplex genotyping of TNF-alpha, IL-1beta, IL-6 and IL-10 polymorphisms associated with transcriptional regulation. Cytokine 1999;11: 789–795.
40 Warzocha K, Ribeiro P, Bienvenu J, et al: Genetic polymorphisms in the tumor necrosis factor locus influence non-Hodgkin's lymphoma outcome. Blood 1998;91:3574–3581.
41 Bone RC: Toward a theory regarding the pathogenesis of the systemic inflammatory response syndrome: what we do and do not know about cytokine regulation. Crit Care Med 1996;24: 163–172.
42 Jaber BL, Pereira BJ: Extracorporeal adsorbent-based strategies in sepsis. Am J Kidney Dis 1997;30(suppl 4):S44–S56.
43 Jaber BL, Rao M, Guo D, et al: Cytokine gene promoter polymorphisms and mortality in acute renal failure. Cytokine 2004;25:212–219.
44 Fekete A, Treszl A, Toth-Heyn P, et al: Association between heat shock protein 72 gene polymorphism and acute renal failure in premature neonates. Pediatr Res 2003;54:452–455.
45 Stallwood MI, Grayson AD, Mills K, et al: Acute renal failure in coronary artery bypass surgery: independent effect of cardiopulmonary bypass. Ann Thorac Surg 2004;77:968–972.

46 Wijeysundera DN, Rao V, Beattie WS, et al: Evaluating surrogate measures of renal dysfunction after cardiac surgery. Anesth Analg 2003;96:1265–1273.

47 Lassnigg A, Schmidlin D, Mouhieddine M, et al: Minimal changes of serum creatinine predict prognosis in patients after cardiothoracic surgery: a prospective cohort study. J Am Soc Nephrol 2004;15:1597–1605.

48 Stafford-Smith M, Podgoreanu M, Swaminathan M, et al: Association of genetic polymorphisms with risk of renal injury after coronary bypass graft surgery. Am J Kidney Dis 2005;45:519–530.

49 Mangano CM, Diamondstone LS, Ramsay JG, et al: Renal dysfunction after myocardial revascularization: risk factors, adverse outcomes, and hospital resource utilization. The Multicenter Study of Perioperative Ischemia Research Group. Ann Intern Med 1998;128:194–203.

50 Conlon PJ, Stafford-Smith M, White WD, et al: Acute renal failure following cardiac surgery. Nephrol Dial Transplant 1999;14:1158–1162.

51 Thakar CV, Arrigain S, Worley S, et al: A clinical score to predict acute renal failure after cardiac surgery. J Am Soc Nephrol 2005;16:162–168.

52 Starkopf J, Zilmer K, Vihalemm T, et al: Time course of oxidative stress during open-heart surgery. Scand J Thorac Cardiovasc Surg 1995;29:181–186.

53 McColl AJ, Keeble T, Hadjinikolaou L, et al: Plasma antioxidants: evidence for a protective role against reactive oxygen species following cardiac surgery. Ann Clin Biochem 1998;35:616–623.

54 Doi K, Suzuki Y, Nakao A, et al: Radical scavenger edaravone developed for clinical use ameliorates ischemia/reperfusion injury in rat kidney. Kidney Int 2004;65:1714–1723.

55 McCord JM: Oxygen-derived free radicals in postischemic tissue injury. N Engl J Med 1985;312:159–163.

56 Andersson LG, Bratteby LE, Ekroth R, et al: Renal function during cardiopulmonary bypass: influence of pump flow and systemic blood pressure. Eur J Cardiothorac Surg 1994;8:597–602.

57 Abbott WM, Austen WG: The reversal of renal cortical ischemia during aortic occlusion by mannitol. J Surg Res 1974;16:482–489.

58 Dada LA, Chandel NS, Ridge KM, et al: Hypoxia-induced endocytosis of Na/K-ATPase in alveolar epithelial cells is mediated by mitochondrial reactive oxygen species and PKC-zeta. J Clin Invest 2003;111:1057–1064.

59 Chello M, Mastroroberto P, Patti G, et al: Simvastatin attenuates leucocyte-endothelial interactions after coronary revascularisation with cardiopulmonary bypass. Heart 2003;89:538–543.

60 Boyle EM Jr, Lille ST, Allaire E, et al: Endothelial cell injury in cardiovascular surgery: atherosclerosis. Ann Thorac Surg 1997;63:885–894.

61 Paparella D, Yau TM, Young E. Cardiopulmonary bypass induced inflammation: pathophysiology and treatment. An update. Eur J Cardiothorac Surg 2002;21:232–244.

62 Partrick DA, Moore EE, Fullerton DA, et al: Cardiopulmonary bypass renders patients at risk for multiple organ failure via early neutrophil priming and late neutrophil disability. J Surg Res 1999;86:42–49.

63 Strittmatter WJ, Saunders AM, Schmechel D, et al: Apolipoprotein E: high-avidity binding to beta-amyloid and increased frequency of type 4 allele in late-onset familial Alzheimer disease. Proc Natl Acad Sci USA 1993;90:1977–1981.

64 Eichner JE, Kuller LH, Orchard TJ, et al: Relation of apolipoprotein E phenotype to myocardial infarction and mortality from coronary artery disease. Am J Cardiol 1993;71:160–165.

65 Corder EH, Saunders AM, Strittmatter WJ, et al: Gene dose of apolipoprotein E type 4 allele and the risk of Alzheimer's disease in late onset families. Science 1993;261:921–923.

66 Corder EH, Saunders AM, Risch NJ, et al: Protective effect of apolipoprotein E type 2 allele for late onset Alzheimer disease. Nat Genet 1994;7:180–184.

67 Tardiff BE, Newman MF, Saunders AM, et al: Preliminary report of a genetic basis for cognitive decline after cardiac operations. The Neurologic Outcome Research Group of the Duke Heart Center. Ann Thorac Surg 1997;64:715–720.

68 Nath KA, Norby SM: Reactive oxygen species and acute renal failure. Am J Med 2000;109:665–678.

69 Cunningham PN, Dyanov HM, Park P, et al: Acute renal failure in endotoxemia is caused by TNF acting directly on TNF receptor-1 in kidney. J Immunol 2002;168:5817–5823.

70 Segerer S, Nelson PJ, Schlondorff D: Chemokines, chemokine receptors, and renal disease: from basic science to pathophysiologic and therapeutic studies. J Am Soc Nephrol 2000;11:152–176.
71 Heyman SN, Rosen S, Darmon D, et al: Endotoxin-induced renal failure. II. A role for tubular hypoxic damage. Exp Nephrol 2000;8:275–282.
72 Macarthur H, Westfall TC, Riley DP, et al: Inactivation of catecholamines by superoxide gives new insights on the pathogenesis of septic shock. Proc Natl Acad Sci USA 2000;97:9753–9758.
73 Bindoli A, Rigobello MP, Deeble DJ: Biochemical and toxicological properties of the oxidation products of catecholamines. Free Radic Biol Med 1992;13:391–405.
74 Stokes AH, Hastings TG, Vrana KE: Cytotoxic and genotoxic potential of dopamine. J Neurosci Res 1999;55:659–665.
75 Bolton JL, Trush MA, Penning TM, et al: Role of quinones in toxicology. Chem Res Toxicol 2000;13:135–160.
76 Gaudino M, Di Castelnuovo A, Zamparelli R, et al: Genetic control of postoperative systemic inflammatory reaction and pulmonary and renal complications after coronary artery surgery. J Thorac Cardiovasc Surg 2003;126:1107–1112.
77 Fishman D, Faulds G, Jeffery R, et al: The effect of novel polymorphisms in the interleukin-6 (IL-6) gene on IL-6 transcription and plasma IL-6 levels, and an association with systemic-onset juvenile chronic arthritis. J Clin Invest 1998;102:1369–1376.
78 Ferrari SL, Ahn-Luong L, Garnero P, et al: Two promoter polymorphisms regulating interleukin-6 gene expression are associated with circulating levels of C-reactive protein and markers of bone resorption in postmenopausal women. J Clin Endocrinol Metab 2003;88:255–259.
79 Chew ST, Newman MF, White WD, et al: Preliminary report on the association of apolipoprotein E polymorphisms, with postoperative peak serum creatinine concentrations in cardiac surgical patients. Anesthesiology 2000;93:325–331.
80 MacKensen GB, Swaminathan M, Ti LK, et al: Preliminary report on the interaction of apolipoprotein E polymorphism with aortic atherosclerosis and acute nephropathy after CABG. Ann Thorac Surg 2004;78:520–526.
81 Weinshilboum RM, Raymond FA: Inheritance of low erythrocyte catechol-o-methyltransferase activity in man. Am J Hum Genet 1977;29:125–135.
82 Floderus Y, Wetterberg L: The inheritance of human erythrocyte catechol-O-methyltransferase activity. Clin Genet 1981;19:392–395.
83 Axelrod J, Tomchick R: Enzymatic O-methylation of epinephrine and other catechols. J Biol Chem 1958;233:702–705.
84 Kalfin RE, Engelman RM, Rousou JA, et al: Induction of interleukin-8 expression during cardiopulmonary bypass. Circulation 1993;88:II401–II406.
85 Schmid FX, Vudattu N, Floerchinger B, et al: Endothelial apoptosis and circulating endothelial cells after bypass grafting with and without cardiopulmonary bypass. Eur J Cardiothorac Surg 2006;29:496–500.

Prof. Rinaldo Bellomo
Department of Intensive Care, Austin Hospital
Melbourne, Vic. 3084 (Australia)
Tel. +61 3 9496 5992, Fax +61 3 9496 3932, E-Mail Rinaldo.Bellomo@austin.org.au

Predictive Capacity of Severity Scoring Systems in the ICU

N. Schusterschitz, M. Joannidis

Medical Intensive Care Unit, Department of Internal Medicine, Division of General Internal Medicine, Medical University Innsbruck, Innsbruck, Austria

Abstract

Severity scoring systems were first introduced to intensive care units (ICUs) in 1980. The basis for their development was the intention to provide information on the prognosis of patients, the efficacy of therapeutic interventions, stratification for clinical studies, workload and benchmarking of ICUs. Despite the appearance of several specialized scoring systems, the general mortality prediction systems such as APACHE, SAPS and MPM scores and their constantly improved successors have become the most popular and widely tested models. The newest development in this field is SAPS III which is the first 'global' model using a data set acquired from 307 ICUs from all over the world.

Copyright © 2007 S. Karger AG, Basel

Several factors influence the prognosis of critically ill patients. The main determinants are the severity of damage to organs and systems and the physiological reserve capacity, mainly determined by age and comorbidities. The use of scoring systems in the intensive care unit (ICU) aims to standardized classifications of illness severity, survival prediction, judgment of the efficacy of therapeutic modalities in certain diseases, quantification of nursing requirements, cost efficiency calculations, stratification of patients when using certain therapeutic interventions or scientific studies, and quality control. Standardized mortality ratios comparing observed hospital mortality with the mortality predicted by statistical models may be used for benchmarking ICUs.

Generally scoring systems can be divided into disease-oriented scores (e.g. sepsis, trauma, burns), patient-related scores (e.g. children, surgical, medical ICU patients), and universally adopted mortality prediction scores (APACHE, SAPS, MPM).

Two major criteria for the validity of a scoring system are discrimination and calibration. Discrimination is the ability of the scoring system to differentiate between survivors and non-survivors, which can be judged by a receiver operating curve (ROC) with an area under the curve (AUC) above 0.9 considered excellent. Calibration reflects the agreement between individual probabilities and actual outcomes and is usually described by the goodness-of-fit statistic (i.e. Hosmer-Lemeshow χ^2 statistic).

Mortality Prediction Scores

Applied Physiology and Chronic Health Evaluation (APACHE) Score

Prognostic scores were introduced in 1981 with the APACHE score [1] which was the first physiology-based mortality model developed for the ICU. The APACHE model focused on seven major physiologic systems: cardiovascular, renal, gastrointestinal, respiratory, hematologic, metabolic and neurologic. Depending on the degree of derangement from normal a severity score from 0 to 4 was applied for each variable. APACHE scores were well correlated with mortality and also performed well within a number of specific cardiovascular, neurological, respiratory and gastrointestinal diagnoses. Importantly, APACHE scores showed no major difference between European and US centers.

The APACHE score was followed by APACHE II in 1985 (fig. 1a) [2] and included 12 physiologic variables (heart rate, mean arterial pressure, temperature, respiratory rate, PaO_2/A-a gradient, hematocrit, white blood cell count, creatinine, sodium, potassium, pH/bicarbonate), the Glasgow Coma Scale, age and previous health status. All 12 variables were considered mandatory and had to be collected during the first 24 h of admission. External validity of APACHE II was investigated by a large number of studies proving good overall discrimination of the system across a variety of comorbidities as well as geography.

In 1991 APACHE III was developed [3] to improve calibration and generalization in specialized populations. APACHE III basically used the variables of APACHE II, the modified Glasgow Coma Scale and added 5 new physiological variables (blood urea nitrogen, urine output, serum albumin, bilirubin and glucose). Furthermore items were weighted by narrowing the range of normal or zero points and increasing the points of extremes of a variable. Although APACHE III showed good discrimination (AUC between 0.8 and 0.9), this system never became as popular as the APACHE II system owing to the fact that some logistic regression coefficients and equations consisted of proprietary information not made publicly available.

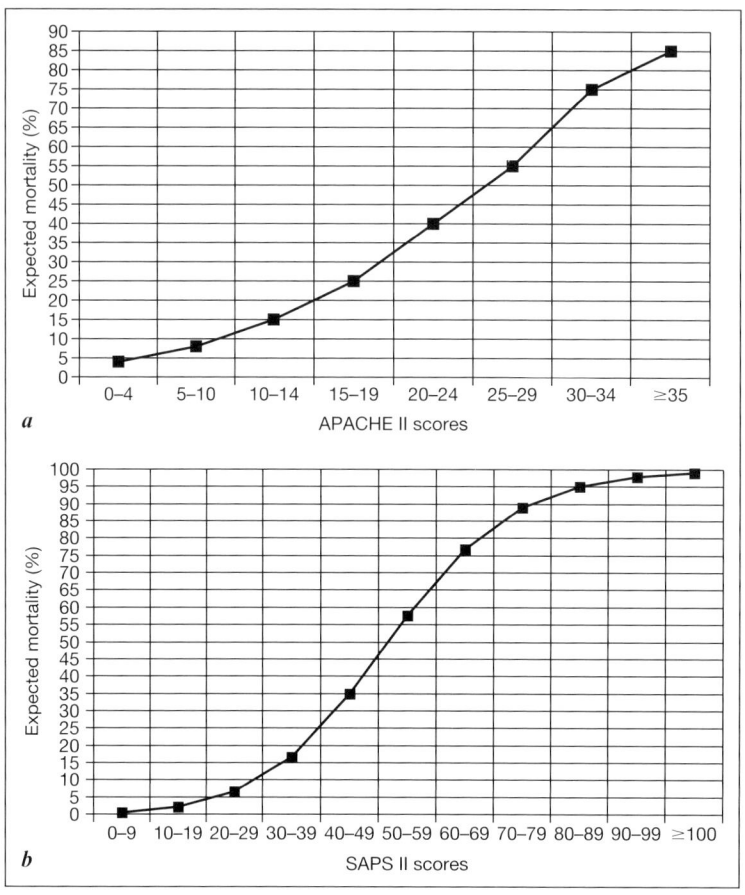

Fig. 1. Expected mortalities as related to mortality prediction scores. *a* APACHE II, modified from Knaus et al. [2]. *b* SAPS II score, modified from Le Gall et al. [5].

Simplified Acute Physiology Score (SAPS)

SAPS was developed by Le Gall et al. [4] in 1984 with the intention to obtain a simpler model but providing the same level of prediction as APACHE. They included 13 physiological variables, the most severe value within the first day of admission was recorded. Discrimination reflected by the obtained AUC of 0.85 was similar to the APACHE score.

SAPS II was developed in 1992 by the same group and was built from a much larger data set of 13,152 patients from 137 ICUs in North America and Europe (fig. 1b) [5]. This model utilized 12 physiologic variables. Furthermore it included age and both type of admission (unscheduled surgical, scheduled surgical and medical) and underlying disease variables (i.e. AIDS, metastatic cancer and hematological malignancy). Although the area under the ROC was 0.88 in the training set and 0.86 in the test set, studies in some European countries showed poor calibration until customization for the local case mix was performed [6].

In 2005 SAPS III, a project of the European Society of Intensive Care Medicine, was introduced [7, 8]. This score was developed on a database including 16,784 patients admitted to 303 ICUs from all over the world. All together 20 variables were selected for SAPS III: ten physiological parameters (Glasgow Coma Scale, total bilirubin, body temperature, creatinine, heart rate, leukocyte count, pH, platelets, systolic blood pressure and PaO_2/FiO_2); five comorbidities (cancer therapy, heart failure (NYHA IV), hematological disorders and cancer, cirrhosis and AIDS), and several admission parameters (length of stay before admission, type of admission [emergency room, other ICU], use of vasoactive drugs, anatomical site of surgery and type of infection on admission).

The reason for the introduction of SAPS III was the development of new major diagnostic and therapeutic methods over the last 10 years which led to shifts towards poor calibration for older models. Secondly SAPS II was developed on a database built exclusively from patients in Europe and North America. SAPS III offers the possibility of referring data to different locations of the world including Australia, Asia as well as Central and South America. Data are acquired within the 1st hour of ICU admission. A recent study including 952 patients with cancer admitted to a specialized ICU in Brazil proved the good performance of SAPS III in these patients [9].

Mortality Prediction Model (MPN)

Eleven noninvasive and easily collectable parameters were the basis of the first version of this model published by Teres et al. [10] in 1987. The model was developed on a dataset including 755 cases collected in 1983. 107 variables were collected on the training set and, after stepwise univariate analysis, the set was reduced to 26 admission variables and 44 variables to be taken at 24 h for patients staying longer in the IUC. After logistic regression the final admission model (MPN_0) contained 7 variables (coma/stupor, emergency admission, cancer, infection, number of organ system failures, age and systolic pressure). Also the final 24-hour model (MPN_{24}) had 7 variables (coma/stupor, infection, mechanical ventilation, shock, emergency admission, age and number of organ system failures).

This model was followed by the MPM II version [11] which again included two versions: MPM_0 II was designed to use parameters taken at the time of admission to the ICU. The admission model contained 15 variables (coma/stupor, non-elective surgery, cancer, age, systolic blood pressure, heart rate, chronic renal insufficiency, cirrhosis, acute renal failure, cardiac dysrhythmia, cerebrovascular accident, gastrointestinal bleeding, intracranial mass effect, cardiopulmonary resuscitation before admission, and mechanical ventilation). The MPN_{24} II was developed for patients staying longer than 24 h included 5 variables from MPM_0 II and eight additional parameters which have to be taken after the first 24 h after admission to the ICU (coma/stupor, infection, mechanical, ventilation, vasoactive drugs, emergency admission, age, cirrhosis, intracranial mass effect, creatinine, PaO_2, prothrombin time and urine output). MPM II appears to be very robust and can be used reliably in medical patients. Although the MPN II system has the advantage of very few quickly determined parameters, it is mainly validated for the US.

Therapeutic Intervention Scoring System (TISS)

This system is one of the oldest scoring systems developed for ICUs [12]. Originally it was intended to correlate the number of required therapeutic interventions with changes in physiological parameters. However, TISS turned out to be insufficient for mortality prediction but has proven to be a reliable measure for nursing workload and for calculations of requirements in nursing staff. Containing 76 items (TISS-76), daily scoring turned out to be time-consuming and many new therapeutic interventions such as positional therapy in acute respiratory distress syndrome were not considered. TISS-28 was developed to provide an updated and simplified version but is still lacking validation by large international databases [13].

Special Severity Scoring Systems

Numerous specialized scoring systems have appeared over the last years, a detailed discussion of those, however, lies outside the scope of this article. Just to mention a few, scores have been developed for pediatric ICUs like the Pediatric Risk of Mortality (PRISM) and the Pediatric Index of Mortality (PIM), scores for trauma patients, e.g. Injury Severity Score (ISS) or for patients with special disease entities such as liver cirrhosis, e.g. Model for End-Stage Liver Disease (MELD).

Sequential organ dysfunction scores were originally introduced into the ICU to describe the individual risk of patients on a day-to-day basis. Because the severity of organ dysfunction has a significant influence on outcome, it was not surprising that organ severity scores predict mortality in individual patients as well. This was shown for Sequential Organ Failure Assessment (SOFA) [14] as well as Multi-Organ Dysfunction Score (MODS) [15].

Table 1. Comparison of popular mortality prediction scores (second and third generation) in terms of discrimination (AUC) [16–30]

	APACHE II	MPM$_0$	MPM$_{24}$	SAPS	APACHE III	SAPS II	MPM II$_0$	MPM II$_{24}$	SAPS III
AUC (mean)	0.82	0.79	0.82	0.77	0.83	0.81	0.8	0.8	0.85
95% CI	(0.79–0.85)	(0.70–0.87)	(0.76–0.88)	(0.64–0.90)	(0.75–0.90)	(0.77–0.86)	(0.72–0.87)	(0.67–0.93)	
Number of studies	15	4	2	3	6	11	4	4	1

Performance of Mortality Prediction Models

Several studies have compared the performance of these models in a variety of ICU populations. Generally spoken the models show good discrimination but a lack of calibration across different countries and populations of ICU patients with special diseases such as malignancies, acute renal failure or liver cirrhosis. Comparison of second-generation models (APACHE II, SAPS, MPN) found APACHE II to have equivalent or superior discrimination. When comparing third-generation models (APACHE III, MPM-II and SAPS II), they all performed equal to or better than the second-generation models (table 1) [16–30]. In terms of calibration the trend was poor across all studies with the only exception of APACHE II.

Limitations of Prediction Models

The development of prediction model is subject to several biases.

First of all, it depends on the parameters collected and then selected for the model. Whereas in the first-generation models variables were originally chosen by subjective judgment, later models used multiple logistic regression for selection of the variable which is more accurate. Secondly, calibration depends on the kinds of complications or syndromes which are known for ICU populations and their representation by a variable at the time of development of a scoring system. For example, when investigating the APACHE score in AIDS patients an under-prediction of mortality in patients with pneumocystis carinii pneumonia was observed. In all other AIDS patients the prediction was good [31]. It is of importance to note that the model was developed before this clinical syndrome had become common, and inclusion of this diagnostic criterion could improve the model's prediction.

Furthermore, in some surgical ICU environments with high trauma admissions, the APACHE II model performed badly due to the few trauma patients included in the data development set.

Special patient groups were not seen in the original test sample (e.g. coronary cardiac surgery in APCHE III).

When comparing standardized mortality ratios in different countries by APACHE II scores, significant differences were noted. This is due to different case mixes in these situations.

An additional unresolved problem is the question of appropriate classification of sedated patients by the Glasgow Coma Scale. SAPS III tries to solve this problem by using an estimated Glasgow Coma Scale before the use of sedation [7, 8].

The question of how to deal with missing values is another problem. Most scoring systems attribute a zero to a missing value which reduces the actual score.

It has repeatedly been shown that ICU scores underestimate mortality in patients with acute renal failure requiring renal replacement therapy [32, 33]. APACHE II suggests a doubling of the measured serum creatinine in case of acute renal failure. Unfortunately this entity is still ill defined leading to incoherent treatment of this variable.

Scoring Systems versus Physician's Judgment

Making a prognosis about patients' outcomes is an essential part of medical decision making. Furthermore, patients and relatives rely on the doctor's prognostic information. When systemically investigating the prognostic capabilities of critical care and primary care physicians, significant differences depending on the level of ICU training were found [34]. However, comparison of physicians' judgment to objective ICU models showed no significant differences in the AUCs between the physicians and APACHE II [35]. One interesting study dividing observed to expected mortality ratios into six severity segments found that the objective model performed better in all segments but the segment with the highest mortality [36]. In an investigation by Christensen et al. [37] attending physicians had the best calibration across all training levels being slightly superior to the objective model prediction.

Finally, the combination of subjective physicians' mortality estimations with a physiologic objective model was found to perform superior to either the subjective or objective model alone [38].

References

1. Knaus WA, Zimmerman JE, Wagner DP, Draper EA, Lawrence DE: APACHE-acute physiology and chronic health evaluation: a physiologically based classification system. Crit Care Med 1981;9:591–597.
2. Knaus WA, Draper EA, Wagner DP, Zimmerman JE: APACHE II: a severity of disease classification system. Crit Care Med 1985;13:818–829.
3. Knaus WA, Wagner DP, Draper EA, Zimmerman JE, Bergner M, Bastos PG, Sirio CA, Murphy DJ, Lotring T, Damiano A: The APACHE III prognostic system. Risk prediction of hospital mortality for critically ill hospitalized adults. Chest 1991;100:1619–1636.
4. Le Gall JR, Loirat P, Alperovitch A, Glaser P, Granthil C, Mathieu D, Mercier P, Thomas R, Villers D: A simplified acute physiology score for ICU patients. Crit Care Med 1984;12:975–977.
5. Le Gall JR, Lemeshow S, Saulnier F: A new Simplified Acute Physiology Score (SAPS II) based on a European/North American multicenter study. JAMA 1993;270:2957–2963.
6. Le Gall JR, Neumann A, Hemery F, Bleriot JP, Fulgencio JP, Garrigues B, Gouzes C, Lepage E, Moine P, Villers D: Mortality prediction using SAPS II: an update for French intensive. Crit Care 2005;9:R645–R652.
7. Metnitz PG, Moreno RP, Almeida E, Jordan B, Bauer P, Campos RA, Iapichino G, Edbrooke D, Capuzzo M, Le Gall JR: SAPS 3 – from evaluation of the patient to evaluation of the intensive. Intensive Care Med 2005;31:1336–1344.
8. Moreno RP, Metnitz PG, Almeida E, Jordan B, Bauer P, Campos RA, Iapichino G, Edbrooke D, Capuzzo M, Le Gall JR: SAPS 3 – From evaluation of the patient to evaluation of the intensive. Intensive Care Med 2005;31:1345–1355.
9. Soares M, Salluh JI: Validation of the SAPS 3 admission prognostic model in patients with cancer in need of intensive care. Intensive Care Med 2006;32:1839–1844.
10. Teres D, Lemeshow S, Avrunin JS, Pastides H: Validation of the mortality prediction model for ICU patients. Crit Care Med 1987;15:208–213.
11. Lemeshow S, Teres D, Klar J, Avrunin JS, Gehlbach SH, Rapoport J: Mortality Probability Models (MPM II) based on an international cohort of intensive care unit patients. JAMA 1993;270:2478–2486.
12. Cullen DJ, Civetta JM, Briggs BA, Ferrara LC: Therapeutic intervention scoring system: a method for quantitative comparison of patient care. Crit Care Med 1974;2:57–60.
13. Miranda DR, de RA, Schaufeli W: Simplified Therapeutic Intervention Scoring System: the TISS-28 items – results from a multicenter study. Crit Care Med 1996;24:64–73.
14. Ferreira FL, Bota DP, Bross A, Melot C, Vincent JL: Serial evaluation of the SOFA score to predict outcome in critically ill patients. JAMA 2001;286:1754–1758.
15. Marshall JC, Cook DJ, Christou NV, Bernard GR, Sprung CL, Sibbald WJ: Multiple organ dysfunction score: a reliable descriptor of a complex clinical outcome. Crit Care Med 1995;23:1638–1652.
16. Beck DH, Smith GB, Pappachan JV, Millar B: External validation of the SAPS II, APACHE II and APACHE III prognostic models in South England: a multicentre study. Intensive Care Med 2003;29:249–256.
17. Capuzzo M, Valpondi V, Sgarbi A, Bortolazzi S, Pavoni V, Gilli G, Candini G, Gritti G, Alvisi R: Validation of severity scoring systems SAPS II and APACHE II in a single-center population. Intensive Care Med 2000;26:1779–1785.
18. Castella X, Gilabert J, Torner F, Torres C: Mortality prediction models in intensive care: acute physiology and chronic health evaluation II and mortality prediction model compared. Crit Care Med 1991;19:191–197.
19. Castella X, Artigas A, Bion J, Kari A: A comparison of severity of illness scoring systems for intensive care unit patients: results of a multicenter, multinational study. The European/North American Severity Study Group. Crit Care Med 1995;23:1327–1335.
20. Del BC, Morelli A, Bassein L, Fasano L, Quarta CC, Pacilli AM, Gunella G: Severity scores in respiratory intensive care: APACHE II predicted mortality better than SAPS II. Respir Care 1995;40:1042–1047.
21. Katsaragakis S, Papadimitropoulos K, Antonakis P, Strergiopoulos S, Konstadoulakis MM, Androulakis G: Comparison of Acute Physiology and Chronic Health Evaluation II (APACHE II)

and Simplified Acute Physiology Score II (SAPS II) scoring systems in a single Greek intensive care unit. Crit Care Med 2000;28:426–432.
22 Livingston BM, MacKirdy FN, Howie JC, Jones R, Norrie JD: Assessment of the performance of five intensive care scoring models within a large Scottish database. Crit Care Med 2000;28:1820–1827.
23 Markgraf R, Deutschinoff G, Pientka L, Scholten T: Comparison of acute physiology and chronic health evaluations II and III and simplified acute physiology score II: a prospective cohort study evaluating these methods to predict outcome in a German interdisciplinary intensive care unit. Crit Care Med 2000;28:26–33.
24 Moreno R, Apolone G, Miranda DR: Evaluation of the uniformity of fit of general outcome prediction models. Intensive Care Med 1998;24:40–47.
25 Nouira S, Belghith M, Elatrous S, Jaafoura M, Ellouzi M, Boujdaria R, Gahbiche M, Bouchoucha S, Abroug F: Predictive value of severity scoring systems: comparison of four models in Tunisian adult intensive care units. Crit Care Med 1998;26:852–859.
26 Patel PA, Grant BJ: Application of mortality prediction systems to individual intensive care units. Intensive Care Med 1999;25:977–982.
27 Rowan KM, Kerr JH, Major E, McPherson K, Short A, Vessey MP: Intensive Care Society's Acute Physiology and Chronic Health Evaluation (APACHE II) study in Britain and Ireland: a prospective, multicenter, cohort study comparing two methods for predicting outcome for adult intensive care patients. Crit Care Med 1994;22:1392–1401.
28 Tan IK: APACHE II and SAPS II are poorly calibrated in a Hong Kong intensive care unit. Ann Acad Med Singapore 1998;27:318–322.
29 Vassar MJ, Lewis FR Jr, Chambers JA, Mullins RJ, O'Brien PE, Weigelt JA, Hoang MT, Holcroft JW: Prediction of outcome in intensive care unit trauma patients: a multicenter study of Acute Physiology and Chronic Health Evaluation (APACHE), Trauma and Injury Severity Score (TRISS), and a 24-hour intensive care unit (ICU) point system. J Trauma 1999;47:324–329.
30 Wilairatana P, Noan NS, Chinprasatsak S, Prodeengam K, Kityaporn D, Looareesuwan S: Scoring systems for predicting outcomes of critically ill patients in northeastern Thailand. Southeast Asian J Trop Med Public Health 1995;26:66–72.
31 Smith RL, Levine SM, Lewis ML: Prognosis of patients with AIDS requiring intensive care. Chest 1989;96:857–861.
32 Levy EM, Viscoli CM, Horwitz RI: The effect of acute renal failure on mortality. A cohort analysis. JAMA 1996;275:1489–1494.
33 Metnitz PG, Krenn CG, Steltzer H, Lang T, Ploder J, Lenz K, Le Gall JR, Druml W: Effect of acute renal failure requiring renal replacement therapy on outcome in critically ill patients. Crit Care Med 2002;30:2051–2058.
34 Poses RM, Bekes C, Copare FJ, Scott WE: The answer to 'What are my chances, doctor?' depends on whom is asked: prognostic disagreement and inaccuracy for critically ill patients. Crit Care Med 1989;17:827–833.
35 Kruse JA, Thill-Baharozian MC, Carlson RW: Comparison of clinical assessment with APACHE II for predicting mortality risk in patients admitted to a medical intensive care unit. JAMA 1988;260:1739–1742.
36 McClish DK, Powell SH: How well can physicians estimate mortality in a medical intensive care unit? Med Decis Making 1989;9:125–132.
37 Christensen C, Cottrell JJ, Murakami J, Mackesy ME, Fetzer AS, Elstein AS: Forecasting survival in the medical intensive care unit: a comparison of clinical prognoses with formal estimates. Methods Inf Med 1993;32:302–308.
38 Marcin JP, Pollack MM, Patel KM, Ruttimann UE: Combining physician's subjective and physiology-based objective mortality risk predictions. Crit Care Med 2000;28:2984–2990.

Prof. Dr. Michael Joannidis
Department of Internal Medicine, Medical Intensive Care Unit
Medical University of Innsbruck, Anichstrasse 35
AT–6020 Innsbruck (Austria)
Tel. +43 512 504 81404, Fax +43 512 504 24199, E-Mail michael.joannidis@i-med.ac.at

Determining the Degree of Immunodysregulation in Sepsis

Jean-Marc Cavaillon, Minou Adib-Conquy

Unit Cytokines & Inflammation, Institut Pasteur, Paris, France

Abstract

During sepsis, the anti-infectious response is closely linked to an overwhelming inflammatory process. The latter is illustrated by the presence in plasma of numerous inflammatory cytokines, markers of cellular stress (e.g. high mobility group box-1 protein), complement-derived compounds (e.g. anaphylatoxin C5a), lipid mediators, and activated coagulation factors. All mediators contribute in synergy to tissue injury, organ dysfunction, and possibly to lethality. To dampen this overzealous process, a counter-regulatory loop is initiated. The anti-inflammatory counterpart involves few anti-inflammatory cytokines (e.g. interleukin-10, transforming growth factor-β), numerous neuromediators (e.g. adrenalin, acetylcholine), and some other factors (e.g. heat shock proteins, ligand of TREM-2, adenosine). These mediators modify the immune status of circulating leukocytes as illustrated by their decreased cell-surface expression of HLA-DR or their reduced ex vivo pro-inflammatory cytokine production in response to Toll-like receptor agonists (e.g. endotoxin, lipoproteins). However, circulating leukocytes remain responsive to whole bacteria and produce normal or even enhanced levels of anti-inflammatory cytokines. Thus, the immune dysregulation observed in sepsis corresponds to a reprogramming of circulating leukocytes.

Copyright © 2007 S. Karger AG, Basel

How to Define Immunodysregulation?

Sepsis is associated with an overwhelming proinflammatory response that is initially aimed at addressing the infectious process. Concomitantly, an anti-inflammatory response is initiated to counteract the inflammatory process. As a consequence, the latter regulatory mechanism has an impact on the immune status of the patients. A long-term alteration in immune status leads to an increased sensitivity to infection and to tumor development. This is illustrated in patients with chronic inflammatory disorders treated with drugs to neutralize tumor necrosis factor (TNF), a key cytokine of innate immunity. These patients have an increased frequency of certain infections (e.g. tuberculosis) or tumors (e.g. lymphoma). It is believed that a short-term alteration in the immune status in septic and critically ill patients further renders them more sensitive to nosocomial infection.

A depressed immune status is illustrated by anergy to skin test antigens, decreased blood cell counts, and reduced expression of surface markers (e.g. major histocompatibility complex class II antigen, CXCR2, etc.). In vitro analysis of circulating leukocytes reveals altered natural killer cell activity, diminished cellular cytotoxicity, reduced antigen presentation, poor proliferation in response to mitogens, enhanced lymphocyte apoptosis, and depressed cytokine production by lymphocytes, neutrophils and monocytes. Accordingly, the terms anergy, immunodepression, or immunoparalysis have been employed to qualify the immune status of septic patients. We have recently demonstrated that these terms are far too excessive, and we proposed that the words 'leukocyte reprogramming' better defines the phenomenon [1].

Biomarkers in Plasma

Markers of Immune Dysregulation

Proinflammatory cytokines play a key role in innate immunity against infection. In contrast, either alone in excess or together in synergy, they contribute to lethality in sepsis (table 1). Most tissues participate in their synthesis during sepsis. In patients, cytokines are produced in excess and are therefore detectable in blood where they are normally absent. High levels of plasma cytokines often correlate with organ dysfunction, length of intensive care stay and poor outcome. However, the circulating cytokines are merely the tip of the iceberg, and cell-associated cytokines can be shown even when plasma levels are undetectable [2]. In addition to cytokines, there are other mediators that have been shown to contribute to sepsis-related lethality. This is the case of some lipid mediators (prostaglandins, leukotrienes, platelet-activating factor), a growth factor (i.e. vascular endothelial growth factor), the unidentified ligand of the triggering receptor expressed on myeloid cells-1 (TREM-1) and anaphylatoxin C5a. High-mobility group box-1 protein is a nuclear protein that is released in sepsis. It is a late mediator and behaves like a cytokine. Like endotoxin, it activates leukocytes through Toll-like receptor 4 (TLR4) and seems to be the mediator that links the occurrence of apoptosis during sepsis and lethality. Other factors are indirectly involved in sepsis-related mortality, and favor the release of proinflammatory cytokines. This is the case with factors generated during coagulation activation (e.g. thrombin, factor Xa) or necrotic cells. Finally, some markers found in large amounts in plasma are only a reflection of the overwhelming process. Many studies have established that procalcitonin is a

Table 1. Immune dysregulation during sepsis is characterized by an exacerbated production of proinflammatory mediators that lead to deleterious effects and lethality, and an exacerbated production of anti-inflammatory mediators that contribute to induce an immune suppressive status

	Mediators that contribute to lethality[a]	Mediators that favor immune suppression
Cytokines	TNF IL-1 IL-12 IL-15 IL-18 IL-27 IFNγ IFNβ Granulocyte-macrophage colony-stimulating factor Leukemia inhibitory factor Macrophage migration inhibitory factor Some chemokines: CXCL8 (IL-8) 　　　　　　　　　CCL5 　　　　　　　　　CXCR1 and 2 ligands 　　　　　　　　　CCR1 ligands 　　　　　　　　　CCR4 ligands[b]	IL-10 IL-13 TGFβ
Growth factors	Vascular endothelial growth factor	—
Cell markers of stress	High mobility group box-1 protein	Heat shock proteins
Plasma factors	Ligand of TREM-1[c] Anaphylatoxin C5a	Ligand of TREM-2
Lipid mediators	Prostaglandins Leukotrienes Platelet-activating factor Oxidized phospholipids	Prostaglandins
Hormones	—	Glucocorticoids
Neuromediators	—	Adrenalin Acetylcholine α-Melanocyte-stimulating hormone Vasoactive intestinal peptide (VIP) Urocortin Adrenomodulin Cortistatin

Table 1. (continued)

	Mediators that contribute to lethality[a]	Mediators that favor immune suppression
Enzymes	Cyclooxygenase-2 5-Lipoxygenase Phospholipase A2 Mast cell dipeptidyl peptidase I	–
Coagulation factor	Tissue factor Thrombin	–
Purine nucleoside	Adenosine (via A_{2A} receptor)	Adenosine (via A_{2A} receptor)

[a] As demonstrated in animal models with the help of specific antibodies, inhibitors or antagonists, or with KO mice.
[b] Either CCL17 or CCL22.
[c] Triggering receptor expressed on myeloid cells.

good marker for the occurrence of an infection in critically ill patients. However, other data have challenged the specificity of plasma procalcitonin elevation as a marker for infection. Soluble TREM-1 was also described as a specific marker for infection [3]. More recent studies suggest that in certain noninfectious clinical settings, levels of soluble TREM-1 are also significantly enhanced [4]. Thus, before using this new marker to diagnose bacterial infections, a large cohort of patients with various noninfectious causes of systemic inflammatory response syndrome (SIRS) should be included to attest to the specific upregulation of TREM-1 during infection.

Molecules Interacting with Lipopolysaccharide

Host serum contains several proteins that interact with lipopolysaccharide (LPS). Some of these proteins are present at homeostasis, such as soluble CD14 and LPS-binding protein (LBP), but their levels are considerably increased during sepsis. Depending on their concentration, these LPS-binding molecules may facilitate LPS interaction with TLR4-bearing cells or, on the contrary, decrease cellular response by transferring cell-bound LPS to plasma lipoproteins as shown with soluble CD14 [5] and LBP [6]. High-density lipoprotein (HDL) and other plasma lipoproteins can bind and neutralize the bioactivity of gram-negative bacterial LPS [7] and gram-positive bacterial lipoteichoic acid [8]. During sepsis, circulating levels of HDL decline dramatically. However, when HDL levels decline in critically ill patients, LPS binds preferentially to low-density and very low-density lipoproteins that maintain their ability to neutralize endotoxin. However, native HDL may enhance the monocyte response to

LPS [9]. This enhancing effect was found in the presence of inhibitory concentrations of LBP. Bactericidal/permeability-increasing (BPI) protein is a cationic protein released by activated or killed neutrophils. This protein shows a high affinity for the lipid A moiety of LPS, and exerts a neutralizing activity and causes bacterial lysis. BPI levels are increased in critically ill patients with bacteremia, and increased circulating BPI is also associated with mortality in patients with ventilator-associated pneumonia [10]. MD-2 is a soluble protein that is associated with TLR4 to form the receptor for LPS. Soluble MD-2 has been detected in the plasma of patients with severe sepsis or septic shock, and in lung edema fluids from patients with acute respiratory distress syndrome [11]. Similarly to soluble CD14, soluble MD-2 may enhance the reactivity of TLR4-positive epithelial cells towards LPS, whereas it would downregulate the reactivity of cells positive for both TLR4 and MD-2, such as monocytes.

Immunosuppressive Markers

An anti-inflammatory process occurs concomitantly to dampen the overzealous anti-infectious response. Anti-inflammatory cytokines and soluble receptors are produced in large amounts during sepsis. They downregulate the production of proinflammatory cytokines and protect animals from sepsis and endotoxin shock. This was shown for interleukin (IL)-10, transforming growth factor-β (TGFβ), IL-4, IL-13, interferon (IFN)α, and IL-6. IL-6 induces a broad array of acute-phase proteins that limit inflammation, such as α_1-acid glycoprotein or C-reactive protein.

The presence of deactivating or immunosuppressive agents within the blood stream may contribute to the hyporeactivity of circulating leukocytes. The fact that 'septic plasma' behaves as an immunosuppressive milieu [12] is illustrated in human volunteers by the capacity of endotoxin to induce plasma inhibitors [13]. Most interestingly, this suppressive effect in septic patients was significantly reduced after passage of plasma through a resin and after incubation with anti-IL-10 antibodies [14]. IL-10 was identified as a major functional deactivator of monocytes in human septic shock plasma [15]. TGFβ was also shown in animal models of hemorrhagic shock or sepsis to be the causative agent of the depressed splenocyte responsiveness [16]. Monocytes from immunocompromised trauma patients seem to be a source of TGFβ [17], and TGFβ released by apoptotic T cells contributes to this immunosuppressive milieu [18]. In addition, there is accumulating evidence for a strong interaction between components of the nervous and the immune systems. Numerous neuromediators behave as immunosuppressors. Interaction of adrenaline with its β_2-adrenergic receptor enhances IL-10 production and decreases TNF production in vitro and in vivo in LPS-challenged healthy volunteers. α-Melanocyte-stimulating hormone also contributes to immunosuppression by inducing IL-10 production by human monocytes [19]. In addition, vasoactive intestinal peptide and pituitary adenylate cyclase-activating polypeptide directly inhibit endotoxin-induced proinflammatory cytokine secretion [20]. Vagal nerve stimulation attenuates hypotension and reduces plasma and liver TNF levels through an interaction between acetylcholine and the α_7 subunit of the nicotinic receptor at the macrophage surface [21]. Sepsis is also associated with an activation of the hypothalamus-pituitary-adrenal axis which leads to the release of glucocorticoids, well known for their potent ability to limit cytokine production. Prostaglandins produced during sepsis also contribute to the downregulation of cytokine production [22]. Finally, adenosine contributes to alter immune status via occupancy of A_{2A} receptor expressed on immune cells [23].

Cellular Markers of Immunodysregulation

Cell Surface Markers on Monocytes

In addition to the quantification of circulating cytokines or other soluble molecules, immunodysregulation in sepsis may be monitored by analyzing the expression of some cell surface molecules. A profound decrease in HLA-DR surface expression on monocytes has been regularly reported in sepsis. Low HLA-DR expression was associated with an increased risk of secondary bacterial infections [24], probably due to a less potent antigen presentation that would not allow an efficient adaptive immunity. The downregulation of HLA-DR is at least partially mediated by the immunosuppressive cytokine IL-10 which was shown to favor the intracellular sequestration of this major histocompatibility complex type II molecule in human monocytes [25]. HLA-DR downregulation was also observed on monocytes of patients with a noninfectious systemic inflammation, such as after pancreatitis, major surgery or trauma.

TREM-1 is a receptor selectively expressed on monocytes/macrophages and neutrophils that mediates cell activation, but its ligand is still unknown. TREM-1 expression is increased on human monocytes upon LPS stimulation in vitro and blockade of TREM-1 protects mice against endotoxic shock [26]. This molecule may also be present in soluble form. Upon LPS injection into healthy volunteers or during sepsis, TREM-1 surface expression was increased on circulating monocytes [27, 28]. An upregulation of TREM-1 has also been found on patients' monocytes after major abdominal surgery [29].

Ex Vivo Leukocyte Reactivity

One of the hallmarks of sepsis is the decreased ex vivo production of proinflammatory cytokines by patients' monocytes in response to LPS stimulation (table 2). This decreased reactivity was also found in many groups of

Table 2. Dysregulation of ex vivo cytokine production by patients' circulating leukocytes during sepsis and systemic inflammatory response syndrome

	Activating agent	Patient group	Cell type studied	Cytokine production	Reference
TLR4 agonist	LPS	Trauma	PBMC	IL-1 ↓, IFNγ ↓	Faist et al: Arch Surg 1988;123:287
	LPS	Sepsis	Monocyte	IL-1 ↓, IL-6 ↓, TNF ↓	Munoz et al: J Clin Invest 1991;88:1747
	LPS	Sepsis	Neutrophil	IL-1 ↓	McCall et al: J Clin Invest 1993;91:853
	LPS	Sepsis	Whole blood	IL-6 ↓, IL-10 ↓, TNF ↓	Marchant et al: J Clin Immunol 1995;15:266
	LPS	Sepsis	Whole blood	IL-1Ra ↑	Van Deuren et al: J infect Dis 1994;169:157
	LPS	Sepsis	Whole blood	IL-1 ↓, IL-6 ↓, TNF ↓	Ertel et al: Blood 1995;85:1341
	LPS	Sepsis	Whole blood	TNF ↓, IL-6 ↓, IL-10 ↓	Marchant et al: J Clin Immunol 1995;15:266
	LPS	Sepsis	Whole blood	IFN-γ ↓, TNF ↓	Mitov et al: Infection 1997;25:206
	LPS	Sepsis	Neutrophil	IL-8 ↓	Marie et al: Blood 1998;91:3439
	LPS	CPB	Neutrophil	IL-8 ↓	Marie et al: Blood 1998;91:3439
	LPS	Trauma	Neutrophil	TNF ↑, IL-8 ↑, IL-1Ra ↔	Zallen et al: J Trauma 1999;46:42
	LPS	Sepsis	Whole blood	G-CSF ↑, TNF ↓	Weiss et al: Cytokine 2001;13:51
	LPS	CPB	Whole blood	TNF ↓	Wilhelm et al: Shock 2002;17:354
	LPS	RCA	Whole blood	IL-1Ra ↑, IL-6 ↓, TNF ↓	Adrie et al: Circulation 2002;106:562
	LPS	Trauma	Whole blood	IL-6 ↓, IL-10 ↓, TNF ↓	Adib-Conquy et al: AJRCCM 2003;168:158
	LPS	Sepsis	PBMC	MIF ↑, TNF ↓	Maxime et al: J Infect Dis 2005;191:138
	LPS	Sepsis	Monocyte	IL-10 ↑, TNF ↓	Adib-Conquy et al: Crit Care Med 2006;34:2377
TLR2 agonist	Pam3CysSK4	Sepsis	Monocyte	IL-10 ↑, TNF ↓	Adib-Conquy et al: Crit Care Med 2006;34:2377
	Pam3CysSK4 LPS from L. interrogans	RCA	Monocyte	IL-10 ↔, TNF ↔	Adib-Conquy et al: Crit Care Med 2006;34:2377
		Trauma	Whole blood	IL-10 ↑, TNF ↔	Adib-Conquy et al: AJRCCM 2003;168:158
TLR9 agonist	CpG	Trauma	Whole blood	IL-6 ↓, IL-10 ↔, TNF ↓	Adib-Conquy et al: AJRCCM 2003;168:158
NOD2 agonist	MDP	Trauma	Monocyte	IL-6 ↔	Szabo et al: J Clin Immunol 1991;11:326
		Trauma	PBMC	IL-10 ↓	Miller-Graziano et al: J Clin Immunol 1995;15:93

Table 2. (continued)

	Activating agent	Patient group	Cell type studied	Cytokine production	Reference
		Sepsis	Monocyte	TNF ↔	Adib-Conquy et al: Crit Care Med 2006;34:2377
		RCA	Monocyte	TNF ↔	Adib-Conquy et al: Crit Care Med 2006;34:2377
G− bacteria	S. typhimurium	Sepsis	Whole blood	IL-6 ↑, TNF ↑	Mitov et al: Infection 1997;25:206
	E. coli	Sepsis	Whole blood	IL-6 ↓, IL-10 ↓, TNF ↓	Haupt et al: J Invest Surg 1997;10:349
	E. coli	Trauma	Whole blood	IL-10 ↑, TNF ↓	Adib-Conquy et al: AJRCCM 2003;168:158
	P. aeruginosa	Sepsis	Whole blood	IL-10 ↔, TNF ↓	Rigato et al: Shock 2003;19:113
G+ bacteria	S. aureus	Sepsis	Whole blood	IL-6 ↑, TNF ↑	Mitov et al: Infection 1997;25:206
	S. aureus	Sepsis	PBMC	IL-10 ↔	Muret et al: Shock 2000;13:169
	S. aureus	Sepsis	Whole blood	TNF ↓	Wihlem et al: Shock 2002;17:354
	S. aureus	RCA	Whole blood	IL-1Ra ↑, IL-6 ↔, TNF ↔	Adrie et al: Circulation 2002;106:562
	S. aureus	CPB	Whole blood	TNF ↓	Wilhelm et al: Shock 2002;17:354
	S. aureus	Trauma	Whole blood	IL-6 ↔, IL-10 ↑, TNF ↔	Adib-Conquy et al: AJRCCM 2003;168:158
	S. aureus	Sepsis	Monocyte	IL-10 ↔, TNF ↔	Adib-Conquy et al: Crit Care Med 2006;34:2377
	S. aureus	RCA	Monocyte	IL-10 ↔, TNF ↔	Adib-Conquy et al: Crit Care Med 2006;34:2377
	S. pyogenes	CPB	Neutrophil	IL-8 ↓	Marie et al: Blood 1998;91:3439
	S. pyogenes	Trauma	Whole blood	IL-6 ↔, IL-10 ↑, TNF ↔	Adib-Conquy et al: AJRCCM 2003;168:158

Cytokine production is indicated as compared to healthy controls: ↑ = increased production; ↓ = decreased production; ↔ = similar production.

CPB = Cardiopulmonary bypass; G− = gram-negative; G+ = gram-positive; LPS = lipopolysaccharide; PBMC = peripheral blood mononuclear cells; RCA = resuscitated cardiac arrest.

patients having noninfectious SIRS (e.g. trauma, major surgery, resuscitation after cardiac arrest; table 2). As observed with LPS, monocytes from septic patients produced significantly less TNF ex vivo in response to a TLR2 agonist. In contrast, we were able to show that this defect did not occur in monocytes from patients with noninfectious SIRS: the production of TNF was comparable to that found in healthy controls. The dysregulation in ex vivo cytokine production was very different when anti-inflammatory cytokines were analyzed. Indeed, in contrast to proinflammatory cytokines, increased production of IL-10 and/or IL-1Ra was found in monocytes of septic and SIRS patients stimulated by TLR4 or TLR2 agonists. Finally, patients' leukocytes were responsive to muramyl dipeptide, the minimal motif of bacterial peptidoglycan, and produced cytokine levels comparable to those obtained in healthy volunteers.

The response to whole bacteria may represent a more relevant pathophysiological approach to monitor the immune status of septic patients. Surprisingly, in contrast to highly specific TLR2 or TLR4 agonists, the production of cytokines by patients' leukocytes gives a more contrasted profile. Cytokine production by septic and SIRS patients in response to heat-killed gram-positive or gram-negative bacteria was usually undiminished or even increased when compared to that obtained with healthy donors (table 2). Thus, the immunodysregulation found during sepsis or SIRS is not a generalized hyporeactivity. Depending upon the cytokine and the stimulus, there may be a decreased, a maintained or an increased response from monocytes.

In conclusion, immune dysregulation during sepsis may be monitored by assessing: (i) circulating cytokines, inflammatory mediators and soluble membrane compounds; (ii) expression of cell surface markers, and (iii) ex vivo monocyte reactivity to LPS. However, one should keep in mind that these modifications are most often similar in patients having sepsis or a non-infectious severe systemic inflammation.

References

1 Cavaillon JM, Adrie C, Fitting C, Adib-Conquy M: Reprogramming of circulatory cells in sepsis and SIRS. J Endotoxin Res 2005;11:311–320.
2 Cavaillon JM, Muñoz C, Fitting C, Misset B, Carlet J: Circulating cytokines: the tip of the iceberg? Circ Shock 1992;38:145–152.
3 Gibot S, Kolopp-Sarda MN, Bene MC, Cravoisy A, Levy B, Faure GC, Bollaert PE: Plasma level of a triggering receptor expressed on myeloid cells-1: its diagnostic accuracy in patients with suspected sepsis. Ann Intern Med 2004;141:9–15.
4 Adib-Conquy M, Monchi M, Goulenok C, Laurent I, Thuong M, Cavaillon JM, Adrie C: Increased plasma levels of soluble triggering receptor expressed on myeloid cells-1 and procalcitonin after cardiac surgery and cardiac arrest without infection, Shock 2007; in press.
5 Kitchens RL, Thompson PA, Viriyakosol S, O'Keefe GE, Munford RS: Plasma CD14 decreases monocyte responses to LPS by transferring cell-bound LPS to plasma lipoproteins. J Clin Invest 2001;108:485–493.

6 Wurfel MM, Kunitake ST, Lichenstein H, Kane JP, Wright SD: Lipopolysaccharide (LPS)-binding protein is carried on lipoproteins and acts as a cofactor in the neutralization of LPS. J Exp Med 1994;180:1025–1035.
7 Cavaillon J, Annane D: Compartmentalization of the inflammatory response in sepsis and SIRS. J Endotoxin Res 2006;12:151–170.
8 Grunfeld C, Marshall M, Shigenaga JK, Moser AH, Tobias P, Feingold KR: Lipoproteins inhibit macrophage activation by lipoteichoic acid. J Lipid Res 1999;40:245–252.
9 Thompson PA, Kitchens RL: Native high-density lipoprotein augments monocyte responses to lipopolysaccharide (LPS) by suppressing the inhibitory activity of LPS-binding protein. J Immunol 2006;177:4880–4887.
10 Froon AH, Bonten MJ, Gaillard CA, Greve JW, Dentener MA, de Leeuw PW, Drent M, Stobberingh EE, Buurman WA: Prediction of clinical severity and outcome of ventilator-associated pneumonia. Comparison of simplified acute physiology score with systemic inflammatory mediators. Am J Respir Crit Care Med 1998;158:1026–1031.
11 Pugin J, Stern-Voeffray S, Daubeuf B, Matthay MA, Elson G, Dunn-Siegrist I: Soluble MD-2 activity in plasma from patients with severe sepsis and septic shock. Blood 2004;104:4071–4079.
12 Cavaillon J-M: 'Septic plasma': an immunosuppressive milieu. Am J Respir Crit Care Med 2002;166:1417–1418.
13 Spinas G, Bloesch D, Kaufmann M, Keller U, Dayer JM: Induction of plasma inhibitors of interleukin 1 and TNF-alpha activity by endotoxin administration to normal humans. Am J Physiol 1990;259:R993–R997.
14 Ronco C, Brendolan A, Lonnemann G, Bellomo R, Piccinni P, Digito A, Dan M, Irone M, La Greca G, Inguaggiato P, Maggiore U, De Nitti C, Wratten ML, Ricci Z, Tetta C: A pilot study of coupled plasma filtration with adsorption in septic shock. Crit Care Med 2002;30:1250–1255.
15 Brandtzaeg P, Osnes L, Øvstebø R, Joø GB, Westwik AB, Kierulf P: Net inflammatory capacity of human septic shock plasma evaluated by a monocyte-based target cell assay: identification of interleukin-10 as a major functional deactivator of human monocytes. J Exp Med 1996;184:51–60.
16 Ayala A, Knotts JB, Ertel W, Perrin MM, Morrison MH, Chaudry IH: Role of interleukin 6 and transforming growth factor-beta in the induction of depressed splenocyte responses following sepsis. Arch Surg 1993;128:89–94.
17 Miller-Graziano CL, Szabo G, Griffey K, Mehta B, Kodys K, Catalano D: Role of elevated monocyte transforming growth factor β production in post-trauma immunosuppression. J Clin Immunol 1991;11:95–102.
18 Chen W, Frank M, Jin W, Wahl S: TGF-beta released by apoptotic T cells contributes to an immunosuppressive milieu. Immunity 2001;14:715–725.
19 Luger TA, Kalden DH, Scholzen TE, Brzoska T: Alpha-melanocyte-stimulating hormone as a mediator of tolerance induction. Pathobiology 1999;67:318–321.
20 Delgado M, Pozo D, Martinez C, Leceta J, Calvo JR, Ganea D, Gomariz RP: Vasoactive intestinal peptide and pituitary adenylate cyclase-activating polypeptide inhibit endotoxin-induced TNF-alpha production by macrophages : in vitro and in vivo studies. J Immunol 1999;162:2358–2367.
21 Wang H, Yu M, Ochani M, Amella CA, Tanovic M, Susarla S, Li JH, Wang H, Yang H, Ulloa L, Al-Abed Y, Czura CJ, Tracey KJ: Nicotinic acetylcholine receptor alpha7 subunit is an essential regulator of inflammation. Nature 2003;421:384–388.
22 Choudhry MA, Ahmad S, Ahmed Z, Sayeed MM: Prostaglandin E2 down-regulation of T cell IL-2 production is independent of IL-10 during Gram negative sepsis. Immunol Lett 1999;67:125–130.
23 Hasko G, Cronstein BN: Adenosine: an endogenous regulator of innate immunity. Trends Immunol 2004;25:33–39.
24 van den Berk JMM, Oldenburger RHJ, van den Berg AP, Klompmaker IJ, Mesander G, van Son WJ, van der Bij W, Slooff MJH, The TH: Low HLA DR expression on monocytes as a prognostic marker for bacterial sepsis after liver transplantation. Transplantation 1997;63:1846–1848.
25 Fumeaux T, Pugin J: Role of interleukin-10 in the intracellular sequestration of human leukocyte antigen-DR in monocytes during septic shock. Am J Respir Crit Care Med 2002;166:1475–1482.
26 Bouchon A, Facchetti F, Weigand MA, Colonna M: TREM-1 amplifies inflammation and is a crucial mediator of septic shock. Nature 2001;410:1103–1107.

27 Knapp S, Gibot S, de Vos A, Versteeg HH, Colonna M, van der Poll T: Cutting edge: expression patterns of surface and soluble triggering receptor expressed on myeloid cells-1 in human endotoxemia. J Immunol 2004;173:7131–7134.
28 Gibot S, Le Renard PE, Bollaert PE, Kolopp-Sarda MN, Bene MC, Faure GC, Levy B: Surface triggering receptor expressed on myeloid cells 1 expression patterns in septic shock. Intensive Care Med 2005;31:594–597.
29 Gonzalez-Roldan N, Ferat-Osorio E, Aduna-Vicente R, Wong-Baeza I, Esquivel-Callejas N, Astudillo-de la Vega H, Sanchez-Fernandez P, Arriaga-Pizano L, Villasis-Keever MA, Lopez-Macias C, Isibasi A: Expression of triggering receptor on myeloid cell 1 and histocompatibility complex molecules in sepsis and major abdominal surgery. World J Gastroenterol 2005;11:7473–7479.

Jean-Marc Cavaillon
Unit Cytokines & Inflammation, Institut Pasteur
28 rue Dr. Roux
FR–75015 Paris (France)
Tel. +33 1 45 68 82 38, Fax +33 1 40 61 35 92, E-Mail jmcavail@pasteur.fr

Metabolism, Electrolytes and Acid-Base Disorders

Nutritional Management in Acute Illness and Acute Kidney Insufficiency

Xavier M. Leverve[a], Noël J.M. Cano[b]

[a]INSERM U884 'Bioénergétique Fondamentale et Appliquée',
Université Joseph Fourier, Grenoble, and [b]Centre Hospitalier Privé Résidence
du Parc, Marseille, France

Abstract

There are now powerful compensatory therapies to counteract kidney deficiency and the prognosis of patients with acute renal failure is mainly related to the severity of the initial disease. Renal failure is accompanied by an increase in both severity and duration of the catabolic phase leading to stronger catabolic consequences. The specificity of the metabolic and nutritional disorders in the most severely ill patients is the consequence of three additive phenomena: (1) the metabolic response to stress and to organ dysfunction, (2) the lack of normal kidney function and (3) the interference with the renal treatment (hemodialysis, hemofiltration or both, continuous or intermittent, lactate or bicarbonate buffer, etc.). As in many other diseases of similar severity, adequate nutritional support in acutely ill patients with ARF is of great interest in clinical practice, although the real improvement as a result of this support is still difficult to assess in terms of morbidity or mortality.

Copyright © 2007 S. Karger AG, Basel

Introduction: Kidney and Whole Body Metabolism

The two kidneys exhibit a high metabolic activity, accounting for 10% of resting energy expenditure while representing only 0.5% of body mass. Kidneys are involved in glucose metabolism by their implication in gluconeogenesis and their role in hormone degradation, mainly insulin and glucagon [1]. Renal glucose release is of the same order of magnitude as liver gluconeogenesis: during the postabsorptive phase, liver glycogenolysis, hepatosplanchnic gluconeogenesis and renal gluconeogenesis, respectively, account for 50, 30 and 20% of whole body glucose release (10–11 μmol/kg body mass/min). In humans, lactate is the main substrate for renal gluconeogenesis, followed by glutamine, and glycerol, while alanine is only a poor substrate for renal glucose synthesis [2]. The kidney is responsible for 50% of whole body lactate gluconeogenesis, therefore it plays a major role in the Cori cycle. Renal glucose synthesis seems more sensitive to hormone action than hepatic gluconeogenesis and has a prominent role during adaptation to various physiological and pathological conditions such as hypoglycemia [3].

The kidney plays a key role in amino acid metabolism and in low molecular weight protein breakdown. It takes up glutamine, proline, citrulline and phenylalanine and significantly releases serine, arginine, tyrosine, taurine, threonine and lysine [4]. Moreover, it plays a crucial role in the control of acid-base regulation by finely tuning the elimination of anions versus cations, thanks to the generation and the elimination of ammonium ions. Glutamine metabolism in proximal tubules produces ammonium and α-ketoglutarate. Acidosis increases ammonium production and excretion, and gluconeogenesis from α-ketoglutarate.

Pathophysiology

Carbohydrate metabolism during renal failure as well as in stress conditions is characterized by insulin resistance, which predominates in peripheral tissues and mainly concerns nonoxidative glucose metabolism [5]. Insulin resistance is attributed to the decrease in glycoregulatory-peptide degradation by the kidney, uremic toxins and acidosis [3]. In addition to insulin resistance, renal failure is also responsible for hypoglycemic manifestations, which may reflect the relative inaptitude of the liver to ensure euglycemia in various severe clinical situations [3]. Hyperinsulinemia together with unadapted glucagon and adrenaline responses to hypoglycemia may explain these manifestations [6]. Indirect calorimetry measurements showed similar resting energy expenditure in control subjects and during real failure and hemodialysis [7]. In spite of abnormalities of plasma lipid transport and clearance, fat is preferentially oxidized after an overnight fast [7], reflecting an accelerated starvation metabolism. Hence, lipids remain a predominant substrate for oxidation in acute renal failure (ARF) as is shown by the low respiratory quotient observed in these patients. Acute patients are also characterized by a low T_3 syndrome and a decrease in testosterone, HGH and IGF levels. In summary, ARF does not much affect specifically the hormonal pattern in comparison with the effect of stress, except for insulin level and insulin resistance since this hormone is degraded mainly in the kidney.

Kidney amino acid exchanges are markedly altered in renal failure [4]. Studies of forearm metabolism during renal failure have shown increased protein synthesis, proteolysis and protein turnover without change in net proteolysis [8]. The effect of acidosis on muscle has been extensively studied [9]: acidosis induces branched-chain AA catabolism and activates the ATP-ubiquitin-dependent cytosolic proteolytic system. Cortisol secretion is activated by acidosis and further stimulates branched-chain AA oxidation and ATP-ubiquitin-dependent proteolysis [10]. Thus, acidosis induces muscle proteolysis and increases renal ammonium excretion, muscle proteolysis being controlled by acid-base balance.

Effect of Renal Replacement Therapy

The metabolic response to acute illness and renal failure can also be affected by some specific alterations related to the replacement therapy. A cascade of events following the contact between the patient's blood and dialysis membrane could be responsible for cytokine and protease activations increasing energy expenditure and protein catabolism [11]. However more biocompatible membranes are often used at present.

Glucose loss during hemodialysis or hemofiltration is similar (between 25 and 50 g) but is higher with hemodiafiltration. The use of 5% dextrose may deliver very large amounts of glucose to the patient when the hemofiltration rate is high (0.5 l/h results in 4,800 cal/day) [12]. Since lipids are circulating only as lipoprotein or are bound to albumin (fatty acids) losses are negligible.

Amino acids are lost with hemodialysis and hemofiltration proportionally to their plasma concentration [13]. When amino acids are infused for nutritional supply, hemofiltration results in about a 10% loss of the total amount infused. Dialysis promotes a net protein catabolism and induces a reduction of protein synthesis. Four to 9 g of free amino acids and 2–3 g of peptide-bound amino acids are removed during each session. The frequent reuse of filters exacerbates the amino acid and albumin loss.

Nutritional Support

Acutely ill patients are highly catabolic and the benefit of adequate nutritional therapy seems important even if it has so far not been possible to show a clear relationship. Limitations of the use of large volumes may be achieved with intermittent dialysis, and continuous hemofiltration or hemodiafiltration facilitates fluid removal and nutritional intakes [14].

The use of the enteral route is recommended whenever possible. Hypermetabolic patients should be fed early and a risk of renal failure should not delay the initiation of nutritional therapy. Conversely, as for the parenteral route, nutrition is indicated only after the acute phase of shock and severe metabolic disorders has been managed. A polymeric diet (1 kcal/ml) is recommended in the majority of clinical situations. In some cases, according to the severity of the disease (acute pancreatitis, gut dysfunction, very high nutritional requirements) more specific diets could be indicated. The addition of glutamine has been proposed since this amino acid plays an important role in the metabolic and immune responses to stress and infection [15].

In hypercatabolic patients with ARF, relatively low levels of energy supply (26 kcal/kg/day) have been associated with a better nitrogen balance compared to higher supplies (35 kcal/kg/day) [16, 17]. High-energy intake (40 kcal/kg/day) has been proposed in patients with high urea nitrogen appearance and very negative nitrogen balance. WHO equations for assessing energy requirement give better estimates than Harris-Benedict equations, and it is important to use the patients' estimated dry weight as ARF patients are often hyperhydrated or have overt edemas. When elevated, heat loss during dialysis should also be taken into consideration in calculating energy requirement. Insulin resistance, reduced glucose tolerance and increased gluconeogenesis, caused by acute uremia itself or acidosis, require a careful monitoring of blood glucose levels by insulin administration in order to avoid hyperglycemic episodes.

In acute and chronic kidney failure, there is a decreased ability to utilize exogenous lipids, and the use of medium-chain triglycerides has not been shown to offer any advantages over long-chain triglycerides [18]. This experimental finding, together with the frequent occurrence of hypertriglyceridcmia, makes it advisable to limit the percentage of lipids to 20–25% of the total energy and to monitor triglyceridemia during treatment. However, lipid intake is important since lipids, besides being a concentrated, low osmolarity source of energy, are carriers of essential fatty acids and the use of fish oil could be proposed [19].

Essential as well as nonessential (histidine, arginine, tyrosine, serine, and cysteine) amino acids become indispensable in ARF, while others, such as phenylalanine and methionine, may accumulate [20]. The use of a mixture of essential amino acids alone must be avoided, because imbalances and severe clinical consequences have been described [21]. Protein requirements in these patients range from 1.0 to 1.5 g/kg/day depending on the severity of catabolism. There is no evidence that increasing the protein intake further results in a better nitrogen balance. However, higher protein/amino acid intake (up to 1.5–2.0 g/kg/day) in more severe ARF patients treated with CVVH, CVVHD, CVVHDF has been advocated. Replacement therapy also produces a considerable loss of amino acids and/or protein with the dialysate, especially with high

Table 1. Expert group recommendations for nutritional treatment of adult patients with renal insufficiency [16, 17]

Clinical condition	Nonprotein energy kcal/kg/day[a]	Protein g/kg/day
Hemodialysis	≥35	1.2–1.4
Peritoneal dialysis	≥35 (glucose absorption from dialysate can account for 25–30% of energy needs)	1.2–1.5 (>50% HBV)
Nephrotic syndrome	≥35	0.8–1.0
Acute renal failure		
Nonoliguric, nonhypercatabolic patients	in most patients: =1.3 BEE	0.55–1.0
Hypercatabolic, dialyzed patients	≥1.3 BEE	1.0–1.5 (or more, see text); NEAA + EAA (ratio >1:1)
Renal transplantation		
Preoperative period	correction of malnutrition	correction of malnutrition
Early postoperative period	30–35	1.3–1.5
Late postoperative period	adapted to maintain an ideal body weight	1.0

BEE = Basal energy expenditure; EAA = essential amino acids; HBV = high biological value; NEAA = nonessential amino acids.
[a] Adapted to individual needs in case of underweight or obesity.

flux dialyzers [16, 17]. This loss should be integrated by artificial nutrition, so an additional amount of protein or amino acids (0.2 g/kg/day) is recommended. The nutritional requirement in patients with kidney insufficiency is summarized in table 1 [16, 17].

Trace elements are excreted mainly by the kidney and parenteral administration to ARF patients requires great care. However, zinc, manganese, copper, selenium and chromium can also be effectively eliminated in the gastroenteric tract. Vitamin A should probably be avoided because of the possibility of accumulation, whereas vitamin C should not exceed 30–50 mg/day, since inappropriate supplementation may result in secondary oxalosis. Even though vitamin D from body storage may provide lasting protection against deficiency, its active renal metabolite (1,25-OH-cholecalciferol) can be rapidly depleted, making repletion necessary. Vitamin K, E, B$_6$ and folate requirements are also increased in ARF [22].

References

1 Cano N: Bench-to-bedside review: glucose production from the kidney. Crit Care 2002;6: 317–321.
2 Gerich JE, Meyer C, Woerle HJ, Stumvoll M: Renal gluconeogenesis: its importance in human glucose homeostasis. Diabetes Care 2001;24:382–391.
3 Adrogue HJ: Glucose homeostasis and the kidney. Kidney Int 1992;42:1266–1282.
4 Tizianello A, De Ferrari G, Garibotto G, Gurreri G, Robaudo C: Renal metabolism of amino acids and ammonia in subjects with normal renal function and in patients with chronic renal insufficiency. J Clin Invest 1980;65:1162–1173.
5 Castellino P, Solini A, Luzi L, Barr JG, Smith DJ, Petrides A, Giordano M, Carroll C, DeFronzo RA: Glucose and amino acid metabolism in chronic renal failure: effect of insulin and amino acids. Am J Physiol 1992;262:F168–176.
6 Castellino P, Bia M, DeFronzo RA: Metabolic response to exercise in dialysis patients. Kidney Int 1987;32:877–883.
7 Schneeweiss B, Graninger W, Stockenhuber F, Druml W, Ferenci P, Eichinger S, Grimm G, Laggner AN, Lenz K: Energy metabolism in acute and chronic renal failure. Am J Clin Nutr 1990;52:596–601.
8 Garibotto G, Russo R, Sofia A, Sala MR, Robaudo C, Moscatelli P, Deferrari G, and Tizianello A: Skeletal muscle protein synthesis and degradation in patients with chronic renal failure. Kidney Int 1994;45:1432–1439.
9 Mitch WE, Price SR, May RC, Jurkovitz C, England BK: Metabolic consequences of uremia: extending the concept of adaptive responses to protein metabolism. Am J Kidney Dis 1994;23: 224–228.
10 Price SR, England BK, Bailey JL, Van Vreede K, Mitch WE: Acidosis and glucocorticoids concomitantly increase ubiquitin and proteasome subunit mRNAs in rat muscle. Am J Physiol 1994;267:C955–C960.
11 Gutierrez A, Alvestrand A, Wahren J, Bergstrom J: Effect of in vivo contact between blood and dialysis membranes on protein catabolism in humans. Kidney Int 1990;38:487–494.
12 Frankenfield DC, Reynolds HN, Badellino MM, Wiles CE 3rd: Glucose dynamics during continuous hemodiafiltration and total parenteral nutrition. Intensive Care Med 1995;21:1016–1022.
13 Davenport A, Roberts NB: Amino acid losses during haemofiltration. Blood Purif 1989;7: 192–196.
14 Bellomo R, Parkin G, Love J, Boyce N: Use of continuous haemodiafiltration: an approach to the management of acute renal failure in the critically ill. Am J Nephrol 1992;12:240–245.
15 Novak I, Sramek V, Pittrova H, Rusavy P, Lacigova S, Eiselt M, Kohoutkova L, Vesela E, Opatrny K Jr: Glutamine and other amino acid losses during continuous venovenous hemodiafiltration. Artif Organs 1997;21:359–363.
16 Toigo G, Aparicio M, Attman PO, Cano N, Cianciaruso B, Engel B, Fouque D, Heidland A, Teplan V, Wanner C: Expert working group report on nutrition in adult patients with renal insufficiency (Part 2 of 2). Clin Nutr 2000;19:281–291.
17 Toigo G, Aparicio M, Attman PO, Cano N, Cianciaruso B, Engel B, Fouque D, Heidland A, Teplan V, Wanner C: Expert working group report on nutrition in adult patients with renal insufficiency (Part 1 of 2). Clin Nutr 2000;19:197–207.
18 Druml W, Fischer M, Sertl S, Schneeweiss B, Lenz K, Widhalm K: Fat elimination in acute renal failure: long-chain vs medium-chain triglycerides. Am J Clin Nutr 1992;55:468–472.
19 Neumayer HH, Heinrich M, Schmissas M, Haller H, Wagner K, Luft FC: Amelioration of ischemic acute renal failure by dietary fish oil administration in conscious dogs. J Am Soc Nephrol 1992;3:1312–1320.

20 Kopple JD: The nutrition management of the patient with acute renal failure. JPEN J Parenter Enteral Nutr 1996;20:3–12.
21 Nakasaki H, Katayama T, Yokoyama S, Tajima T, Mitomi T, Tsuda M, Suga T, Fujii K: Complication of parenteral nutrition composed of essential amino acids and histidine in adults with renal failure. JPEN J Parenter Enteral Nutr 1993;17:86–90.
22 Gilmour ER, Hartley GH, Goodship TH: Trace elements and vitamins in renal disease; in Mitch WE, Klahr S (eds): Nutrition and the Kidney. Boston, Little, Brown, 1993, pp 114–131.

Pr. Xavier Leverve
Bioénergétique Fondamentale et Appliquée INSERM U884
Université Joseph Fourier, BP 53X
FR–38041 Grenoble Cedex (France)
Tel. +33 476 51 43 86, Fax +33 476 51 42 18, E-Mail Xavier.Leverve@ujf-grenoble.fr

Fundamentals of Oxygen Delivery

James Yassin, Mervyn Singer

Bloomsbury Institute of Intensive Care Medicine, University College London, London, UK

Abstract

Oxygen is vital to life. A series of steps are needed to transport oxygen from the lungs to the mitochondrion where the bulk of it is used for generation of energy. Understanding this pathway, which still remains to be properly characterized, will greatly aid both diagnosis and management of the hypoxic patient.

Copyright © 2007 S. Karger AG, Basel

What Is Oxygen and Why Do We Need It?

In 1774 Joseph Priestley discovered oxygen whilst acting as librarian for the Earl of Sherbourne. He performed a series of experiments with mercury and noticed that mercuric oxide, when heated, produced a gas that made a candle burn much faster, and that supported respiration. He called this gas 'dephlogisticated air' [1]. The name oxygen was coined by the French chemist Antoine Lavoisier, from the Greek 'acid producer' as it was mistakenly felt at the time that all acids contained oxygen [2].

Oxygen (O_2) is utilized by most eukaryotic cells to produce energy, in the form of adenosine triphosphate (ATP), by the process of aerobic metabolism that occurs in mitochondria. There is some debate as to why this happens. In 1883 Andreas Schimper [3] put forward the theory of endosymbiosis, later refined by Margulis [4], whereby some 3–4 billion years ago, blue-green algae respired anaerobically and produced oxygen as a by-product of photosynthesis. To deal with this toxic molecule, new bacteria (probably *Rickettsia* spp.) developed aerobic metabolism. Over time, these bacteria were engulfed by anaerobic cells, the two becoming symbiotic. The host cell produced nutrients for the ingested aerobic cell that, in turn, produced energy for the host as a mitochondrion.

Conclusive proof is obviously lacking, though this theory is strongly supported by the composition of mitochondrial membranes closely resembling that of bacteria [4], by both mitochondrial and bacterial DNA being circular, and containing a higher-than-expected percentage of guanine-cytosine [4].

Cellular utilization of O_2 is the last step in a long chain of reactions that occur in the cell cytoplasm and within the mitochondrion. The aim of aerobic metabolism is to enhance energy generation from available substrate. This is then utilized by all cellular functions, from housekeeping duties of membrane stabilization and maintenance of the resting membrane potential, to the highly specialized functions that different organs provide, such as contractile function of cardiac myocytes or the many complex functions of the renal tubules. Failure of aerobic metabolism, either due to insufficient delivery of O_2 or reduced cellular utilization, results in a greater dependence on anaerobic metabolism (glycolysis) in the cellular cytoplasm. While glycolytic activity can upregulate to a certain degree, this has the disadvantage of a greatly reduced energy generation per unit of substrate. Enhanced glycolysis will also increase production of by-products, such as hydrogen ions (H^+), carbon dioxide (CO_2) and lactate. Though traditionally perceived as negative, there are adaptive roles served by these by-products that may assist cell function and integrity in critical illness, for example, a right shift of the oxyhaemoglobin dissociation curve, alternative substrate provision through lactate, and vasodilatation through activation of targeted ion channels.

Aerobic and Anaerobic Metabolism

Figure 1 describes energy generation from carbohydrate metabolism under aerobic conditions. Per molecule of glucose, a net two moles of ATP are produced by glycolysis, a further two are formed in the Krebs cycle and, at cytochrome oxidase, the last step of the electron transport chain where O_2 is first introduced, a further 28 or so molecules of ATP are produced. This contribution emphasizes the importance of oxidative phosphorylation toward energy generation in most of the body's cells, being responsible for $\geq$95% of ATP production under healthy, resting conditions. Notable exceptions include the erythrocyte, which contains no mitochondria, and neutrophils which have a much larger glycolytic component.

Figure 2 shows the production of by-products that occur when the cell respires anaerobically. The lack of oxygen stimulates production of molecules that reduce the conversion of pyruvate to acetyl-CoA by pyruvate dehydrogenase [5]; under these conditions there is a relative increase in production of lactate, CO_2 and H^+ ions for less ATP.

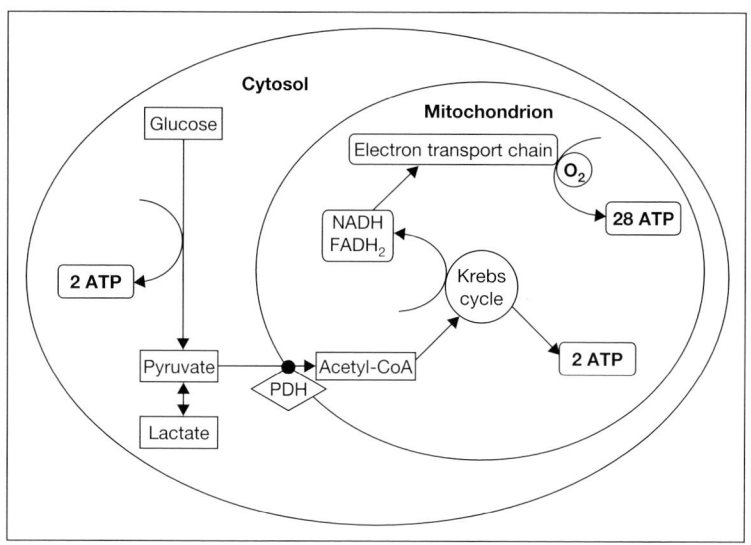

Fig. 1. Energy generation from carbohydrate metabolism.

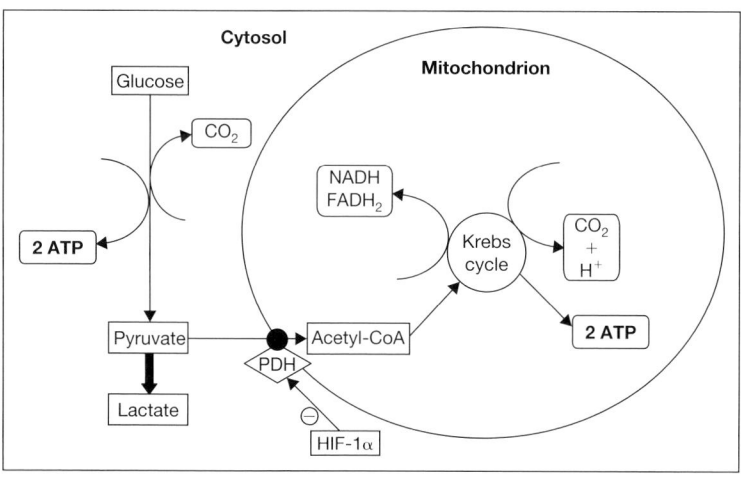

Fig. 2. Byproducts of glucose metabolism.

Lactate

A reduced O_2 delivery often but not always results in lactic acidosis. Conversely, lactic acidosis, a common phenomenon in the critically ill, may occur in the absence of tissue hypoxia. At neutral pH, lactic acid is almost totally dissociated into lactate and H^+. Under normal conditions, the hydrogen ion is used in oxidative phosphorylation to produce ATP. Thus, when stressed, the body does not become acidotic unless oxidative phosphorylation is impaired, for example during sepsis [6]. As can be seen in figure 2, more lactate is produced under anaerobic compared to aerobic conditions.

Traditionally, hyperlactataemia has been divided into type A and type B. Tissue hypoxia causes type A, while the causes of type B are non-hypoxic and include biguanide therapy and also renal replacement therapy, which may use lactate as a buffer for the dialysate. In this situation, the serum lactate concentration may be expected to increase, though it is not accompanied by an acidosis, and then stabilizes at a new and higher level provided liver function is normal.

The ratio of lactate to pyruvate can be measured but is not performed clinically as pyruvate is very unstable, and quickly breaks down. The lactate:pyruvate ratio in health is approximately 10:1. Under hypoxic conditions, as pyruvate dehydrogenase is inhibited, pyruvate is converted into lactate and so the ratio may be expected to rise to around 40:1 [7].

Oxygen Delivery

Oxygen delivery (DO_2) can be considered as occurring in a number of stages: (1) passage of O_2 to the alveolus, (2) transfer of O_2 from the alveolus to the red cell, (3) movement from alveolar capillary to the pulmonary vein, (4) transport of O_2 to the tissues and (5) cellular uptake of O_2.

Figure 3 shows the 'oxygen cascade'. O_2 tension falls progressively from inspired air to the level of the mitochondria (approximately 0.1–1 kPa), with shunt and diffusion being represented within the lungs.

Alveolar Oxygen

The partial pressure of O_2 in the alveolus (P_AO_2) is influenced by a number of factors. At sea level, PO_2 is approximately 21 kPa, that is to say 21% of atmospheric pressure (P_{atm}). Under normal physiological conditions, a tidal breath passes to the carina and the air is warmed and humidified. As a result of the addition of saturated vapour pressure (6 kPa), the PO_2 drops to [21 × (100−6)] = 19.7 kPa. This is hardly significant at sea level, but becomes very important if breathing rarefied air at the summit of Mount Everest. Up to the

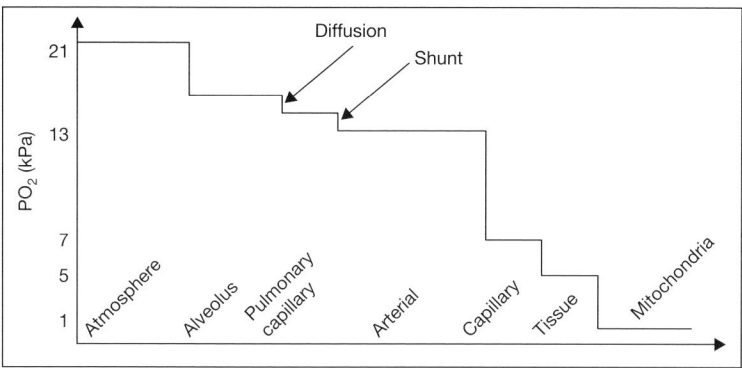

Fig. 3. The oxygen cascade from air to mitochondrion.

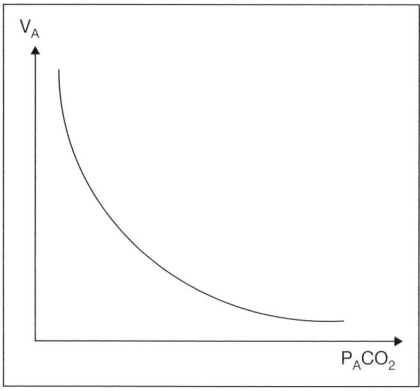

Fig. 4. The relationship between alveolar ventilation and alveolar PCO_2.

8th generation of conducting airways, O_2 travels by convection, and by the 12th generation this bulk flow decreases with diffusion becoming increasingly important [8].

Under conditions of constant O_2 utilization and CO_2 production, P_AO_2 is influenced primarily by P_ACO_2, as the partial pressure of the other constituents of air (mostly N_2) is relatively stable. An increase in P_ACO_2 as a result of reduced alveolar ventilation (V_A) (fig. 4), a relative increase in physiological dead space, or increased CO_2 production, results in a corresponding decrease in P_AO_2 (fig. 5). The relationship between O_2 and CO_2 is thus linear at physiological gas tensions. The simplified alveolar gas equation (equation 1) allows prediction of P_AO_2 if $PaCO_2$ is known and its near match for P_ACO_2 assumed.

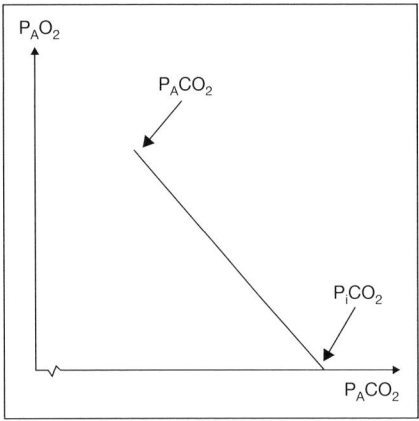

Fig. 5. The relationship between alveolar PO_2 and PCO_2.

$$P_AO_2 = P_iO_2 - \frac{PaCO_2}{R} \quad (1)$$

where P_AO_2 is the alveolar PO_2, P_iO_2 is the partial pressure of inspired O_2 (allowing for vapour pressure), $PaCO_2$ is the arterial PCO_2 and R is the respiratory quotient (usually held to be 0.8). Substituting in some normal values (equation 2) we can see that the P_AO_2 is usually 13 kPa.

$$P_AO_2 = [(100-6) \times 0.21] - (5.3/0.8) = 13.1 \text{ kPa} \quad (2)$$

From the alveolar gas equation the importance of $PaCO_2$ becomes clear. The biggest influence on P_ACO_2 is alveolar minute volume, i.e. the volume of gas reaching the alveolus in 1 min. Thus, respiratory depression, usually either from drugs or a neurological cause, leads to an increase in CO_2 and thus a decrease in P_AO_2.

From Alveolus to Red Cell

The passage of O_2 from alveolus to erythrocyte occurs in two stages: firstly, diffusion of the gas across the basement membrane into the capillary and, secondly, binding of O_2 to haemoglobin.

Oxygen is not a particularly soluble gas compared to CO_2 so its transfer to capillary blood is aided by the small distance it has to travel. It only has to traverse the alveolar basement membrane and capillary endothelium (around 0.6 μm) [9]. The speed of diffusion of oxygen across the lung is dependent on oxygen's molecular weight and solubility, as well as the area available for diffusion,

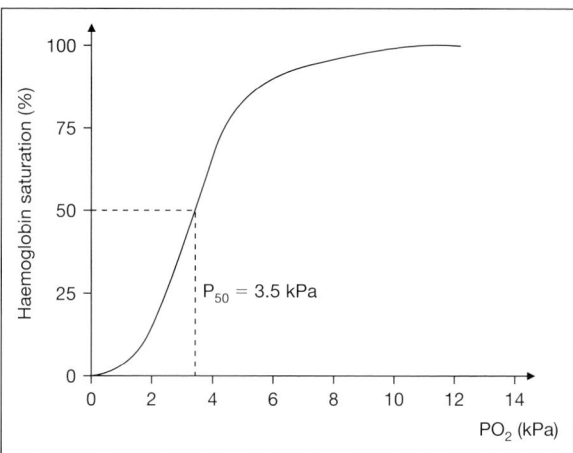

Fig. 6. The oxyhaemoglobin dissociation curve.

the distance it must travel and the partial pressure difference across its path. This is described by Fick's law of diffusion (equation 3):

$$V_{oxygen} \propto \frac{A}{T} \cdot D \cdot (P_1 - P_2) \qquad (3)$$

where V_{oxygen} is the rate of diffusion, A is the area of the lung, T is the thickness of the alveolar-capillary boundary, D is the diffusion constant (dependent on molecular weight and solubility), and $P_1 - P_2$ is the partial pressure difference across the membrane.

Under normal resting physiological circumstances, haemoglobin is fully bound by the time it is a third of the way down the alveolar capillary [10]. Thus, haemoglobin binding is *perfusion* limited. If the cardiac output is increased, for example during exercise, full binding will be delayed due to the increased blood velocity, but will still occur before the end of the capillary. In disease, anything that increases the difficulty of diffusion, such as pulmonary oedema, lung fibrosis or basement membrane disease, will result in an inability to load haemoglobin by the time it leaves the capillary. This is known as *diffusion* limitation.

Oxygen binds to the four haem moieties within the haemoglobin molecule in a homotropic fashion. As each molecule binds, the affinity of haemoglobin for oxygen increases [11]. This gives the characteristic shape of the oxygen dissociation curve (ODC) (fig. 6). The binding affinity of haemoglobin for O_2 is also heterotropic, and is influenced by H^+ concentration, temperature, 2,3-diphosphoglycerate concentration, and CO_2 [12]. The presence of H^+ and CO_2

shifts the ODC to the right and so facilitates O_2 release (Bohr effect) [13], whilst binding of O_2 leads to the release of CO_2 (Haldane effect) [14].

As well as binding to haemoglobin, a small amount of O_2 also dissolves in blood. At 1 atm, even when given supplementary O_2, this dissolved gas is insignificant when compared with the amount carried on haemoglobin.

From Capillary to Pulmonary Vein

With blood leaving the alveolar capillary is fully saturated with oxygen, it passes to the pulmonary vein and thence to the left atrium. Along this path it is mixed with blood that has not been exposed to an alveolus and thus the overall saturation of blood in the left atrium is lower than would be expected. This desaturated blood is the result of physiological and pathological shunt (*true* shunt) and ventilation/perfusion mismatch (*effective* shunt). The desaturated blood is called *venous admixture*, and is the theoretical volume of desaturated blood that would need to be added to fully saturated arterial blood in order to produce the observed arterial SaO_2.

Physiological shunt allows desaturated venous blood into the systemic circulation via normal vascular anatomy. In the coronary circulation, venous blood drains directly into the left ventricle via the thebesian veins. Also, blood that perfuses the bronchial tree drains directly into the pulmonary veins. This shunt is usually a small fraction of cardiac output, and thus reduces the arterial PO_2 still further.

Pathological shunt may be caused by pulmonary arteriovenous malformations, or right to left cardiac septal defects. Again, this will depress the arterial PO_2, but importantly for both physiological and pathological shunts, neither can be improved by the administration of 100% O_2, as nowhere in its path will the blood be exposed to the oxygen, and the increased P_IO_2 will have little effect on the already fully saturated blood passing through the alveoli.

Under normal physiological conditions in the standing subject, both the ventilation and perfusion of the lung are approximately matched. The result is that alveoli with a good blood supply are well ventilated, and vice versa. Both ventilation and perfusion are greater in the dependent parts of the lung. This effect is due, in part, to the action of gravity on the weight of the lung tissue itself and the column of blood [15]. Anything that alters the ratio of ventilation and perfusion may result in unsaturated blood reaching the systemic circulation.

Transfer of O_2 to the Tissues

The amount of oxygen delivered to the tissues is a product of the oxygen content per unit volume of blood and the cardiac output. The O_2 content (CaO_2) depends on the amount of haemoglobin, and how well saturated that haemoglobin is. There is also a small addition for the dissolved O_2 as discussed earlier. This is described in equation 4:

$$CaO_2 = (Hb \times SaO_2 \times 1.39) + (PaO_2 \times 0.02) \qquad (4)$$

where CaO_2 is the arterial O_2 content in ml/100 ml blood, Hb is the haemoglobin concentration in g/dl, SaO_2 are the arterial haemoglobin saturation, 1.39 is Huffner's constant, and refers to the amount of O_2 in milliliters 1 g of haemoglobin can hold, and P_aO_2 is the arterial PO_2 in kPa.

For example, a healthy subject breathing room air at 1 atm, saturating at 97% with a haemoglobin concentration of 15 g/dl, we can see that:

$$CaO_2 = (15 \times 0.97 \times 1.39) + (13 \times 0.02) = 20.2 + 0.26 = 20.5 \text{ ml}/100 \text{ ml}$$

Therefore every litre of blood carries 200 ml of O_2. If this is then multiplied by cardiac output, the global DO_2 can be calculated (equation 5).

$$DO_2 = CaO_2 \times \text{cardiac output} = 200 \times 5 = 1{,}000 \text{ ml/min} \qquad (5)$$

1,000 ml O_2 is ejected from the left ventricle every minute. This represents the global DO_2, but does not indicate the delivery to individual vascular beds, i.e. splanchnic, renal, or cerebral. Each of these beds can regulate their own blood flow [16–18], and, with the exception of the coronary system, can alter the amount of O_2 removed from that regional circulation [19].

Cellular Uptake of O_2

The PO_2 within mitochondria is extremely low, being in the order of 0.5–2.7 kPa [20]. This varies between cell types and is intimately related to cytosolic PO_2. In order to enter the cell, the O_2 molecules must first dissociate from the erythrocyte haemoglobin. This process is aided by the presence of CO_2 and H^+ ions in the capillary, as described above [21]. Once released, O_2 can then diffuse down its 'tension gradient' into the cell, and thus be available to the mitochondria.

With respect to the capillary, these gradients are both longitudinal and radial, and were first described by Krogh [22] and Erlangen with the 'Krogh cylinder model' in 1919 (fig. 7).

The exact site of O_2 transfer into the tissue is the subject of current debate and research. Investigators have measured a large drop in PO_2 across the terminal arteriole, with less change down the capillary than might be expected and, finally, an increase in PO_2 in the postcapillary venule [23]. The reasons for this

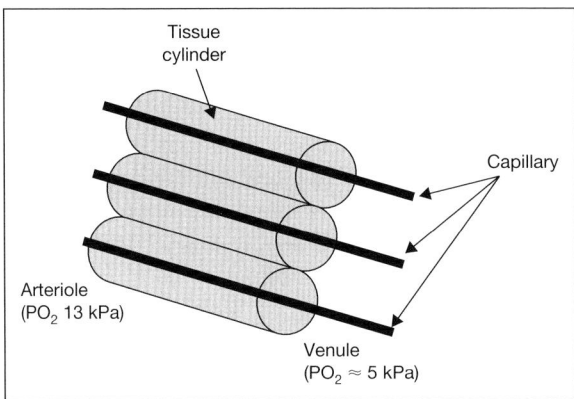

Fig. 7. The Krogh cylinder model of oxygen diffusion.

are unclear, but suggestions include protection of the capillary bed from hyperoxia (and thus increased free radical production) by a 'countercurrent' transfer of O_2 from the arteriole and venule [24]. The countercurrent transport of CO_2 also allows the capillary pH to remain low, and so increase the efficiency of O_2 offloading from haemoglobin for a given PO_2 [25]. This process may be affected by anything that prevents O_2 unloading from the red cell, increased distance from the capillary to the mitochondria, or any inability of the mitochondria to use the O_2 for aerobic respiration.

Hypoxia

Hypoxia is defined as the point at which aerobic respiration ceases to continue, and further metabolism continues anaerobically. This may occur with a 'normal' PaO_2. Hypoxaemia is defined as a low arterial oxygen tension, using an arbitrary cut-off, of 8 kPa in room air. The two terms are not interchangeable, and here we shall be discussing hypoxia and its causes.

Using the system outlined above, the causes of hypoxia can be categorized into: hypoxic hypoxia, anaemic hypoxia, stagnant hypoxia, and cytopathic dysoxia.

Hypoxic hypoxia is the result of either a failure of gas exchange, and/or an increase in venous admixture. The causes of these are many, and include chronic obstructive pulmonary disease, collapse/consolidation, lung fibrosis and intracardiac shunts. This may respond to administration of oxygen provided that shunt is not a predominant feature.

Table 1. Common causes of hypoxia

Causes of hypoxia	
↓ *O_2 content*	
↓ P_AO_2	↓ P_iO_2, ↓ MV
↓ diffusion	pulmonary oedema, fibrosis
↓ binding to Hb	abnormal Hb, alkalosis
↑ admixture	↑ venous admixture
Anaemia	
↓ cardiac output	hypovolaemia, heart failure
↓ cellular utilization	sepsis, drugs, cyanide poisoning, carbon monoxide poisoning

P_AO_2 = Alveolar PO_2; P_iO_2 = inspired PO_2; MV = minute volume; Hb = haemoglobin.

Anaemic hypoxia is self-explanatory and may be treated by an infusion of red cells. The haemoglobin concentration at which oxygen delivery is insufficient depends on oxygen requirements. In healthy volunteers a concentration of 3 g/dl [26] may be tolerated. However, in the critically unwell population, a target of 7 g/dl has been shown to be sufficient for most patients [27]. However, uncertainty remains about the optimal value in patients with cardiorespiratory failure. An outcome study investigating early, goal-directed resuscitation of patients with early sepsis targeted a haematocrit value of 30% [28].

Stagnant hypoxia refers to the patient with an inadequate cardiac index. This may be ventricular filling-related, in which case either blood or another intravenous fluid should be administered. Low cardiac output due to poor ventricular function (either left or right) may respond to inotropes, but a cause should be identified and, if possible, treated, for example myocardial infarction or cardiac tamponade.

Cytopathic dysoxia is the last and most recently discovered cause of hypoxia. This is the result of an inability of the electron transport chain within the mitochondria to utilize available O_2 from within the cytoplasm of the cell. A common cause of this is sepsis [29, 30], where the electron transport chain is inhibited by nitric oxide [6]. Other causes include cyanide or carbon monoxide poisoning, where the cyanide or CO molecule binds to the ferric ion of mitochondrial cytochrome oxidase and stops its ability to respire aerobically [31].

Table 1 gives a summary of the common causes of hypoxia.

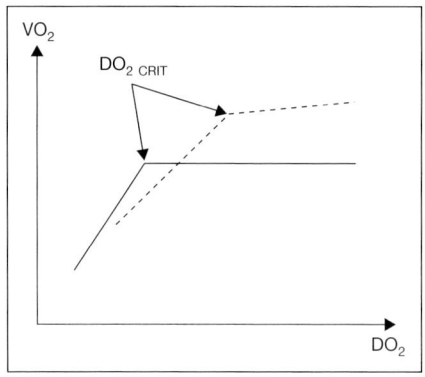

Fig. 8. The stylized relationship between oxygen delivery and consumption. — = Relationship in health; ---- = relationship in sepsis.

Oxygen Supply Dependence

Oxygen consumption (VO_2) is determined by mitochondrial O_2 requirements. Under normal conditions there is a relationship between DO_2 and VO_2 (fig. 8, solid line), such that an increase in O_2 requirement results in an increase in oxygen extraction, thus maintaining aerobic metabolism independent of DO_2 [32]. If DO_2 decreases, a level will be reached where VO_2 cannot be maintained by increasing extraction alone. This is known as the critical DO_2 ($DO_{2\ CRIT}$) [33]. Different organs under varying conditions of stress will have a differing ability to extract oxygen, and so figure 8 represents a global picture. Critical illness changes the relationship between DO_2 and VO_2 (fig. 10, dashed line) such that the tissue may continue to utilize O_2 as its delivery increases, so called *supply dependency* [34].

Conclusion

A series of steps are needed to transport oxygen from the atmosphere to the mitochondrion where the bulk of it is used for generation of energy. Understanding this pathway greatly aids diagnosis and management of the hypoxic patient. Cellular uptake of O_2 is as yet incompletely characterized and further research will elucidate the physiological control mechanisms and processes that occur during pathological states.

References

1. Priestley J: Experiments and Observations on Different Airs. London, Royal Society, 1775.
2. Lavoisier A: Considérations Générales sur la Nature des Acides. Acides, 1778.
3. Schimper AFW: Über die Entwicklung der Chlorophyllkörner und Farbkörper. Bot Ztg 1883;41:105–162.
4. Margulis L: Symbiosis in cell evolution. Freeman, San Francisco, 1981, pp 206–227.
5. Kim JW, Tchernyshyov I, Semenza GL, Dang CV: HIF-1-mediated expression of pyruvate dehydrogenase kinase: a metabolic switch required for cellular adaptation to hypoxia. Cell Metab 2006;3:150–151.
6. Brealey D, Brand M, Hargreaves I, Heales S, Land J, Smolenski R, Davies NA, Cooper CE, Singer M: Association between mitochondrial dysfunction and severity and outcome of septic shock. Lancet 2002;360:219–223.
7. Levy B, Sadoune LO, Gelot AM, et al. Evolution of lactate/pyruvate and arterial ketone body ratios in the early course of catecholamine-treated septic shock. Crit Care Med 2000;28:114–119.
8. Wilson TA, Lin K: Convection and diffusion in the airways and the design of the bronchial tree; in Bouhuys A (ed): Airway Dynamics, Physiology and Pharmacology. Springfield, Thomas, 1970, pp 5–20.
9. Hogan J, Smith P, Heath D, Harris P: The thickness of the alveolar capillary wall in the human lung at high and low altitude. Br J Dis Chest 1986;80:13–18.
10. West JB: Respiratory Physiology – the Essentials. Baltimore, Williams & Wilkins, 2000, p 23.
11. Monod J, Wyman J, Changeux J-P: On the nature of allosteric transitions: a plausible model. J Mol Biol 1965;12:88–118.
12. Imai K: Allosteric Effects in Haemoglobin. Cambridge, Cambridge University Press, 1982.
13. Bohr C, Hasselbalch KA, Krogh A: Über einen in biologischer Beziehung wichtigen Einfluss, den die Kohlensäurespannung des Blutes auf dessen Sauerstoffbindung übt. Skand Arch Physiol 1904;16:402–412.
14. Christiansen J, Douglas CG, Haldane JS: The absorption and dissociation of carbon dioxide by human blood. J Physiol (Lond) 1914;48:244–271.
15. West JB, Dollery CT: Distribution of blood flow and ventilation-perfusion ratio in the lung, measured with radioactive CO_2. J Appl Physiol 1960;15:405–410.
16. Miller DJ, Stanek A, Langfitt TW: Concepts of cerebral perfusion pressure and vascular compression during intracranial hypertension. Prog Brain Res 1972;31:411–432.
17. Navar LG: Renal autoregulation: perspectives from whole kidney and single nephron studies. Am J Physiol Renal Physiol 1978;234:F357–F370.
18. Granger DN, Richardson PDI, Kvietys PR, Mortillaro NA: Intestinal blood flow. Gastroenterology 1980;78:837–863.
19. Allela A, Williams F, Bolene-Williams C, Katz L: Interrelation between cardiac oxygen consumption and coronary flow. Am J Physiol 1955;183:570–582.
20. Nunn JF: Oxygen; in Nunn JF (ed): Nunn's Applied Respiratory Physiology. Oxford, Butterworth-Heinemann, 1993, pp 247–305.
21. Pittman RN, Dulling BR: Effects of altered carbon dioxide tension on haemoglobin oxygenation in the hamster cheek pouch microvessels. Microvasc Res 1977;13:211–224.
22. Krogh A: The number and the distribution of capillaries in muscle with the calculation of the oxygen pressure necessary for supplying tissue. J Physiol (Lond) 1919;52:409–515.
23. Kerger H, Torres Filho IP, Rivas M, Winslow RM, Intaglietta M: Systemic and subcutaneous oxygen tension in conscious Syrian golden hamsters. Am J Physiol Heart Circ Physiol 1995;267:H802–H810.
24. Kobayashi H, Takizawa N: Oxygen saturation and pH changes in cremaster microvessels of the rat. Am J Physiol Heart Circ Physiol 1996;270:H1453–H1461.
25. Wiebel ER: The Pathway for Oxygen. Cambridge, Harvard University Press, 1984.
26. Fontana JL, Welborn L, Mongan PD, Sturm P, Martin G, Bünger R: Oxygen consumption and cardiovascular function in children during profound intraoperative normovolemic hemodilution. Anesth Analg 1995;80:219–225.

27 Hébert PC, Wells G, Blajchman MA, Marshall J, Martin C, Pagliarello G, Tweeddale M, Schweitzer I, Yetisir E: A multimember, randomized, controlled clinical trial of transfusion requirements in critical care. N Engl J Med 1999;340:409–417.
28 Rivers E, Bryant MPH, Havstad S: Early goal directed therapy in the treatment of severe sepsis and septic shock. N Engl J Med 2001;345:1368–1377.
29 Astiz M, Rackow EC, Weil MH, Schumer W: Early impairment of oxidative metabolism and energy production in severe sepsis. Circ Shock 1988;26:311–320.
30 Boekstegers P, Weidenhofer S, Kapsner T, Werden K: Skeletal muscle partial pressure of oxygen in patients with sepsis. Crit Care Med 1994;22:640–650.
31 Kerns W, Isom G, Kirk MA: Cyanide and hydrogen sulfide; in Goldfrank LR, Flomenbaum NE, Lewin NA, et al (eds): Toxicologic Emergencies, ed 7. New York, McGraw-Hill, 2002, pp 1493–1513.
32 Leach RM, Treacher DF: The pulmonary physician in critical care 2: oxygen delivery and consumption in the critically ill. Thorax 2002;57:170–177.
33 Schumacher PT, Cain SM: The concept of a critical DO_2. Intensive Care Med 1987;13:223.
34 Bihari D, Smithies M, Gimson A, et al: The effect of vasodilatation with prostacyclin on oxygen delivery and uptake in critically ill patients. N Engl J Med 1987;317:397–403.

Prof. Mervyn Singer
Bloomsbury Institute of Intensive Care Medicine, University College London
Cruciform Building
Gower St., London WC1E 6BT (UK)
Tel. +44 207 679 6714, Fax +44 207 679 6952, E-Mail m.singer@ucl.ac.uk

Principals of Hemodynamic Monitoring

Patricio M. Polanco[a], *Michael R. Pinsky*[b]

[a]Division of Trauma, Department of Surgery and [b]Department of Critical Care Medicine, University of Pittsburgh School of Medicine, Pittsburgh, Pa., USA

Abstract

Background/Aims: Hemodynamic monitoring is the cornerstone of patient management in the intensive care unit. However, to be used effectively its applications and limitations need to be defined and its values applied within the context of proven therapeutic approaches. **Methods:** Review of the physiological basis for monitoring and a review of the literature on its utility in altering patient outcomes. **Results:** Most forms of monitoring are used to prevent cardiovascular deterioration or restore cardiovascular wellness. However, little data support the generalized use of aggressive resuscitation protocols in all but the most acutely ill prior to the onset of organ injury. Outcomes improve with aggressive resuscitation in some patients presenting with early severe sepsis and in postoperative high-risk surgical patients. **Conclusions:** Monitoring should be targeted to meet the specific needs of the patient and should not be applied in a broad fashion and whenever possible it should be used as part of a treatment protocol of proven efficacy.

Copyright © 2007 S. Karger AG, Basel

Hemodynamic monitoring is a cornerstone in the care of the critically ill; high acuity units, such as the emergency department, intensive care unit (ICU) and operating room and acute treatment centers, such as dialysis units, monitor the cardiovascular status of their patients to both identify new cardiovascular insufficiency, diagnose its etiology and monitor response to resuscitation therapy. Modern medicine has witnessed an impressive degree of recent medical technological advances, allowing monitoring, display, and assessment of an almost unimaginable number of physiological variables. Yet the utility of most hemodynamic monitoring is unproven, whereas it is the commonly available technologies where clinical studies have demonstrated relevance. Furthermore, despite the many options available, most acute care centers monitor and display only blood pressure, heart rate and pulse oximetry (SpO_2), as they have done for the last 20 years. Furthermore, with few exceptions, such monitoring does not drive treatment protocols but rather serves as automated vital signs recorded to trigger further attention. It is hard to validate the utility of monitoring when it is used in this fashion because no hemodynamic monitoring device will improve outcome unless coupled to a treatment which, itself, improves outcome. Thus, the effectiveness of hemodynamic monitoring to improve outcome is limited to specific patient groups and disease processes where proven effective treatments exist. Although, like most of medicine, the utility of hemodynamic monitoring is not well documented, a primary rationale for the use of hemodynamic monitoring is to identify cardiovascular instability and guide therapy.

Rationale for Hemodynamic Monitoring

The arguments to justify the use of specific types of monitoring techniques can be roughly grouped into three levels of defense based on their level of validation [1]. At the basic level, the specific monitoring technique can be defended based on historical controls. At this level, prior experience using similar monitoring was traditionally used and presumed to be beneficial. The mechanism by which the benefit is achieved need not be understood. The second level of defense comes through an understanding of the pathophysiology of the process being treated. This physiological argument can be stated as 'knowledge of how a disease process creates its effect and thus preventing the process from altering measured bodily functions should prevent the disease process from progressing or injuring remote physiological functions'. Most of the rationale for hemodynamic monitoring resides at this level. It is not clear from recent clinical studies in critically ill patients that this argument is valid, primarily because knowledge of the actual processes involved in the expression of disease and tissue injury is often inadequate. The third level of defense comes from documentation that the monitoring device, by altering therapy in otherwise unexpected ways, improves outcome in terms of survival and quality of life. In reality, few therapies done in medicine can claim benefit at this level. Thus, we are left with the physiological rationale as the primary defense of monitoring critically ill patients.

The Physiological Basis for Hemodynamic Monitoring

Monitoring of critically ill patients usual serves a dual function. First, it is used to document hemodynamic stability and the lack of a need for acute interventions and second, it is used to monitor when measured variables vary from their defined baseline values thus defining disease. Accordingly, knowing the limits to which such monitoring reflects actual physiological values is an essential aspect of their utility.

One the physical side, hemodynamic monitoring can be invasive or noninvasive, and continuous or intermittent. Monitoring devices can measure physiological variables directly or derive these variables through signal processing. Signal processing does not minimize the usefulness of physiological variable analysis; it just separates the output data from the patient by the use of the data processor. The most common signal processing physiological variables measured clinically is the electrocardiogram. Although it is a highly processed signal bearing no simple relationship to the surface electrical potential on the heart, the electrocardiogram is a highly sensitive and specific measure of arrhythmias and ischemic injury.

Variables that can be measured noninvasively include body temperature, heart rate, systolic and diastolic arterial blood pressure, and respiratory frequency. Although central venous pressure (CVP) can be estimated as jugular venous distention and hepatojugular reflux, these signs can be ambiguous and are not usually applied in routine clinical practice today. Processed noninvasive variables include the electrocardiogram, transcutaneous SpO_2, expired CO_2, transthoracic echocardiography, tissue O_2 saturation (StO_2) and noninvasive respiratory plethysmography, although the latter two devices are not readily available. Invasive monitoring reflects intravascular catheter insertion, transesophageal echocardiographic probe insertion and blood component analysis. Invasive hemodynamic monitoring of vascular pressures is usually performed by the percutaneous insertion of a catheter into a vascular space and transducing the pressure sensed at the distal end. This allows for the continual display and monitoring of these complex pressure waveforms. Similar intrapulmonary vascular catheters can be used to derive thermal signals and mixed venous O_2 saturation (SvO_2), central venous O_2 saturation ($ScvO_2$), and cardiac output. How useful this hemodynamic information is to diagnosis, treatment and prognosis is a function of its reliability, established treatment protocols and guidelines and the expertise of the operator. What follows is an anatomical survey of the various monitoring devices and a discussion of their utility.

Arterial Pressure Monitoring

After pulse rate, arterial pressure is the most common hemodynamic variable monitored and recorded. Blood pressure is usually measured noninvasively using a sphygmomanometer and the auscultation technique. Importantly, very large and obese subjects in whom the upper arm circumference exceeds the width limitations of a normal blood pressure cuff will record pressures that are higher than they actually are. In such patients using the large thigh blood pressure cuff usually resolves this problem. Blood pressure can be measured automatically using computer-driven devices (e.g. Dynamat®) that greatly reduce nursing time. Sphygmomanometer-derived blood pressure measures display slightly higher systolic and lower diastolic pressures than simultaneously measured indwelling arterial catheters, but the mean arterial pressure is usually similar and the actual systolic and diastolic pressure differences are often small except in the setting of increased peripheral vasomotor tone. If perfusion pressure of the finger is similar to arterial pressure, then both blood pressure and the pressure profile may be recorded noninvasively and continuously using the optical finger probe (Fenapres). However, finger perfusion is often compromised during hypovolemic shock and hypothermia, limiting this monitoring technique to relatively well-perfused patients.

Accurate and continuous measures of arterial pressure can be done through arterial catheterization of easily accessible arterial sites in the arm (axillary, brachial or radial arterial) or groin (femoral arterial). Rarely are upper extremity sites other than the radial artery used because of fear of vascular compromised distal to the catheter insertion site, although data supporting these fears are nonexistent. Arterial catheterization displaying continuous arterial pressure waveforms lends itself to arterial waveform analysis, essential in calculation pulse pressure and pulse pressure variations and cardiac output.

The Physiological Significance of Arterial Pressure

Arterial pressure is the input pressure for organ perfusion. Organ perfusion is usually dependent on organ metabolic demand and perfusion pressure. With increasing tissue metabolism, organ blood flow proportionally increases by selective local vasodilation of the small resistance arterioles. If cardiac output cannot increase as well, as is the case with heart failure, then blood pressure decreases limiting the ability of local vasomotor control to regulate organ blood flow. If local metabolic demand remains constant, however, changes in arterial pressure are usually matched by changes in arterial tone so as to maintain organ blood flow relatively constant. This local vasomotor control mechanism is referred to as autoregulation. Cerebral blood flow over the normal autoregulatory range of 65 to 120 mm Hg is remarkably constant. Although autoregulation occurs in many organs, like the brain, liver, skeletal muscle and skin, it is not a universal phenomenon. For example, coronary flow increases with increasing arterial pressure because the myocardial O_2 demand increases as the heart ejects into a higher arterial pressure circuit. Furthermore, renal blood flow increases in a pressure-dependent fashion over its entire pressure for similar reasons. As renal flow increases, so does renal filtrate flow into the tubules, increasing renal metabolic demand. Thus, a normal blood pressure does not mean that all organs have an adequate amount of perfusion, because increases in local vasomotor tone and mechanical vascular obstruction can still induce asymmetrical vascular ischemia.

Determinants of Arterial Pressure

Arterial pressure is a function of both vasomotor tone and cardiac output. The local vasomotor tone also determines blood flow distribution, which itself is usually determined by local metabolic demands. For a constant vasomotor tone, vascular resistance can be described by the relation between changes in both arterial pressure and cardiac output. The body defends organ perfusion pressure above all else in its autonomic hierarchy through alterations in α-adrenergic tone, mediated though baroreceptors located in the carotid sinus and aortic arch. This supremacy of arterial pressure in the adaptive response to circulatory shock exists because both coronary and cerebral blood flows are dependent only on perfusion pressure. The cerebral vasculature has no α-adrenergic receptors; the coronary circulation has only a few. Accordingly, hypotension always reflects cardiovascular embarrassment, but normotension does not exclude it. Hypotension decreases organ blood flow and stimulates a strong sympathetic response that induces a combined α-adrenergic (increased vasomotor tone) and β-adrenergic (increased heart rate and cardiac contractility) effect and causes a massive ACTH-induced cortisol release from the adrenal glands. Thus, to understand the determinants of arterial pressure one must also know the level of vasomotor tone.

In the ICU setting, arterial tone can be estimated at the bedside by calculating systemic vascular resistance. Using Ohm's law, resistance equals the ratio of the pressure to flow, usually calculated as the ratio of the pressure gradient between aorta and CVP to cardiac output. Arterial tone can also be calculated as total peripheral resistance, which is the ratio of mean arterial pressure to cardiac output. Regrettably, neither systemic vascular resistance nor total peripheral resistance faithfully describes arterial resistance. Arterial resistance is the slope of the arterial pressure-flow relation. The calculation of systemic vascular resistance using CVP as the backpressure to flow has no physiological rationale and the use of systemic vascular resistance for clinical decision making should be abolished. Regrettably, both systemic vascular resistance and total peripheral resistance are still commonly used in hemodynamic monitoring because they allow for the simultaneous assessment of both pressure and flow, while the actual measure of arterial tone is more difficult to estimate.

The determinants of arterial pressure can simplistically be defined as systemic arterial tone and blood flow. Since blood flow distribution will vary amongst organs relative to their local vasomotor tone and arterial pressure is similar for most organs, measures of peripheral resistance, by any means or formula, reflect the lump parameter of all the vascular beds, and thus, describe no specific vascular bed completely. If no hemodynamic instability alters normal regulatory mechanisms, then local blood flow will also be proportional to local metabolic demand. Within this construct, the only reason cardiac output becomes important is to sustain an adequate and changing blood flow to match changes in vasomotor tone such that arterial input pressure remains constant. Since cardiac output is proportional to metabolic demand there is no level of cardiac output that reflects normal values in the unstable and metabolically active patient. However, as blood pressure decreases below a mean of 60 mm Hg and/or cardiac indices decrease below 2.0 liters/min/m^2, organ perfusion usually becomes compromised, and if sustained it will lead to organ failure and death. Presently, only one clinical trial examined the effect of increasing mean arterial pressure on tissue blood flow. When patients with circulatory shock were resuscitated with volume and vasopressors to a mean arterial pressure range of 60–70, 70–80 or 80–90 mm Hg, no increased organ blood flow could be identified above a mean arterial pressure of 65 mm Hg. Clearly, subjects with prior hypertension will have their optimal perfusion pressure range increased over normotensive patients. Thus, there is no firm data supporting any one limit of arterial pressure or cardiac output values or therapeutic approaches based on these values that have proven more beneficial than any other has. Accordingly, empiricism is the rule regarding target values of both mean arterial pressure and cardiac output. At present, the literature suggests that maintaining a nonpreviously hypertensive patient's mean arterial pressure >65 mm Hg by the use of fluid resuscitation and subsequent vasopressor therapy, as needed, is an acceptable target. Previously hypertensive subjects will need a higher mean arterial pressure to insure the same degree of blood flow [2]. There is no proven value in forcing either arterial tone or cardiac output to higher levels to achieve a mean arterial pressure above this threshold. In fact, data suggest that further resuscitative efforts using vasoactive agents markedly increase mortality [3], and the relatively new concept of 'delayed' and 'hypotensive resuscitation' for traumatic hemorrhagic shock on the other hand had shown improved outcome in some clinical and experimental studies [4–6]. However, these studies were done in trauma patients with penetrating wounds and no immediate access to surgical repair. Once a patient is in the hospital and the sites of active bleeding addressed, then aggressive fluid and pressor resuscitation is indicated.

Arterial Pressure Variations during Ventilation
The majority of the critically ill surgical patients treated in the ICU are usually on mechanical ventilation. Ventilation-induced arterial pressure variations have been described since antiquity as pulsus paradoxus. Inspiratory decreases in arterial pressure were used to monitor both the severity of bronchospasm in asthmatics and their inspiratory efforts [7].

Recently renewed interest in the hemodynamic significance of heart-lung interactions has emerged. The commonly observed variations in arterial pressure and aortic flow seen during positive pressure ventilation have been analyzed as a measure of preload responsiveness [8]. The rationale for this approach is that positive-pressure ventilation-induced changes in either systolic arterial pressure (used to describe pulsus paradoxus), arterial pulse pressure or stroke volume can predict in which subjects cardiac output will increase in response to fluid resuscitation. Ventilation-induced changes in systolic arterial pressure (pulsus paradoxus) and arterial pulse pressure are easy to measure from arterial pressure recordings. The greater the degree of systolic arterial pressure or pulse pressure variation over the respiratory cycle, the greater the increase in cardiac output in response to a defined fluid challenge. Recently, measuring the mean change in aortic blood flow during passive leg raising in spontaneous breathing patients has also proven accurate in predicting preload responsiveness [9].

Although arterial pressure variations are a measure of preload responsiveness [10] the 'traditional' preload measures, such as right atrial pressure (Pra), pulmonary artery occlusion pressure (Ppao), right ventricular (RV) end-diastolic volume and intrathoracic blood volume, poorly reflect preload responsiveness [11]. In essence, preload is not preload responsiveness.

Indications for Arterial Catheterization

The arterial catheter is frequently inserted as a 'routine' at the admission of patients to the ICU for continuous monitoring blood pressure and repetitive measurements of blood gases. There is no evidence to support this exaggerated clinical practice. Although probably the only proven indication for arterial catheterization is to synchronize the intra-aortic balloon of counterpulsation, there are some others indications where the information obtained is valuable in the assessment and treatment of the patient, such as cardiovascular instability or the use of vasopressors or vasodilators during resuscitation. The probable indications for arterial catheterization are summarized in table 1. Although arterial catheterization is an invasive procedure that is not free of complications, a recent systematic review of a large number of cases showed that most of the complications were minor, including temporary vascular occlusion (19.7%) and hematoma (14.4%). Permanent ischemic damage, sepsis and pseudoaneurysm formation occurred in less than 1% of cases [12].

CVP Monitoring

Methods of Measuring CVP

CVP is the pressure in the large central veins proximal to the right atrium relative to atmosphere. In the ICUs the CVP is usually measured using a fluid-filled

Table 1. Arterial catheterization

Indications for arterial catheterization
- As a guide to synchronization of intra-aortic balloon counterpulsation

Probable indications for arterial catheterization
- Guide to management of potent vasodilator drug infusions to prevent systemic hypotension
- Guide to management of potent vasopressor drug infusions to maintain a target mean arterial pressure
- As a port for the rapid and repetitive sampling of arterial blood in patients in whom multiple arterial blood samples are indicated
- As a monitor of cardiovascular deterioration in patients at risk of cardiovascular instability

Useful applications of arterial pressure monitoring in the diagnosis of cardiovascular insufficiency
- Differentiating cardiac tamponade (pulsus paradoxus) from respiration-induced swings in systolic arterial pressure
  Tamponade reduces the pulse pressure but keeps diastolic pressure constant; respiration reduces systolic and diastolic pressure equally, such that pulse pressure is constant
- Differentiating hypovolemia from cardiac dysfunction as the cause of hemodynamic instability
  Systolic arterial pressure decreases more following a positive pressure breath as compared to an apneic baseline during hypovolemia; systolic arterial pressure increases more during positive pressure inspiration when LV contractility is reduced

---

catheter (central venous line or Swan-Ganz catheter) with the distal tip located in the superior vena cava connected to a manometer or more often to a pressure transducer of a monitor, displaying the waveform in a continuous fashion. CVP can also be measured noninvasively as jugular venous pressure by the height of the column of blood distending the internal and external jugular veins, when the subjects are sitting in a semireclined position, such that the small elevations in CVP will be reflected in a persistent jugular venous distention.

Determinants of CVP

Starling demonstrated the relationship between cardiac output, venous return and CVP showing that increasing the venous return (and preload) increases the stroke volume (and cardiac output) until a plateau is reached. Although the CVP is clearly influenced by the volume of blood in the central compartment and its venous compliance, there are several physiological and anatomical factors that can influence its measurement and waveform such as the vascular tone, RV function, intrathoracic pressure changes, tricuspid valve disease, arrhythmias, and both myocardial and pericardial disease (table 2).

Monitoring CVP

CVP has been used as a monitor of central venous blood volume and an estimate of the right atrial pressure for many years, being wrongly used as a parameter and sometimes goal for replacement of intravascular volume in

Table 2. Factors affecting the measured CVP

Central venous blood volume
   Venous return/cardiac output
   Total blood volume
   Regional vascular tone

Compliance of central compartment
   Vascular tone
   RV compliance
      Myocardial disease
      Pericardial disease
      Tamponade

Tricuspid valve disease
   Stenosis
   Regurgitation

Cardiac rhythm
   Junctional rhythm
   Atrial fibrillation
   A-V dissociation

Reference level of transducer
   Positioning of patient

Intrathoracic pressure
   Respiration
   Intermittent positive pressure ventilation
   PEEP
   Tension pneumothorax

shock patients. The validity of this measure as an index of RV preload is nonexistent across numerous studies. It has been shown that CVP has a poor correlation with cardiac index, stroke volume, left ventricular (LV) end-diastolic volume and RV end-diastolic volume [13–15].

Although a very high CVP demands a certain level of total circulating blood volume, one may have a CVP of 20 mm Hg and still have an underfilled LV that is fluid responsive. For example, in the setting acute RV infarction CVP can be markedly elevated, whereas cardiac output often increases further with volume loading. In reported series, some patients with low CVP failed to respond to fluids and some patients with high CVP responded to challenge of fluids [16]. Based on this and the poor correlations described above it is impossible to define ideal values of CVP. However there is some evidence that volume loading in patients with CVP > 12 mm Hg is very unlikely to increase cardiac output [17]. Thus, the only usefulness of CVP is to define relative hypervolemia, since an elevated CVP only occurs in disease. Two clinical studies

Table 3. Physiological measures derived from invasive monitoring and their physiological relevance

Arterial pressure
  Mean arterial pressure
    Organ perfusion inflow pressure
  Arterial pulse pressure and its variation during ventilation
    LV stroke volume changes and pulsus paradoxus
    Preload responsiveness (if assessed during intermittent positive pressure ventilation)
  Arterial pressure waveform
    Aortic valvulopathy, input impedance and arterial resistance
    Used to calculate stroke volume by pulse contour technique

Central venous pressure (CVP)
  Mean CVP
    If elevated, the effective circulating blood volume is not reduced
  CVP variations during ventilation
    Tricuspid insufficiency, tamponade physiology
    Preload responsiveness (if assessed during spontaneous breathing)

Pulmonary arterial pressure (Ppa)
  Mean Ppa
    Pulmonary inflow pressure
  Systolic Ppa
    RV pressure load
  Diastolic Ppa and pulse pressure and their variations during ventilation
    RV stroke volume, PVR
    Diastolic pressure tract changes in intrathoracic pressure during ventilation

Pulmonary artery occlusion pressure (Ppao)
  Mean Ppao
    Left atrial and LV intraluminal pressure and by inference, LV preload
    Backpressure to pulmonary blood flow
  Ppao waveform and its variation during occlusion and ventilation
    Mitral valvulopathy, atrial or ventricular etiology of arrhythmia, accuracy of mean
      Ppao to measure intraluminal LV pressure, and pulmonary capillary pressure (Ppc)

showed a potential benefit in specific groups of surgical patients (hip replacement and renal transplant patients) [18, 19] in whom CVP was used to guide therapy; however, there is no clinical evidence that CVP monitoring improves outcome in critically ill patients and attempts to normalize CVP in early goal-directed therapy during resuscitation do not display any benefit [20].

Pulmonary Artery Catheterization and Its Associated Monitored Variables

Pulmonary arterial catheterization allows the measurement of many clinically relevant hemodynamic variables (table 3) and, in combination with measures of

Table 4. Physiological variables derived from invasive monitoring

---

Calculated using multiple measured variables including cardiac output by thermodilution (COtd), arterial and mixed venous blood gases and end-tidal CO_2 (PetCO$_2$)

Vascular resistances
  Total peripheral resistance = MAP/COtd
  Systemic vascular resistance = (MAP – CVP)/COtd
  Pulmonary arterial resistance = (mean Ppa – Ppc)/COtd
  Pulmonary venous resistance = (Ppc – Ppao)/COtd
  Pulmonary vascular resistance = (mean Ppa – Ppao)/COtd

Vascular pump function
  Left ventricular stroke volume (SVIv) = COtd/HR
  Left ventricular stroke work (SWIv) = (MAP – Ppao)/SVIv
  Preload-recruitable stroke work = SWIv/Ppao

Oxygen transport and metabolism
  Global oxygen transport or delivery (DO$_2$) = CaO$_2$/COtd
  Global oxygen uptake (VO$_2$) = (CaO$_2$ – CvO$_2$)/COtd
  Venous admixture
  Ratio of dead space to total tidal volume (Vd/Vt) = PaCO$_2$/(PaCO$_2$ – PetCO$_2$)

RV function using RV ejection fraction (EFrv) catheter-derived data
  RV end-diastolic volume (EDVrv) = SV/EFrv
  RV end-systolic volume (ESVrv) = EDVrv – SV

HR = Heart rate; MAP = mean arterial pressure.

arterial and mixed venous blood gases, many relevant calculated parameters (table 4). One can measure the intrapulmonary vascular pressures including CVP, pulmonary arterial pressure (Ppa), and by intermittent balloon occlusion of the pulmonary artery, Ppao and pulmonary capillary pressure (Ppc). Furthermore, by using the thermodilution technique and the Stewart-Hamilton equation one can estimate cardiac output and RV ejection fraction, global cardiac volume and intrathoracic blood volume. Finally, one can measure SvO$_2$ either intermittently by direct sampling of blood from the distal pulmonary arterial port or continuously via fiberoptic reflectometry. Assuming one knows the hemoglobin concentration and can tract arterial O$_2$ saturation (SaO$_2$), easily estimated noninvasively by SpO$_2$, one can calculate numerous derived variables that describe well the global cardiovascular state of the patient. These derived variables include total O$_2$ delivery (DO$_2$), whole body O$_2$ consumption (VO$_2$), venous admixture (as an estimate of intrapulmonary shunt), pulmonary and systemic vascular resistance, RV end-diastolic and end-systolic volumes, and both RV and LV stroke work index.

Pulmonary Artery Pressure

The determinants of Ppa are the volume of blood ejected to the pulmonary artery during systole, the resistance of the pulmonary vascular bed and the downstream left atrial pressure. The pulmonary vascular bed is a low resistance circuit with a large reserve that allows increases of cardiac output with minor changes in the Ppa. On the other hand, increases in the downstream venous pressure (e.g. LV failure) or in the flow resistance (e.g. lung diseases) rises the Ppa. Although increases in cardiac output alone do not cause pulmonary hypertension, having an increased vascular resistance the Ppa can be increased due to changes in cardiac output. Based on these considerations the Ppa should not be used as a reliable parameter of ventricular filling in several lung diseases that conditioned changes in the vascular tone and cardiac output. The normal range of values for Ppa are systolic 15–30 mm Hg, diastolic 4–12 mm Hg, and mean 9–18 mm Hg [21].

Ppao Monitoring

Methods of Measuring Ppao

Numerous studies of physicians have demonstrated that the ability to accurately measure Ppao from a strip chart recording or a freeze frame snapshot of the monitor screen is poor. Many initiatives have been put into place to educate physicians and nurses, but the reality is that since the pressure measured also reports changes in intrathoracic pressure, a value which is always changing, the accuracy of Ppao measures is likely to remain poor in graduates of all educations programs.

The Ppao value is thought to reflect the LV filling because of the unique characteristic of the pulmonary circulation. Balloon inflation of the pulmonary artery catheter forces the tip to migrate distally into smaller vessels until the tip occludes a medium-sized (1.2-cm-diameter) pulmonary artery. This occlusion stops all blood flow in that vascular tree distal to the occlusion site until such time as other venous branches reconnect downstream to this venous draining bed. The point where such parallel pulmonary vascular beds anastomose is at a point about 1.5 cm from the left atrium. Thus, if a continuous column of blood is present from the catheter tip to the left heart, then Ppao measures pulmonary venous pressure at this first junction, or J-1 point, of the pulmonary veins [22]. As downstream pulmonary blood flow ceases, distal Ppa falls in a double exponential fashion to a minimal value, reflecting the pressure downstream in the pulmonary vasculature from the point of occlusion. The Ppa value where the first exponential pressure decay is overtaken by the second longer exponential pressure decay reflects Ppc measures, useful in calculating pulmonary arterial

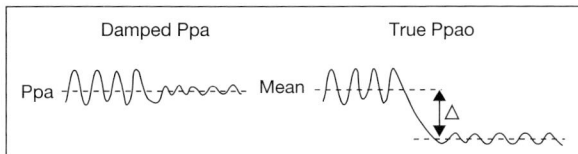

Fig. 1. Two examples of a Ppa waveform before and then during balloon occlusion to measure pulmonary artery occlusion pressure (Ppao). Note the left-sided occlusion tracing has a higher pressure than diastolic Ppa indicating an inaccurate estimate of Ppao, whereas the right-sided tracing indicates a more accurate estimate of Ppao.

and venous resistances. Importantly, the column of water at the end of the catheter is now extended to include the pulmonary vascular circuit up to this J-1 point of blood flow. Since the vasculature is compliant relative to the catheter, vascular pressure signals dampen relative to the nonoccluded Ppa signal. Thus, the two primary aspects of Ppao measures that are used to identify an occluded pressure are the decrease in diastolic pressure values to less than diastolic Ppa and the dampening of the pressure signal (fig. 1). If one needed further validation that the catheter is actually in an occluded vascular bed, then one could measure the pH, pCO_2 and pO_2 of blood sampled from the occluded distal tip of the catheter. Since the sampled blood will be from the stagnant pool of blood, its removal will make it be pulled back into the pulmonary artery catheter (PAC) from the pulmonary veins. Since the blood sampled will have crossed the alveolar capillaries twice, its pCO_2 will be lower than arterial pCO_2 and its pO_2 higher, due to the law of mass action.

Pleural Pressure and Ppao

Although one may measure Ppao accurately relative to atmosphere, the heart and large vessel pulmonary vasculature live in an intrathoracic compartment and sense pleural pressure (Ppl) as their surrounding pressure. Ventilation causes significant swings in Ppl. Pulmonary vascular pressures, when measured relative to atmospheric pressure, will reflect these respiratory changes in Ppl. To minimize this 'respiratory artifact' on intrathoracic vascular pressure recordings, measures are usually made at end-expiration. During quiet spontaneous breathing, end-expiration occurs at the highest vascular pressure values, whereas during passive positive-pressure breathing, end-expiration occurs at the lowest vascular pressure values. With assisted ventilation or with forced spontaneous ventilation, it is often difficult to define end-expiration [23]. These limitations are the primary reasons for inaccuracies in estimating Ppao at the bedside.

Even if measures of Ppao are made at end-expiration and Ppao values reflecting a continuous column of fluid from the catheter tip to the J-1 point, these Ppao measures may still overestimate Ppao if Ppl is elevated at end-expiration. Hyperinflation, due to air trapping, dynamic hyperinflation or the use of extrinsic positive end-expiratory pressure (PEEP) will all increase end-expiratory Ppl to a varying degree as a function of airway resistance and lung and chest wall compliance. It is not possible to predict with accuracy the degree to which increases in PEEP will increase Ppl. Since differences in lung and chest wall compliance exist among patients and in the same patient over time, one cannot assume a fixed relation between increases in Paw and Ppl [24].

Why Measure Ppao?

Ppao is used most often in the bedside assessment of: (1) pulmonary edema, (2) pulmonary vasomotor tone, (3) intravascular volume status and LV preload, and (4) LV performance. These points were summarized recently and will be restated below [25].

Pulmonary Edema

Pulmonary edema can be caused by either elevations of Ppc, referred to as hydrostatic or secondary pulmonary edema, or increased alveolar capillary or epithelial permeability, referred to as primary pulmonary edema. Usually hydrostatic pulmonary edema requires a pulmonary capillary increase to >18 mm Hg. However, if capillary or alveolar cell injury is present, alveolar flooding can occur at much lower Ppc. Furthermore, in the setting of chronic pulmonary vascular congestion, increased pulmonary lymphatic flow and increased respiratory excursions promote a rapid clearance of lung interstitial fluid minimizing edema formation. Still, measures of Ppao are commonly used to determine the cause of pulmonary edema. Ppao values <18 mm Hg suggest a nonhydrostatic cause, whereas values >20 mm Hg suggest a hydrostatic cause of pulmonary edema [22]. However, many exceptions to this rule exist. As mentioned above if increased lung permeability is present then fluid-resuscitation-induced pulmonary edema may occur at Ppao values much below 18 mm Hg, and treatment strategies aimed at reducing Ppao will further reduce pulmonary edema formation. Similarly, if pulmonary venous resistance is increased, then Ppc may be much higher than the measured Ppao inducing hydrostatic pulmonary edema despite no increased lung permeability and a low Ppao. Similarly, Ppao may be >20 mm Hg without any evidence of hydrostatic pulmonary edema, either because Ppl is also elevated or because of increased pulmonary lymphatic flow.

Pulmonary Vasomotor Tone

The pulmonary circulation normally has a low resistance, with pulmonary arterial diastolic pressure only slightly higher than Ppao and mean

Ppa a few millimeters Hg higher than Ppao. Pulmonary vascular resistance (PVR) can be estimated using Ohm's law as the ratio of the pulmonary vascular pressure gradient (mean pulmonary artery pressure minus Ppao) and cardiac output [i.e. PVR = (mean Ppa – Ppao)/cardiac output]. Normal PVR is between 2 and 4 mm Hg $\times$ l/min/m^2. Usually these values are multiplied by 80 to give a normal PVR range of 150–250 dyn s/cm^5. Either an increased PVR or a passive pressure build-up from the pulmonary veins can induce pulmonary hypertension. If pulmonary hypertension is associated with an increased PVR then the causes are primarily within the lung. Diagnoses such as pulmonary embolism, pulmonary fibrosis, essential pulmonary hypertension and pulmonary venoocclusive disease need to be excluded. Whereas if PVR is normal then LV dysfunction is the more likely cause of pulmonary hypertension [26]. Since the treatments for these two groups of diseases is quite different despite similar increases in Ppa, the determination of PVR in the setting of pulmonary hypertension is very important. Regrettably, PVR poorly reflects true pulmonary vasomotor tone in lung disease states and during mechanical ventilation, especially with the application of PEEP. Alveolar pressure (Palv) can be the backpressure to pulmonary blood flow in certain lung regions during positive-pressure ventilation and in the presence of hyperinflation because Palv exceeds left atrial pressure. Furthermore, since lung disease is usually nonhomogeneous, pulmonary blood flow is preferentially shifted from compressed vessels in West Zone 1 and 2 conditions (i.e. Ppao < Palv and Ppa = Palv, respectively) to those circuits with the lowest resistance (West Zone 3, i.e. Ppao > Palv), thus making the lung vascular pathology appear less than it actually is.

LV Preload

Ppao is often taken to reflect LV filling pressure, and by inference, LV end-diastolic volume. Patients with cardiovascular insufficiency and a low Ppao are presumed to be hypovolemic and initially treated with fluid resuscitation, whereas patients with similar presentations but an elevated Ppao are presumed to have an impaired contractile function. Although there are no accepted high and low Ppao values for which LV underfilling is presumed to occur or not, Ppao values <10 mm Hg are usually used as presumed evidence of a low LV end-diastolic volume, whereas values >18 mm Hg suggest a distended LV [27]. Unfortunately, there is very little data to support this approach and almost no data to defend this logic. There are multiple documented reasons for this observed inaccuracy that relate to individual differences in LV diastolic compliance and contractile function [28]. First, the relation between Ppao and LV end-diastolic volume is curvilinear and is often very different among subjects and within subjects over time. Thus, neither absolute values of Ppao or changes in

Ppao will define a specific LV end-diastolic volume or its change [29]. Second, Ppao is not the distending pressure for LV filling. It is only the internal pressure of the pulmonary veins relative to atmospheric pressure. Assuming Ppao approximated left atrial pressure, it will poorly reflect LV end-diastolic pressure because it poorly follows the late diastolic pressure rise induced by atrial contraction and does not measure pericardial pressure, which is the outside pressure for LV distention. With lung distention Ppl increases increasing pericardial pressure. Although we can estimate Ppl using esophageal balloon catheters, pericardial pressure is often different. Changes in pericardial pressure will alter LV end-diastolic volume independent of Ppao. Finally, even if one knew pericardial pressure and Ppao did accurately reflect LV end-diastolic pressure, LV diastolic compliance can vary rapidly changing the relation between LV filling pressure and LV end-diastolic volume. Myocardial ischemia, arrhythmias, and acute RV dilation can all occur over a few heartbeats. Thus, it is not surprising that Ppao is a very poor predictor of preload responsiveness. Thus, it is not recommended to use Ppao to predict response to fluid resuscitation in critically ill patients.

LV Performance

The four primary determinants of LV performance are preload (LV end-diastolic volume), afterload (maximal LV wall stress), heart rate and contractility. Ppao is often used as a substitute for LV end-diastolic volume when constructing Starling curves (i.e. relationship between changing LV preload and ejection phase indices). Usually one plots Ppao versus LV stroke work (LV stroke volume × developed pressure). Using this construct, patients with heart failure can be divided into four groups depending on their Ppao (> or <18 mm Hg) and cardiac index values (> or <2.2 liters/min/m^2) [27]. Those patients with low cardiac indices and high Ppao are presumed to have primary heart failure, and a low cardiac output and low Ppao, on the other hand, reflect hypovolemia. Those with high cardiac indices and high Ppao are presumed to be volume overloaded, and having high cardiac output and low Ppao reflect increased sympathetic tone. Although this maybe a useful construct for determining diagnosis, treatment and prognosis of patients with acute coronary syndrome, it poorly predicts cardiovascular status in other patient groups. However, as described above, if LV end-diastolic volume and Ppao do not trend together in response to fluid loading or inotropic drug infusion, then inferences about LV contractility based on this Ppao/LV stroke work relation may be incorrect. This is not a minor point. Volume loading may induce acute RV dilation markedly reducing LV diastolic compliance, such that Ppao will increase as LV stroke work decreases. However, the relationship between LV end-diastolic volume and stroke work need not have changed at all. Similarly, inotropic drugs, like dobutamine, may reduce biventricular volumes by decreasing venous return, decreasing LV diastolic compliance, even if the heart is not responsive to inotropic therapy. Thus, the same limitations on the use of Ppao in assessing LV preload must be considered when using it to assess LV performance.

Cardiac Output Monitoring

Measuring Cardiac Output

Cardiac output can be estimated by many techniques, including invasive hemodynamic monitoring. Pulmonary blood flow, using a balloon floatation PAC equipped with a distal thermistor, and transpulmonary blood flow, using an arterial thermistor, both with a central venous cold volume injection, can be used. Similarly, minimally invasive echo Doppler techniques can measure blood flow at the aortic value and descending aortic flow using esophageal Doppler monitoring. Cardiac output can be measured intermittently by bolus cold injection or continuously by cold infusion. The advantage of the continuous cardiac output technique and the transpulmonary technique is that neither is influenced greatly by the ventilation-induced swings in pulmonary blood flow. Measurement of cardiac output by intermittent pulmonary artery flow measures using bolus cold indicator and monitoring the thermal decay curve is the most common method to measure cardiac output at the bedside. However, such intermittent measures will show profound ventilatory cycle-specific patterns [30]. By making numerous measures at random with the ventilatory cycle and then averaging all measures with proper thermal decay profiles, regardless of their values, one can derive an accurate measure of pulmonary blood flow [31].

Recently, a renewed interest in pulse contour analysis to estimate LV stroke volume, and therefore cardiac output, from the arterial pressure profile over ejection has acquired its own set of supporters [32]. Arterial pressure and arterial pulse pressure are a function of rate of LV ejection, LV stroke volume and the resistance, compliance and inertance characteristics of the arterial tree and blood. If the arterial components of tone remain constant, then changes in pulse pressure most proportionally reflect changes in LV stroke volume. Thus, it is not surprising that the aortic flow variation parallels arterial pulse pressure variation [33], and pulse contour-derived estimates of stroke volume variation can be used to determine preload responsiveness [34, 35]. Caution must be applied to using the pulse contour method because it has not been validated in subjects with rapidly changing arterial tone, as often occurs in subjects with hemodynamic instability. Furthermore, it requires the application of abnormally large tidal volumes [34–36]. Thus, at the present time, the pulse contour-derived stroke volume variation technique represents a potentially great but still unproven clinical decision tool [37].

Currently three commercial devices that use pulse contour analysis of an arterial line waveform to obtain continuous cardiac output are approved for clinical use (PiCCO™, LIDCO™ and VigileoEdwards™ systems). The benefit of being minimally invasive and the correlation shown with 'standard' methods of measuring cardiac output in some clinical and experimental studies make them a promising tool for hemodynamic monitoring [38, 39].

Mixed Venous SO_2 Monitoring

Measuring SvO_2

SvO2 reflects the pooled SvO_2 and is an important parameter in the assessment of the adequacy of DO_2 and its relation with VO_2. A decrease in SvO_2 could be explained by a decrease in DO_2 or any of the parameters that determine this like saturation (SaO_2), cardiac output and hemoglobin, and also by an increase in VO_2. A decrease of DO_2 will be followed by stable VO_2 with a consequent decrease of the SvO_2 until a critical value of DO_2 is reached where the tissues are not longer able to compensate having a constant VO_2, and VO_2 becomes dependent on DO_2 in an almost linear relation. At this level SvO_2, though continuing to decrease, becomes less sensitive to changes of tissue perfusion.

SvO2 measured from blood drawn from the distal tip of a PAC represents the true mixed venous value of the blood blended in the right ventricle. Care must be taken to withdraw blood slowly so that it does not get aspirated from the downstream pulmonary capillaries. Validation of true mixed venous blood requires documentation that the measured $PvCO_2$ is greater than $PaCO_2$, because blood drawn over the capillaries sees alveolar gas twice and will have a lower PCO_2 than arterial blood. Continuous measures of SvO_2 can be made using fiberoptic reflectance spectroscopy. Two techniques are commercially available. Both use one fiberoptic line to send a light signal and another to receive the reflected light at a different wavelength. However, only one catheter (Abbott™) uses the Shaw technique of also measuring hemoglobin reflectance and thus remains accurate of wide changes in hemoglobin concentration. The other catheter (Edward™) requires recalibration if hemoglobin levels vary by more than 1 g/dl. Both techniques are valuable to monitoring SvO_2 trends as either cardiac output, arterial O_2 content or metabolic demand varies.

$ScvO_2$

Recent interest in central venous O_2 saturation ($ScvO_2$) has evolved over the past years with the positive results of the study of Rivers et al. [40].

Rivers et al. demonstrated that in patients with septic shock or severe sepsis admitted to the Emergency Department an early and aggressive resuscitation guided by $ScvO_2$, CVP and mean arterial pressure reduced 28-day mortality from 46.5 to 30.5%. However, measures of SvO_2 remain the gold standard to reflect minimal DO_2. This is because although $ScvO_2$ and SvO_2 covary and seem to follow a parallel tracking, their differences can exceed 5%. Furthermore, during dynamic changes in cardiac output as occur in shock states, $ScvO_2$ may exceed SvO_2 by 5% or more or be less than SvO_2 by 5% or more [41]. Thus, using a defined threshold value for $ScvO_2$ to identify when to start or stop resuscitation in a critically ill patient is fundamentally flawed. Still, a low $ScvO_2$ (<65%) is invariably associated with a low SvO_2 (<72%), making it less sensitive but still clinically useful at lower threshold values.

The Meaning of Cardiac Output and SvO_2 as End Points of Resuscitation

Although one may potentially measure cardiac output accurately at the bedside, there is no such thing as a normal cardiac output. Cardiac output is either adequate for the needs of the body or it is not. For example, the same cardiac output and DO_2 that is adequate at rest may be grossly inadequate and not associated with life during periods of increased metabolic demand. Since the primary goal of the cardiorespiratory system is to continuously maintain adequate amounts of O_2 (DO_2) to meet the metabolic demands of the tissues (VO_2), neither cardiac output nor mean arterial pressure are sensitive or specific measures of the adequacy of cardiovascular function. Clearly, the best measures of the adequacy of blood flow are the continued maintenance of normal end-organ function without evidence of excessive anaerobic metabolism. Normal urine output, gut activity, mentation, normal blood lactate levels and spontaneous voluntary muscular activities reflect the most easily validated measures of body health [42]. Regrettably, many patients present with coexistent organ system dysfunction, either preexistent or due to the insult. Furthermore, organ function cannot be monitored quickly enough to allow for titration of care. Thus, one cannot rely on these absolute markers to direct therapy [43]. Perhaps a more functional marker of adequacy of DO_2 to the tissues is SvO_2 [44]. Although values of SvO_2 >70% do not insure that all vascular beds are adequately perfused, SvO_2 values <60% are associated with oxidative impairment of tissues with a high metabolic rate and values <50% are uniformly associated with evidence of anaerobic metabolism in some vascular beds [45]. Thus, as a negative predictive marker, preventing SvO_2 from decreasing below 50% and hopefully keeping it above 70% by fluid resuscitation, sedation and ancillary support (e.g. mechanical ventilation to reduce the work cost of breathing), all may improve DO_2 to metabolically active tissues.

If the metabolic demand changes, cardiac output should covary with it [46]. Since this puts an added variable on the analysis of hemodynamic stability, a common approach in the cardiovascular management of the critically ill patient is to minimize the extraneous metabolic demands of the patients during intervals in which therapeutic interventions and diagnostic processes are being performed so as to maintain stable baseline O_2 consumption for comparison. Thus, minimizing the work cost of breathing by using mechanical ventilation, and reducing sympathetic responses by infusion of sedative agents, reflect stabilizing processes that allow for accurate hemodynamic assessment. This is often more difficult to achieve than imagined [31]. Even a sedated and mechanically ventilated subject can be expending much effort assisting or resisting the ventilator-derived breaths. Muscular activities, such as moving in bed or being turned, 'fighting the ventilator', and breathing spontaneously can easily double resting VO_2 [47]. O_2 supply and demand must covary as a normal and expected aspect of homeostasis under almost all conditions. In cardiovascular insufficiency states, such as cardiogenic shock or hypovolemic shock, total cardiac output is often limited and cannot increase enough in response to increasing metabolic demand to match the demand. Under these severe conditions VO_2 tends to remain constant by varying the extraction of O_2 in the tissues rather than by varying total blood flow. Thus, measures of SvO_2 can be used to identify patients in circulatory shock. Furthermore, resuscitation efforts that increase SvO_2 to >70% should be associated with improved end-organ function.

The Controversy of the PAC

One would think that the clinical use of the pulmonary catheter in the management of the hemodynamically unstable patient would be invaluable. However, this utility has not been documented. Although there are no proven indications for the insertion of PAC, there are potential indications (yet not proven) for its use based on the need to assess cardiac function, global DO_2, intravascular volume status and pulmonary pressures as summarized in table 4.

The controversy over the use of the pulmonary arterial catheter in the management of critically ill patients continues to rage. Proponents of its use cite physiological rationale to diagnosis and titration of complex treatments that may otherwise be detrimental. Opponents of its use cite the almost total lack of data showing that its use in the management of critically ill patients improves outcome. Still, one truth remains: no catheter will improve outcome unless coupled to a treatment that itself improves outcome.

Despite some exciting initial uncontrolled reports of markedly improved outcome in high-risk surgery patients [48, 49], further well-controlled studies in both high-risk surgical patients [50] and trauma patients [51, 52] failed to document that any improved survival when patients were treated based on pulmonary arterial catheter-derived data. In fact, the patients resuscitated aggressively to force DO_2 into these survivor levels suffered a much higher mortality rate that did the control group treated conservatively [3]. Interestingly, as mentioned above, using only arterial pressure and superior vena caval O_2 saturation, but with a defined physiology-based treatment algorithm Rivers et al. [40] demonstrated a markedly improved survival in septic shock patients without the need of PAC. On the other hand, a recent statistical analysis that includes over 50,000 patients of the National Trauma Data Bank showed for the first time a decrease in mortality in a selective group of trauma patients (severely injured, elderly, who arrived in shock) with the use of PAC [53].

Because of this nonclear benefit of the use of PAC, the fact of it being an invasive monitoring procedure with potential serious complications acquires a major relevance when deciding on the risk-benefit indicating its use. Two recent large prospective multicenter studies showed an incidence of 5 and 10% of complications [54, 55]. The most frequent complications described in this series were hematomas, arterial puncture, arrhythmias, and PAC-related infections, although a long list of complications has been described. No deaths attributable to PAC were found in this series but other authors had previously reported mortality generally due to right heart and pulmonary artery perforation [56, 57].

Beyond the controversial use of the PAC, two recent randomized clinical trials using active protocols of hemodynamic monitoring and algorithms of goal-directed therapy guided by esophageal Doppler flowmetry [58] and pulse contour analysis for cardiac output [59] in postoperative surgical patients had shown a decreased duration of hospital stay and morbidity. Thus, the literature suggests that the generalized use of hemodynamic monitoring and aggressive goal-directed therapy could improve outcome but that one does not need to use a PAC to achieve these goals. However, the fact that these entire arguments miss the point of the utility of hemodynamic monitoring is relative, namely, that no monitoring device, no matter how accurate, safe and simple to use, will improve outcome unless coupled to a treatment, which itself, improves outcome. Thus, the question should not be, 'Does the PAC improve outcome?' but rather, 'Do treatment protocols that require information only attainable from pulmonary arterial catheterization improve outcome?' Furthermore, the treatment protocol itself should also be shown to improve outcome prior to the study, because otherwise if the trail shows no difference in outcome with or without a PAC, the results may well reflect the fact that there was no benefit for the protocol in either arm of the study.

Conclusion

All surgical patients require monitoring to assess cardiovascular stability and sometimes may benefit from optimization of their hemodynamic status. Therefore, all surgeons require a basic understanding of physiological underpinnings of hemodynamic monitoring. The physiological rationale is still the primary level of defense for monitoring critically ill patients.

Arterial catheterization to monitor arterial pressure is a safe procedure with a low complication rate. However, it should be used only when clear indications exist. There is no evidence that achieving pressures over 65 mm Hg increases organ perfusion or favors outcome. The analysis of pulse pressure variation is a useful method to assess preload responsiveness and a potential tool for resuscitation. CVP has being wrongly used as a parameter of goal for replacement of intravascular volume in shock patients. Volume loading in patients with CVP >12 mmHg is unlikely to increase cardiac output and attempts to normalize CVP in early goal-directed therapy during resuscitation has no proven benefit. The use of PAC provides direct access to several physiological parameters, both as raw data and derived measurements (CO, SvO_2, DO_2). At the present targeting specific levels of DO_2 have proven effective only in high-risk surgery patients in the perioperative time. Ppao is often used as bedside assessment of pulmonary edema, pulmonary vasomotor tone, intravascular volume status and LV preload, and LV performance. Several publications have explored the potential indications and benefits of the PAC in goal-directed therapies. Beyond this controversy there is a trend to less invasive methods of hemodynamic monitoring and current data support protocols of monitoring and goal-directed therapy that could improve outcome in selected groups of surgical patients.

Acknowledgment

This work was supported by grant federal funding HL67181 and HL0761570.

References

1. Bellomo R, Pinsky MR: Invasive monitoring; in Tinker J, Browne D, Sibbald W (eds): Critical Care – Standards, Audit and Ethics. Oxford, University Press, 2006, pp 82–104.
2. Ledoux D, Astiz ME, Carpati CM, Rackow EC: Effects of perfusion pressure on tissue perfusion in septic shock. Crit Care Med 2000;28:2729–2732.
3. Hayes MA, Timmins AC, Yau EH, Palazzo M, Hinds CJ, Watson D: Elevation of systemic oxygen delivery in the treatment of critically ill patients. N Engl J Med 1994;330:1717–1722.
4. Bickell WH, Wall MJ Jr, Pepe PE, et al: Immediate versus delayed fluid resuscitation for hypotensive patients with penetrating torso injuries. N Engl J Med 1994;331:1105–1109.

5 Capone AC, Safar P, Stezoski W, Tisherman S, Peitzman AB: Improved outcome with fluid restriction in treatment of uncontrolled hemorrhagic shock. J Am Coll Surg 1995;180:49–56.
6 Kowalenko T, Stern S, Dronen S, Wang X: Improved outcome with hypotensive resuscitation of uncontrolled hemorrhagic shock in a swine model. J Trauma 1992;33:349–353.
7 Rebuck AS, Read J: Assessment and management of severe asthma. Am J Med 1971;51:788–798.
8 Michard F, Boussat S, Chemla D, et al: Relation between respiratory changes in arterial pulse pressure and fluid responsiveness in septic patients with acute circulatory failure. Am J Respir Crit Care Med 2000;162:134–138.
9 Monnet X, Rienzo M, Osman D, et al: Passive leg raising predicts fluid responsiveness in the critically ill. Crit Care Med 2006;34:1402–1407.
10 Gunn SR, Pinsky MR: Implications of arterial pressure variation in patients in the intensive care unit. Curr Opin Crit Care 2001;7:212–217.
11 Michard F, Teboul JL: Predicting fluid responsiveness in ICU patients: a critical analysis of the evidence. Chest 2002;121:2000–2008.
12 Scheer B, Perel A, Pfeiffer UJ: Clinical review: complications and risk factors of peripheral arterial catheters used for haemodynamic monitoring in anaesthesia and intensive care medicine. Crit Care 2002;6:199–204.
13 Michard F, Alaya S, Zarka V, Bahloul M, Richard C, Teboul JL: Global end-diastolic volume as an indicator of cardiac preload in patients with septic shock. Chest 2003;124:1900–1908.
14 Godje O, Peyerl M, Seebauer T, Lamm P, Mair H, Reichart B: Central venous pressure, pulmonary capillary wedge pressure and intrathoracic blood volumes as preload indicators in cardiac surgery patients. Eur J Cardiothorac Surg 1998;13:533–539.
15 Buhre W, Weyland A, Schorn B, et al: Changes in central venous pressure and pulmonary capillary wedge pressure do not indicate changes in right and left heart volume in patients undergoing coronary artery bypass surgery. Eur J Anaesthesiol 1999;16:11–17.
16 Magder S, Georgiadis G, Tuck C: Respiratory variations in right atrial pressure predict response to fluid challenge. J Crit Care 2004;7:76–85.
17 Bafaqeeh F, Magder S: CVP and volume responsiveness of cardiac output (abstract). Am J Respir Crit Care Med 2004;169:A343.
18 Venn R, Steele A, Richardson P, Poloniecki J, Grounds M, Newman P: Randomized controlled trial to investigate influence of the fluid challenge on duration of hospital stay and perioperative morbidity in patients with hip fractures. Br J Anaesth 2002;88:65–71.
19 Thomsen HS, Lokkegaard H, Munck O: Influence of normal central venous pressure on onset of function in renal allografts. Scand J Urol Nephrol 1987;21:143–145.
20 Shoemaker WC, Kram HB, Appel PL, Fleming AW: The efficacy of central venous and pulmonary artery catheters and therapy based upon them in reducing mortality and morbidity. Arch Surg 1990;125:1332–1337.
21 Sharkey SW: Beyond the wedge: clinical physiology and the Swan-Ganz catheter. Am J Med 1987;83:111–122.
22 Swan HJ, Ganz W, Forrester J, Marcus H, Diamond G, Chonette D: Catheterization of the heart in man with use of a flow-directed balloon-tipped catheter. N Engl J Med 1970;283:447–451.
23 Hoyt JD, Leatherman JW: Interpretation of the pulmonary artery occlusion pressure in mechanically ventilated patients with large respiratory excursions in intrathoracic pressure. Intensive Care Med 1997;23:1125–1131.
24 Pinsky M, Vincent JL, De Smet JM: Estimating left ventricular filling pressure during positive end-expiratory pressure in humans. Am Rev Respir Dis 1991;143:25–31.
25 Pinsky MR: Clinical significance of pulmonary artery occlusion pressure. Intensive Care Med 2003;29:175–178.
26 Abraham AS, Cole RB, Green ID, Hedworth-Whitty RB, Clarke SW, Bishop JM: Factors contributing to the reversible pulmonary hypertension of patients with acute respiratory failure studies by serial observations during recovery. Circ Res 1969;24:51–60.
27 Forrester JS, Diamond G, Chatterjee K, Swan HJ: Medical therapy of acute myocardial infarction by application of hemodynamic subsets (first of two parts). N Engl J Med 1976;295:1356–1362.
28 Raper R, Sibbald WJ: Misled by the wedge? The Swan-Ganz catheter and left ventricular preload. Chest 1986;89:427–434.

29 Kumar A, Anel R, Bunnell E, et al: Pulmonary artery occlusion pressure and central venous pressure fail to predict ventricular filling volume, cardiac performance, or the response to volume infusion in normal subjects. Crit Care Med 2004;32:691–699.

30 Jansen JR, Bogaard JM, Versprille A: Extrapolation of thermodilution curves obtained during a pause in artificial ventilation. J Appl Physiol 1987;63:1551–1557.

31 Synder JV, Powner DJ: Effects of mechanical ventilation on the measurement of cardiac output by thermodilution. Crit Care Med 1982;10:677–682.

32 Wesseling K, Wit BD, Weber J, Smith NT: A simple device for the continuous measurement of cardiac output. Adv Cardiovasc Physiol 1983;5:16–52.

33 Feissel M, Michard F, Mangin I, Ruyer O, Faller JP, Teboul JL: Respiratory changes in aortic blood velocity as an indicator of fluid responsiveness in ventilated patients with septic shock. Chest 2001;119:867–873.

34 Reuter DA, Felbinger TW, Schmidt C, et al: Stroke volume variations for assessment of cardiac responsiveness to volume loading in mechanically ventilated patients after cardiac surgery. Intensive Care Med 2002;28:392–398.

35 Reuter DA, Felbinger TW, Kilger E, Schmidt C, Lamm P, Goetz AE: Optimizing fluid therapy in mechanically ventilated patients after cardiac surgery by on-line monitoring of left ventricular stroke volume variations. Comparison with aortic systolic pressure variations. Br J Anaesth 2002;88: 124–126.

36 Berkenstadt H, Margalit N, Hadani M, et al: Stroke volume variation as a predictor of fluid responsiveness in patients undergoing brain surgery. Anesth Analg 2001;92:984–989.

37 Pinsky MR, Payen D: Functional hemodynamic monitoring. Crit Care 2005;9:566–572.

38 Linton R, Band D, O'Brien T, Jonas M, Leach R: Lithium dilution cardiac output measurement: a comparison with thermodilution. Crit Care Med 1997;25:1796–1800.

39 Kurita T, Morita K, Kato S, Kikura M, Horie M, Ikeda K: Comparison of the accuracy of the lithium dilution technique with the thermodilution technique for measurement of cardiac output. Br J Anaesth 1997;79:770–775.

40 Rivers E, Nguyen B, Havstad S, et al: Early goal-directed therapy in the treatment of severe sepsis and septic shock. N Engl J Med 2001;345:1368–1377.

41 Reinhart K, Kuhn HJ, Hartog C, Bredle DL: Continuous central venous and pulmonary artery oxygen saturation monitoring in the critically ill. Intensive Care Med 2004;30:1572–1578.

42 Marik PE: Gastric intramucosal pH. A better predictor of multiorgan dysfunction syndrome and death than oxygen-derived variables in patients with sepsis. Chest 1993;104:225–229.

43 Pinsky MR: Beyond global oxygen supply-demand relations: in search of measures of dysoxia. Intensive Care Med 1994;20:1–3.

44 Kandel G, Aberman A: Mixed venous oxygen saturation. Its role in the assessment of the critically ill patient. Arch Intern Med 1983;143:1400–1402.

45 Miller MJ, Cook W, Mithoefer J: Limitations of the use of mixed venous pO_2 as an indicator of tissue hypoxia. Clin Res 1979;27:401A.

46 Pinsky MR: The meaning of cardiac output. Intensive Care Med 1990;16:415–417.

47 Weissman C, Kemper M, Damask MC, Askanazi J, Hyman AI, Kinney JM: Effect of routine intensive care interactions on metabolic rate. Chest 1984;86:815–818.

48 Tuchschmidt J, Fried J, Astiz M, Rackow E: Elevation of cardiac output and oxygen delivery improves outcome in septic shock. Chest 1992;102:216–220.

49 Boyd O, Grounds RM, Bennett ED: A randomized clinical trial of the effect of deliberate perioperative increase of oxygen delivery on mortality in high-risk surgical patients. JAMA 1993;270: 2699–2707.

50 Sandham JD, Hull RD, Brant RF, et al: A randomized, controlled trial of the use of pulmonary-artery catheters in high-risk surgical patients. N Engl J Med 2003;348:5–14.

51 McKinley BA, Kozar RA, Cocanour CS, et al: Normal versus supranormal oxygen delivery goals in shock resuscitation: the response is the same. J Trauma 2002;53:825–832.

52 Velmahos GC, Demetriades D, Shoemaker WC, et al: Endpoints of resuscitation of critically injured patients: normal or supranormal? A prospective randomized trial. Ann Surg 2000;232:409–418.

53 Friese RS, Shafi S, Gentilello LM: Pulmonary artery catheter use is associated with reduced mortality in severely injured patients: a National Trauma Data Bank analysis of 53,312 patients. Crit Care Med 2006;34:1597–1601.
54 Binanay C, Califf RM, Hasselblad V, et al: Evaluation study of congestive heart failure and pulmonary artery catheterization effectiveness: the ESCAPE trial. JAMA 2005;294:1625–1633.
55 Harvey S, Harrison DA, Singer M, et al: Assessment of the clinical effectiveness of pulmonary artery catheters in management of patients in intensive care (PAC-Man): a randomised controlled trial. Lancet 2005;366:472–477.
56 Ducatman BS, McMichan JC, Edwards WD: Catheter-induced lesions of the right side of the heart. A one-year prospective study of 141 autopsies. JAMA 1985:253:791–795.
57 Kearney TJ, Shabot MM: Pulmonary artery rupture associated with the Swan-Ganz catheter. Chest 1995;108:1349–1352.
58 McKendry M, McGloin H, Saberi D, Caudwell L, Brady AR, Singer M: Randomised controlled trial assessing the impact of a nurse delivered, flow monitored protocol for optimisation of circulatory status after cardiac surgery. BMJ 2004;329:258.
59 Pearse R, Dawson D, Fawcett J, Rhodes A, Grounds RM, Bennett ED: Early goal-directed therapy after major surgery reduces complications and duration of hospital stay. A randomised, controlled trial [ISRCTN38797445]. Crit Care 2005;9:R687–R693.
60 Schlichtig R, Kramer DJ, Boston JR, Pinsky MR: Renal O_2 consumption during progressive hemorrhage. J Appl Physiol 1991;70:1957–1962.

Michael R. Pinsky, MD
606 Scaife Hall
3550 Terrace Street
Pittsburgh, PA 15261 (USA)
Tel./Fax +1 412 647 5387, E-Mail pinsky@pitt.edu

Acid-Base Disorders and Strong Ion Gap

John A. Kellum

Department of Critical Care Medicine, University of Pittsburgh Medical Center, Pittsburgh, Pa., USA

Abstract

The application of modern quantitative physical chemical techniques to clinical acid-base has yielded important new information about the nature and clinical significance of metabolic acid-base disorders. Abnormalities identified by the strong ion gap appear to be common in critically ill patients and are associated with increased mortality especially when identified early in the course of critical illness. Attempts to identify the exact chemical nature of ions identified by the strong ion gap have only been of limited success and further study is needed.

Copyright © 2007 S. Karger AG, Basel

Acid-base balance is among the most tightly regulated variables in human physiology. Acute changes in blood pH induce powerful effects at the level of the cell, organ, and organism [1]. Yet the mechanisms responsible for local, regional and systemic acid-base control are incompletely understood and controversy exists in the literature as to what methods should be used to understand them [2]. Once basic categorization into respiratory and metabolic acid-base disorders is accomplished, evaluation of ion balance is undertaken in order to identify complex acid-base disorders and to help narrow the differential diagnosis. Evaluation of ion balance can be done using the familiar anion gap (AG) or the slightly more complex strong ion gap (SIG). Both techniques are based on the principle of electrical neutrality which dictates that in macroscopic aqueous solutions (like blood plasma) the sum of all positive charges (cations) must equal the sum of all negative charges (anions).

Anion Gap

The AG is calculated, or rather estimated, from the differences between the routinely measured concentrations of serum cations (Na^+ and K^+) and anions

(Cl^- and HCO_3^-). Since there can be no actual difference (electrical neutrality must be preserved), the measured difference reflects missing, or 'unmeasured' ions. Normally, this difference or 'gap' is filled primarily by the ionized portion of the weak acids (A^-) principally albumin, and, to a lesser extent, phosphate. Sulfate and lactate also contribute a small amount to the normal AG, typically less than 2 mEq/l. However, there are also unmeasured cations such as Ca^{2+} and Mg^{2+} and these tend to offset the effects of sulfate and lactate except when either is abnormally increased. Plasma proteins other than albumin can be either positively or negatively charged but in the aggregate tend to be neutral [3] except in rare cases of abnormal paraproteins such as in multiple myeloma.

When the AG is greater than that produced by albumin and phosphate, other anions (e.g. lactate, ketones) must be present in higher than normal concentrations. In this way, the AG can be used to narrow the differential diagnosis in the case of a metabolic acidosis. Furthermore, the magnitude of the AG reflects the concentration of the offending acid and can therefore provide a means of monitoring when measurement of the acid is difficult (e.g. ketoacidosis). Finally the AG may provide a clue to the presence of life-threatening conditions such as poisonings.

In practice the AG is calculated as follows:

$$AG = (Na^+ + K^+) - (Cl^- + HCO_3^-)$$

Because of its low and narrow extracellular concentration, K^+ is often omitted from the calculation. Respective normal values with relatively wide ranges reported by most laboratories are 12 ± 4 mEq/l (if K^+ is considered) and 8 ± 4 mEq/l (if K^+ is not considered). The value of a 'normal AG' has decreased in recent years following the introduction of more accurate methods for measuring Cl^- concentration [4, 5]. However, the various measurement techniques available mandate that each institution reports its own expected normal AG.

Importantly, the concept of a normal AG is based on the premise that A^- is normal, which requires albumin and phosphate, the two major constituents of the A^-, to be normal, both in concentration and in charge. As it turns out this is usually the case in healthy subjects but rarely in critically ill patients [6]. Dehydration may induce a parallel increment in the apparent AG by increasing the concentration of all the ions. Conversely, severe hypoalbuminemia causes a decrease in the AG and it has been recommended to 'correct' the AG for the prevailing albumin concentration since each gram per deciliter decline in serum albumin reduces the apparent AG by 2.5–3 mEq/l [7].

Some authors have cast doubt on the diagnostic value of the AG in certain situations [6, 8]. For example, Salem and Mujais [6] found routine reliance on the AG to be 'fraught with numerous pitfalls'. The primary problem with the

AG is its reliance on the use of a normal range produced by albumin and to a lesser extent phosphate as discussed above. These constituents may be grossly abnormal in patients with critical illness leading to a change in the normal range for these patients. This has prompted some authors to adjust the 'normal range' for the AG by the patient's albumin [7] or even phosphate [9] concentration. Because these anions are not strong anions their charge will be altered by changes in pH. Each gram per deciliter of albumin has a charge of 2.8 mEq/l at pH 7.4 (2.3 mEq/l at 7.0 and 3.0 mEq/l at 7.6) and each milligram per deciliter of phosphate has a charge of 0.59 mEq/l at pH 7.4 (0.55 mEq/l at 7.0 and 0.61 mEq/l at 7.6). Thus, under physiological conditions, the variance is reasonably small. A convenient way to estimate the normal AG for a given patient is by use of the following formula [9]:

normal AG = 2 (albumin g/dl) + 0.5 (phosphate mg/dl)

or for international units:

normal AG = 0.2 (albumin g/l) + 1.5 (phosphate mmol/l)

When this patient-specific normal range was used to examine the presence of unmeasured anions in the blood of critically ill patients, the accuracy of this method improved from 33% with the routine AG (normal range = 12 mEq/l) to 96% [9].

Alternatively, the estimated charge coming from albumin and phosphate can be added with Cl^- and HCO_3^- as total anions. Lactate can also be considered and the resultant 'corrected AG' (cAG) should be close to zero.

cAG = $(Na^+ + K^+) - [Cl^- + HCO_3^- + 2$ (albumin) $+ 0.5$ (phosphate) $+$ lactate]

or for international units:

cAG = $(Na^+ + K^+) - (Cl^- + HCO_3^- + 0.2$ (albumin) $+ 1.5$ (phosphate) $+$ lactate)

Either technique is only accurate within about 5 mEq/l. When more accuracy is desired a slightly more complicated method of estimating A^- is required [10].

Strong Ion Gap

About a decade ago [10], our laboratory applied newly published data by Figge et al. [3] concerning the net charge on the surface of albumin to ideas proposed by Stewart [11] a decade earlier. The idea was to compare two methods of estimating the total charge difference between plasma cations and anions known as the strong ion difference (SID). The first method, known as the 'apparent' SID (SIDa), was to simply measure as many strong (completely or near completely dissociated) cations and anions as possible and sum their charges. The second was to estimate the SID from the partial pressure of CO_2 (from which HCO_3^- and CO_3^{2-} can be estimated) and the concentration of the weak acids (mostly albumin and phosphate as globulins are both cationic and anionic and in healthy humans their net charge is near zero). This second estimate of SID is termed the 'effective' SID (SIDe). Neither estimate is exact. While considering all of the usual electrolytes and lactate, the SIDa will 'miss' strong ions such as ketones and sulfate because they are not measured. Similarly, the SIDe is only an accurate estimate of SID if there are not significant amounts of unmeasured weak acids (e.g. proteins other than albumin or normal globulins) and if albumin and globulins are themselves normal in charge, conformation and composition. Neither of these seem very likely in critically ill patients so neither SIDa nor SIDe should be assumed to be equal to the SID. However, both SIDa and SIDe should equal SID and hence be equal to each other in healthy plasma. When SIDa is not equal to SIDe, some unmeasured anions or cations must be present. We termed this difference the SIG to distinguish it from the AG [10]. By convention, SIDa – SIDe = SIG and hence SIG is 'positive' when unmeasured anions are present in excess of unmeasured cations, and negative when unmeasured cations exceed unmeasured anions. Unfortunately, the name SIG might seem to imply that strong ions are involved. Yet, as detailed above, either strong or weak ions, or both, may produce a 'gap' between these two complementary estimates of SID. The SIG cannot tell us which.

$$SIDa = (Na^+ + K^+ + Ca^{2+} + Mg^{2+}) - (Cl^- + lactate)$$

$$SIDe = 2.46 \cdot 10^{-8} \cdot PCO_2/10^{-pH} + [albumin] \cdot (0.123 \cdot pH - 0.631) + [PO_4^{2-}] \cdot (0.39 \cdot pH - 0.469)$$

$$SIG = SIDa - SIDe$$

Interpreting the Gaps

The utility of the AG/SIG comes primarily from its ability to quickly and easily limit the differential diagnosis in a patient with metabolic acidosis. If an increased AG/SIG is present, the explanation will often be found among five disorders: ketosis, lactic acidosis, poisoning, renal failure and sepsis. Table 1 provides a list, including other disorders associated with an increased AG/SIG.

A number of factors may influence the AG apart from those corrected for above. Respiratory and metabolic alkalosis are associated with an increase of up to 3–10 mEq/l in the apparent AG following an enhanced lactate production

Table 1. Causes of an increased AG and SIG

Common causes
  Renal failure
  Ketoacidosis
    Diabetic
    Alcoholic
    Starvation
    Metabolic errors
  Lactic acidosis[1]
  Toxins
    Methanol
    Ethylene glycol
    Salicylates
    Paraldehyde
    Toluene

Rare causes
  Dehydration
  Sodium salts
    Sodium lactate[1]
    Sodium citrate
    Sodium acetate
    Sodium PCN ($>50\,mU/day$)
    Carbenicilin ($>30\,g/day$)
  Decreased unmeasured cation
  Hypomagnesemia[1]
  Hypocalcemia[1]
  Alkalemia

[1]Already accounted for by SIG.

(from stimulated phosphofructokinase enzymatic activity), the reduction in the ionized weak acids (A^-) and possibly, the additional effect of dehydration (with its own impact on AG calculation). Low Mg^{2+} concentration with associated low K^+ and Ca^{2+} concentrations, as well as the administration of sodium salts of poorly reabsorbable anions (such as β-lactam antibiotics) are known causes of an increased AG [12]. Certain parenteral nutrition formulations, such as those containing acetate, may increase both the AG and the SIG and citrate may rarely have the same effect in the setting of multiple blood transfusions particularly if massive doses of banked blood are used, such as during liver transplantation [13]. None of these rare causes, however, will increase the AG or SIG significantly, and they are usually easily identified.

We expected to find very little, if any, unmeasured ions in the blood of normal humans and by applying our methodology to a published dataset of healthy exercising subjects [14] we found, or rather, did not find very much at all. The total unmeasured anions in the blood of these subjects was a mere 0.3 ± 0.6 mEq/l [10]. However, unlike healthy exercising subjects or normal laboratory animals [15, 16], critically ill patients seem to have much higher SIG values [17–22]. Recently, there has been controversy as to what constitutes a 'normal' SIG and as to whether an abnormal SIG is associated with adverse clinical outcomes. Reports from the United States [17, 18, 22] and from Holland [19] have found that the SIG was close to 5 mEq/l in critically ill patients while studies from England and Australia [20, 21] have found much higher values. One might speculate that the difference may lie with the use of gelatins (an exogenous source of unmeasured ions [23]) for resuscitation in these countries [24]. In this scenario, the SIG is likely to be a mixture of endogenous and exogenous anions. Interestingly, these two studies involving patients receiving gelatins [20, 21] have failed to find a correlation between SIG and mortality while studies in patients not receiving gelatins [17, 18, 25] a positive correlation between SIG and hospital mortality has been found. Indeed Kaplan and Kellum [18] have recently reported that preresuscitation SIG predicts mortality in injured patients better than blood lactate, pH or injury severity scores. Dondorp et al. [25] had similar results with preresuscitation SIG as a strong mortality predictor in patients with severe malaria. More recently, Durward et al. [26] report yet another instance when SIG and mortality correlate in patients not receiving gelatins. These authors also found that SIG, at admission to the ICU, was superior to lactate and other acid-base variables in terms of predicting subsequent hospital survival. In this study SIG had equal predictive accuracy to the pediatric index of mortality score, a risk prediction tool that comprises eight variables including base excess.

Etiology of SIG

While numerous studies have identified clinical conditions associated with unmeasured anions, the exact chemical nature of these substances is unknown. Unmeasured anions have been reported in the blood of patients with sepsis [8, 27] and liver disease [10, 28] and in experimental animals given endotoxin [15]. These anions may be the source of much of the unexplained acidosis seen in patients with critical illness. However, the very idea that something is happening during cardiac surgery [26], during the early stages of major vascular injury [18] and during malarial sepsis [25] as well during other types of critical illness [17] that results in the release of anions that correlate with subsequent mortality is astonishing, especially since we do not know what these anions are. Given that individual patients may have SIG values of more than 10–15 mEq/l, it seems unlikely that any strong ion could be present in the plasma at these concentrations and be unknown to us. Yet, it seems stranger still, for weak acids such as proteins to be the cause given that they are, in fact, weak. In healthy subjects the total charge concentration of plasma albumin is only about 10–12 mEq/l. For a similarly charged protein to affect a SIG of 15 mEq/l, it would need to be present in very large quantities indeed. Recent attempts to determine the etiology of SIG in critically ill patients reveal that although certain low molecular weight anions usually associated with intermediary metabolism are found to be significantly elevated in the plasma obtained from patients with metabolic acidosis [29] the overall concentration of these molecules explains less than 50% of the observed SIG.

The answer, probably, is that the identity of the SIG in these patients is multifactorial. Endogenous strong ions such a ketones and sulfate are added to exogenous ones such as acetate and citrate. Reduced metabolism of these and other ions owing to liver [15] and kidney [30] dysfunction likely exacerbates this situation. The release of a myriad of acute phase proteins, principally from the liver, in the setting of critical illness and injury likely adds to the SIG. Furthermore, the systemic inflammatory response is associated with the release of a substantial quantity of proteins including cytokines and chemokines some of which, like high-mobility group B1, have been linked to mortality [31]. The cumulative effect of all of these factors may well be a reflection of both organ injury and dysfunction. It is perhaps not surprising that there is a correlation between SIG and mortality.

However, whatever the source of SIG, it appears that its presence in the circulation, especially early in the course of illness or injury, portends a poor prognosis. While the prognostic significance of SIG is reduced (or abolished) when exogenous unmeasured anions are administered (e.g. gelatins), a SIG acidosis seems to be far worse than a similar amount of hyperchloremic acidosis and more like lactic acidosis in terms of significance [26, 32]. Although it is possible that saline-based resuscitation fluids contaminate the prognostic value of hyperchloremia the same way gelatins appear to confound SIG, there remains strong evidence that not all metabolic acidoses are the same.

References

1 Kellum JA: Diagnosis and treatment of acid-base disorders; in Grenvik A, Ayres SM, Holbrook PR, Shoemaker WC (eds): Textbook of Critical Care. Philadelphia, Saunders, 2000, pp 839–853.
2 Severinghaus JW: Siggard-Andersen and the 'great trans-Atlantic acid-base debate'. Scand J Clin Lab Invest 1993;53(suppl 214):99–104.

3 Figge J, Mydosh T, Fencl V: Serum proteins and acid-base equilibria: a follow-up. J Lab Clin Med 1992;120:713–719.
4 Sadjadi SA: A new range for the anion gap. Ann Intern Med 1995;123:807–808.
5 Winter SD, Pearson R, Gabow PG, Schultz A, Lepoff RB: The fall of the serum anion gap. Arch Intern Med 1990;150:3113–3115.
6 Salem MM, Mujais SK: Gaps in the anion gap. Arch Intern Med 1992;152:1625–1629.
7 Gabow PA: Disorders associated with an altered anion gap. Kidney Int 1985;27:472–483.
8 Mecher C, Rackow EC, Astiz ME, Weil MH: Unaccounted for anion in metabolic acidosis during severe sepsis in humans. Crit Care Med 1991;19:705–711.
9 Kellum JA: Determinants of blood pH in health and disease. Crit Care 2000;4:6–14.
10 Kellum JA, Kramer DJ, Pinsky MR: Strong ion gap: a methodology for exploring unexplained anions. J Crit Care 1995;10:51–55.
11 Stewart P: Modern quantitative acid-base chemistry. Can J Physiol Pharmacol 1983;61:1444–1461.
12 Whelton A, Carter GG, Garth M: Carbenicillin-induced acidosis and seizures. JAMA 1971;218:1942–1944.
13 Kang Y, Aggarwal S, Virji M: Clinical evaluation of autotransfusion during liver transplantation. Anesth Analg 1991;72:94–100.
14 Lindinger MI, Heigenhauser GJF, McKelvie RS, Jones NL: Blood ion regulation during repeated maximal exercise and recovery in humans. Am J Physiol 1992;262:R126–R136.
15 Kellum JA, Bellomo R, Kramer DJ, Pinsky MR: Hepatic anion flux during acute endotoxemia. J Appl Physiol 1995;78:2212–2217.
16 Kellum JA, Bellomo R, Kramer DJ, Pinsky MR: Etiology of metabolic acidosis during saline resuscitation in endotoxemia. Shock 1998;9:364–368.
17 Balasubramanyan N, Havens PL, Hoffman GM: Unmeasured anions identified by the Fencl-Stewart method predict mortality better than base excess, anion gap, and lactate in patients in the pediatric intensive care unit. Crit Care Med 1999;27:1577–1581.
18 Kaplan L, Kellum JA: Initial pH, base deficit, lactate, anion gap, strong ion difference, and strong ion gap predict outcome from major vascular injury. Crit Care Med 2004;32:1120–1124.
19 Moviat M, van Haren F, van der Hoeven H: Conventional or physicochemical approach in intensive care unit patients with metabolic acidosis. Crit Care 2003;7:R41–R45.
20 Cusack RJ, Rhodes A, Lochhead P, Jordan B, Perry S, Ball JAS, et al: The strong ion gap does not have prognostic value in critically ill patients in a mixed medical/surgical adult ICU. Intensive Care Med 2002;28:864–869.
21 Rocktaschel J, Morimatsu H, Uchino S, Bellomo R: Unmeasured anions in critically ill patients: can they predict mortality? Crit Care Med 2003;31:2131–2136.
22 Gunnerson KJ, Roberts G, Kellum JA: What is normal strong ion gap (SIG) in healthy subjects and critically ill patients without acid-base abnormalities (abstract). Crit Care Med 2003;31(12 suppl):A111.
23 Hayhoe M, Bellomo R, Liu G, Kellum JA, McNicol L, Buxton B: Role of the splanchnic circulation in acid-base balance during cardiopulmonary bypass. Crit Care Med 1999;27:2671–2677.
24 Kellum JA: Closing the gap on unmeasured anions. Crit Care 2003;7:219–220.
25 Dondorp AM, Chau TT, Phu NH, Mai NT, Loc PP, Chuong LV, et al: Unidentified acids of strong prognostic significance in severe malaria. Crit Care Med 2004;32:1683–1688.
26 Durward A, Tibby SM, Skellett S, Austin C, Anderson D, Murdoch IA: The strong ion gap predicts mortality in children following cardiopulmonary bypass surgery. Pediatr Crit Care Med 2005;6:281–285.
27 Gilfix BM, Bique M, Magder S: A physical chemical approach to the analysis of acid-base balance in the clinical setting. J Crit Care 1993;8:187–197.
28 Kirschbaum B: Increased anion gap after liver transplantation. Am J Med Sci 1997;313:107–110.
29 Forni LG, McKinnon W, Lord GA, Treacher DF, Peron JM, Hilton PJ: Circulating anions usually associated with the Krebs cycle in patients with metabolic acidosis. Crit Care 2005;9:R591–R595.
30 Rocktaschel J, Morimatsu H, Uchino S, Goldsmith D, Poustie S, Story D, et al: Acid-base status of critically ill patients with acute renal failure: analysis based on Stewart-Figge methodology. Crit Care 2003;7:R60–R66.

31 Wang H, Bloom O, Zhang M, Vishnubhakat JM, Ombrellino M, Che J, et al: HMG-1 as a late mediator of endotoxin lethality in mice. Science 1999;285:248–251.
32 Gunnerson KJ, Saul M, He S, Kellum JA: Lactate versus non-lactate metabolic acidosis: a retrospective outcome evaluation of critically ill patients. Crit Care 2006;10:R22–R26.

John A. Kellum, MD
Department of Critical Care Medicine, University of Pittsburgh Medical Center
3550 Terrace Street
Pittsburgh, PA 15216 (USA)
Tel. +1 412 647 6966, Fax +1 412 647 8060, E-Mail kellumja@upmc.edu

Fluid Resuscitation and the Septic Kidney: The Evidence

Elisa Licari[a], Paolo Calzavacca[a], Claudio Ronco[b], Rinaldo Bellomo[a]

[a]Department of Intensive Care, Austin Hospital, Melbourne, Vic., Australia;
[b]Department of Nephrology, St. Bortolo Hospital, Vicenza, Italy

Abstract

Acute kidney injury (AKI) is a common complication of severe sepsis. Severe sepsis is the most common cause of AKI in ICU. The widely accepted and practiced initial cornerstone of treatment for septic AKI is fluid resuscitation. The biological rationale for fluid resuscitation in septic AKI is based on the assumption that septic AKI is an ischemic form of AKI and that increasing renal perfusion and oxygen delivery by means of fluid resuscitation will protect the kidney. Whether this is true, however, remains uncertain. In this paper, we discuss salient pathophysiological aspects of AKI, review the evidence available on the need for fluid resuscitation, the amount and the type of fluid that might be best suited to AKI and discuss all major aspects of fluid resuscitation for septic AKI in humans and experimental animals.

Copyright © 2007 S. Karger AG, Basel

Sepsis accounts for a similar number of death per year in the United States as myocardial infarction [1]. It is the most frequent cause of vasodilatory shock accounting for more than 200,000 cases per year in the United States [2]. Occurrence of admission in ICU is caused by sepsis in 10% of cases, with mortality ranging from 25 to 80% [3].

Sepsis is the most common factor predisposing to acute kidney injury (AKI) in critically ill patients [4, 5]. It accounts for around 50% of cases [1, 6]. Patients with AKI due to sepsis have a worse prognosis than those with nonseptic AKI [1]. Unfortunately, until recently, the definition of AKI has been a confounding element in its epidemiology. This is because the diagnosis of AKI is complex and involves data obtained from history, biochemical analysis, body size, sex, hematological information and imaging. The second International Consensus Conference of the Acute Dialysis Quality Initiative Group in 2002 proposed a classification scheme for AKI to facilitate communication and research in this field. It produced the so-called RIFLE criteria for the classification and definition of AKI [7]. These criteria use creatinine and urine output and consider changes from baseline creatinine value in reaching a classification [7]. The RIFLE classification has now been applied to understanding the epidemiology of AKI and septic AKI in ICU and the findings confirm the high incidence of AKI in ICU, its high mortality and strong association with sepsis [8]. Thus septic AKI is perhaps the biggest physiological and therapeutic challenge in critical care nephrology. In order to develop a rational approach of its treatment, one needs to understand some aspects of its pathophysiology.

The Pathophysiology of Septic AKI

A decrease in renal blood flow (RBF) causing renal ischemia has been proposed to be central in the pathogenesis of septic AKI [1, 9, 10]. Sepsis induces increased nitric oxide synthase activity and generates oxygen radicals. Nitric oxide is believed, in turn, to cause systemic vasodilatation, peroxynitrite-related tubular injury and downregulation of renal endothelial nitric oxide synthase [1]. The vasodilation that results from the systemic changes induced by severe sepsis shifts blood flow from the renal bed to peripheral vascular beds causing decreased RBF, subsequent ischemia and acute tubal necrosis.

This paradigm mainly relies on animal data because the measurement of RBF in man is extremely difficult and requires invasive techniques. However, a recent systematic review [11] of human and animal studies concluded that the primary determinant of RBF in sepsis is cardiac output (CO). If the CO is increased, RBF is typically either increased or preserved. CO in human sepsis is usually elevated. Only three studies conducted in septic ICU patients were found in which RBF was measured [12–14]. All showed either preserved or increased RBF. Recent animal experiments [15–18] using a hyperdynamic model of sepsis also found that RBF in sepsis was increased due to vasodilation of renal circulation of both afferent and efferent arteriole. Thus, the pathophysiology of septic AKI might not necessarily involve ischemia (acute tubular necrosis), especially in hyperdynamic sepsis.

These observations have important repercussions on our understanding and biologic rationale for specific kidney protective interventions, especially in the field of fluid resuscitation.

The Biologic Rationale for Fluid Resuscitation in Septic AKI

Prompt and aggressive fluid resuscitation is considered a cornerstone for renal protection and preservation of renal function [19]. The rationale for fluid resuscitation is that septic systemic vasodilation and capillary leak cause relative hypovolemia [2]. Relative hypovolemia would then cause decreased vital organ perfusion and, therefore, decreased oxygen delivery, which, in turn, would cause organ dysfunction. This would appear to be particularly relevant to the kidney, which is considered highly sensitive to hypovolemia.

Correction of septic hypovolemia may require continuous and large volumes of fluid administration to maintain renal oxygen delivery above a critical threshold and to increase or maintain mean arterial pressure to a level that allows appropriate distribution of CO and adequate organ perfusion to the kidney [20, 21].

In addition, although the kidney seeks to autoregulate its own perfusion, this ability may be impaired in severe sepsis [22]. While these considerations support the use of fluid resuscitation, more recent insights into the pathophysiology of septic AKI, as described above, challenge the notion that ischemia is responsible for septic AKI and, by implication, also challenge the notion that fluid resuscitation would be of major benefit.

The Evidence of the Renal Benefits of Fluid Resuscitation in Septic AKI

Although it is widely recognized as a common practice to start cardiovascular resuscitation of septic patients with a fluid challenge to maintain organ perfusion and thus renal function, no randomized control trial has ever evaluated the effect of fluid challenge or fluid resuscitation in general on kidney function in septic AKI [23]. A recent evidenced-based review [24] stated the need to perform a fluid challenge as soon as a hypovolemic state is supported by only grade E evidence. This means that no evidence has been found to support this statement, which, therefore, relies on the expert opinion. On the other hand, recent evidence has started raising caution about the liberal use of fluid in the management of critically ill patients [25–29].

For example, the ARDS network enrolled 1,000 patients with acute lung injury in a randomized trial. The patients were randomly assigned to a strategy involving either conservative or liberal use of fluids. Over a period of 7 days, the conservative-strategy group received a fluid balance of -136 ± 491 ml as compared with $6,992 \pm 502$ ml in the liberal-strategy group ($p < 0.001$). More than 80% of patients enrolled in this study were septic and, therefore, at high risk of septic AKI due to relative hypovolemia related not only to sepsis but also to mechanical ventilation and high levels of PEEP. Within the first 60 days, there were no significant differences in the percentage of patients receiving renal replacement therapy (10% in the conservative-strategy group vs. 14% in

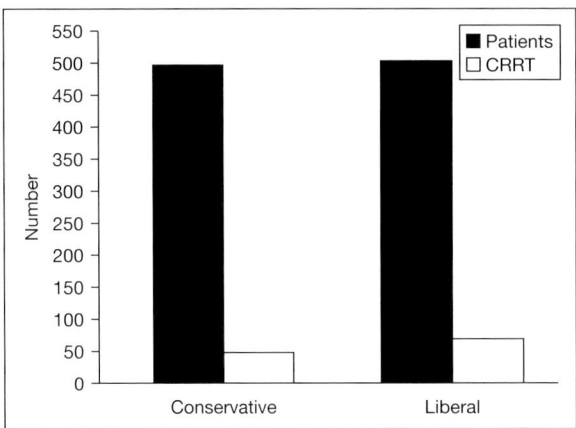

Fig. 1. Histogram presenting the incidence of the need for renal replacement therapy in ARDS patients treated with liberal versus conservative fluid strategy in a large multicenter trial.

the liberal-strategy group, p = 0.06) (fig. 1) although the trend was clearly in favor of a conservative approach. In conclusion the conservative strategy improved lung function and shortened the duration of mechanical ventilation and intensive care, and, possibly, even offered some protection from severe AKI.

In the SAFE study, a randomized control trial comparing albumin 4% to normal saline, 6,997 patients were enrolled. Of these, 3,497 were assigned to receive albumin and 3,500 assigned to receive saline. The 4 days' fluid balance was 3 liters in the albumin group and 4 liters in the saline group. Unfortunately no data about renal function were provided, making it uncertain whether the different choice of fluids and fluid balance affected the likelihood of AKI.

The early goal-directed therapy study [30] is a randomized controlled trial of resuscitation strategies, which enrolled 263 patients. Of these, 130 were randomly assigned to early goal-directed therapy (EGDT) and 133 to standard therapy. The initial fluids administered over the first 6 h were greater with EGDT (5 l in EGDT vs. 3.5 l for standard therapy; p < 0.001). This difference might be expected to translate to improved renal function and protection from AKI. Unfortunately, no data about renal function are provided to tell us whether the EGDT approach affects the likelihood of AKI.

Thus, in the three major clinical trials of fluid therapy and sepsis, no conclusive data are available in relation to the septic kidney. However, it is possible that the type of fluids used by clinicians might affect the likelihood and course of septic AKI.

Type of Fluid

Colloids and crystalloids might affect renal function differently. Colloids have long been used in the care of septic patients to raise oncotic pressure [31]. This approach has the rationale of minimizing edema formation [31]. Unfortunately little is known [32] about their impact on the septic kidney. Some evidence exists, however, on three major subtypes of colloid fluids: albumin, starches and gelatins.

Albumin

The use of albumin in intensive care setting is still debated, many studies having found very conflicting data. However the SAFE trial (saline vs. albumin fluid evaluation) [33], a multicentric, randomized, controlled trial that enrolled nearly 7,000 patients, found no differences in new organ failure, urine output or in the number of renal replacement therapy days in patients treated with albumin or saline for fluid resuscitation. This suggests that there is no intrinsic advantage or disadvantage in terms of AKI from using albumin or saline.

Starches

There are different types of starches available in different countries. Pharmacokinetics [34] and pharmacodynamics of starches depend on molecular weight (MW) and rate of substitution. All these macromolecules are metabolized by serum amylases and excreted by the kidney. Hydroxyethylstarch can be classified according to the MW in high MW (450–480 kDa), medium MW (200 kDa) and low MW (70 kDa), or according to the rate of substitution (high 0.6–0.7 or low 0.4–0.5), or according to the C2/C6 ratio (high >8, low <8), or according to the concentration (high 10%, low 6%). Although definitive studies addressing safety and efficacy of hydroxyethylstarch in preventing alterations of renal function in septic patients are not conclusive, some evidence exists that at least high MW starch preparations might contribute to AKI.

In a multicenter randomized study by Schortgen et al. [35] published in 2001, 129 patients received medium MW hydroxyethylstarch 6%. These investigators found that starch was associated with an increased risk of developing AKI or need for renal replacement therapy when compared to a 3% gelatin (OR 2.57). This observation is consistent with data from autopsy findings reporting osmotic nephrosis-like lesions [36] in 80% of kidneys from donors treated with starch. In a prospective randomized study in patients undergoing renal transplantation, Cittanova et al. [37] found osmotic nephrosis-like lesions in the kidneys of donors treated with high MW hydroxyethylstarch and a statistically significant increase in the need for hemodiafiltration and serum creatinine level compared to kidneys from donors treated with gelatin only. These findings are consistent with a recent trial [23] (VISEP trial, Efficacy of Volume Substitution and Insulin Therapy in Severe Sepsis), a randomized comparison of crystalloid (Ringer's lactate) and colloid (10% HES) fluid therapy in critically ill patients with severe sepsis. Preliminary data indicate that the use of starch resulted in a significantly higher incidence of AKI.

Gelatin

Although some case reports [32] exist suggesting gelatin can adversely affect kidney function, there is no controlled trial comparing gelatin solutions to crystalloid. Comparisons to starches indicate that gelatin solutions might be safer (see above).

Hypertonic Saline

A review on the use of hypertonic saline for resuscitation [38] suggested a favorable effect of this therapy. Unfortunately no randomized controlled trial in septic humans using hypertonic saline has been done [39].

Saline

No trial has ever been performed comparing saline versus no treatment in humans; neither has a study comparing saline and other crystalloids (Hartmann's, lactated solution) ever been performed in the field of septic AKI. The only randomized control trial in septic critically ill patients addressing saline is the SAFE trial as discussed above. Although it does not give us any insight into the effectiveness of saline resuscitation in the treatment of septic AKI, we can state that there is no evidence in that study that saline is either better or worse than albumin in resuscitating septic patients.

Lactated Solutions

We could not find any trial of crystalloid solution other than normal saline in sepsis, so there is no evidence available at the moment on whether lactated solutions can deliver significant advantages or disadvantages in septic AKI.

In conclusion, insufficient human evidence exists that colloids (except for high MW starch) are better or worse than crystalloids or that a particular type of crystalloid is better or worse than another. Thus, we can only rely on data generated from experimental studies of sepsis and extrapolate from the physiological effects of fluid resuscitation on systemic and regional hemodynamics and renal function.

Animal Studies

Given the limitations of the human data available, a review of the animal data on the effect of fluids in septic models or septic AKI might be helpful. Unfortunately many biases are present when dealing with animal studies: animal size, consciousness of animals, time from septic insult, methods of inducing sepsis, and CO, so that it is very hard to compare different studies and draw conclusions that apply to humans [11].

Only one randomized, unblinded [40] trial compared isotonic or hypertonic fluid therapy with control in a porcine model of endotoxemic shock. In 24 Landrace pigs a hypodynamic endotoxin shock was induced. The animals were then randomly divided into three groups: a control group, a group treated with hypertonic (7.5%) saline-6% dextran 70 (HSD group) at 4 ml/kg and a group treated with isotonic (0.9%) saline-6% dextran 70 (ISD) at 4 ml/kg. The mortality rate at 300 min was 67% in the control group, 14% in the HSD group and 75% in ISD group. This gives a relative risk of death in the HSD group when compared with the control group of 0.17 and a relative risk of death in the ISD group of 1.2 when compared to control. The findings that isotonic saline was associated with the highest mortality raise some concerns. More work is required before any conclusions can be drawn from these unusual observations. Importantly, for the point of view of fluid therapy in septic AKI, there were no data on renal function in this study.

Unfortunately there is no other study specifically addressing the need for fluid resuscitation in order to sustain renal function in septic animals. Some experimental studies have analyzed the effect of crystalloid resuscitation on organ blood flow and compared their efficacy versus colloid resuscitation.

Bressack et al. [41] published a study comparing normal saline and 5% albuminated saline in piglets. They found that RBF was related only to CO and that organ edema formation occurred only in the saline-treated group.

Wan et al. [42] in 2006 published a randomized controlled crossover animal study in which they demonstrate that resuscitation with normal saline increased central venous pressure, CO, urine output, creatinine clearance and fractional excretion of sodium in sheep but had no effect on RBF. All these findings were transient (<1 h).

In a recent study, Garrido et al. [43] compared Ringer's lactate and hypertonic saline in the resuscitation of a dog model of hypodynamic shock. After 30 min from the start of an infusion of *Escherichia coli* in mongrel dogs, they

Table 1. Summary of all available animal studies

Authors, year	Study performed	Main finding
Crystalloid vs. nothing		
Oi et al., 2000 [40]	No fluid vs. isotonic or hypertonic infusion in a porcine model of endotoxemia	OR 1.2 for death in untreated animals
Wan et al., 2006 [42]	Normal saline vs. control	Normal saline transiently increases CVP, CO, mesenteric flow, UO, CC and FeNa in sheep, no effect on RBF
Crystalloid vs. colloid		
Bressack et al., 1987 [41]	Normal saline vs. 5% albuminated saline in piglets	Organ edema in the saline-treated group
Garrido Adel et al., 2006 [43]	Ringer lactate and hypertonic saline in resuscitation of dog in hypodynamic shock	Both transient improvement in systemic and regional blood flow after infusion; in the hypertonic group significant and sustained reduction of systemic and mesenteric oxygen extraction

CC = Creatinine clearance; CVP = central venous pressure; UO = urine output.

randomized 7 animals to receive lactated Ringer at 32 ml/kg over 30 min or 7.5% hypertonic saline solution 4 ml/kg over 5 min. They then observed (120 min) the effects on hemodynamic parameters and on regional perfusion markers. Both infusion regimens produced a transient improvement in systemic and regional blood flow. However, no specific information on the renal effects was reported. All available animal studies are summarized in table 1.

Unfortunately, no study so far has specifically used animal models of sepsis or septic AKI to study the optimal amount of fluid resuscitation or the prevention of AKI. Also no studies have compared resuscitation with fluids versus resuscitation with vasopressors alone or the combination of the two. In addition, many different hemodynamic targets are used in titrating fluid resuscitation but no controlled trial has been performed to address this specific issue.

The above observations from animal studies related to septic AKI and fluid therapy highlight the dearth of information that exists in this field and the need for further investigations.

Conclusions

Fluid resuscitation is a common empirical practice in the treatment of patients with septic AKI. The rationale for this is to attempt to restore an adequate RBF in an environment of suspected decreased oxygen delivery due to ischemia. However, more recent studies challenge this paradigm. In addition, recent evidence suggests that liberal fluid administration may be dangerous to both kidney and patient and that high MW starches may be nephrotoxic. Animal studies remain inadequate in helping us understand what might be the best approach to fluid therapy in septic AKI. Much more research is needed in this important field of critical care nephrology.

References

1. Schrier RW, Wang W: Acute renal failure and sepsis. N Engl J Med 2004;351:159–169.
2. Landry DW, Oliver JA: The pathogenesis of vasodilatory shock. N Engl J Med 2001;345:588–595.
3. National Nosocomial Infections Surveillance System (NNIS) report, October 2005. www.cdc.gov/ncidod/dhqp/nnis_pubs.html.
4. Uchino S, Kellum JA, Bellomo R: Acute renal failure in critically ill patients: a multinational, multicenter study. JAMA 2005;294:813–818.
5. Cole L, Bellomo R, Silvester W: A prospective, multicenter study of the epidemiology, management, and outcome of severe acute renal failure in a 'closed' ICU system. Am J Respir Crit Care Med 2000;162:191–196.
6. De Mondança A, Vincent JL, Suther PM, et al: Acute renal failure in the ICU: risk factors and outcome evaluated by the SOFA score. Intensive Care Med 2000;26:915–921.
7. Bellomo R, Ronco C, Kellum JA, et al: Acute renal failure-definition, outcome measures, animal models, fluid therapy and information technology needs: the Second International Consensus Conference of the Acute Dialysis Quality Initiative (ADQI) Group. Crit Care 2004;8:R204–R212.
8. Bellomo R, Kellum JA, Ronco C: Defining and classifying acute renal failure: from advocacy to consensus and validation of the RIFLE criteria. Intensive Care Med 2007;33:409–413.
9. Badr KF: Sepsis-associated renal vasoconstriction: potential targets for future therapy. Am J Kidney Dis 1992;20:207–213.
10. De Vriese AS, Bourgeois M: Pharmacologic treatment of acute renal failure in sepsis. Curr Opin Crit Care 2003;9:474–480.
11. Langenberg C, Bellomo R, May C, Li W, Moritoki E, Morgera S: Renal blood flow in sepsis. Crit Care 2005;9:R363–R374.
12. Brenner M, Schaer GL, Mallory DL, et al: Detection of renal blood flow abnormalities in septic and critically ill patients using a newly designed indwelling thermodilution renal vein catheter. Chest 1990;98:170–179.

13 Lucas CE, Rector FE, Werner M, et al: Altered renal homeostasis with acute sepsis. Clinical significance. Arch Surg 1973;106:444–449.
14 Rector F, Goyal S, Rosemberg IK, et al: Sepsis: a mechanism for vasodilatation in the kidney. Ann Surg 1973;178:222–226.
15 Langenberg C, Wan L, Egi M, May CN, Bellomo R, et al: Renal blood flow in experimental septic acute renal failure. Kidney Int 2006;69:1996–2002.
16 Heemskerk AE, Huisman E, van Lambalgen AA, et al: Gram-negative shock in rats depends on the presence of capsulated bacteria and is modified by laparotomy. Shock 1996;6:418–425.
17 Heemskerk AE, Huisman E, van Lambalgen AA, et al: Laparotomy and renal function during endotoxin shock in rats. Shock 1996;6:410–417.
18 Heemskerk AE, Huisman E, van Lambalgen AA, et al: Renal function and oxygen consumption during bacteremia and endotoxemia in rats. Nephrol Dial Transplant 1997;12:1586–1594.
19 Dellinger RP, Carlet JM, Masur H: Surviving sepsis campaign guidelines for management of severe sepsis and septic shock. Crit Care Med 2004;32:858–873.
20 Landry DW, Oliver JA: The pathogenesis of vasodilatory shock. N Engl J Med 2001;23:588–595.
21 Ince C, Sinaasappel M: Microcirculatory oxygenation and shunting in sepsis and shock. Crit Care Med 1999;27:1369–1377.
22 Bersten AD, Holt AW: Vasoactive drugs and the importance of renal perfusion pressure. New Horiz 1995;3:650–661.
23 Bagshaw S, Bellomo R: Fluid resuscitation and the septic kidney. Curr Opin Crit Care 2006;12:527–530.
24 Vincent JL, Gherlac H: Fluid resuscitation in severe sepsis and septic shock: an evidence-based review. Crit Care Med 2004;32(suppl 11):s451–s454.
25 Van Biesen W, Yegenaga I, Vanholder R: Relationship between fluid status and its management on acute renal failure AKI in intensive care unit ICU patients with sepsis: a prospective analysis. J Nephrol 2005;18:54–60.
26 Wiedemann HP, Wheeler AP, Bernard GR: Comparison of two fluid-management strategies in acute lung injury. N Engl J Med 2006;354:2564–2575.
27 Sakr Y, Vincent JL, Reinhart K: High tidal volume and positive fluid balance are associated with worse outcome in acute lung injury. Chest 2005;128:3098–3108.
28 Simmons RS, Berdine GG, Seidenfeld JJ: Fluid balance and the adult respiratory distress syndrome. Am Rev Respir Dis 1987;135:924–929.
29 Mehta RL, Clark WC, Schetz M: Techniques for assessing and achieving fluid balance in acute renal failure. Curr Opin Crit Care 2002;8:535–543.
30 Rivers E, Nguyen B, Havstad S: Early goal-directed therapy in the treatment of severe sepsis and septic shock. N Engl J Med 2001;345:1368–1377.
31 Rackow EC, Falk JL, Fein IA: Fluid resuscitation in circulatory shock: a comparison of the cardiorespiratory effects of albumin, hetastarch, and saline solutions in patients with hypovolemic and septic shock. Crit Care Med 1983;11:839–850.
32 Davidson IJ: Renal impact of fluid management with colloids: a comparative review. Eur J Anaesthesiol 2006;23:721–738.
33 Finfer F, Bellomo R, Boyce N: A comparison of albumin and saline for fluid resuscitation in the intensive care unit. N Engl J Med 2004;350:2247–2256.
34 Treib J, Baron JF, Grauer MT, Strauss RG: An international view of hydroxyethyl starches. Intensive Care Med 1999;25:258–268.
35 Schortgen F, Lacherade JC, Bruneel F, et al: Effects of hydroxyethylstarch and gelatin on renal function in severe sepsis: a multicentre randomized study. Lancet 2001;357:911–916.
36 Legendre C, Thervet E, Page B, et al: Hydroxyethylstarch and osmotic-nephrosis-like lesions in kidney transplantation. Lancet 1993;342:248.
37 Cittanova ML, Leblanc I, Legendre C, et al: Effect of hydroxyethylstarch in brain-dead kidney donors on renal function in kidney transplant recipients. Lancet 1996;348:1620–1622.
38 Wade CE: Hypertonic saline resuscitation in sepsis. Crit Care 2002;6:397–398.
39 Oliveira RP, Velasco I, Soriano FG, Friedman G: Clinical review: hypertonic saline resuscitation in sepsis. Crit Care 2002;6:418–423.

40 Oi Y, Aneman A, Svensson M, et al: Hypertonic saline-dextran improves intestinal perfusion and survival in porcine endotoxin shock. Crit Care Med 2000;28:2843–2850.
41 Bressack MA, Morton NS, Hortop J: Group B streptococcal sepsis in the piglet: effects of fluid therapy on venous return, organ oedema, and organ blood flow. Circulation 1987;61:659–669.
42 Wan L, Bellomo R, May CN: The effect of normal saline resuscitation on vital organ blood flow and septic sheep. Intensive Care Med 2006;32:1238–1242.
43 Garrido Adel P, Cruz Junior RJ, Poli de Figueiredo LF, et al: Small volume of hypertonic saline as the initial fluid replacement in experimental hypodynamic sepsis. Crit Care 2006;10:R62–R71.

Prof. Rinaldo Bellomo
Department of Intensive Care, Austin Health
Melbourne, Vic. 3084 (Australia)
Tel. +61 3 9496 5992, Fax +61 3 9496 3932, E-Mail rinaldo.bellomo@austin.org.au

Factors Affecting Circuit Patency and Filter 'Life'

Ian Baldwin

Department of Intensive Care, Austin Hospital, and Department of Nursing and Health Sciences, RMIT University, Melbourne, Vic., Australia

Abstract

Frequent clotting applying continuous renal replacement therapy means treatment can be inadequate and with increased costs for circuits and nursing time. Patency of the extracorporeal circuit is commonly achieved using anticoagulants such as heparin. When anticoagulants are not used, or clotting occurs within a few hours of use, with anticoagulation, blood flow failure is a likely cause. The blood pump can fail to deliver without operator awareness. Clotting within the membrane and/or venous 'air-trap' chamber is common where resistance to blood flow is high with stasis and turbulence. The design of the venous chamber allows the blood fill level to oscillate and form a clot, with a blood filter at the exit of the chamber also causing clot development. Several practices attempt to prevent clotting, however most without evidence. Adding heparin to the circuit during the preparation phase, ensuring that the access catheter is not obstructed, a blood flow setting of ≥200 ml/min, and administration of substitution fluids before the membrane (predilution) can be useful strategies for increasing circuit patency. An audit of filter life is useful and necessary feedback to nursing staff training strategies. This promotes safety and, when circuit patency is poor, may reflect poor troubleshooting ability.

Copyright © 2007 S. Karger AG, Basel

In this article we will review issues affecting the patency of the extracorporeal circuit (EC) and filter 'life' in relation to continuous renal replacement therapy (CRRT) by first discussing where clotting commonly occurs. This will be followed by a review of factors affecting circuit patency including consideration of heparin during preparation, the access catheter, blood flow, the membrane, the administration of substitution fluids, the venous 'air-trap' chamber, and the importance of training and education for staff.

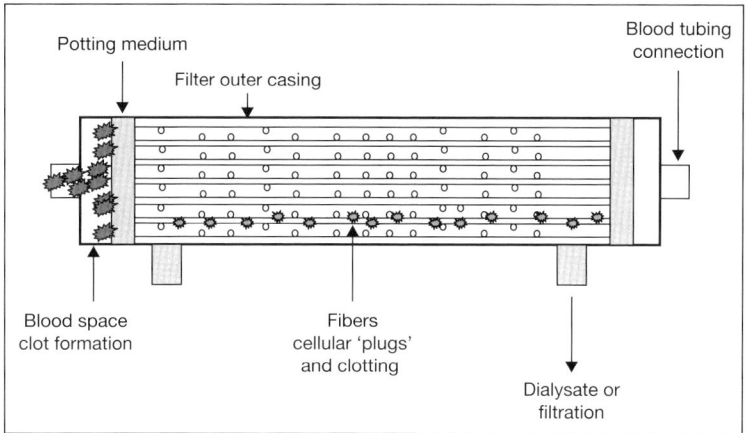

Fig. 1. Schematic drawing of the CRRT membrane indicating sites of clotting: before the potting medium at blood entry, and within fibers by cellular protein plugs.

Circuit Patency, Where Clotting Occurs

Clotting occurs most commonly in the hemofilter (membrane) and or the 'venous' air-trap chamber [1, 2]. Figure 1 is a drawing of a hemofilter schematically highlighting that clot formation can initially occur as blood enters the membrane, and resistance is high prior to the 'potting' material for the filter fibers. Further clotting and clogging of the fibers with proteins then possibly occurs as a secondary additional process. Figure 2 is a drawing of a venous chamber schematically highlighting flow resistance factors at the chamber exit due to the chamber blood filter. If the venous chamber clots, this can obstruct blood flow completely with blood not easily returned to the patient, an undesirable event. This event is often misdiagnosed as 'filter clotting'. Therefore the venous chamber is an important factor in circuit and filter patency.

Anticoagulants

Anticoagulant drugs prevent or delay the formation of clots in the EC, and many methods are used with different levels of evidence supporting their success [3]. Anticoagulants are the focus of another article in this book and will be not discussed other than to suggest that circuit clotting may not always be due to insufficient or incorrect anticoagulation. Maintaining blood flow can be as important in preventing clotting. These flow 'mechanics' are particularly

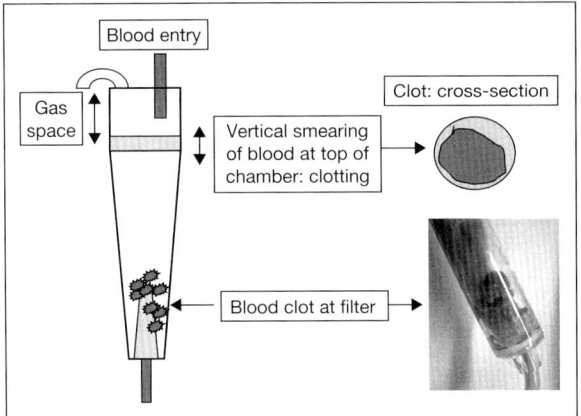

Fig. 2. Schematic drawing of the venous air-trap chamber showing where clot formation commonly occurs: top of the chamber by blood oscillation building a clot (cross-section indicated) and the bottom around the outlet filter (picture included).

important when no anticoagulation is used in patients at risk of bleeding, and useful to understand for all treatments and for circuit patency in every patient.

Access Catheter

The veno-venous access catheter used for CRRT is not a common site for clot formation, however the access catheter may be associated with formation of clots in the EC. Increasing negative ('arterial') and positive ('venous') pressure in the EC reflects access lumen obstruction, causing blood pump failure and reduction in prescribed output. This can then cause slowing of blood flow in the membrane, and clotting as ultrafiltration continues regardless of blood flow indicator speed [4]. This access catheter failure may be unrecognized with reduced blood flow over long periods causing membrane clotting [4].

The insertion site (e.g. femoral, subclavian, internal jugular) of the catheter may also be implicated in circuit patency with anecdotal experience suggesting that poor blood flow can occur when repositioning patients during nursing care. Nursing care and physical therapy must be managed with caution and a response to changes in arterial and venous circuit pressures indicating catheter obstruction is often needed. It is an important aspect of CRRT nursing knowledge to recognize catheter obstruction, and modify the patient position to maintain blood flow without excessive 'arterial' ($\geq 100\,mm\,Hg$) or 'venous' ($>150\,mm\,Hg$) pressures. Sometimes subtle patient position changes can be sufficient to alleviate such pressure changes and facilitate correct blood flow.

Blood Pump – Flow Speed

Blood flow controlled by the blood pump setting may influence the development of EC clot formation [5, 6]. In theory: the faster the blood flow, the less clot formation. A blood flow rate of 200 ml/min is adequate for all modes of CRRT, however hemoconcentration of blood can occur in the membrane if the ultrafiltrate and blood flow are not correctly mixed (filtration fraction error). This highlights a possibly important relationship between clearance mode or technique (diffusion or convection) and the potential for clotting in the EC. Whilst there is minimal evidence, some clinicians do suggest that convective clearance may have a higher potential for clotting in comparison to diffusive clearance [7, 8].

Membrane and Circuit Patency

The addition of heparin into a circuit during or after priming in order to 'coat' the membrane is suggested by many authors to be a useful strategy to prevent clotting [9–12]. There is no good evidence to suggest this has any effect on all membranes, however plastic and membrane surfaces do take up heparin particularly after treating to neutralize negative charge [13, 14]. Therefore, unless contraindicated, heparin can be added to a circuit after priming whilst awaiting connection to the patient, with this heparinized saline pumped around the circuit to promote the coating effect. Different fiber types may also influence clotting [15–17]. Although synthetic membranes are considered biocompatible, premature clotting when exposed to an acrylonitrile membrane may be reduced when changed to polysulfone, with some differences in racial and genetic disposition to clotting for different membrane exposures a possible association [18]. Finally, larger surface area membranes offer less resistance to blood flow [19] and may increase circuit patency and functional life. Increasing membrane size from 1.0 to 1.4 m^2 is of minimal clinical consequence in the adult patient, and often at no increased cost, but may increase filter life.

Circuit Patency and Administration of Substitution Fluids

There is evidence that use of predilution reduces membrane clotting in comparison to postdilution in pure convective modes [20]. The amount of predilution volume required to achieve this affect is not clear, however an alteration in hematocrit is the likely mechanism. Citrate anticoagulation can be performed by addition of the citrate to predilution replacement fluids in a convective therapy. This combines the effect of anticoagulation and predilution [21].

Venous Chamber

An important safety feature of any EC is an air-trap chamber placed in the EC prior to the blood returning to the patient, however this EC component is often a site for clotting [1, 5, 9, 10, 22].

The chamber allows turbulent flow to occur with the blood level in the chamber oscillating or rising and falling. This oscillation is consistent with the pulsatile flow generated by the blood pump and the varying resistance at the venous lumen of the access catheter. The chamber cannot be completely full as a pocket of gas (air $\pm$ CO^2) above the blood level acts as a medium for pressure readings to a transducer, and prevents blood entering the transducer line. The oscillating blood level causes a constant smearing and cell deposition on the inside of the chamber eventually developing a ring of deposited cells inside the chamber, building a clot, as indicated in figure 2. Attempts have been made to prevent this clotting by adding heparin into the chamber before and during use [23], adding fluids to the chamber (postdilution), and use of a tubing design such that incoming blood enters under the blood chamber level. This last approach can create a cell–plasma separation with a small layer of plasma separating to the top of the chamber. This provides a plasma layer protecting the cells from exposure to the gas and reduces cell smearing.

While evidence is lacking, circuit patency may be enhanced by keeping the venous chamber level close to full with a minimal gas pocket, adjusting the level down when a ring of clot begins to form, adding postdilution fluids into this chamber when used, and adding heparin into the chamber during the priming procedure – heparin coating.

Staff Training, Education

Safe and skilled use of CRRT machines requires nursing education and training activities with theoretical and practical components. Inability to manage and/or correct simple alarm events may be associated with poor circuit patency and/or serious patient harm [24].

Current day machines with rigid priming and preparation sequences, troubleshooting prompts, and automated alarms are a safety net for use, but are not absolute in respect to the human–machine interface. An inability to manage an alarm event where the blood pump stops can be the cause of circuit clotting and failure. No blood flow for 3–5 min or less can cause cell and plasma separation, clotting, and no recovery.

There are many strategies to train nurses for these events using a simulation set up of the machine and EC, interactive video activities of these alarms, and simple tutorial activities [10, 25]. All of these are useful towards providing safe and successful CRRT, and are another strategy for maintaining circuit patency.

With regard to training programs, bedside records of 'filter life' are useful and important data to review. This type of circuit patency or filter life audit provides useful feedback to teachers, particularly when circuit 'life' is poor (e.g. <3–4 h). Repeated events of this can reflect the adequacy of nursing education and training needs.

Conclusion

Clotting in the circuit during CRRT can be delayed or prevented by both the administration of anticoagulants and prevention of blood stasis and resistance in the circuit. The access catheter, blood flow, membrane and venous chamber are in a relationship and this has an association with circuit patency. The use of heparin during preparation by coating the circuit and administration of substitution as predilution are also useful strategies to prevent clotting. Training and education with audit of each circuit or filter 'life' are useful data to collect and provide feedback. These strategies promote safety and successful use where circuit patency is one measure of success.

References

1 Davenport A: The coagulation system in the critically ill patient with acute renal failure and the effect of an extracorporeal circuit. Am J Kidney Dis 1997;30(suppl 4):s20–s27.
2 Keller F, Seeman J, Preuschof L, Offermann G: Risk factors of system clotting in heparin free hemodialysis. Nephrol Dial Transplant 1990;5:802–807.
3 Oudemans-Van Straaten HM, Wester JPJ, de Pont ACJM, Schetz MRC: Anticoagulation strategies in continuous renal replacement therapy: can the choice be evidence based ? Intensive Care Med 2006;32:188–202.
4 Baldwin I, Bellomo R: The relationship between blood flow, access catheter and circuit failure during CRRT; a practical review; in Ronco C, Bellomo R, Brendolan A (eds): Sepsis, Kidney and Multiple Organ Failure. Contrib Nephrol. Basel, Karger, 2004, vol 144, pp 203–213.
5 Webb AR, Mythen, MG, Jacobsen D, Mackie IJ: Maintaining blood flow in the extracorporeal circuit: hemostasis and anticoagulation. Intensive Care Med 1995;21:84–93.
6 Bellomo R, Ronco C: Circulation of the continuous artificial kidney: blood flow, pressures, clearance and the search for the best; in Artigas A, Bellomo R, Ronco C (eds): Circulation in Native and Artificial Kidneys. Karger, Basel, 1997, pp 138–149.
7 Mitchell A, Daul AE, Beiderlinden M, Schafers RF, Heemann U, Kribben A, et al: A new system for regional citrate anticoagulation in continuous veno-venous hemodialysis (CVVHD). Clin Nephrol 2003;59:106–114.
8 Kutsogiannis DJ, Mayers I, Chin WD, Gibney RT: Regional citrate anticoagulation in continuous veno-venous hemodiafiltration. Am J Kidney Dis 2000;35:802–811.

9 Schetz M: Anticoagulation for continuous renal replacement therapy. Curr Opin Anaesthesiol 2001;14 143–149.
10 Baldwin IC, Elderkin TD: CVVH in intensive care. Nursing perspectives. New Horiz 1995;3:738–747.
11 Baldwin IC, Bridge NP, Elderkin TD: Nursing issues, practices, and perspectives for the management of continuous renal replacement therapy in the intensive care unit; in Bellomo R, Ronco C (eds): Critical Care Nephrology. Dordrecht, Kluwer Academic, 1998, pp 1309–1327.
12 Davenport A: Extracorporeal anticoagulation for intermittent and continuous forms of renal replacement therapy in the intensive care unit; in Murray PT, Brady HR, Hall JB (eds): Intensive Care Nephrology. London, Taylor & Francis, 2006, pp 165–180.
13 Lavaud S, Canivet E, Wuillai A, Maheut H, Randoux C, Bonnet J M, Renaux JL, Chanard J: Optimal anticoagulation strategy in haemodialysis with heparin coated polyacrylonitrile membrane. Nephrol Dial Transplant 2003;18:2097–2094.
14 Lavaud S, Paris B, Maheut H, Randoux C, Renaux JL, Rieu P, Chanard J: Assessment of the heparin-binding AN69 ST hemodialysis membrane: II. Clinical studies without heparin administration. ASAIO J 2005;51:348–351.
15 Jones CH: Continuous renal replacement therapy in acute renal failure: membranes for CRRT. Artif Organs 1998;22:2–7.
16 Salmon J, Cardigan R, Mackie I, Cohen SL, Machin S, Singer M: Continuous venovenous haemofiltration using polyacrylonitrile filters does not activate contact system and intrinsic coagulation pathways. Intensive Care Med 1997;23:38–43.
17 Frank RD, Weber J, Dresbach H, Thelan H, Weiss C, Floeg J: Role of contact system activation in hemodialyzer-induced thrombogenicity. Kidney Int 2001;60:1972–1981.
18 Hadef S, Raoudha G, Saoussen A, Imen K, Henda E, Jalel G, Abdelaziz H: A prospective study of the prevalence of heparin-induced antibodies and other associated thromboembolic risk factors in pediatric patients undergoing hemodialysis. Am J Hematol 2006;81:328–334.
19 Jenkins RD: The extra-corporeal circuit: physical principles and monitoring; in Ronco C, Bellomo R (eds): Critical Care Nephrology, ed 1. Dordrecht, Kluwer Academic, 1998, pp 1189–1197.
20 Uchino S, Fealy N, Baldwin I, Morimatsu H, Bellomo R: Pre-dilution vs. post-dilution during continuous veno-venous hemofiltration: impact on filter life and azotemic control. Nephron Clin Pract 2003;94:94–98.
21 Egi M, Naka T, Bellomo R, Cole L, French C, Trethewy C, Wan L, Langenberg CC, Fealy N, Baldwin I: A comparison of two citrate anticoagulation regimens for continuous veno-venous hemofiltration. J Artif Organs 2005;28:1211–1218.
22 Keller F, Seeman J, Preuschof L, Offermann G: Risk factors of system clotting in heparin free hemodialysis. Nephrol Dial Transplant 1990;5:802–807.
23 Baldwin I, Tan HK, Bridge N, Bellomo R: Possible strategies to prolong circuit life during hemofiltration: three controlled studies. Ren Fail 2002;24:839–848.
24 Ronco C: Fluid balance in CRRT: a call to attention! Int J Artif Organs 2005;28:763–764.
25 Baldwin I: Training management and credentialing for CRRT in critical care. Am J Kidney Dis 1997;30(suppl 4):S112–S116.

Prof. Ian Baldwin
Department of Intensive Care, Austin Hospital
Melbourne, Vic. 3084 (Australia)
Tel. +61 3 9496 5000, Fax +61 3 9496 3932, E-Mail ian.baldwin@austin.org.au

Starting Up a Continuous Renal Replacement Therapy Program on ICU

Wilfried De Becker

Department of Intensive Care, University Hospital Gasthuisberg, Leuven, Belgium

Abstract

Background/Aim: The questions as to which treatment is the most effective for the replacement of renal function in critically ill patients with acute renal failure and the qualifications needed by nurses to manage the continuous renal replacement therapy (CRRT) device are part of an ongoing debate between nephrologists and intensivists, between nurses of the renal ward and the ICU. **Methods:** The keys to a successful CRRT program are a well-balanced and practical education program, a user-friendly dialysis machine, and technical support 24 h/day. A computerized data management system will diminish the workload to an acceptable level. **Results:** Intensive care nurses on our ICUs are well trained to execute CRRT without the involvement of nephrology nurses. On the ICU, the 24-hour presence of an intensivist is an additional advantage to solve medical problems involving CRRT. The daily cost of CRRT is only dependent on the devices and independent of human resources. **Conclusion:** Initiating and maintaining a CRRT program is a great challenge for the ICU nurse. The possible problems remain within the ICU staff's ability to solve if they follow an education program. If the workload for the nurses is well monitored, extra personnel can be avoided.

Copyright © 2007 S. Karger AG, Basel

Introduction

The questions as to which treatment is the most effective for the replacement of renal function in critically ill patients with acute renal failure and the qualifications needed by the nurses to manage the continuous renal replacement therapy (CRRT) device is an ongoing debate between nephrologists and intensivists as well as between nurses in the renal ward and the ICU [2, 3, 8]. Some centers create a specialized CRRT team with nurses and physicians from both disciplines [3, 6]. Our experience is that all ICU nurses are well educated and well trained to perform all actions of the CRRT.

In 1980 continuous arteriovenous hemofiltration was introduced to our ICU. The intensivist installed the circuit and the nurse managed the fluid balance. A few years later we used continuous veno-venous hemofiltration CVVH with a roller pump device (BSM22™ by Gambro™). In the beginning the intensivists still installed the filter but step-by-step the nurses took over this procedure. The Prisma™ (Gambro) was introduced in 1994. All nurses were trained to do the installation and the physician only managed the settings and electrolyte balance. The responsibility and the workload for the ICU nurse increased and more education was needed. The physiology of renal failure, CRRT techniques, care plan and practical training were included in the education program.

The introduction of the new Prismaflex™ device (Gambro) was preceded by an update lesson about renal failure (1 h) by a member of the medical staff and one about the prevention of filter failure (1 h) by a staff nurse. A 2-hour practical training was organized for every nurse in small sessions of 6 people by a representative of Gambro.

Our Intensive Care Department has 56 ICU beds divided into 4 units: 2 general ICUs with 16 beds each; a pediatric ICU with 10 beds, and a burn unit with 14 beds. Each unit has a head nurse and 2 assistant head nurses. One of them has a special interest in CRRT and is also an instructor in the education program.

The nursing staff rate is 2.8 full time equivalents for each ICU bed. In practice every nurse takes care of 2 patients during every shift (morning, evening and night).

The medical staff members are intensivists who are well trained in the critical care of kidney failure.

Nurses and intensivists work complementarily in the handling of all CRRT tasks. So no dialysis nurse or nephrologist is needed in this setting. Only intermittent hemodialysis (IHD) is performed by staff from the dialysis department of our hospital.

The incidence of CVVH in 2006 was 4.9/day (0–9) and IHD was 2/day.

Keys to a Successful CRRT Program [5]

Physicians well trained in intensive care medicine and renal failure must be available on the ward 24 h/day so that nurses can easily ask their advice, even for minor problems involving the CRRT.

Intensive care nurses must be well trained in care plans for renal failure and CRRT techniques, and the CRRT devices must be user-friendly.

Table 1. The 4 modules of the education program

Module		Lessons	In-practice training
I	General intensive care basic I	85 h	121 h on general ICU
II	General intensive care basic II	80 h	121 h on general ICU
III	Emergency care	80 h	167 h on emergency ward
IV	Pediatric intensive care	36 h	121 h on PICU or NICU

Module II includes several topics about renal failure: physiology of kidney failure (1 h); epuration techniques (1 h); nursing care plan for CRRT (1.5 h); practical laboratory installation on the Prismaflex™ (1.5 h), and practice on an ICU in the presence of qualified nurses.

A CRRT education program using the 'Bachelor after bachelor in intensive care and emergency care' program [1] gives a theoretical background and large practical experience is offered to future ICU nurses.

The ICU nurse is supported by experienced staff nurses who are able to organize additional training on a regular basis.

A computerized monitoring system manages all pressures and fluid values in the patient's file. Full technical support must be offered by the manufacturer (24 h/day).

Education [1]

The education starts at nursing school and contains: anatomy and physiology of the kidney (4 h); 4 h of theory and 6 h of case study regarding pathology and nursing of patients with renal failure, especially chronic renal failure with intermittent hemodialysis and peritoneal dialysis, and acid-base balance (2 h).

Nurses who have completed the nursing school have a Bachelor Diploma in Nursing and they can continue with the Bachelor after Bachelor program.

The nursing high school which is affiliated with our university hospital organizes a Bachelor after Bachelor education program in 4 modules as shown in table 1.

This education program takes a whole year; in-practice training included, and is completed after an oral jury examination on the basis of cases from the work field. All the lecturers are physicians and nurses from our hospital, and most of them work on the intensive care or emergency ward of our hospital. After this course the nurses can easily find work on any ICU, PICU, NICU or emergency ward.

Equipment

The choice of an appropriate CRRT device is very important for the ICU nurse. Before changing from Prisma to Prismaflex, other manufacturers were also invited to demonstrated with their latest devices which we tested at the bedside with a small dedicated group of nurses. The test focused on the ergonomic aspects of the device and how easy it was to learn the filter set installation. The beta-trial of the Prismaflex device was also performed by this group. They all choose this device to succeed the Prisma because of the recognizable installation procedure and the new ergonomic aspects such as higher scales. The Prismaflex is a good example of a user-friendly machine: easy to learn, and an easy to install 'all-in-one' set with a 'step-by-step' explanation on the color display. Many tasks run automatically. All alarms and possible actions are clearly explained on the display. No handout is needed. All these advantages save time in the workload of the ICU nurse. If a dual lumen catheter is already in place, therapy can be installed and running 30 min after the decision to start CRRT is made.

Our department has 6 devices on standby and has the ability to extend this with 2 devices from the medical ICU. When the demand is larger we still have 2 Prisma devices and on exception we can call for support from Gambro to rent more devices.

Computerized Monitoring

Prismaflex is connectable to our patient data management system (PDMS; MetaVision™ of iMDsoft™). The data from settings, pressure monitoring, fluids, scales and therapy status plus additional fluid balance calculations and laboratory results are displayed every minute on the bedside computer screen. Pop-up warning menus are shown when some actions must be done. The run time of every filter is monitored and can be used as a quality parameter for CRRT. This advantage improves the quality of collecting data and significantly diminishes the workload for the nurse.

Tasks for the Nurses

The tasks of the nurses are the following: priming the circuit using 1 l of normal saline containing 5,000 IU heparin, an initial bolus of heparin is not used; installation of the circuit, input of the prescribed settings of CVVH and starting up; dialysis machine resetting after any trouble; replacement of exhausted fluids (substitution and anticoagulant); return blood before filter is clotted and re-initiate CRRT day and night; circuit removal and catheter refilling with anticoagulant after circuit clotting or major machine troubles day and night; monitoring pressures and fluid loss with a PDMS; fluid balance control, automatically calculated by the PDMS every hour; blood sampling for coagulation (aPTT control) and electrolytes every 4 h; catheter care, and assist the intensivist while placing the catheter.

Standard CVVH Settings

We simultaneously use two 5-liter bags of a commercialized electrolyte solution hanging on the dialysate and substitution scales. The effluent is captured in 5-liter bags. If heparin is prescribed we use 20-ml syringes (standard 2,000 IU/20 ml) with the infusion pump of the machine. The filter is a M100 (AN69 membrane), and the blood flow rate is 200–300 ml/min, the substitution rate is 2,000–3,500 ml/h, predilution is 30%, postdilution is 70%, and a 13-french dual-lumen venous catheter (Hemo-Access™, Gambro) is used.

Discussion

CVVH does not require extra personnel cost because the cost of an ICU nurse and intensivists are already calculated in the total budget of an ICU. 16% of the total cost can be saved [7]. When IHD therapy is needed, a dialysis nurse from the renal ward will do the whole procedure during 2–4 h. Sometimes they do two IHD therapies simultaneously.

The affinity of the ICU nurse to their patient is greater and this affects morbidity in a positive way [4].

Most ICU nurses have great technical skills with all kinds of ICU equipment. The more they use a CVVH device by installing the set, the better they can anticipate all kinds of alarms. This affects filter survival.

The down-time between two filters can be kept very low [9, 10] by changing the filter immediately after clotting by the ICU nurses themselves. They do not have to wait for a renal nurse and they can organize their own work better.

References

1 Deneire M: Bachelor after bachelor in the intensive care and emergency care: education program of the Catholic High School for Nursing Leuven (Belgium). European Credit Transfer System form, 2006.

2 Craig MA, Depner TA, Tweedy RL, Hokana L, Newby-Lintz M: Implementing a continuous renal replacement therapies program. Adv Ren Replace Ther 1996;3:348–350.
3 Gilbert RW, Caruso DM, Foster KN, Canulla MV, Nelson ML, Gilbert EA: Development of a continuous renal replacement program in critically ill patients. Am J Surg 2002;184:526–533.
4 Politoski G, Mayer B, Davy T, Swartz MD: Continuous renal replacement therapy. A national perspective AACN/NKF. Crit Care Nurs Clin North Am 1998;10:171–177.
5 Willis J, Hodge KS, Dwyer J: Keys for a successful CRRT educational program. Blood Purif 2004;22:229–247.
6 Hodge KS, Willis J, Dwyer J: Development of a consolidated CRRT program. Blood Purif 2004;22:229–247.
7 Vitale C, Bagnis C, Marangella M, Belloni G, Lupo M, Spina G, Bondonio P, Ramello A: Cost analysis of blood purification in intensive care units: continuous versus intermittent hemodiafiltration. J Nephrol 2003;16:572–579.
8 Ronco C, Zanella M, Brendolan A, Milan M, Canato G, Zamperetti N, Bellomo R: Management of severe acute renal failure in critically ill patients: an international survey in 245 centers. Nephrol Dial Transplant 2001;16:230–237.
9 Uchino S, Fealy N, Baldwin I, Morimatsu H, Bellomo R: Continuous is not continuous: the incidence and impact of circuit 'down-time' on uraemic control during continuous veno-venous haemofiltration. Intensive Care Med 2003;29:575–578.
10 Baldwin I, Bellomo R, Koch B: Blood flow reductions during continuous renal replacement therapy and circuit life. Intensive Care Med 2004;30:2074–2079.

Wilfried De Becker, RN, CCRN
Department of Intensive Care, University Hospital Gasthuisberg
Herestraat 49
BE–3000 Leuven (Belgium)
Tel. +32 16 344778, Fax +32 16 344015, E-Mail wilfried.debecker@uz.kuleuven.ac.be

Is There a Need for a Nurse Emergency Team for Continuous Renal Replacement Therapy?

Ian Baldwin

Department of Intensive Care, Austin Hospital, and Department of Nursing and Health Sciences, RMIT University, Melbourne, Vic., Australia

Abstract

The use of an emergency response team for unwell patients has provided an improvement in hospital care standards by reducing medical and postoperative adverse outcomes. Use of a nurse emergency team for patients treated with continuous renal replacement therapy (CRRT) also has potential to reduce adverse outcomes with CRRT, where staff may lack experience or find troubleshooting CRRT difficult in an ICU with many critically ill patients in their care. Differing nursing models are used to provide CRRT in the ICU, and all of these could benefit from a nursing response team at some time. The response must be immediate, with suitably available and CRRT-experienced nurses. As with medical emergency team use, the nursing emergency team for CRRT would be called when a deviation from a standard criterion list occurs. The list could include: prolonged blood pump stoppage ($\sim$2 min); air detection alarm; blood leakage; sudden circuit pressure changes – transmembrane pressure ($>$200 mm Hg) or venous pressure ($>$200 mm Hg) or arterial pressure negative ($\geq$100 mm Hg); the need to override a fluid balance alarm 3 times in 5 min; patient hypotension; cardiac arrest or similar event, or the nurse is concerned that the machine is malfunctioning. The 'human resource' is the biggest challenge to developing a suitable response team 24/7, however where ICU and nephrology nurses work in a collaborative approach for CRRT, a response team would be more easily established and may not be required continuously.

Copyright © 2007 S. Karger AG, Basel

The use of a rapid response team, with acute care skills, can better manage the unwell patient in a hospital ward or subacute area by preventing cardiac and respiratory arrest [1]. The team is called usually via the hospital public address system or a pager text when patient carers detect changes in vital signs reflecting a deterioration and/or they are simply 'worried' about the patient [2]. This approach (Medical Emergency Team, MET) is being adopted in hospitals to prevent the traditional 'crisis' intervention associated with cardiac or respiratory arrest [3].

CRRT Emergency Team

Some parallels may be made between the unwell patient on a hospital ward and the continuous renal replacement therapy (CRRT) machine when in use on a critically ill patient. Both have a blood pump system with measurable pressures reflecting a normal and abnormal state. Changes in 'vital signs' can reflect a deterioration that, if not corrected, may cause the 'pump' to stop and death to occur.

In the context of CRRT, prolonged blood pump stoppage in response to acute circuit pressure changes may cause a clotting event and an extracorporeal circuit 'death' [4–6]. This will mean a loss of treatment, a need to replace the circuit and membrane, and commonly the loss of patient blood. All of these are undesirable, reflect inefficiency, and are potentially preventable.

If subtle and less acute changes in measured circuit pressures are identified and corrected, this event may not occur; the cardiac arrest is prevented. Is there a need for a similar response team to better manage CRRT?

Nursing Models and CRRT

Different nursing models are used to apply CRRT to the critically ill patient in the ICU. Nursing management and care of the CRRT can be provided by ICU nurses alone, a mixed or collaborative arrangement with the ICU and nephrology unit nurses, or by nephrology nurses only [7, 8].

The continuous application of the treatment requires a high percentage of staff to be trained and skilled, and in some centers this has been impossible [9]. With increasing use of CRRT in the ICU it can be difficult to provide enough ICU nurses with CRRT training, for a 24-hour, 7-days/week use of CRRT with multiple patients being treated. This can also be a challenge for the collaborative models. In contrast, when a small number of patients are treated, knowledge and skills can be difficult to acquire and maintain.

Therefore regardless of nursing models, situations can occur where inexperienced staff are managing a patient on CRRT. In addition, skilled nurses may be caring for an acutely unstable patient or more than one critically ill patient, and their attention to CRRT may be compromised.

CRRT Machines and Alarm Events

The design of new CRRT machines has made preparation routines more simple and automated, alarm systems that set and adjust automatically, and user interfaces with messages and prompts for potential malfunctions [10].

However, when the blood pump stops with an alarm event, rapid and skilled intervention may be required. This may be due to a circuit pressure change, an air bubble detected, or fluid flow problem [5]. However if the alarm is 'latched' the blood pump will stop and not restart until the situation is corrected and the alarm reset [5]. Without a rapid response, secondary alarms may then begin such that multiple alarms are occurring making troubleshooting the task for an expert.

In addition to blood pump stoppage, other acute alarm events are important to manage appropriately as these can cause serious harm to the critically ill patient, e.g. continued override and ignoring a fluid balance alarm may cause hypovolemia and cardiac arrest [11, 12].

CRRT Nurse Emergency Team

The intentions of the MET are to provide skilled help quickly to an immediate need, where an acute deterioration is thought to be imminent. In addition, the MET provides an education role, is collaborative and makes useful human links for a better hospital where knowledge is disseminated outside the closed walls of a specialized area [2]. It is important to acknowledge that a MET service is not to reduce knowledge of carers or take over their skill development. The MET is not a forum for grandstanding and promoting superiority [13], creating a culture of hesitation where staff feel they will be criticized after making a call [14].

The use of a similar service for CRRT could benefit patients and nurses in many centers. The criteria for calling the team may vary depending on the setting, however table 1 is a suggested list potentially applicable to all. The main focus of the criteria would be to prevent blood pump stoppage and serious adverse patient events related to fluid balance and circuit function. Routine preparation, checking, settings and disconnection routines are not activities where a CRRT emergency team would be called. These skills may require a

Table 1. Nurse emergency team CRRT: criteria for calling

Call for the CRRT nurse emergency team if:
- Blood pump stoppage and unable to restart before ~2 min
- Significant air in the circuit
- Blood leakage anywhere
- Sudden rise in either transmembrane pressure (>200 mm Hg) or venous pressure (>200 mm Hg) or arterial pressure negative (≥100 mm Hg)
- Override 'fluid balance' alarm 3 times in 5 min
- Significant patient hypotension
- At cardiac arrest or similar
- You are concerned or worried the machine is malfunctioning

support and education service, and constitute areas for basic and ongoing learning with CRRT in a classroom simulation or similar.

Challenges to a Nurse Emergency Team for CRRT

A MET service relies on suitably qualified people who respond quickly to help 24 h, 7 days/week. The nursing or medical responders must be readily available to attend the call, and have a high level of knowledge. This suggests that they either do not have a patient allocation or can be backed up immediately during their absence for potentially long periods. Is this possible for a CRRT emergency team? Which nurses would fit this demand? In addition, would they have the knowledge necessary? For example, if a nephrology nursing group becomes the CRRT response team, they would be required to leave a patient, during dialysis, be backed up by others while they were absent, and then have the knowledge for a different machine technology used for CRRT in the ICU.

Yes, this is possible in some settings providing the human resources are provided, and appropriate training and expertise is obtained by the nurse. Where, collaborative models are currently used [8], this would be more easily achieved as the human resources and knowledge required may already exist; nephrology nurses have a role in the ICU and experience with the CRRT machine. Where an ICU manages CRRT alone, these factors may be more challenging to achieve, suggesting that a response team may be very difficult and/or not worthwhile. This is possibly why many ICU areas have adopted a daily dialysis (SLED) approach during the day as this reduces the training requirement of CRRT and allows a more achievable collaborative situation with nephrology nurses [15–17], particularly when machine technology is very similar in both areas.

Conclusion

The use of a nurse emergency team has potential to apply CRRT in the ICU with more success, reduce circuit failure and prevent mistakes or serious adverse events. A similar approach to the MET established for unwell patients on a general ward would ask suitable nurses to rapidly attend the ICU and assist others managing CRRT. This could be to a criteria or list including contextual factors of a nurses 'concern' and/or aberrations of measured circuit pressures and/or alarms suggesting incorrect function. Such a team would need to respond quickly, being able to leave their area and have the knowledge required on arrival. This concept would be more easily applied in hospitals where collaborative models of nursing for CRRT in the ICU already exist. The availability of nurses or the human resource is the most important key to success.

References

1 Buist MD, Moore GE, Bernard SA, et al: Effects of a medical emergency team on reduction in incidence of and mortality from unexpected cardiac arrest in hospital: preliminary study. BMJ 2002;324:387–390.
2 Bellomo R, Goldsmith D, Uchino S, Buckmaster J, et al: Prospective controlled trial of the effect medical emergency team on postoperative morbidity and mortality rates. Crit Care Med 2004;32:916–921.
3 DeVita M: Medical emergency teams: deciphering clues to crises in hospitals. Crit Care 2005;9:325–326.
4 Webb AR, Mythen MG, Jacobsen D, Mackie IJ: Maintaining blood flow in the extracorporeal circuit: haemostasis and anticoagulation. Intensive Care Med 1995;21:84–93.
5 Baldwin IC, Elderkin TD: CVVH in intensive care. Nursing perspectives. New Horiz 1995;3: 738–747.
6 Baldwin IC, Bridge NP, Elderkin TD: Nursing issues, practices, and perspectives for the management of continuous renal replacement therapy in the intensive care unit; in Bellomo R, Ronco C (eds): Critical Care Nephrology. Dordrecht, Kluwer Academic, 1998, pp 1309–1327.
7 Giuliano K, Pysznik E: Renal replacement therapy in critical care: implementation of a unit-based continuous venovenous hemodialysis program. Crit Care Nurse 1998;18:40–51.
8 Dirkes S: Continuous renal replacement therapy: dialytic therapy for acute renal failure in intensive care. Nephrol Nurs J 2000;27:6:581–586.
9 Kihara M, Ikeda Y, Shibata K, Masumori S, Fujita H, Ebira H, Toya Y, Takagi N, Shionoiri H, Umemura S, Ishii M: Slow hemodialysis performed during the day in managing renal failure in critically ill patient. Nephron 1994;67:36–41.
10 Ronco C, Brendolan A, Dan M, Piccinni P, Bellomo R: Machines for continuous renal replacement therapy; in Ronco C, Bellomo R, La Greca G (eds): Blood Purification in Intensive Care. Contrib Nephrol. Basel, Karger, 2000, vol 132, pp 323–334.
11 Schultz D: Gambro Prisma Continuous Renal Replacement System. FDA Updated Public Health Notification. 2005: Retrieved December 19th, 2006, from http:www.fda.gov/cdrh/safety/022706 – gambro.html

12 Ronco C: Fluid balance in CRRT: a call to attention! Int J Artif Organs 2005;28:8:763–764.
13 Foraida MI, DeVita MA, Braithwaite S, et al: Improving the utilization of medical crisis teams (condition C) at an urban tertiary care hospital. J Crit Care 2003;19:87–94.
14 Jones D, Baldwin I, McIntyre T, Story D, Mercer I, Miglic A, Goldsmith D, Bellomo R: Nurses attitudes to a medical emergency team service in a teaching hospital. Qual Saf Health Care 2006;15:427–432.
15 Schlaeper C, Amerling R, Manns M, Levin N: High clearance continuous renal replacement therapy with a modified dialysis machine. Kidney Int 1999;56(suppl 72):S20–S23.
16 Kumar VA, Craig M, Depner TA, Yeun JY: Extended daily dialysis: a new approach to renal replacement for acute renal failure in the intensive care unit. Am J Kidney Dis 2000;36:294–300.
17 Marshall MR, Golper TA, Shaver MJ, Chatoth DK: Hybrid renal replacement modalities for the critically ill; in Ronco C, Bellomo R, La Greca G (eds): Blood Purification in Intensive Care. Contrib Nephrol. Basel, Karger, 2001, vol 132, pp 252–257.

Prof. Ian Baldwin
Department of Intensive Care, Austin Hospital
Melbourne, Vic. 3084 (Australia)
Tel. +61 3 9496 5000, Fax +61 3 9496 3932, E-Mail ian.baldwin@austin.org.au

Information Technology for CRRT and Dose Delivery Calculator

Zaccaria Ricci[a,b], *Claudio Ronco*[b]

[a]Department of Pediatric Cardiology and Cardiac Surgery, Bambino Gesù Hospital, Rome, and [b]Department of Nephrology, Dialysis and Transplantation, St. Bortolo Hospital, Vicenza, Italy

Abstract

Background: The application of information technology (IT) to the field of critical care nephrology is a process that may reduce errors in care delivery, improve monitoring, decrease unintentional practice variation, increase the quality and accuracy of delivered treatments. **Methods:** This review presents some examples of potential applications of recent IT achievements to clinical practice. **Results:** The adequacy calculator for continuous therapy dose prescription was recently shown to accurately predict urea clearance. When clearances above 60 ml/min where prescribed, the calculator tended to overestimate effective clearances; this overestimation generally remained within an error of 15%. Nevertheless, the delivered Kt/V in 24 h will always approach the target value of 1.2. The use of the calculator enabled strict monitoring of treatments. Furthermore, the so-called 'next generation' machines have technical characteristics in common that allow the highest safety and accuracy levels: some of these aspects are addressed and commented on in the present review. **Conclusion:** IT is having and will likely have a significant impact on patient safety, practice variation, patient assessment and monitoring, and documentation of the demographics of acute renal failure and dialysis. One of the most recent and potentially interesting aspects of IT implementation on acute dialysis might be the renal replacement dose monitoring and calculation: close control of the therapy delivery and, eventually, prescription adjustments might be optimized.

Copyright © 2007 S. Karger AG, Basel

Introduction

Information technology (IT) in the medical field is a synonym for technical modernization and improvement in medical equipment. Over the last few years, the application of IT to the field of critical care nephrology has been an ongoing process and, when properly utilized, it may result in both a better understanding of the disease as well as in improvements in patient outcomes. Some fundamental targets and priorities of IT have recently been shown by the Acute Dialysis Quality Initiative (ADQI) [1]. These aspects can be summarized by four points: (i) current and future IT innovations should be used to reduce errors in care delivery which could potentially lead to patient harm; (ii) IT should improve monitoring of the current practice of acute dialysis; (iii) IT should be implemented to reduce unintentional practice variation without limiting practice preferences, and (iv) advances in technology should be applied to acute dialysis pump systems in order to improve the quality of acute dialysis care delivery.

Application of IT to Dialysis Dose Calculation

Software called the 'Adequacy Calculator for ARF' has recently been described [2]. This is a simple Microsoft Excel-based program [3] that calculates urea clearance and estimates fractional clearance and Kt/V_{urea} for all continuous renal replacement therapy (CRRT) modalities. The calculator works on the assumption that the urea sieving coefficient is 1 for convective therapies and that complete saturation of spent dialysate occurs during continuous dialysis. In a pilot evaluation of a small cohort of patients, the value of clearance predicted by the calculator correlated significantly to the value obtained from direct blood and dialysate determination during the first 24 h of treatment, regardless of the renal replacement modality used. Renal replacement therapy (RRT), in fact, consists of various modalities which differ in many features including the continuity of treatment (intermittent vs. continuous), vascular access (arteriovenous vs. veno-venous), and the mechanism of solute removal (diffusion vs. convection). Accordingly, it is difficult to find an ideal marker and an universal method to compare the doses of these different treatments. Using urea as a marker molecule, as is done in chronic hemodialysis, treatment dose of CRRT can be defined by various aspects such as efficiency, intensity, frequency, and clinical efficacy [4]. Of course, urea simply represents a surrogate of the low molecular weight toxin. The major shortcoming of the traditional solute marker-based approach to dialysis dose in acute renal failure (ARF) lies well beyond any methodological critique of single-solute kinetics-based thinking. In patients with ARF, the majority of whom are in intensive care, a restrictive (solute-based only) concept of dialysis dose seems grossly inappropriate. In these patients the therapeutic needs that can/need to be affected by the 'dose' of RRT are much more than the simple control of small solutes as represented by urea. They include control of acid-base, tonicity, potassium, magnesium, calcium and phosphate, intravascular volume, extravascular volume and temperature;

furthermore, the avoidance of unwanted side effects associated with the delivery of solute control must be considered [4].

Nonetheless, the possibility of implementing a simple software in the routine management of CRRT prescription is an ideal example of IT: the calculator can be used to reduce errors in care delivery (mostly, undertreatment or fluid balance errors); it would improve monitoring of the actual delivery of RRT sessions; it could truly reduce unintentional practice variation without limiting practice preferences, and if directly implemented in dialysis machine monitors, the calculator would certainly improve the quality of acute dialysis care delivery. This tool or its future developments would finally help operators in the field of prescription, delivery and monitoring of RRT, would improve standardization of dose selection, and could potentially facilitate dialysis prescription in a large scale trial on RRT dose.

In chronic hemodialysis, treatment dose of RRT is defined as a fractional clearance, Kt/V, where K is the instantaneous clearance, t is treatment time and V is the volume of distribution of the marker molecule. This is a dimensionless parameter that represents the *efficacy* of treatments, and allows comparison among different therapies and among different patients. In fact, different instantaneous clearances, representing treatment *efficiency*, can yield to comparable results in terms of efficacy only if correlated to treatment time and patient total body weight. A Kt/V value of 1.2 is an established marker of adequacy shown to correlate with morbidity and mortality in patients with end-stage kidney disease [5, 6]. Kt/V has not yet been validated as a marker of adequacy in ARF patients but it seems that a good rationale exists for its use in continuous therapies. Theoretically speaking, in its original concept, clearance was signed to evaluate renal function among disparate individuals where, however, function was operating 24 h/day and blood levels were at steady state. This is the reason why, after some days of CRRT, patients' urea levels approach a real steady state (never obtained in the intermittent dialysis population) and post-dialysis rebound is not present. It is finally reasonable to consider urea distribution equivalent in total body weight, as in the case of a single pool kinetic model (spKt/V). Ronco et al. [7] demonstrated an improved outcome with post-dilution hemofiltration delivered at $35\,\text{ml}\cdot\text{h}^{-1}\cdot\text{kg}^{-1}$ in a population of 450 patients. Setting a spKt/V threshold that could guide clinicians towards adequate treatments, we could possibly meet the target of $35\,\text{ml}\cdot\text{h}^{-1}\cdot\text{kg}^{-1}$ which, delivered as 24-hour treatment, may translate into a spKt/V of 1.4 independent of the RRT modality.

We tested the adequacy calculator [2] and found that it was able to accurately predict the delivered urea clearance, apart from which the CRRT modality was selected; the correlation between prediction and effective delivery remained high in a range of time of 24 h. When clearances above 60 ml/min

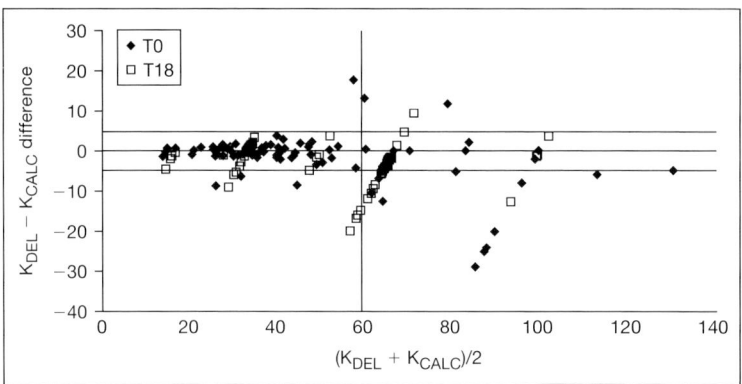

Fig. 1. Bland-Altman analysis illustrates correlation between urea clearance obtained by the two methods. Urea clearance calculated using the described software (K_{CALC}) and urea clearance obtained by direct measurement on pre-filter blood and effluent samples (K_{DEL}). The mean $K_{CALC} - K_{DEL}$ value for each treatment is presented on the x axis; the difference between both methods is shown on the y axis. Parallel lines indicate the standard deviation. The difference between K_{CALC} and K_{DEL} tends to increase for mean $K_{CALC} - K_{DEL}$ values above 60 ml/min (vertical line). It is possible to distinguish the correlations between K_{CALC} and K_{DEL} at the start of therapy (T0) and after 18–24 h of uninterrupted therapy (T18).

where prescribed, the calculator showed a tendency to overestimate effective clearances: this overestimation remained generally within an error of 15% (fig. 1). Considering our results and the dissociation between treatment delivery and calculator estimation when high clearances are involved, as could be the case of slow efficiency extended dialysis, a slight correction to prevent overestimation of effective treatment delivery is strongly advised. Nevertheless, even in the presence of an error of up to 15%, the delivered Kt/V in 24 h will always approach the target value of 1.2. The use of the calculator allowed us to strictly monitor our treatments during the study period, and described an average 10.7% ($p < 0.05$) reduction in therapy delivery, when compared to the prescribed dose. In our population, this delivery reduction was mainly due to operative treatment time, often shorter than the prescribed treatment time (during the substitution of bags and filter change, treatment is not administered). Our observation is consistent with a recent large retrospective analysis [8]. In this setting, when a 'standardized' downtime is foreseen, treatment prescription might be adjusted to correct for the time of zero clearance.

However, all these considerations must be seen in the light of an absolute lack of any previous attempt to adjust treatment dose to specific target levels. Furthermore, a clear understanding of the adequate levels of RRT has still to be achieved. Despite all the uncertainty surrounding the meaning of adequate

CRRT dose and the gross shortcomings related to its accuracy in patients with ARF, the idea that there might be an optimal dose of solute removal continues to have a powerful hold in the literature. According to some authors this concept seems optimistic [4]: nonetheless, a dose prescription should be made after the indication for an extracorporeal blood purification technique is given and the treatment dose delivered should be carefully monitored.

Application of IT to Dialysis Accuracy and Safety

A new generation of CRRT equipment is being developed more than 30 years after the first continuous treatment was instituted. Today technological evolution is supporting new clinical indications, new safety and accuracy standards while also trying to reduce the workload for the operators. The so-called 'next generation' machines [9] have in common technical characteristics of high-level safety requirements, pressure measurements of all crucial segments of the circuit (catheter inlet and outlet, filter inlet and outlet, ultrafiltrate and dialysate ports), accurate ultrafiltration and therapy delivery control, obtained by four or more roller volumetric pumps (blood, replacement, dialysate and effluent), and ultra-precise scales. The accuracies of these systems allow errors in fluid administration/ultrafiltration of a few grams per hour. Another important feature of this integrated systems is a user-friendly operator interface. These monitors perform a complete range of therapies including slow continuous ultrafiltration, continuous veno-venous hemofiltration, continuous veno-venous hemodialysis, continuous veno-venous hemodiafiltration, and plasma exchange. Complete monitoring of fluid balance and all important events is provided by continuous recording of the history of the last 24 h (or more) of treatment. When an alarm occurs a message on the screen suggests the most appropriate intervention required. Blood flow ranges from 10 to >400 ml/min. Hemofiltration and dialysate range from 0 to about 10 l/h. The possibility of making fluid balance errors during CRRT has been recognized from the beginning of this treatment strategy. The possibility of errors is continuous, and hardware and circuits are highly challenged by the uninterrupted utilization. The advent of automated machines and the implementation of IT has partially overcome this problem. Nevertheless, there are conditions and operation modes in which the potential for errors is still present. In particular, a fluid balance error can lead to fatal outcomes [10]. Features and alarms, even the safest, can be manipulated by operators creating the opportunity for serious errors. Physicians and nurses involved in the prescription and delivery of CRRT should have precise protocols and defined procedures in relation to machine alarms to prevent major clinical problems. There is no solution to the unwise utilization of a perfect system.

Conclusion

IT is having and will likely continue to have a significant impact on patient safety, practice variation, patient assessment and monitoring, and documentation of the demographics of ARF and dialysis. Continuing work in these areas is necessary together with repeated assessment of the targets achieved in order to augment IT effects on routine practice. While increased system accuracy and strict safety controls are already a reality, dialysis dose monitoring and calculation might be an interesting implementation in future developments of CRRT machines: treatment prescription could theoretically be guided by a 'calculator' in successive steps before RRT is started. Such semi-automated treatment should implement a close control of therapy delivery and eventually, in a closed loop, suggest prescription adjustments when the target of the dose is not reached.

References

1 Savage B, Marquardt GW, Paolini F, Schlaeper C: Information technology and acute dialysis. Curr Opin Crit Care 2002;8:544–548.
2 Ricci Z, Salvatori G, Bonello M, Bolgan I, D'Amico G, Dan M, Piccinni P, Ronco C: In vivo validation of the adequacy calculator for continuous renal replacement therapies. Crit Care 2005;9:R266–R273.
3 Pisitkun T, Tiranathanagul K, Poulin S, et al: A practical tool for determining the adequacy of renal replacement therapy in acute renal failure patients; in Ronco C, Bellomo R, Brendolan A (eds): Sepsis, Kidney and Multiple Organ Dysfunction. Contrib Nephrol. Basel, Karger, 2004, vol 144, pp 329–349.
4 Ricci Z, Bellomo R, Ronco C: Dose of dialysis in acute renal failure. Clin J Am Soc Nephrol 2006;1:380–388.
5 NKF/DOQI Clinical practice guidelines for haemodialysis adequacy: updater 2000. Am J Kidney Dis 2001;37(suppl 1):S7–S64.
6 Owen WF Jr, Chertow GM, Lazarus JM, Lowrie EG: Dose of haemodialysis and survival: differences by race and sex. JAMA 1998;280:1764–1768.
7 Ronco C, Bellomo R, Homel P, Brendolan A, Dan M, Piccini P, La Greca G: Effects of different doses in continuous veno-venous haemofiltration on outcomes of acute renal failure: a prospective randomised trial. Lancet 2000;355:26–30.
8 Venkataraman R, Kellum JA, Palevsky P: Dosing patterns for CRRT at a large academic medical center in the United States. J Crit Care 2002;17:246–250.
9 Ricci Z, Salvatori G, Bonello M, Ratanarat R, Andrikos E, Dan M, Piccinni P, Ronco C: A new machine for continuous renal replacement therapy: from development to clinical testing. Expert Rev Med Devices 2005;2:47–55.
10 Ronco C, Ricci Z, Bellomo R, Baldwin I, Kellum J: Management of fluid balance in CRRT: a technical approach. Int J Artif Organs 2005;28:765–776.

Zaccaria Ricci
Department of Pediatric Cardiology and Cardiac Surgery, Bambino Gesù Hospital
Piazza S. Onofrio
IT–00100 Rome (Italy)
Tel. +39 06 6859 3333, E-Mail z.ricci@libero.it

Emerging Biomarkers of Acute Kidney Injury

Prasad Devarajan

Nephrology and Hypertension, Cincinnati Children's Hospital Medical Center, University of Cincinnati, Cincinnati, Ohio, USA

Abstract

Background: Acute kidney injury (AKI) is a major clinical problem with a rising incidence and high mortality rate. The lack of early biomarkers has resulted in an unacceptable delay in initiating therapies. **Methods:** Here we will update the reader on promising new blood and urinary biomarkers that have recently emerged through the application of innovative technologies such as functional genomics and proteomics to human and animal models of AKI. **Results:** The most promising biomarkers of AKI for clinical use include a plasma panel (NGAL and cystatin C) and a urine panel (NGAL, Il-18 and KIM-1). **Conclusions:** As they represent tandem biomarkers, it is likely that the AKI panels will be useful for timing the initial insult and assessing the duration and severity of AKI. Based on the differential expression of the biomarkers, it is also likely that the AKI panels will distinguish between the various types and etiologies of AKI. It will be important in future studies to validate the sensitivity and specificity of these biomarker panels in clinical samples from large cohorts and from multiple clinical situations.

Copyright © 2007 S. Karger AG, Basel

Acute kidney injury (AKI), previously referred to as acute renal failure (ARF), is a significant and devastating problem in clinical medicine [1–4]. The incidence of AKI varies from 5% in hospitalized patients to 30–50% of patients in intensive care units, and there is now substantial evidence that the incidence is rising at an alarming rate. Despite significant improvements in therapeutics, the mortality and morbidity associated with AKI remain dismally high. Outstanding advances in basic research have illuminated the pathogenesis of AKI and have paved the way for successful therapeutic approaches in animal models. However, translational research efforts in humans have yielded extremely disappointing results. A major reason for this is the lack of early markers for AKI, akin to troponins in acute myocardial disease, and hence an unacceptable delay in initiating therapy [5–7]. In current clinical practice, AKI is typically diagnosed by measuring serum creatinine. Unfortunately, creatinine is an unreliable indicator during acute changes in kidney function [8]. First, serum creatinine concentrations may not change until about 50% of kidney function has already been lost. Second, serum creatinine does not accurately depict kidney function until a steady state has been reached, which may require several days. However, animal studies have shown that while AKI can be prevented and/or treated by several maneuvers, these must be instituted very early after the insult, well before the rise in serum creatinine. The lack of early biomarkers of AKI in humans has hitherto crippled our ability to launch potentially effective therapies in a timely manner. Indeed, human investigations have now clearly established that earlier intervention improves the chance of ameliorating renal dysfunction [5]. The lack of early biomarkers has negatively impacted on a number of landmark clinical trials investigating highly promising therapies for AKI [9, 10].

In addition to aiding in the early diagnosis and prediction, biomarkers may serve several additional purposes in AKI. Thus, biomarkers are also needed for: (a) discerning AKI subtypes (pre-renal, intrinsic renal, or post-renal); (b) identifying AKI etiologies (ischemia, toxins, sepsis, or a combination); (c) differentiating AKI from other forms of acute kidney disease (urinary tract infection, glomerulonephritis, interstitial nephritis); (d) predicting the AKI severity (risk stratification for prognostication as well as to guide therapy); (e) monitoring the course of AKI, and (f) monitoring the response to AKI interventions. Furthermore, AKI biomarkers may play a critical role in expediting the drug development process. The Critical Path Initiative issued by the FDA in 2004 stated that 'Additional biomarkers (quantitative measures of biologic effects that provide informative links between mechanism of action and clinical effectiveness) and additional surrogate markers (quantitative measures that can predict effectiveness) are needed to guide product development'. Identification of novel AKI biomarkers has been designated as a top priority by the American Society of Nephrology [11]. The concept of developing a new toolbox for earlier diagnosis of disease states is also prominently featured in the NIH Road Map for biomedical research [12].

Desirable characteristics of clinically applicable AKI biomarkers include: (a) they should be noninvasive and easy to perform at the bedside or in a standard clinical laboratory, using easily accessible samples such as blood or urine; (b) they should be rapidly and reliably measurable using a standardized assay platform; (c) they should be highly sensitive to facilitate early detection, and with a wide dynamic range and cutoff values that allow risk stratification, and (d) they should be highly specific for AKI, and allow the identification of AKI subtypes and etiologies. This will almost certainly involve a combination of a panel of biomarkers, along with clinical information.

Table 1. Current status of promising AKI biomarkers in various clinical situations

Biomarker name	Sample source	Cardiac surgery	Contrast nephropathy	Sepsis or ICU	Kidney transplant	Commercial test?
NGAL	Plasma	Early	Early	Early	Early	Biosite[a]
Cystatin C	Plasma	Intermediate	Intermediate	Intermediate	Intermediate	Dade-Behring
NGAL	Urine	Early	Early	Early	Early	Abbott[a]
IL-18	Urine	Intermediate	Absent	Intermediate	Intermediate	None
KIM-1	Urine	Intermediate	Not tested	Not tested	Not tested	None

[a] In development.

The quest for AKI biomarkers is an area of intense contemporary research [13–15]. Conventional urinary biomarkers such as casts and fractional excretion of sodium have been insensitive and nonspecific in the clinical setting of AKI. Other traditional urinary biomarkers such as filtered high molecular weight proteins and tubular proteins or enzymes have also suffered from a lack of specificity and a dearth of standardized assays [15]. Fortunately, the application of innovative technologies such as cDNA microarrays and proteomics to human and animal models of AKI has uncovered several novel genes and gene products that are emerging as biomarkers [13, 16]. The most promising of these are outlined in table 1, and their current status in human AKI is chronicled in this article.

Novel AKI Biomarkers under Evaluation in Humans

Neutrophil Gelatinase-Associated Lipocalin

Human neutrophil gelatinase-associated lipocalin (NGAL) was originally identified as a 25-kDa protein covalently bound to gelatinase from neutrophils. NGAL is normally expressed at very low levels in several human tissues, including kidneys, lungs, stomach, and colon. NGAL expression is markedly induced in injured epithelia. For example, NGAL concentrations are elevated in the serum of patients with acute bacterial infections, the sputum of subjects with asthma or chronic obstructive pulmonary disease, and the bronchial fluid from the emphysematous lung [17]. NGAL was recently identified by microarray analysis as one of the earliest and most robustly induced genes and proteins in the kidney after ischemic or nephrotoxic injury in animal models, and NGAL protein was easily detected in the blood and urine soon after AKI [18–22].

These findings have spawned a number of translational studies to evaluate NGAL as a novel biomarker of human AKI.

In a cross-sectional study, human adults in the intensive care unit with established ARF (defined as a doubling of the serum creatinine in less than 5 days) secondary to sepsis, ischemia, or nephrotoxins displayed a greater than 10-fold increase in plasma NGAL and more than a 100-fold increase in urine NGAL by Western blotting when compared to normal controls [21]. Both plasma and urine NGAL correlated highly with serum creatinine levels. Kidney biopsies in these patients showed intense accumulation of immunoreactive NGAL in 50% of the cortical tubules. These results identified NGAL as a widespread and sensitive response to established AKI in humans.

In a prospective study of children undergoing cardiopulmonary bypass, AKI (defined as a 50% increase in serum creatinine) occurred in 28% of the subjects, but the diagnosis using serum creatinine was only possible 1–3 days after surgery [23]. In marked contrast, NGAL measurements by Western blotting and ELISA revealed a robust 10-fold or more increase in the urine and plasma within 2–6 h of surgery in patients who subsequently developed AKI. Both urine and plasma NGAL were powerful independent predictors of AKI, with an outstanding area under the curve (AUC) of 0.998 for the 2-hour urine NGAL and 0.91 for the 2-hour plasma NGAL measurement [23]. Thus, plasma and urine NGAL emerged as sensitive, specific, and highly predictive early biomarkers of AKI after cardiac surgery in children. It should be emphasized that the patients in this study were primarily children with congenital heart disease, who lacked many of the common comorbid conditions (such as diabetes, hypertension, and atherosclerosis) that are frequently encountered in adults. Nevertheless, these findings have now been confirmed in a prospective study in adults who developed AKI after cardiac surgery, in whom urinary NGAL was significantly elevated 1–3 h after the operation [24]. AKI, defined as a 50% increase in serum creatinine, did not occur until the 3rd postoperative day. However, patients who did not encounter AKI also displayed a significant increase in urine NGAL in the early postoperative period, although to a much lesser degree than in those who subsequently developed AKI. The AUC reported in the adult study was 0.74 for the 3-hour NGAL and 0.80 for the 18-hour NGAL, which is perhaps reflective of the confounding variables one typically accumulates as we age.

NGAL has also been evaluated as a biomarker of AKI in kidney transplantation. Biopsies of kidneys obtained 1 h after vascular anastomosis revealed a significant correlation between NGAL staining intensity and the subsequent development of delayed graft function [25]. In a prospective multicenter study of children and adults, urine NGAL levels in samples collected on the day of transplant clearly identified cadaveric kidney recipients who subsequently developed delayed graft function and dialysis requirement (which typically occurred 2–4 days later). The receiver-operating characteristic curve for prediction of delayed graft function based on urine NGAL at day 0 showed an AUC of 0.9, indicative of an excellent predictive biomarker [26]. Urine NGAL has also been shown to predict the severity of AKI and dialysis requirement in a multicenter study of children with diarrhea-associated hemolytic uremic syndrome [27]. Preliminary results also suggest that plasma and urine NGAL measurements represent predictive biomarkers of AKI following contrast administration [28–30] and in the intensive care setting [31].

In summary, NGAL is emerging as a center-stage player in the AKI field, as a novel predictive biomarker. However, it is acknowledged that the studies published thus far are small, and NGAL appears to be most sensitive and specific in relatively uncomplicated patient populations with AKI. NGAL measurements may be influenced by a number of coexisting variables such as preexisting renal disease [32] and systemic or urinary tract infections [15, 17]. Large multicenter studies to further define the predictive role of plasma and urine NGAL as a member of the putative 'AKI panel' have been initiated, robust assays for commercialization are nearly complete, and the results are awaited with optimism.

Cystatin C

Cystatin C is a cysteine protease inhibitor that is synthesized and released into the blood at a relatively constant rate by all nucleated cells. It is freely filtered by the glomerulus, completely reabsorbed by the proximal tubule, and not secreted. Since blood levels of cystatin C are not significantly affected by age, gender, race, or muscle mass, it is a better predictor of glomerular function than serum creatinine in patients with chronic kidney disease [33]. Urinary excretion of cystatin C has been shown predict the requirement for renal replacement therapy about 1 day earlier in patients with established AKI, with an AUC of 0.75 [34]. In the intensive care setting, a 50% increase in serum cystatin C predicted AKI 1–2 days before the rise in serum creatinine, with an AUC of 0.97 and 0.82, respectively [7].

A recent prospective study compared the ability of serum cystatin C and NGAL in the prediction of AKI following cardiac surgery [35]. Of 129 patients, 41 developed AKI (defined as a 50% increase in serum creatinine) 1–3 days after cardiopulmonary bypass. In AKI cases, serum NGAL levels were elevated 2 h after surgery, whereas serum cystatin C levels increased only after 12 h. Both NGAL and cystatin C levels at 12 h were strong independent predictors of AKI, but NGAL outperformed cystatin C at earlier time points.

Thus, both NGAL and cystatin C represent promising tandem biomarker candidates for inclusion in the blood 'AKI panel'. An advantage of cystatin C is the commercial availability of a standardized immunonephelometric assay, which is automated and provides results in minutes. Additionally, routine clinical storage conditions, freeze/thaw cycles, the presence of interfering substances, and the etiology of the AKI do not affect serum cystatin C measurements.

Kidney Injury Molecule-1

Kidney injury molecule-1 (KIM-1) is a transmembrane protein that is highly overexpressed in dedifferentiated proximal tubule cells after ischemic or nephrotoxic AKI in animal models [36, 37], and a proteolytically processed domain is easily detected in urine [38]. In a small human cross-sectional study, KIM-1 was found to be markedly induced in proximal tubules in kidney biopsies from patients with established AKI (primarily ischemic), and urinary KIM-1 distinguished ischemic AKI from pre-renal azotemia and chronic renal disease [36]. Patients with AKI induced by contrast did not have increased urinary KIM-1.

Recent preliminary studies have expanded the potential clinical utility of KIM-1 as a predictive AKI biomarker. In a cohort of 103 adults undergoing cardiopulmonary bypass, AKI (defined as a 0.3-mg/dl increase in serum creatinine) developed in 31% in whom the urinary KIM-1 levels increased by about 40% 2 h after surgery and by more than 100% at the 24-hour time point [39]. In a small case-control study of 40 children undergoing cardiac surgery, 20 with AKI (defined as a 50% increase in serum creatinine) and 20 without AKI, urinary KIM-1 levels were markedly enhanced, with an AUC of 0.83 at the 12-hour time point [40].

Thus, KIM-1 represents a promising candidate for inclusion in the urinary 'AKI panel'. An advantage of KIM-1 over NGAL is that it appears to be more specific to ischemic or nephrotoxic kidney injury, and not significantly affected by chronic kidney disease or urinary tract infections. It is likely that NGAL and KIM-1 will emerge as tandem biomarkers of AKI, with NGAL being most sensitive at the earliest time points and KIM-1 adding significant specificity at slightly later time points.

Interleukin-18

Interleukin (IL)-18 is a proinflammatory cytokine that is induced and cleaved in the proximal tubule, and subsequently easily detected in the urine following ischemic AKI in animal models [41]. In a cross-sectional study, urine IL-18 levels were markedly increased in patients with established AKI, but not in subjects with urinary tract infection, chronic kidney disease, nephritic syndrome, or pre-renal failure [42]. Urinary IL-18 levels displayed a sensitivity and specificity of >90% for the diagnosis of established AKI. In addition, IL-18 in urine obtained on the day of kidney transplantation was significantly increased in patients who subsequently developed delayed graft function, with an AUC of 0.95. Urinary IL-18 was significantly upregulated up to 48 h prior to the increase in serum creatinine in patients with acute respiratory distress syndrome who develop AKI, with an AUC of 0.73, and represented an independent predictor of mortality in this cohort [43].

Urinary NGAL and IL-18 were recently shown to represent early, predictive, sequential AKI biomarkers in children undergoing cardiac surgery [44]. In patients who developed AKI 2–3 days after surgery, urinary NGAL was induced within 2 h and peaked at 6 h whereas urine IL-18 levels increased around 6 h and peaked at over 25-fold 12 h after surgery (AUC 0.75). Both NGAL and IL-18 were independently associated with the duration of AKI among cases. Urine NGAL and IL-18 have also emerged as predictive biomarkers for delayed graft function following kidney transplantation [26]. In a prospective multicenter study of children and adults, both NGAL and IL-18 in urine samples collected on the day of transplant predicted delayed graft function and dialysis requirement with AUC of 0.9.

Thus, IL-18 also represents a promising candidate for inclusion in the urinary 'AKI panel'. IL-18 is more specific to ischemic AKI, and not affected by nephrotoxins, chronic kidney disease or urinary tract infections. It is likely that NGAL, IL-18 and KIM-1 will emerge as sequential urinary biomarkers of AKI.

Conclusions

The tools of modern science have provided us with promising novel biomarkers for AKI, with potentially high sensitivity and specificity. These include a plasma panel (NGAL and cystatin C) and a urine panel (NGAL, IL-18, and KIM-1). Since they represent tandem biomarkers, it is likely that the AKI panels will be useful for timing the initial insult and assessing the duration of AKI (analogous to the cardiac panel for evaluating chest pain). Based on the differential expression of the biomarkers, it is also likely that the AKI panels will help distinguish between the various types and etiologies of AKI. However, they have hitherto been tested only in small studies and in a limited number of clinical situations. It will be important in future studies to validate the sensitivity and specificity of these biomarker panels in clinical samples from large cohorts and from multiple clinical situations. Such studies will be markedly facilitated by the availability of commercial tools for the reliable and reproducible measurement of biomarkers across different laboratories.

References

1. Lameire N, Van Biesen W, Vanholder R: Acute renal failure. Lancet 2005;365:417–430.
2. Schrier RW, Wang W, Poole B, Mitra A: Acute renal failure: definitions, diagnosis, pathogenesis, and therapy. J Clin Invest 2004;114:5–14.
3. Uchino S, Kellum JA, Bellomo R, Doig GS, Morimatsu H, Morgera S, Schetz M, Tan I, Bouman C, Macedo E, Gibney N, Tolwani A, Ronco C; Beginning and Ending Supportive Therapy for the Kidney (BEST Kidney) Investigators: Acute renal failure in critically ill patients: a multinational, multicenter study. JAMA 2005;294:813–818.
4. Devarajan P: Update on mechanisms of ischemic acute kidney injury. J Am Soc Nephrol 2006;17:1503–1520.
5. Schrier RW: Need to intervene in established acute renal failure. J Am Soc Nephrol 2004;15:2756–2758.
6. Hewitt SM, Dear J, Star RA: Discovery of protein biomarkers for renal diseases. J Am Soc Nephrol 2004;15:1677–1689.
7. Herget-Rosenthal S, Marggraf G, Hüsing J, Goring F, Pietruck F, Janssen O, Phillip T, Kribben A: Early detection of acute renal failure by serum cystatin C. Kidney Int 2004;66:1115–1122.
8. Bellomo R, Kellum JA, Ronco C: Defining acute renal failure: physiological principles. Intensive Care Med 2004;30:33–37.
9. Allgren RL, Marbury TC, Rahman SN, Weisberg LS, Fenves AZ, Lafayette RA, Sweet RM, Genter FC, Kurnik BR, Conger JD, Sayegh MH: Anaritide in acute tubule necrosis. Auriculin Anaritide Acute Renal Failure Study Group. N Engl J Med 1997;336:828–834.
10. Hirschberg R, Kopple J, Lipsett P, Benjamin E, Minei J, Albertson T, Munger M, Metzler M, Zaloga G, Murray M, Lowry S, Conger J, McKeown W, O'Shea M, Baughman R, Wood K, Haupt M, Kaiser R, Simms H, Warnock D, Summer W, Hintz R, Myers B, Haenftling K, Capra W, Pike M, Guler H-P: Multicenter clinical trial of recombinant human insulin-like growth factor 1 in patients with acute renal failure. Kidney Int 1999;55:2423–2432.
11. American Society of Nephrology: American Society of Nephrology Renal Research Report. J Am Soc Nephrol 2005;16:1886–1893.
12. Zerhouni E: The NIH roadmap. Science 2003;302:63–65.
13. Devarajan P, Mishra J, Supavekin S, Patterson LT, Potter SS: Gene expression in early ischemic renal injury: clues towards pathogenesis, biomarker discovery, and novel therapeutics. Mol Genet Metab 2003;80:365–376.
14. Han WK, Bonventre JV: Biologic markers for the early detection of acute kidney injury. Curr Opin Crit Care 2004;10:476–482.
15. Zhou H, Hewitt SM, Yuen PST, Star RA: Acute kidney injury biomarkers – needs, present status, and future promise. Nephrology Self Assessment Program, American Society of Nephrology. NephSAP 2006;5:63–71.
16. Nguyen M, Ross G, Dent C, Devarajan P: Early prediction of acute renal injury using urinary proteomics. Am J Nephrol 2005;25:318–326.
17. Xu S, Venge P: Lipocalins as biochemical markers of disease. Biochim Biophys Acta 2000;1482:298–307.
18. Supavekin S, Zhang W, Kucherlapati R, Kaskel FJ, Moore LC, Devarajan P: Differential gene expression following early renal ischemia-reperfusion. Kidney Int 2003;63:1714–1724.
19. Mishra J, Ma Q, Prada A, Zahedi K, Yang Y, Barasch J, Devarajan P: Identification of neutrophil gelatinase-associated lipocalin as a novel urinary biomarker for ischemic injury. J Am Soc Nephrol 2003;4:2534–2543.
20. Mishra J, Mori K, Ma Q, Kelly C, Barasch J, Devarajan P: Neutrophil Gelatinase-Associated Lipocalin (NGAL): a novel urinary biomarker for cisplatin nephrotoxicity. Am J Nephrol 2004;24:307–315.
21. Mori K, Lee HT, Rapoport D, Drexler I, Foster K, Yang J, Schmidt-Ott KM, Chen X, Li JY, Weiss S, Mishra J, Cheema FH, Markowitz G, Suganami T, Sawai K, Mukoyama M, Kunis C, D'Agati V, Devarajan P, Barasch J: Endocytic delivery of lipocalin-siderophore-iron complex rescues the kidney from ischemia-reperfusion injury. J Clin Invest 2005;115:610–621.

22 Schmitt-Ott KM, Mori K, Kalandadze A, Li J-Y, Paragas N, Nicholas T, Devarajan P, Barasch J: Neutrophil gelatinase-associated lipocalin-mediated iron traffic in kidney epithelia. Curr Opin Nephrol Hypertens 2005;15:442–449.
23 Mishra J, Dent C, Tarabishi R, Mitsnefes MM, Ma Q, Kelly C, Ruff SM, Zahedi K, Shao M, Bean J, Mori K, Barasch J, Devarajan P: Neutrophil gelatinase-associated lipocalin (NGAL) as a biomarker for acute renal injury following cardiac surgery. Lancet 2005;365:1231–1238.
24 Wagener G, Jan M, Kim M, Mori K, Barasch JM, Sladen RN, Lee HT: Association between increases in urinary neutrophil gelatinase-associated lipocalin and acute renal dysfunction after adult cardiac surgery. Anesthesiology 2006;105:485–491.
25 Mishra J, Ma Q, Kelly C, Mitsnefes M, Mori K, Barasch J, Devarajan P: Kidney NGAL is a novel early marker of acute injury following transplantation. Pediatr Nephrol 2006;21:856–863.
26 Parikh CR, Jani A, Mishra J, Ma Q, Kelly C, Barasch J, Edelstein CL, Devarajan P: Urine NGAL and IL-18 are predictive biomarkers for delayed graft function following kidney transplantation. Am J Transplant 2006;6:1639–1645.
27 Trachtman H, Christen E, Cnaan A, Patrick J, Mai V, Mishra J, Jain A, Bullington N, Devarajan P: Urinary neutrophil gelatinase-associated lipocalin in D+HUS: A novel marker of renal injury. Pediatr Nephrol 2006;21:989–994.
28 Bachorzewska-Gajewska H, Malyszko J, Sitniewska E, Malyszko JS, Dobrzycki S: Neutrophil gelatinase-associated lipocalin and renal function after percutaneous coronary interventions. Am J Nephrol 2006;26:287–292.
29 Malyszko J, Bachorzewska-Gajewska H, Malyszko JS, Dobrzycki S: Could NGAL predict renal function after percutaneous coronary interventions? J Am Soc Nephrol 2006;17:406A.
30 Devarajan P, Hirsch R, Dent C, Pfriem H, Allen J, Beekman R, Mishra J, Ma Q, Kelly C, Mitsnefes M: NGAL is an early predictive biomarker of acute kidney injury following contrast administration. J Am Soc Nephrol 2006;17:48A.
31 Zappitelli M, Washburn K, Arikan AA, Mishra J, Loftis L, Devarajan P, Goldstein SL: Urine NGAL is an early predictive biomarker of acute kidney injury in critically ill children. J Am Soc Nephrol 2006;17:404A.
32 Mitsnefes M, Kathman T, Mishra J, Kartal J, Khoury P, Nickolas T, Barasch J, Devarajan P: Serum NGAL as a marker of renal function in children with chronic kidney disease. Pediatr Nephrol 2007;22:101–108.
33 Dharnidharka VR, Kwon C, Stevens G: Serum cystatin C is superior to serum creatinine as a marker of kidney function: a meta-analysis. Am J Kidney Dis 2002;40:221–226.
34 Herget-Rosenthal S, Poppen D, Husing J, Marggraf G, Pietruck F, Jakob HG, Phillip T, Kribben A: Prognostic value of tubular proteinuria and enzymuria in nonoliguric acute tubular necrosis. Clin Chem 2004;50:552–558.
35 VandeVoorde RG, Katlman TI, Ma Q, Kelly C, Mishra J, Dent CA, Mitsnefes MM, Devarajan P: Serum NGAL and cystatin C as predictive biomarkers for acute kidney injury. J Am Soc Nephrol 2006;17:404A.
36 Han WK, Bailly V, Abichandani R, Thadani R, Bonventre JV: Kidney injury molecule-1 (KIM-1): a novel biomarker for human renal proximal tubule injury. Kidney Int 2002;62:237–244.
37 Ichimura T, Hung CC, Yang SA, Stevens JL, Bonventre JV: Kidney injury molecule-1: a tissue and urinary biomarker for nephrotoxicant-induced renal injury. Am J Physiol Renal Physiol 2004;286: F552–F563.
38 Vaidya VS, Ramirez V, Ichimura T, Bobadilla NA, Bonventre JV: Urinary kidney injury molecule-1: a sensitive quantitative biomarker for early detection of kidney tubular injury. Am J Physiol Renal Physiol 2006;290:F517–F529.
39 Liangos O, Han WK, Wald R, Perianayagam MC, Mackinnon RW, Dolan N, Warner KG, Symes JF, Bonventre JV, Jaber BL: Urinary kidney injury molecule-1 level is an early and sensitive marker of acute kidney injury following cardiopulmonary bypass. J Am Soc Nephrol 2006;17:403A.
40 Han WK, Waikar SS, Johnson A, Curhan GC, Devarajan P, Bonventre JV: Urinary biomarkers for early detection of acute kidney injury. J Am Soc Nephrol 2006;17:403A.
41 Melnikov VY, Ecder T, Fantuzzi G, Siegmund B, Lucia MS, Dinarello CA, Schrier RW, Edelstein CL: Impaired IL-18 processing protects caspase-1 deficient mice from ischemic acute renal failure. J Clin Invest 2001;107:1145–1152.

42 Parikh CR, Jani A, Melnikov VY, Faubel S, Edelstein CL: Urinary interleukin-18 is a marker of human acute tubular necrosis. Am J Kidney Dis 2004;43:405–414.
43 Parikh CR, Abraham E, Ancukiewicz M, Edelstein CL: Urine IL-18 is an early diagnostic marker for acute kidney injury and predicts mortality in the intensive care unit. J Am Soc Nephrol 2005;16:3046–3052.
44 Parikh CR, Mishra J, Thiessen-Philbrook H, Dursun B, Ma Q, Kelly C, Dent C, Devarajan P, Edelstein CL: Urinary IL-18 is an early predictive biomarker of acute kidney injury after cardiac surgery. Kidney Int 2006;70:199–203.

Prasad Devarajan, MD
Nephrology and Hypertension, MLC 7022, Cincinnati Children's Hospital Medical Center
3333 Burnet Avenue
Cincinnati, OH 45229–3039 (USA)
Tel. +1 513 636 4531, Fax +1 513 636 7407, E-Mail prasad.devarajan@cchmc.org

Diagnosis of Acute Kidney Injury: From Classic Parameters to New Biomarkers

Joseph V. Bonventre

Renal Division, Brigham and Women's Hospital and Department of Medicine, Harvard Stem Cell Institute, Harvard Medical School and Harvard-Massachusetts Institute of Technology, Division of Health Sciences and Technology, Boston, Mass., USA

Abstract

A change in serum creatinine is the standard metric used to define and monitor the progression of acute kidney injury (AKI). This marker is inadequate for a number of reasons including the fact that changes in serum creatinine are delayed in time after kidney injury and hence creatinine is not a good indicator to use in order to target therapy in a timely fashion. There is an urgent need for early biomarkers for the diagnosis of AKI. There is also a need for biomarkers that will be predictive of outcome and which can be used to monitor therapy. There are a limited number of biomarkers that are being validated by a number of groups and from this list clinically useful reagents are likely to be derived over the next few years. In this article the status of 5 potential urinary biomarkers for AKI are discussed: kidney injury molecule-1, N-acetyl-β-D-glucosaminidase, neutrophil gelatinase-associated lipocalin, cystatin C, and interleukin-18. Considerable progress has been made although much continues to be needed to validate these markers for routine clinical use. Armed with these new tools the future will look much brighter for the patient with AKI as it is likely that early diagnosis and better predictors of outcome will lead to new therapies which can be introduced earlier in the course of disease.

Copyright © 2007 S. Karger AG, Basel

Acute kidney injury (AKI) is a common condition that is associated with a high mortality rate. It has been recognized that routinely used measures of renal function, such as blood urea nitrogen and serum creatinine concentrations, significantly increase only after substantial kidney injury occurs and then with a time delay. In addition multiple factors can effect blood urea nitrogen and serum creatinine. Serum creatinine is known to be secreted by the renal tubule and this secretion can be modified by pharmacological agents. Serum levels can be modified by changes in volume status which often occurs especially in postoperative patients who are at risk of AKI and for whom early diagnosis of AKI would be desirable. Creatinine production varies greatly among individuals and consequently the changes in serum creatinine will vary greatly from individual to individual even in the context of an equivalent change in renal function. There are also technical problems with the assay as there are urine components which can interfere.

The insensitivity of tests used to detect injury to the kidney delays the diagnosis in humans, making it particularly challenging to administer putative therapeutic agents in a timely fashion. While significant attempts have been made to increase the utility of serum creatinine in the steady state with the introduction of equations to calculate an 'effective GFR' [1], this does not alleviate the problem with this biomarker of renal function when kidney function is changing rapidly as it does with AKI. Furthermore, the insensitivity of traditional markers of kidney damage affects the evaluation of toxicity in preclinical studies by allowing drug candidates, which have low but nevertheless important nephrotoxic side effects in animals, to pass the preclinical safety criteria only to be found to be clinically nephrotoxic at great human costs. In this brief review I will summarize the importance of better biomarkers for kidney injury and discuss the current status of specific biomarkers to detect preclinical and clinical renal injury and potentially predict outcome of AKI in humans.

Urgent Need for Early Biomarkers in the Management of AKI

One of the disappointments in the care of patients with AKI is that little has been shown to be effective in preventing the syndrome or arresting its progression. We have support therapies that can provide correction of metabolic consequences of kidney failure but despite this the mortality rates with AKI remain very high, especially in patients in the ICU. It has recently been recognized that even small changes in kidney function lead to marked increases in mortality [2, 3]. Thus it is particularly important to diagnose injury early. Furthermore progress in the field has been significantly impaired by inadequate clinical studies of therapeutic agents because treatment protocols are adversely effected by late diagnosis. By the time creatinine increases and the study investigators confirm this increase, the acute event resulting in the AKI is long past. This is analogous to initiating treatment in patients with acute myocardial infarction 48 h after coronary occlusion.

What Are Biomarkers?

Biomarkers can be any parameter in a patient that can be quantitated and provides useful information about a normal or pathobiological state. We will limit our discussion primarily to proteins or other molecules which are found in blood or urine. A biomarker may be used to help stratify patients at risk of AKI, or diagnose or predict the natural history of a disease process. A biomarker might be used as an indicator to guide the timing and type of therapy, to predict mortality and monitor the response to therapy. Response to therapy can be quantitated with a biomarker. The best biomarkers are measured noninvasively, and are easy to determine in a timely fashion so that clinical decision making can be facilitated. Ideally the measurement is made at the bedside. The biomarker should not be affected by changes in the composition of the fluid in which it is measured, especially as patients with kidney injury have marked changes in the composition of blood and urine. It is possible that no single biomarker will be ideally suited to satisfy all these needs. It is possible that a set of biomarkers will ultimately be used.

To validate biomarkers it would be ideal if there were a 'gold standard' to which the marker is being compared. Unfortunately this is not a trivial task as we have already indicated that creatinine is not such a standard. Pathology would be very useful but we routinely have few AKI patients in whom renal biopsies have been obtained. Urinary casts and changes in the fractional excretion of sodium may be helpful but both of these markers have their own inherent insensitivity and problems with specificity, especially with quantitation of injury. In this context it is very useful to consider animal data because in the animal pathology is more easily obtained and a detailed time course can be evaluated relative to a very precise time definition of the insult. Newly proposed biomarkers should be simultaneously and quantitatively compared with those used in past.

Potential Biomarkers for AKI

There are a limited number of biomarkers that are being validated by various groups and from this list clinically useful reagents are likely to derive over the next few years. Of these biomarkers, kidney injury molecule-1 (KIM-1), N-acetyl β-D-glucosaminidase (NAG), neutrophil gelatinase-associated lipocalin (NGAL), cystatin C, and interleukin-18 (IL-18) will be discussed.

Kidney Injury Molecule-1
KIM-1 (or Kim-1 in rodents) encodes a type I cell membrane glycoprotein containing, in its extracellular portion, a novel six-cysteine immunoglobulin-like domain and a threonine/serine and proline-rich domain characteristic of mucin-like O-glycosylated proteins, suggesting its potential involvement in cell–cell and/or cell–matrix interactions [4] (fig. 1). We found the mRNA that encodes this protein to be upregulated in the kidney more than any other mRNA

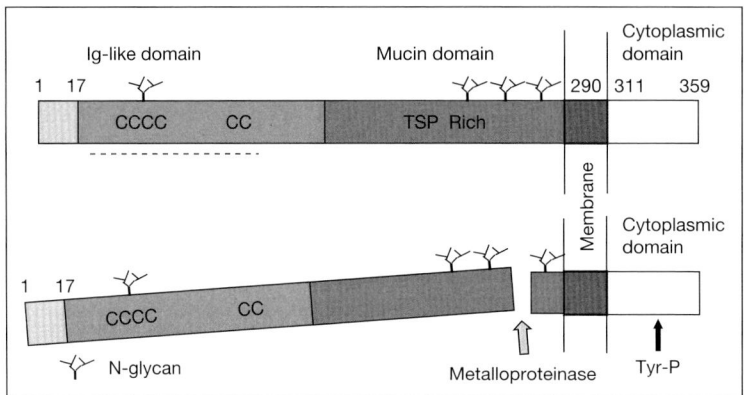

Fig. 1. Drawing showing the structure of KIM-1. The protein is a type-1 membrane protein with most of the protein made up of an extracellular domain that consists of a signal peptide, and Ig domain and a mucin domain. There is a short cytoplasmic domain with at least one important tyrosine phosphorylation domain. The protein is cleaved by a metalloproteinase and the ectodomain appears in the urine of rodents and humans with AKI.

in response to experimental AKI in rodents. This was confirmed using the nephrotoxin, cisplatin, in an unbiased genomic approach taken by a pharmaceutical consortium [5]. After proximal tubular kidney injury, the ectodomain of KIM-1 is shed from proximal tubule cells in vivo into the urine in rodents [6–8] and humans [9]. In preclinical and clinical studies urinary Kim-1 serves as an early diagnostic indicator of kidney injury. In rodents Kim-1 is increased earlier than any of the conventional biomarkers, e.g. plasma creatinine, blood urea nitrogen, glycosuria, increased proteinuria or increased urinary NAG levels [8, 9].

N-Acetyl-β-D-Glucosaminidase

NAG is a proximal tubular brush border-specific lysosomal enzyme which has been used as an indicator of renal proximal tubule injury. In our hands in humans and rodents this biomarker performs well [8, 10], although not as well as Kim-1 in rodents where it is less sensitive. Appropriate comparisons are not complete in humans. It is know that some metals and other nephrotoxicants can directly inhibit NAG activity, and therefore in such cases NAG cannot be used as a biomarker [11, 12].

Neutrophil Gelatinase-Associated Lipocalin

NGAL is a 25-kDa protein bound to gelatinase which is expressed and secreted by hepatocytes, neutrophils, and kidney epithelial cells during inflammation and AKI [13]. NGAL is involved in iron shuttling from extracellular to intracellular compartments. NGAL was identified as being one of the seven genes whose expression was upregulated >10-fold within the first few hours after ischemic renal injury in a mouse model. It has recently been reported that on day 0 both urine NGAL and IL-18 (see below) predicted the trend in serum creatinine in the post-transplant period after adjusting for effects of age, gender, race, urine output, and cold ischemia time [14].

Cystatin C

Cystatin C is a 13-kDa cysteine protease inhibitor and one of the 12 members of human cystatin family [15]. Cystatin C is unique amongst all cystatins as it seems to be produced by all human nucleated cells. It has been argued by some that serum cystatin C should be used to replace creatinine clearance in patients with chronic kidney disease. For example, while a meat meal increases serum creatinine and reduces the calculated effective GFR significantly, there is no significant change in the serum cystatin concentration [16]. Herget-Rosenthal et al. [17] showed that urinary excretion of cystatin C at entry to the study in patients with non-oliguric acute tubular necrosis had a higher sensitivity and specificity than β_1-microglobulin, NAG, and the Liano severity of illness score [18] in predicting the requirement for renal replacement therapy.

Interleukin-18

IL-18 is a cytokine whose levels have been reported to be elevated in the urine of patients with AKI and delayed graft function compared with normal subjects and patients with pre-renal azotemia, urinary tract infection, chronic renal insufficiency, and nephritic syndrome [19]. Parikh et al. [20] have reported that IL-18 is an early, predictive biomarker of AKI after cardiopulmonary bypass, and that NGAL and IL-18 are increased in tandem after cardiopulmonary bypass.

Conclusions

Acute kidney injury is a complex entity with multiple causes. It is possible that no single marker will provide the levels of sensitivity and specificity necessary to be clinically useful across the full spectrum of AKI. It is possible that a panel of markers will be optimal to satisfy all the previously mentioned requirements. Laboratories carrying out biomarker studies are collaborating with clinical investigators performing prevention and treatment studies. A network should be developed within the biomarker field. Multiple biomarkers should be studied simultaneously. For both the discovery of new biomarkers and validation of the ones at hand, it is necessary to have close interactions between clinician scientists and laboratory scientists. Armed with these new tools the future will look much brighter for the patient with AKI as it is likely that clinical trials will lead to new therapies which can be introduced earlier in the course of disease and be more effective in leading to mitigation of the severity of disease and potentiation of recovery of this fundamentally reversible condition.

Acknowledgments

This work was supported by the National Institutes of Health (grants DK 39773, DK54741, DK72381).

References

1. Levey AS, Coresh J, Balk E, Kausz AT, Levin A, Steffes MW, Hogg RJ, Perrone RD, Lau J, Eknoyan G: National Kidney Foundation practice guidelines for chronic kidney disease: evaluation, classification, and stratification. Ann Intern Med 2003;139:137–147.
2. Chertow GM, Burdick E, Honour M, Bonventre JV, Bates DW: Acute kidney injury, mortality, length of stay, and costs in hospitalized patients. J Am Soc Nephrol 2005;16:3365–3370.
3. Lassnigg A, Schmidlin D, Mouhieddine M, Bachmann LM, Druml W, Bauer P, Hiesmayr M: Minimal changes of serum creatinine predict prognosis in patients after cardiothoracic surgery: a prospective cohort study. J Am Soc Nephrol 2004;15:1597–1605.
4. Bailly V, Zhang Z, Meier W, Cate R, Sanicola M, Bonventre JV: Shedding of kidney injury molecule-1, a putative adhesion protein involved in renal regeneration. J Biol Chem 2002;277: 39739–39748.
5. Amin RP, Vickers AE, Sistare F, Thompson KL, Roman RJ, Lawton M, Kramer J, Hamadeh HK, Collins J, Grissom S, et al: Identification of putative gene based markers of renal toxicity. Environ Health Perspect 2004;112:465–479.
6. de Borst MH, van Timmeren MM, Vaidya VS, de Boer RA, van Dalen MB, Kramer AB, Schuurs TA, Bonventre JV, Navis G, van Goor H: Induction of kidney injury molecule-1 in homozygous Ren2 rats is attenuated by blockade of the renin-angiotensin system or p38 MAP kinase. Am J Physiol Renal Physiol 2007;292:F313–F320.
7. Ichimura T, Hung CC, Yang SA, Stevens JL, Bonventre JV: Kidney injury molecule-1:a tissue and urinary biomarker for nephrotoxicant-induced renal injury. Am J Physiol Renal Physiol 2004;286:F552–F563.
8. Vaidya VS, Ramirez V, Ichimura T, Bobadilla NA, Bonventre JV: Urinary kidney injury molecule-1: a sensitive quantitative biomarker for early detection of kidney tubular injury. Am J Physiol Renal Physiol 2006;290:F517–F529.
9. Han WK, Bailly V, Abichandani R, Thadhani R, Bonventre JV: Kidney injury molecule-1 (KIM-1): a novel biomarker for human renal proximal tubule injury. Kidney Int 2002;62:237–244.
10. Liangos O, Perianayagam MC, Vaidya VS, Han WK, Wald R, Tighiouart H, Mackinnon RW, Li L, Balakrishnan VS, Pereira BJ, et al: Urinary N-acetyl-{beta}-(D)-glucosaminidase activity and kidney injury molecule-1 level are associated with adverse outcomes in acute renal failure. J Am Soc Nephrol 2007;Epub ahead of print.
11. Wiley RA, Choo HY, Traiger GJ: The effect of nephrotoxic furans on urinary N-acetylglucosaminidase levels in mice. Toxicol Lett 1982;14:93–96.
12. Vaidya VS, Shankar K, Lock EA, Bucci TJ, Mehendale HM: Renal injury and repair following S-1,2 dichlorovinyl-L-cysteine administration to mice. Toxicol Appl Pharmacol 2003;188:110–121.

13 Schmidt-Ott KM, Mori K, Li JY, Kalandadze A, Cohen DJ, Devarajan P, Barasch J: Dual action of neutrophil gelatinase-associated lipocalin. J Am Soc Nephrol 2007;18:407–413.
14 Parikh CR, Jani A, Mishra J, Ma Q, Kelly C, Barasch J, Edelstein CL, Devarajan P: Urine NGAL and IL-18 are predictive biomarkers for delayed graft function following kidney transplantation. Am J Transplant 2006;6:1639–1645.
15 Filler G, Bokenkamp A, Hofmann W, Le Bricon T, Martinez-Bru C, Grubb A: Cystatin C as a marker of GFR–history, indications, and future research. Clin Biochem 2005;38:1–8.
16 Preiss DJ, Godber IM, Lamb EJ, Dalton RN, Gunn IR: The influence of a cooked-meat meal on estimated glomerular filtration rate. Ann Clin Biochem 2007;44:35–42.
17 Herget-Rosenthal S, Poppen D, Husing J, Marggraf G, Pietruck F, Jakob HG, Philipp T, Kribben A: Prognostic value of tubular proteinuria and enzymuria in nonoliguric acute tubular necrosis. Clin Chem 2004;50:552–558.
18 Liano F, Junco E, Pascual J, Madero R, Verde E: The spectrum of acute renal failure in the intensive care unit compared with that seen in other settings. The Madrid Acute Renal Failure Study Group. Kidney Int Suppl 1998;66:S16–S24.
19 Parikh CR, Jani A, Melnikov VY, Faubel S, Edelstein CL: Urinary interleukin-18 is a marker of human acute tubular necrosis. Am J Kidney Dis 2004;43:405–414.
20 Parikh CR, Mishra J, Thiessen-Philbrook H, Dursun B, Ma Q, Kelly C, Dent C, Devarajan P, Edelstein CL: Urinary IL-18 is an early predictive biomarker of acute kidney injury after cardiac surgery. Kidney Int 2006;70:199–203.

Joseph V. Bonventre, MD, PhD
Brigham and Women's Hospital, Renal Division, Harvard Institutes of Medicine
4 Blackfan Circle
Boston, MA 02115 (USA)
Tel. +1 617 525 5960, Fax +1 617 525 5965, E-Mail joseph_bonventre@hms.harvard.edu

Endotoxin and Cytokine Detection Systems as Biomarkers for Sepsis-Induced Renal Injury

Steven M. Opal

Infectious Disease Division, Brown Medical School, Providence, R.I., USA

Abstract

Background: A reliable biomarker as an indicator of the presence of severe sepsis is an unmet medical need. **Methods:** Review of recent literature on this topic focusing upon endotoxin and cytokine assays. **Results:** The ideal biomarker for sepsis would be readily available, technically easy to perform with a quick turn-around time, inexpensive, highly specific, very sensitive, and preferably highly correlated in quantitative terms with disease severity. Such a test would provide early diagnostic accuracy, prognostic information, and indicate responsiveness to treatment interventions. Regrettably no such biomarker exists for sepsis at present, and it is not likely that such an ideal assay will be developed in the foreseeable future. **Conclusions:** Despite their shortcomings, a number of existing and candidate biomarker assays are available and can provide some useful information to the clinician caring for septic patients. The relative merits of endotoxin measurement, interleukin-6 levels and a variety of other sepsis markers are reviewed. Full implementation of these biomarkers may improve diagnostic accuracy over the standard clinical criteria for sepsis.

Copyright © 2007 S. Karger AG, Basel

One of the major unmet medical needs in the management of sepsis is the development of reliable methods to differentiate patients who are becoming septic from those who have a physiological systemic response to infection. As Machiavelli noted over 500 years ago in his famous treatise *The Prince*, 'Hectic fever at its inception is difficult to recognize but easy to treat, left untended it becomes easy to recognize but too difficult to treat'. Regrettably this same statement could be made about sepsis even today. Despite intense clinical and laboratory monitoring, early recognition of the early phases of sepsis remains a fundamental challenge for clinicians and investigators alike. When does the systemic host response to infection convert from a controlled and advantageous

Table 1. Summary of some of the currently available biomarkers for the recognition of sepsis

Biomarker	Advantages	Disadvantages
Endotoxin (LPS)	High levels predictive of outcome, approved assays available	Difficult assays and not readily available, not highly specific
Interleukin-6	Easily measured and high levels predictive of outcome	Not available in clinical laboratories, highly variable, nonspecific
Procalcitonin	Well-documented predictive value, reliable, easily measured	Not available in many clinical laboratories
C-Reactive protein	Readily available assays, easily measured acute phase protein	Nonspecific assay, limited predictive value in sepsis
aPTT wave form analysis	Easy, rapid assay if equipment is available, highly predictive	Limited clinical experience, mechanism of action unclear
Coagulation markers (protein C, platelets, D-dimer, TAT, F1.2, etc.)	Standard assays are available, levels correlated with outcome	Not readily available as rapid assays, indirect and nonspecific
HLA-DR expression	Clear pathophysiological link with sepsis-induced immune suppression	Not readily available, less valuable for early sepsis
Soluble TREM-1	Highly predictive in early clinical studies	Limited experience, not readily available clinically

aPTT = Activated partial thromboplastin time; F1.2 = prothrombin fragment 1.2; HLA = human leukocyte antigen; LPS = lipopolysaccharide; TAT = thrombin antithrombin complex; TREM = triggering receptor expressed on myeloid cells.

host defense response to a deleterious pathophysiological process capable of diffuse tissue injury with potentially lethal consequences?

There is no single diagnostic test that defines the clinical syndrome that we now refer to as sepsis, and clinical criteria alone are notoriously inaccurate in determining the precise time when an injurious and dysfunctional septic host response has developed. We can readily measure evidence of end-organ injury and tissue hypoperfusion (defined as severe sepsis), but these are late findings of established organ dysfunction. Waiting to recognize that a patient was septic in retrospect is often too late for successful intervention to prevent further tissue injury. For these reasons, a biomarker assay has been much sought after to assist the clinician in the detection of the early signs of sepsis. The most promising biomarkers have been detection assays for microbial mediators or early host response mediators indicative of a pathologic state of an injurious host response. Some of these candidate biomarkers will be briefly described in the following paragraphs. Their relative merits as each biomarker are summarized in table 1.

Endotoxin (LPS) Assays

Bacterial endotoxin, or lipopolysaccharide (LPS), is an intrinsic component of the outer membrane of gram-negative bacteria and is essential for the viability of enteric bacteria [1]. Endotoxin functions as an alarm molecule alerting the host to the presence of microbial invasion by gram-negative bacteria. The potentially lethal consequences following endotoxin release into the circulation is attributable to the exaggerated host response to the endotoxin, rather than the endotoxin molecule itself.

LPS is a di-phosphorylated, polar macromolecule that contains hydrophobic elements within its lipid A core structure, and hydrophilic elements in its repeating polysaccharide surface components. LPS forms microaggregates (micelles). LPS signaling is mediated by interactions with a hepatically derived, acute-phase plasma protein known as LPS-binding protein (LBP) [2, 3]. LBP functions as a shuttle carrier protein transferring LPS monomers to CD14. CD14 is a glycosyl phosphatidylinositol-linked protein found primarily on the cell surfaces of myeloid cells. After docking to membrane-bound CD14, LPS is transferred to the essential soluble adaptor protein MD2. This LPS-MD2 complex is then presented to the extracellular domain of TLR4 where they aggregate on lipid rafts on the cell surface to activate intracellular signaling. Through a well-characterized series of activation steps by specific threonine/serine kinases, intracellular signaling ultimately leads to phosphorylation, ubiquinylation and degradation of inhibitory κB (IκB). IκB degradation releases nuclear factor κB (NFκB) from its cytoplasm stores, NFκB then translocates into the nucleus activating a myriad of transcriptional programs including clotting elements, complement, acute phase proteins, cytokines, chemokines and nitric oxide synthase genes. The outpouring of these inflammatory mediators is central to the pathogenesis of septic shock induced by gram-negative bacteria [1, 2].

High levels of endotoxin in patients with severe sepsis correlated with the presence of hypotension and is associated with worse prognosis [3, 4]. The problems with endotoxin assays are that the measurable levels are affected by endotoxin-binding proteins such as LBP, soluble CD14, and high- and low-density lipoproteins. The levels of these endotoxin-binding proteins are highly variable in septic patients. Moreover, the pathophysiological impact of a given amount of circulating endotoxin is highly dependent upon the responsiveness of host tissues to this microbial mediator. As an example, older patients tolerate endotoxemia poorly when compared with younger patients [3].

Measurement of endotoxin in the blood has been performed by a technically difficult and time-intensive bioassay, the Limulus amebocyte assay. Recently, a new rapid endotoxin activity assay has been developed that may prove to be of greater clinical utility. This assay is based upon the degree of priming of the circulating neutrophil population by endotoxin exposure. In a study of patients admitted to a mixed surgical-medical ICU, 58% of the patients had elevated endotoxin levels by this assay. This percentage increased to 85% in patients with severe sepsis [4]. This new endotoxin detection method correlated with excess ICU mortality, septic shock and gram-negative bacteremia.

IL-6 as a Biomarker for Cytokine Networks

Inflammatory cytokines play a pivotal role in the pathogenesis of sepsis. The major proinflammatory cytokines, tumor necrosis factor (TNF)-α and interleukin (IL)-1β, induce their hemodynamic and metabolic effects in concert with an expanding group of host-derived inflammatory mediators that work in a coordinated fashion to produce the systemic inflammatory response. The multitude of inflammatory cytokines and chemokines found in excess quantities in the bloodstream in patients with septic shock is impressive and is only matched by an equally daunting group of anti-inflammatory mediators. The proinflammatory mediators tend to predominate locally and in the early phases of sepsis, whereas the endogenous anti-inflammatory components often prevail systemically in the later phases of sepsis. Monocyte-macrophage-generated cytokines and chemokines primarily drive the early septic process; whereas, the lymphocyte-derived cytokines and interferons become important in the later phases of sepsis [5, 6].

Cytokine levels are notoriously variable in the blood and have proven rather difficult to assess from routine blood samples. The most reliable and widely utilized cytokine measure continues to be IL-6 [2–6]. This GP130 receptor ligand is found and easily measured in the majority of septic patients but these levels are capricious and vary by orders of magnitude in patients with similar clinical presentations. IL-6 has both anti-inflammatory and proinflammatory actions and is not lethal by itself, as is TNF or IL-1, when injected into experimental animals. IL-6 is viewed as an indicator cytokine during periods of excess cytokine synthesis in acute severe illness. Its measurement has proven to be of significant but limited clinical relevance as a biomarker for severe sepsis [5, 6].

Monocyte Deactivation and Immunodepression as Marker for Sepsis

The sepsis-induced immune suppressive phenomenon is part of a general, compensatory, host defense mechanism designed to limit the potentially injurious impact of ongoing systemic inflammation. This occurs primarily at the

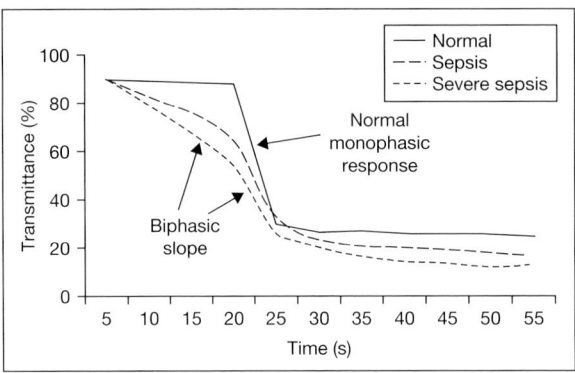

Fig. 1. Hypothetical appearance of optical transmission of coagulation events in the activated partial thromboplastin time. The normal response is a monophasic, sudden drop in percent transmittance when fibrin polymerization occurs (solid line). In septic plasma, complexes of CRP and lipoproteins form immediately and result in a biphasic decrease in transmittance (dotted lines). The slope of the initial decline in percent of transmittance is correlated with disease severity in sepsis.

transcriptional level, with downregulation of genes encoding for proinflammatory cytokines and other acute phase proteins. Stress hormones (i.e. catecholamines, corticosteroids) and anti-inflammatory cytokines such as IL-10 are upregulated. Excess CD4+ lymphocyte apoptosis and a shift to a TH2-type cytokine response are common concomitants following an initial septic insult [7]. This relative immune refractory state places the patient at increased risk of secondary bacterial or fungal infection. Methods to detect this immunosuppressed state and restore immune competence are under active clinical investigation. Patients with depressed expression of MHC class II antigens (e.g., HLA-DR) on the cell surface of macrophages may be in a functionally immunosuppressed state [8, 9]. The level of expression of monocyte HLA-DR is measurable by a rapid bioassay to assess immune function and prognosis in sepsis [8–10].

aPTT Biphasic Wave Form Analysis as a Marker of Sepsis

A remarkably specific abnormality in the optical transmission waveform obtained during measurement of the activated partial thromboplastin time on some photometric hemostasis autoanalyzers deserves mention as a potential biomarker for sepsis [11]. A biphasic waveform abnormality is related to a complex of C-reactive protein and very low-density lipoprotein associated with the clinical diagnosis of disseminated intravascular coagulation and severe sepsis (fig. 1).

Several studies have confirmed the association of a biphasic waveform and the diagnosis of severe sepsis [11–13]. The diagnostic accuracy of the abnormal waveform for severe sepsis appears to be comparable to procalcitonin with the advantage of a higher negative predictive value. This simple and rapidly available test may prove to be of added diagnostic value during the workup for common coagulation abnormalities in sepsis. An early procoagulant state is almost uniformly generated in the initial stages of sepsis and simple measurement of coagulation parameters have provided consistent prognostic value in many sepsis studies [12].

Procalcitonin in Sepsis

Perhaps the most extensively studied biomarker for sepsis is procalcitonin, the propeptide of calcitonin. In septic patients, procalcitonin is generated by numerous extrathyroidal tissues and possesses many favorable attributes as a diagnostic test for sepsis. It has a long half-life (approximately 24 h), and blood levels increase from undetectable to over 100 ng/ml during the course of septic shock. Procalcitonin levels do not become elevated as rapidly as the cytokine IL-6 and more reliably distinguish between non-infectious versus bacterial causes of inflammation than do cytokine measures in numerous studies [5, 6, 14, 15]. Recent studies indicate that procalcitonin levels could be utilized to determine the appropriate use of antibacterial agents for bacterial respiratory infections versus supportive care alone for viral infections in community-acquired pneumonia [15]. Procalcitonin is approved for risk assessment for patients with sepsis. Assay time is less than 20 min and results are available in 1 h.

Soluble TREM-1 as a Potential Biomarker for Sepsis

Triggering receptor expressed on myeloid cells (TREM-1) is a member of the immunoglobulin superfamily and is expressed on the cell surface of neutrophils and monocytes. TREM-1 is upregulated in the setting of bacterial and fungal infection and, after binding with its as yet unknown ligand, acts synergistically with LPS to induce cytokine production. A soluble form of TREM is released during infection and at a cutoff of 60 ng/ml in the blood, soluble TREM has an excellent sensitivity and specificity in differentiating systemic inflammatory response syndrome from sepsis [16]. The ultimate diagnostic utility of soluble TREM-1 measurement awaits further testing.

References

1 Calvano SE, Wenzhong X, Rihards DR, Felciano RM, Baker HV, Cho RJ, Chen RO, Brownstein BH, Perren J, Tschoeke SK, Miller-Graziano C, Moldawer LL, Mindrinos MN, Davis RW, Tompkins RG, Lowry SF; Inflammatory Host Responses to Injury Large Scale Collaborative Research Program: A network-based analysis of systemic inflammation in humans. Nature 2005;437:1032–1037.
2 Opal SM: The clinical relevance of endotoxin in human sepsis: a critical analysis. J Endotoxin Res 2002;8:473–476.
3 Opal SM, Scannon PJ, Vincent J-L, White M, Carroll SF, Palardy JE, Parejo NA, Pribble JP, Lemke JH: Relationship between plasma levels of lipopolysaccharide (LPS) and LPS-binding protein in patients with severe sepsis and septic shock. J Infect Dis 1999;180:1584–1589.
4 Marshall JC, Foster D, Vincent JL Cook D, Cohen J, Dellinger P, Opal SM: Diagnostic and prognostic implications of endotoxemia in critical illness: results of the Medic trial. J Infect Dis 2004;190:527–534.
5 Mokart D, Merlin M, Sannini A, Brun JP, Delpero JR, Houvenaeghel G, Moutardier V, Blache JL: Procalcitonin, interleukin 6 and systemic inflammatory response syndrome (SIRS): early markers of postoperative sepsis after major surgery. Br J Anaesth 2005;94:767–773.
6 Fraunberger P, Want Y, Holler E, Parhofer KG, Nagel D, Walli AK, Seidel D: Prognostic value of interleukin 6, procalcitonin, and C-reactive protein levels in intensive care unit patients during first increase of fever. Shock 2006;26:10–12.
7 Hotchkiss RS, Karl IE: The pathophysiology and treatment of sepsis. N Engl J Med 2003;348: 138–150.
8 Ng PC, Li G, Chui KM, Chu WC, Li K, Wong RP, Fok TF: Quantitative measurement of monocyte HLA-DR expression in the identification of early onset neonatal infection. Biol Neonate 2006;89:75–81.
9 Le Tulzo Y, Pangault C, Amiot L, Guilloux V, Tribut O, Arvieux C, Camus C, Fauchet R, Thomas R, Drénou B: Monocyte human leukocyte antigen-DR transcriptional downregulation by cortisol during septic shock. Am J Respir Crit Care Med 2004;169:1144–1151.
10 Perry SE, Mostafa SM, Wenstone R, Shenkin A, McLaughlin PJ: Is low monocyte HLA-DR expression helpful to predict out come in severe sepsis? Intensive Care Med 2003;29:1245–1252.
11 Toh CH, Samis J, Downey C, et al: Biphasic transmittance waveform in the APTT coagulation assay is due to the formation of Ca($++$)-dependent complex of C-reactive protein and very-low density lipoprotein and is a novel marker of impending disseminated intravascular coagulation. Blood 2002;100:2522–2529.
12 Levi M, Opal SM: Coagulation abnormalities in critically ill patients. Crit Care 2006;10:222–229.
13 Chopin N, Floccard B, Sobas F, et al: Activated partial thromboplastin time waveform analysis: a new tool to detect infection? Crit Care Med 2006;34:1654–1660.
14 Aikawa N, Fujishima S, Endo S, Sekine I, Kogawa K, Yamamoto Y, Kushimoto S, Yukioka H, Kato N, Totsuka K, Kikuchi K, Ideda T, Ideda K, Harada K, Satomura S: Multicenter prospective study of procalcitonin as an indicator of sepsis. J Infect Chemother 2005;11:152–159.
15 Christ-Crain M, Stolz D, Bingisser R, et al: Procalcitonin guidance of antibiotic therapy in community-acquired pneumonia: a randomized trial. Am J Respir Crit Care Med 2006;174:84–91.
16 Gibot S, Kolopp-Sarda MN, Benie MC, et al: Plasma level of a triggering receptor expressed on myeloid cells-1: its diagnostic accuracy in patients with suspected sepsis. Ann Intern Med 2004; 141:9–15.

Steven M. Opal
Infectious Disease Division, Memorial Hospital of Rhone Island
111 Brewster Street
Pawtucket, RI 02860 (USA)
Tel. +1 401 729 2545, Fax +1 401 729 2795, E-Mail Steven_Opal@brown.edu

Quantifying Dynamic Kidney Processes Utilizing Multi-Photon Microscopy

Bruce A. Molitoris, Ruben M. Sandoval

Department of Medicine, Division of Nephrology, Indiana University School of Medicine, and Indiana Center for Biological Microscopy, Indianapolis, Ind., USA

Abstract

Multi-photon microscopy and advances in optics, computer sciences, and the available labeling fluorophores now allow investigators to study the dynamic events within the functioning kidney with subcellular resolution. This emerging technology, with improved spatial and temporal resolution and sensitivity, enables investigators to follow complex heterogenous processes in organs such as the kidney. Repeated determinations within the same animal are possible minimizing their use and inter-animal variability. Furthermore, the ability to obtain volumetric data (3D) makes quantitative 4D (time) analysis possible. Finally, use of up to three fluorophores concurrently allows three different or interactive processes to be observed simultaneously. Therefore, this approach compliments existing molecular, biochemical, and pharmacologic techniques by advancing data analysis and interpretation to subcellular levels for molecules without the requirement for fixation.

Copyright © 2007 S. Karger AG, Basel

New imaging technologies, such as multi-photon microscopy, have equipped researchers with extremely powerful tools to uniquely address biologically important questions that can only be accomplished in whole organ studies [1–3]. In parallel with this, advances in fluorophores with increased quantum yields and ease of labeling [4], molecular and transgenic approaches, and new delivery techniques [5] have enabled the development of intravital studies that can follow and quantify events with enhanced spatial and temporal resolution. Furthermore, exponential developments in computer sciences, specifically with applications to imaging, have removed many of the obstacles previously limiting the ability to utilize microscopy to study and quantify dynamic cellular processes. These imaging technologies enable the measurement of dynamic 4-dimensional (3D plus time) structure and function in organs and tissues [6], the measurement of chemical and biochemical composition of tissues, the expression of fluorescently labeled molecular agents including drugs and proteins, quantification of the rates of physiological processes such as microvascular perfusion rates, glomerular permeability and the mechanism of cellular uptake and intracellular trafficking [1, 2, 7].

The kidney is an extremely complex heterogeneous organ consisting of vascular and epithelial components functioning in a highly coordinated fashion that enables the regulation of a myriad of interdependent processes. Over the years talented and creative individuals have developed novel experimental approaches that enable the isolation, understanding and integrating the unique structure–function relationships that occur. Investigators have developed model systems to enable enhanced manipulation and isolation of specific variables of interest. This has lead to a mechanistic understanding of cellular processes and the identification of alterations in these processes under defined conditions that attempt to mimic either physiologic or disease states. However these models often lack the organ-specific complexity, the dynamic nature of cellular processes and cell–cell interactions necessary for adequate understanding of the process under study. For example, proximal tubule cells in cell culture undergo dedifferentiation resulting in reduced metabolic rates and alterations in many cellular processes. This has limited our ability to test therapeutic approaches and has resulted in difficulties translating preclinical data into therapeutic advances.

Multi-photon microscopy offers the investigator a minimally invasive high resolution technique, with increased depth of penetration and markedly reduced phototoxicity, for visualization of cell–cell and intracellular events intravitally. The reduction in phototoxicity with multi-photon microscopy results as fluorescence excitation occurs only at the focal point, thereby eliminating out-of-focus fluorescent excitation within the tissue as would occur with confocal microscopy. The genesis of these advances was covered in a previous article [3] and in our previous publications [1, 2, 8]. Second, improved detectors with increased sensitivity, enhanced software and faster hardware, and new computational algorithms for 3D analysis and quantification have enabled more rapid, sensitive and accurate data gathering, visualization and interpretation [9]. Finally, the revolution in fluorophores capable of reporting on a growing number of cellular processes has markedly improved the capabilities available to the investigator. This is especially true for the biotech industry where small proteins and oligonucleotides can be labeled without affecting the pharmacokinetics or pharmacologic effects of the agent. In addition, fluorescent labeling of proteins, either by genetic or chemical means of attachment using a wide spectrum of colors, makes simultaneous multi-colored imaging of different cellular processes possible [1–7].

Table 1. Investigational uses for multi-photon microscopy

Glomerular
- Size/volume
- Permeability/filtration
- Fibrosis/sclerosis

Microvasculature
- Blood flow rate
- Endothelial permeability
- WBC adherence/rolling
- Vasoconstriction

Cellular uptake
- Cell type specific uptake
- Site – apical vs. basolateral membrane
- Mechanism – endocytosis vs. carrier/transporter mediated

Cellular trafficking
- Intracellular organelle distribution
- Cytosol localization

Cellular metabolism
- Fluorescence decay over time

Cell toxicity
- Cell injury in necrosis, apoptosis
- Surface membrane/blebbing
- Mitochondrial function

Applications

Table 1 lists the types of data that can be obtained using multi-photon microscopy of the kidney. Both dynamic structural and functional observations are possible. Therefore, one can observe and correlate cell–cell interactions, structural changes, glomerular filtration, permeability, reabsorption, cellular metabolism, microvascular flow and the functional effects of a substance being administered. Since tissue fixation is not necessary these structural-functional studies can be undertaken with a large number of small as well as large molecules that cannot be fixed in tissues. In figure 1a we show the use of multiple fluorescent molecules to follow several processes simultaneously. Glomerular capillary blood flow is easily seen and can be quantified in the capillary network of Munich Wistar rat surface glomeruli. At the same time filtration of a small molecular weight (MW) Texas Red dextran is seen in Bowman's space. We have used this approach to quantify glomerular permeability, show that

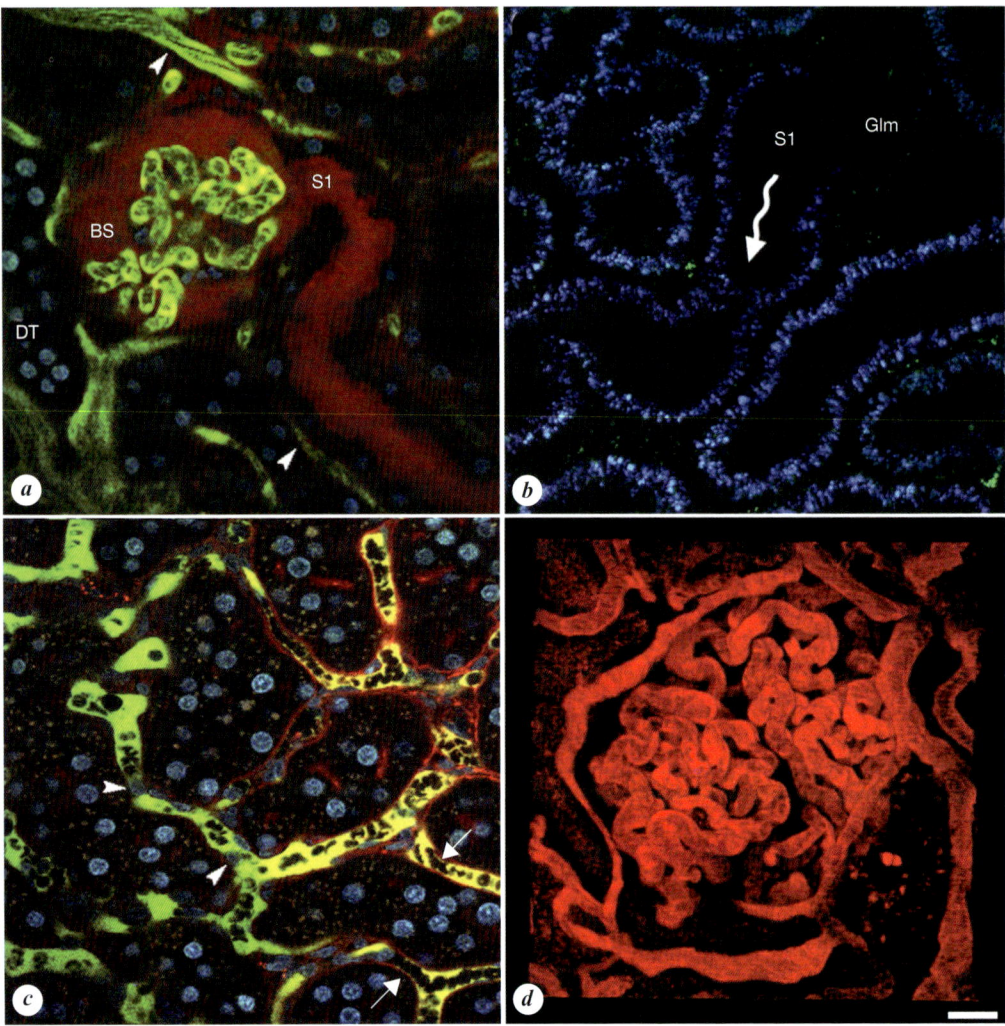

Fig. 1. Physiologic and morphologic parameters in the superficial rat kidney identified by intravital 2-photon microscopy. *a* Glomerular filtration and movement along a nephron segment. A large, non-filtering 500,000 MW fluorescein dextran (green) is retained within the capillary loops in the glomerulus (center) and the microvasculature (arrowheads). Within both of these structures circulating red blood cells (RBCs) appear as black oblong streaks because of exclusion of the large MW dextran. A small, freely filtered 3,000 MW Texas Red dextran (red) is seen filtering into Bowman's Space (BS) around the glomerulus and down into the S1 segment. A variation in nuclear morphology between cells types, labeled with Hoechst 33342 (cyan), can be seen between proximal tubule cells (S1), distal tubule cells (DT), and podocytes around the glomerulus (center). *b* Endocytic uptake by proximal tubule cells. A small, freely filtered 10,000 MW Cascade Blue dextran (blue) was given 24 h prior to imaging. The bulk of

molecular charge is not a determinant of filtration and the dissociation of protein-bound molecules prior to filtration [2, 10]. In figure1b we show the cell-specific uptake of a Cascade Blue-labeled dextran probe within the kidney. This agent is not bound to proteins, is freely filtered and rapidly taken up by proximal tubule cells across their apical membrane via endocytosis. No other cell type within the kidney either bound or internalized the dextran as shown by the lack of fluorescence in endothelial cells or in distal tubule cells. Within seconds of intravenous injection there was filtration and rapid binding to the apical membrane of proximal tubule cells. With increased time there was enhanced cellular accumulation, especially in lysosomal structures. Using total integrated fluorescence it is possible to quantify the extent of cellular uptake and to even partition it into apical binding and cellular accumulation [1, 2, 7]. Additional studies with folic acid (FA)-FITC revealed rapid loss of intracellular fluorescence following cellular uptake. These data did not indicate that FA was rapidly catabolized, but rather that FA was taken up into acidic compartments resulting in FITC quenching and loss of its fluorescent signal [11]. In fact, using the R-FA probe we documented transcytosis as a mechanism for reclamation of the reabsorbed FA [11]. Finally, recent data indicate that serum albumin undergoes glomerular filtration at a much greater level than previously believed [12]. Thereafter, it is rapidly reabsorbed by PTC and a fraction undergoes transcytosis, a process of reclaiming without intracellular catabolism.

the dextran can be seen localized within lysosomes (large punctate structures) at the basal portion of proximal tubule cells. The leading S1 proximal tubule segment with the open connection to Bowman's space is seen here adjacent to an unlabeled glomerulus (Glm). The arrow indicates the direction of flow. *c* Microvascular injury following exposure to endotoxin. With the same dyes used in (*a*), alterations in microvascular dynamics, primarily RBC flow are seen. The oblong streaks seen in (*a*) become more defined as RBCs due to the reduced flow and the presence of obstruction causing Rouleaux formations (arrows), and white cells adhering to the microvascular walls (arrowheads). The heterogeneity of this alteration is seen by the color profile of the dextrans in the blood. On the left half of the image, the blood, although slow, has circulated sufficiently that the small 3,000 MW dextran (red) has been cleared and only the large non-filtering 500,000 MW dextran (green) remains in the plasma. On the right half of the image, stagnation of flow in those vessels has prevented that pool of blood from reaching a glomerulus to filter out the 3,000 MW dextran (red). As a result, the color profile is yellow, a combination of the large (green) and small (red) dextrans occupying the same space in the plasma. Also visible is the vasoconstriction of the vessels to some degree, and small MW dextran leaking into the interstitial space around the microvasculature. *d* A 3D rendering of a surface glomerulus. Using VOXX software (developed at the Indiana Center for Biological Microscopy), focal planes taken at 1-μm intervals were rendered to produce a solid appearing composite. The complex inter-weaving of the capillary loops is readily seen along with the surrounding microvasculature. Bar = 20 μm.

Figure 1c shows the renal cortical microvasculature following endotoxin injection. Again, quantitation of erythrocyte flow, permeability alterations, and white blood cell (WBC) rolling, adherence and infiltration are possible. WBC within the microvasculature can be identified using Hoechst 33342 as a nuclear marker. This is also true for all nucleated cells within the tissue. Since the staining intensity is different for PTC and distal tubule nuclei we can use this to identify tubular segments. Rouleau formation can be noted. This is a common occurrence in several types of acute kidney injury including ischemia, sepsis, lipopolysaccharide and radiocontrast (unpublished observations, B.A.M.).

Figure 1d shows a 3D volume reconstruction of multiple z-axis sections of a cortical section containing surface glomeruli. This enables quantitative analysis of structural changes over time as multiple volumes can be collected in a time series. Finally, using powerful software programs like Amira one can segment out and quantify individual cells or areas of interest within cells [13, 14].

Numerous other approaches to understanding protein function and gene regulation have been developed. This is a rapidly growing field that will continue to make use of molecular and transgenic advances to selectively label individual proteins. These approaches will enable the study of specific proteins, compartments and processes in a dynamic fashion.

The use of fluorescent rationing to study events within the kidney was recently advanced by Yu et al. [10]. Their studies have outlined specific ways to evaluate glomerular permeability, glomerular sieving coefficients, and tubular reabsorption of different compounds as affected by size and charge selectivity using a generalized polarity concept. The use of this ratiometric concept has several advantages including minimizing errors secondary to the effect of intensity attenuation by tissue depth or fluctuations in excitation intensity or detector sensitivity on the overall quantitative process. As variations in fluorescence intensity can be a major problem in quantification, the use of ratiometric techniques is of great importance.

Once the fluorescent compound is within a cell, it then becomes possible to quantify its intracellular distribution and metabolism. Furthermore, it is quite possible to follow the intracellular accumulation and subcellular distribution over time in the same animal, and to undertake repeated observations in that animal at varying intervals over days to weeks. Furthermore, analysis of volumetric data, obtained by collecting images along the z axis, with quantitative software such as VOXX [9] or Amira [13] can yield additional information regarding cellular uptake and intracellular distribution. These studies can be particularly helpful in the pharmacokinetic understanding of drug delivery and metabolism at the individual cell level. However, one must remember that the fluorescence half life and biologic half life may vary and that specific studies are required to relate these to important parameters.

Specific intracellular organelles can be studied utilizing fluorescent dyes that have been developed to selectively label these organelles [1]. For example rhodamine-123 is utilized to label the mitochondria of tubular epithelial cells. As the fluorescence of this compound is directly related to the potential difference across the mitochondrial membrane, one can then develop quantitative assays for mitochondrial function for both acute and chronic studies. It is also possible to selectively label the mitochondria of endothelial cells and circulating WBCs utilizing rhodamine B hexyl ester which stays within the microvascular compartment. Again, quantitative assays can be developed to look at the individual number and fluorescence potential of mitochondria. This is an area where development of an internal standard for ratiometric imaging would be advantageous.

It is also possible to use the DNA fluorescent marker Hoechst to specifically evaluate intranuclear uptake of other fluorescent compounds and to identify specific cell types based upon their nuclear morphology. Identifying apoptosis in vivo, utilizing standard nuclear condensation criteria, can also be done following an acute injection of Hoechst 33342 [15]. As is shown in figure 1, different cellular nuclei have different morphologies allowing one to identify podocyte nuclei, distal tubule nuclei, proximal tubule nuclei and endothelial nuclei.

Challenges and Future Opportunities

Many challenges remain to maximize the ability to study and quantify cell–cell and subcellular processes at the cellular level within the kidney using multi-photon microscopy. Two major areas include image acquisition rates and the depth of penetration within the kidney. The studies presented in this review were recorded at approximately one frame per second. One can increase the acquisition rate by limiting the area of study, but this often sacrifices other important data. The depth of imaging possible, although 4–5 times greater than confocal imaging, remains limited to less than 200 μm for the kidney. Thus, we are unable to visualize the cortical-medullary area from the surface of the kidney. Perhaps new external detectors, access to longer wave length light sources, and lenses specifically designed for multi-photon microscopes will allow enhanced depth of penetration. Phototoxicity does remain a potential problem at the focal point of excitation and this must always be considered during study design. Quantifying the recorded results is also a major area under development. Continuing improvement in software and hardware has a goal of automation of data collection, segmentation and analysis. Finally, cost remains an obstacle for the individual PI, with core imaging facilities and expertise generally required [1].

In summary, recent developments in intravital multi-photon studies within the kidney now allow investigators to utilize unique techniques and fluorescent probes to visualize the functioning kidney and characterize cellular and subcellular events in a dynamic fashion. This approach will lead to enhanced understanding of renal physiology and the pathophysiology of disease processes and their therapy. This will result in more efficient and effective translation of preclinical data into therapeutic advances.

Acknowledgements

This work was made possible by National Institutes of Health grants P50 DK-61594, PO1 DK-53465 and RO1 DK-069408, a Veterans Affairs Merit Review (to B.A.M.), and an INGEN (Indiana Genomics Initiative) grant from the Lilly Foundation to the Indiana University School of Medicine.

References

1 Molitoris BA, Sandoval RM: Intravital multiphoton microscopy of dynamic renal processes. Am J Physiol Renal Physiol 2005;288:F1084–F1089.
2 Molitoris BA, Sandoval RM: Pharmacophotonics: utilizing multi-photon microscopy to quantify drug delivery and intracellular trafficking in the kidney. Adv Drug Deliv Rev 2006;58:809–823.
3 Zipfel WR, Williams RM, Webb WW: Nonlinear magic: multiphoton microscopy in the biosciences. Nat Biotechnol 2003;21:1369–1377.
4 Miyawaki A, Sawano A, Kogure T: Lighting up cells: labelling proteins with fluorophores. Nat Cell Biol 2003;suppl:S1–S7.
5 Ashworth SL, Tanner GA: Fluorescent labeling of renal cells in vivo. Nephron Physiol 2006;103: p91–p96.
6 Gerlich D, Ellenberg J: 4D imaging to assay complex dynamics in live specimens. Nat Cell Biol 2003;suppl:S14–S19.
7 Peti-Peterdi J: Multiphoton imaging of renal tissues in vitro. Am J Physiol Renal Physiol 2005;288: F1079–F1083.
8 Dunn KW, Sandoval RM, Kelly KJ, Dagher PC, Tanner GA, Atkinson SJ, Bacallao RL, Molitoris BA: Functional studies of the kidney of living animals using multicolor two-photon microscopy. Am J Physiol Cell Physiol 2002;283:C905–C916.
9 Clendenon JL, Phillips CL, Sandoval RM, Fang S, Dunn KW: Voxx: a PC-based, near real-time volume rendering system for biological microscopy. Am J Physiol Cell Physiol 2002;282: C213–C218.
10 Yu W, Sandoval RM, Molitoris BA: Quantitative intravital microscopy using a Generalized Polarity concept for kidney studies. Am J Physiol Cell Physiol 2005;289:C1197–C1208.
11 Sandoval RM, Kennedy MD, Low PS, Molitoris BA: Uptake and trafficking of fluorescent conjugates of folic acid in intact kidney determined using intravital two-photon microscopy. Am J Physiol Cell Physiol 2004;287:C517–C526.
12 Russo LM, Sandoval RM, McKee M, Osicka TM, Collins AB, Brown D, Molitoris BA, Comper WD: The normal kidney filters nephrotic levels of albumin retrieved by proximal tubule cells: Retrieval is disrupted in nephrotic states. Kidney Int 2007; Epub ahead of print.
13 Clendenon JL, Byars JM, Hyink DP: Image processing software for 3D light microscopy. Nephron Exp Nephrol 2006;103:e50–e54.

14 Phillips CL, Gattone VH 2nd, Bonsib SM: Imaging glomeruli in renal biopsy specimens. Nephron Physiol 2006;103:p75–p81.
15 Kelly KJ, Sandoval RM, Dunn KW, Molitoris BA, Dagher PC: A novel method to determine specificity and sensitivity of the TUNEL reaction in the quantitation of apoptosis. Am J Physiol Cell Physiol 2003;284:C1309–C1318.

Bruce A. Molitoris, MD
Indiana Center for Biological Microscopy, Indiana University School of Medicine
950 W. Walnut St., R2–202C
Indianapolis, IN 46202 (USA)
Tel. +1 317 274 5287, Fax +1 317 274 8575, E-Mail bmolitor@iupui.edu

Diuretics in the Management of Acute Kidney Injury: A Multinational Survey

Sean M. Bagshaw[a,g], *Anthony Delaney*[b,c], *Daryl Jones*[d], *Claudio Ronco*[f], *Rinaldo Bellomo*[a,e]

[a]Department of Intensive Care, Austin Hospital, Melbourne, Vic., [b]Department of Intensive Care, Royal North Shore Hospital, Sydney, [c]Northern Clinical School, University of Sydney, St. Leonards, [d]Department of Epidemiology and Preventive Medicine, Monash University, Melbourne, Vic., [e]Department of Medicine, Melbourne University, Melbourne, Vic., Australia, and [f]Department of Nephrology, St. Bortolo Hospital, Vicenza, Italy, [g]Division of Critical Care Medicine, University of Alberta Hospital, University of Alberta, Edmonton, Alta., Canada

Abstract

Background: Diuretics are a common intervention in critically ill patients with acute kidney injury (AKI). However, there is no information that describes the practice patterns of diuretic use by clinicians. **Methods:** Multinational, multicenter survey of intensive care and nephrology clinicians that utilized an 18-question self-reported questionnaire. **Results:** The survey generated 331 responses from 16 countries. Academic institutions comprised 77.5%, with the remaining being from metropolitan, regional or private hospitals. The use of furosemide was most common (67.1%), delivered primarily intravenously (71.9%) and by bolus dosing (43.3%). Other diuretics were infrequently used. The majority rated current serum creatinine (73.6%) and urine output (73.4%), blood pressure (59.7%), central venous pressure (65.2%) and risk of toxicity (62.4%) important when deciding on a dose. Pulmonary edema was a prime physiologic indication for diuretic use (86.3%). Diuretic use was also common with rhabdomyolysis (55.6%), major surgery (56%), and cardiogenic shock (56.2%), and sepsis (49.5%). Diuretic use was most commonly given either prior to (57.7%) or during recovery (33.9%) after renal replacement therapy (RRT). Most (76.6%) targeted a diuresis of $\geq$0.5–1.0 ml/kg/h. The majority did not believe that diuretics could reduce mortality (74.3%), reduce need for (50.8%) or duration of RRT (57.8%) or improve renal recovery (68.2%), however, many stated uncertainty. Most (85.1%) would be willing to participate in a randomized trial (RCT) of diuretics in AKI with 72.4% believing it ethically acceptable to allocate patients to placebo. **Conclusion:** Diuretics are frequently used in AKI. Clinicians are most familiar with furosemide given intravenously and titrated to a physiologic endpoint of urine output. Most clinicians believe an RCT on diuretic use in AKI is justified and ethical. This survey confirms clinical agreement and a need for higher quality evidence on diuretic use in AKI.

Copyright © 2007 S. Karger AG, Basel

Acute kidney injury (AKI) is a common complication in critical illness which portends a poor prognosis [1, 2]. Few interventions have been proven to impact the clinical course and outcome once AKI is established [3–5].

Diuretics, in particular loop diuretics, are a common intervention used in the management of critically ill patients with AKI [6, 7]. Loop diuretics act at the medullary thick ascending loop of Henle to inhibit the $Na^+/K^+/Cl^-$ pump on the luminal cell membrane surface and reduce oxygen demand [8, 9]. Theoretically, the timely use of loop diuretics might attenuate the severity of AKI. Diuretics may further play a vital role in managing extravascular volume overload by augmenting urine output and aid in acid-base and potassium homeostasis.

Small clinical studies have suggested that diuretics might diminish the severity of injury by converting 'oliguric' to 'non-oliguric' AKI, shorten the duration of AKI, improve the rate of renal recovery, and perhaps delay or ameliorate the need for renal replacement therapy (RRT) [10–15]. However, improvements in survival or renal recovery have yet to be shown by high quality research evidence [6, 16]. Accordingly, there is controversy regarding whether diuretics can impact clinical outcomes in AKI [17–20]. Thus, while there appears to be a biological rationale for their use, there is a limited understanding of how and when diuretics are used. Similarly, there are no data describing the beliefs, attitudes and practice patterns (i.e. indications, dosing regimens and end points for efficacy) used by clinicians routinely involved in caring for critically ill patients with AKI. Lastly, there is uncertainty on whether clinicians have a genuine need for a randomized controlled trial (RCT) assessing diuretics in AKI.

Therefore, as part of a larger initiative to understand the potential therapeutic role of diuretics in AKI, we have conducted a multinational multicenter survey of intensive care specialists, nephrologists and other clinicians who routinely manage critically ill patients with AKI in order to gain insight into the current patterns of practice and potentially aid in the future design of a RCT.

Methods

Target Population
The target population for this survey were intensivists, nephrologists and other clinicians who routinely provide care for critically ill patients with AKI.

Survey Objectives
This survey was undertaken first to explore the beliefs, attitudes and practice patterns of clinicians regarding the use of diuretics in the management of AKI. Specifically, the survey was designed to investigate several aspects about the use of diuretics in AKI including:

(1) frequency; (2) timing in the course of AKI; (3) indications; (4) class of diuretic (i.e. loop, proximal, potassium sparing, osmotic); (5) methods of administration; (6) dosing regimens; (7) what endpoints are used by clinicians to determine a response, and (8) whether clinicians believe the use of diuretics in AKI can impact outcome. In addition, the survey inquired on: (1) whether clinicians need a future RCT on diuretic use in AKI; (2) the clinical/physiologic characteristics on patients they would enroll, and (3) whether they had interest in participating in such a trial.

Questionnaire Development

The initial questionnaire included 18 questions. The initial 3 questions were focused on obtaining basic demographic information from the respondents (i.e. specialty, years of experience, practice type). The next 14 questions were designed to be uni-dimensional and included closed format questions that used either a 5-point Likert-type agreement scale or a dichotomous response (i.e. yes/no). Finally, the last question was open-ended and provided space for additional comments.

Pilot Testing of the Survey

The questionnaire was initially piloted using a focus group, comprised of clinicians with different levels of experience, to evaluate for clarity, comprehension and interpretation. The focus group was moderated by the principal investigators (S.M.B., D.J. and R.B.). All participants of the focus group contributed and provided feedback. During this process, the wording of several questions was modified. The survey was then piloted in a small sample of intensivists from two centers for additional feedback.

Survey Administration

The survey was intended to capture responses across a broad range of practices, in multiple centers and in multiple countries. The survey was sent to member centers of the Australia and New Zealand Intensive Care Society Clinical Trials Group, to member centers of a large multinational consortium of investigators, the Beginning and Ending Supportive Therapy for kidney investigators with an interest in AKI, to members of the Australia and New Zealand Society of Nephrology, and several member centers of the Canadian Critical Care Trials Group. Contacts at each center were then requested to circulate the survey to all members of their respective departments. A second reminder notice was sent to the same groups and centers at 4 weeks.

Clinicians were provided with three methods to respond to the survey. The first was electronic completion of the survey and return to the principle investigator (S.M.B.) as an e-mail attachment. The second was paper completion and return by either mail or fax. The final method was completion of an online version of the survey posted on Survey Monkey.com (available at the website: http://www.surveymonkey.com). Clinicians were provided with a direct link via e-mail to the survey and were able to complete the survey online. A total of 169 responses (51%) were completed online. The survey was conducted from September to November 2006.

Data Management and Analysis

Questionnaires returned by e-mail/fax/letter were manually entered into a master Excel spreadsheet (Microsoft Corp, Richmond, USA). Questionnaires completed on SurveyMonkey.com were exported in another Excel spreadsheet, reconfigured, and merged with the master Excel

Table 1. Distribution of countries and cities surveyed

Country	Cities	Responses
Australia	21	155 (48.9%)
Canada	7	46 (14.5%)
United States	3	32 (10.1%)
New Zealand	5	18 (5.7%)
China	7	14 (4.4%)
Japan	2	13 (4.1%)
Netherlands	3	11 (3.5%)
Belgium	3	11 (3.5%)
Italy	5	8 (2.5%)
Germany	1	2 (0.6%)
Brazil	1	1 (0.3%)
Czech Republic	1	1 (0.3%)
Russia	1	1 (0.3%)
Sweden	1	1 (0.3%)
United Kingdom	1	1 (0.3%)
Uruguay	1	1 (0.3%)
Unknown	–	1 (0.3%)
Total 16	63	317 (100%)

database. No assumptions were made about missing data and missing fields were not replaced. Categorical data are presented as proportions and compared using Fisher's exact test. Continuous data are presented as means (range) and compared using a Student's t test. Analysis was performed using STATA 8.2 (Stata Corp, College Station, USA).

Ethical Approval

This survey was approved by the Human Research Ethics Committee at the Austin Hospital prior to commencement.

Results

Characteristics of the Sampling Frame

The survey generated 331 responses. Of these, 14 (4.2%) were incomplete and provided no usable data. All these responses were generated from the online version of the survey. These 14 were excluded from the analysis.

The surveys were generated from 63 cities in 16 countries (table 1). Australia, New Zealand, Canada and the United States comprised 79.4% of responses. Most were from academic or tertiary institutions (77.5%) with smaller proportions from metropolitan, regional, rural, or private hospitals (fig. 1).

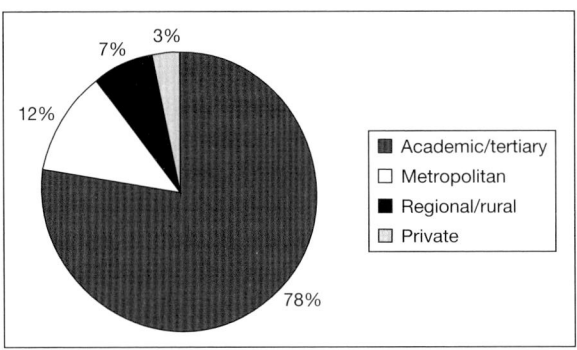

Fig. 1. Summary of center types sampled for the survey.

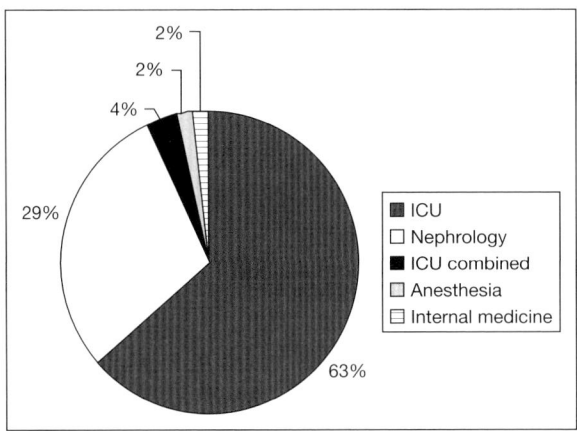

Fig. 2. Summary of clinical specialties sampled for the survey.

Demographics

Intensivists or those with combined ICU training represented 68.9%, whereas nephrology comprised 29.5% (fig. 2). All had generally been in practice for a median (intraquartile range) of 10 (5–18) years. The vast majority (94.9%) practiced adult medicine.

Details of How Diuretics Are Administered

Loop diuretics, specifically furosemide, was the most commonly used diuretic, with 67.1% using it 'frequently' or 'almost always'. Additional loop

Table 2. Summary of responses pertaining to factors influencing the dose of diuretic administered

Factor	Very unimportant, %	Unimportant %	Uncertain %	Important %	Very important, %
Patient weight (n = 311)	11.3	32.2	17	32.8	6.8
Serum creatinine (n = 314)	6.4	13.4	6.7	54.8	18.8
Urine output (n = 302)	3.2	14.6	8.9	55.6	17.8
Toxicity (n = 314)	4.1	20.1	13.4	49.7	12.7
Cardiac output (n = 315)	7.6	25.4	20.3	39.4	7.3
Blood pressure (n = 315)	6.7	21	12.7	46.7	13
CVP (n = 313)	5.8	16.6	12.5	47.9	17.3
PAOP (n = 313)	11.2	24.3	22.7	32	9.9
MVO_2 (n = 314)	11.5	31.9	32.2	21	3.5
PaO_2/FiO_2 ratio (n = 313)	8	29.1	30.4	29.4	3.2
MV (n = 312)	10.9	38.8	31.1	18.3	1

CVP = Central venous pressure; MV = mechanical ventilation; MVO_2 = mixed venous oxygen saturation; PAOP = pulmonary artery occlusion pressure.

diuretics such as torsemide, ethacrynic acid, and bumetanide were rarely used. Use of other classes/types of diuretics was also much less common. The majority responded either 'almost never' or 'infrequently' to use of hydrochlorozide (79.5%), spirolactone (79.7%), metolazone (81.1%), acetazolamide (88.8%), and mannitol (84.7%).

The majority of respondents deemed several factors as 'important' or 'very important' when determining the dose of diuretic to administer (table 2). These factors included current serum creatinine (73.6%), current urine output (73.4%), blood pressure (59.7%), central venous pressure (65.2%), and risk of toxicity (62.4%). Only a few respondents commented that patient age, baseline kidney function, and serum albumin also influenced the dose of diuretic to be administered.

Diuretics are primarily administered by the intravenous (IV) route (71.9% responded 'frequently' or 'almost always'). Diuretics are rarely given by the oral route in the setting of AKI (75.7% reporting 'almost never' or 'infrequently'). While both IV bolus and infusion are common, IV bolus appears to be used more frequently than an IV infusion.

A protocol to guide diuretic therapy was reported by only 5.3% (n = 16) of the respondents.

Indications and Timing of Administration

Pulmonary edema was the only reported physiologic indication where the vast majority (86.3%) responded to use of diuretics either 'frequently' or

Table 3. Summary of the responses pertaining to physiologic indications for use of diuretics in the management of ARF

Factor	Almost never, %	Infrequently %	Sometimes %	Frequently %	Almost always, %
Increasing SCr (n = 313)	29.4	21.1	31.6	14.4	3.5
Oliguria when SCr not known (n = 314)	29.9	22	22.9	19.1	6.1
Oliguria when SCr is increasing (n = 313)	16	19.8	32.3	21.4	10.5
Pulmonary edema (n = 314)	0.32	1	12.4	38.9	47.5
Metabolic acidosis (n = 314)	34.7	31.2	29.3	4.1	0.64
Hyperkalemia (n = 315)	7	16.5	38.1	25.1	13.3

SCr = Serum creatinine.

Table 4. Summary of the responses pertaining to the clinical scenarios for use of diuretics in the management of ARF

Factor	Almost never, %	Infrequently %	Sometimes %	Frequently %	Almost always, %
CIN (n = 315)	36.2	26	22.2	13	8
Rhabdomyolysis (n = 313)	22	15.7	28.8	26.8	6.7
ACS (n = 313)	39	26.5	22.4	8.6	3.5
Postoperative (n = 314)	18.5	21	28.3	27.7	4.5
Sepsis (n = 315)	26.4	23.2	29.2	16.5	4.8
Cardiogenic shock (n = 315)	19.1	16.8	28.3	27.9	7.9
Hemorrhagic shock (n = 314)	56.7	23.3	13.1	5.4	1.6

ACS = Abdominal compartment syndrome; CIN = contrast-induced nephropathy.

'almost always' (table 3). An increasing serum creatinine, oliguria where the serum creatinine is yet to be determined, and metabolic acidosis were seldom indications for diuretic use for most respondents. For both oliguria where the serum creatinine is increasing and hyperkalemia, diuretic use appears common, with most largely reporting such use either 'sometimes' or 'frequently'. Additionally, a few respondents also commented on the use of diuretics for metabolic alkalosis (post-hypercapnic), hypercalcemia, hypertension due to hypervolemia, as well as a single challenge to assess diuretic responsiveness.

The majority responded either 'infrequently' or 'almost never' to administering diuretics for AKI associated with hemorrhagic shock (80%), contrast-induced nephropathy (62.2%), and abdominal compartment syndrome (65.5%;

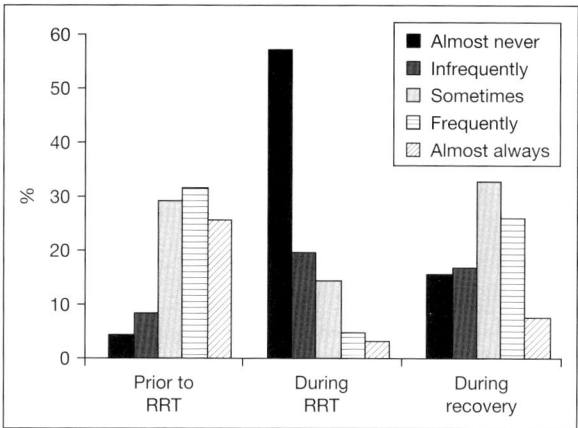

Fig. 3. Summary of responses pertaining to the timing of diuretic use in the management of ARF.

table 4). On the contrary, their use was more common in AKI associated with rhabdomyolysis, major surgery, and cardiogenic shock with respondents reporting 'sometimes' or 'frequently' in 55.6, 56, and 56.2%, respectively. In septic AKI, 49.6% responded either 'infrequently' or 'almost never', however 45.7% reported using diuretics 'sometimes' or 'frequently'. Other reported clinical indications also included hepatorenal syndrome and post-prostatectomy for clot retention.

The majority administer diuretics most commonly prior to initiation of RRT (57.7% responded 'frequently' or 'almost always') whereas only 8% responded similarly for use during RRT. Diuretics are also commonly used during the recovery phase with 33.9% having responded 'frequently' or 'almost always' (fig. 3).

Assessment of Clinical Response

A urine output of ≥0.5 or ≥1.0 ml/kg/h in response to diuretics was targeted by 76.6% of respondents. Only 11.5% set a target urine output of ≥1.5–2.0 ml/kg/h. A minority (8.3%) reported not using urine output to assess the response to diuretics in AKI. An additional 13.7% reported a variety of other targets (i.e. 80–100 ml/h, >1 l over 2 h, >1–1.5 l/day, any improvement in urine output and a sufficient increase to maintain a target fluid balance) and numerous other factors that contribute to their determination of a response to diuretics in AKI (i.e. volume status, serum electrolytes, pulmonary edema).

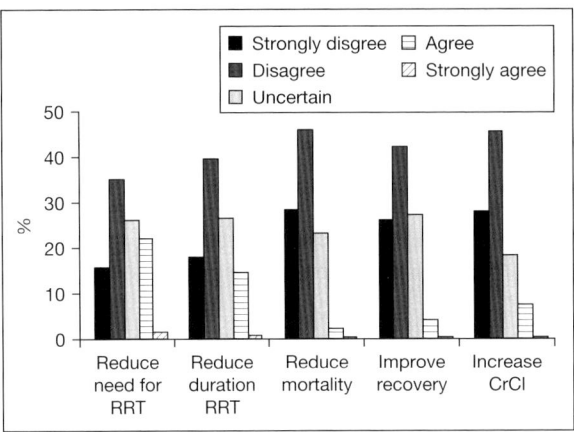

Fig. 4. Summary of responses pertaining to the beliefs about outcomes with the use of diuretics in the management of ARF.

While 17.8% responded that they do not specifically target a given fluid balance when using diuretics in AKI, 16.2% reported that they aim for a neutral balance and 35.9% targeted a negative daily balance in the range of 0.5–1 l. Another 30.2% reported that fluid balance goals were dependent on additional factors (i.e. present fluid balance status, presence of pulmonary or peripheral edema, non-renal organ dysfunction, presence of oliguria, and phase of AKI). The majority of respondents (65.8%) do not routinely use the change in serum creatinine or urea to assess whether there has been a response to diuretics. A few respondents used serum creatinine as a marker for toxicity or commented that a rise in serum creatinine coupled with a negative fluid balance may indicate either over-diuresis or relative hypovolemia.

Clinical Outcomes

The majority of respondents did not believe (either 'strongly disagree' or 'disagree') that the use of diuretics in AKI could reduce mortality (74.3%) or improve renal recovery (68.2%; fig. 4). However, 23.2 and 27.3% still reported uncertainty about whether diuretics could impact these outcomes. Similarly, most reported disagreement that diuretics could reduce the need for RRT (50.8%) or the duration of RRT (57.8%), however again, a significant proportion (26 and 26.7%) reported uncertainty.

In total, 72.4% of respondents would consider it ethical to give patients a placebo in an RCT of diuretics in AKI, whereas 15.7% believed it would be unethical. Another 11.9% provided additional comments such as it would depend on the protocol or trial design, and on the presence or absence of dangerous fluid overload or pulmonary edema. Overall, 85.1% stated they would be interested in participating in an RCT of diuretics in critically ill patients with AKI. Most responded that they would be willing to enroll patients with a variety of clinical features including: oliguria for 2 (57.3%) or 6 h (84%); an elevated serum creatinine (84.1%); septic shock (65.9%), and those likely to need RRT (88.3%).

Additional Comments

Additional comments were made by 78 (24.6%) respondents. While the comments varied, they were grouped into 4 broad themes. The majority of comments (73.1%) provided further details on beliefs about the clinical impact of diuretics, and when and for what indications diuretics are should be used in AKI. The majority mentioned diuretics in the context of fluid overload, improving urine output, enabling adequate nutrition, or simply that they do not use diuretics in AKI.

Another 19.2% provided feedback on issues related to RCT design. For example, remarks suggested that a trial protocol should include: provisions for exclusion of post-renal etiologies; provisions to ensure patients were not volume depleted; provisions to allow attending clinicians to place patients on RRT or give a rescue dose of diuretics if indicated, and finally suggestions for the enrollment of particular subgroups (i.e. cardiac surgical patients).

There were 3 (3.9%) comments that focused on the ethics of potentially withholding diuretics (i.e. allocated to placebo) and uncertainty about clinical equality in the context of the available literature. Finally, only 3 (3.9%) respondents remarked on the survey, in particular on the wording of questions and how the questions appeared to focus on the potential benefit of diuretics in AKI rather than harm.

Discussion

We have conducted a multinational multicenter survey of intensive care specialists, nephrologists and other clinicians routinely involved in the management of critically ill patients with AKI. Our principal objective was to gain insight into the existing beliefs, attitudes and self-reported practice patterns of diuretic use in patients with AKI.

There are a number of important findings from this survey. First, we found that furosemide is by far the most common and universally administered diuretic and that diuretics are generally given intravenously as a bolus, however, they are also commonly given by continuous IV infusion. A urine output in the range of 0.5–1.0 ml/kg/h is most commonly targeted. When deciding on the dose, clinicians consider several factors important, in particular, the baseline serum creatinine, current urine output, hemodynamics and the risk for toxicity. Second, we found that the presence of pulmonary edema was the most widely accepted indication for diuretics use. Moreover, clinicians would be reluctant to potentially enroll patients into a trial where placebo would be allocated to a patient with evidence of pulmonary edema unless there were clear provisions in the study protocol for rescue therapy if necessary. Third, the majority of respondents do not believe that use of diuretics in ARF will directly contribute to improved outcomes such as reduced mortality or need for RRT or increased renal recovery. On the other hand, a significant proportion of respondents were unsure, in particular on whether diuretics could reduce either the need for or the duration of RRT. Finally, we found that most believed that enrollment in an RCT, where patients could be allocated to placebo, would be considered ethical. What is more, most would be willing to participate in such an RCT and enroll patients with a variety of clinical presentations.

So far clinical trials of loop diuretics have been small, confounded by co-interventions (i.e. mannitol and dopamine), and characterized by delayed or late intervention (i.e. prolonged periods of oligo-anuria or already on RRT at enrollment). This is potentially important, as delay to appropriate therapy in AKI has been identified as a potential contributor to increased mortality and reduced the likelihood of renal recovery [21, 22]. Likewise, in these trials, furosemide was often given in large IV bolus doses, where no specific titration of therapy to physiologic endpoints such as urine output was performed. In addition, these trials failed to include critically ill patients. The evidence from these trials largely forms the basis for the prevailing view on whether diuretics can potentially impact outcome in AKI. Yet, interestingly, the pattern of practice as described in this survey would appear somewhat contradictory with this evidence, in particular when considering how and when diuretics are given. The majority of respondents in this survey administered diuretics early in the course of AKI, and 76.6 and 52.1% titrated them to urine output and fluid balance targets.

While only a minority reported not using urine output or fluid balance targets nor using diuretics at all during the course of AKI, many further clarified that they reserved the use of diuretics for patients with clinically important fluid overload. This may become a greater issue now considering that the incidence of AKI has recently been shown to be as high as 67% in critically ill patients [1]. More importantly, such AKI regularly occurs in the context of additional organ failure such as acute lung injury or sepsis that prompts the need for mechanical ventilation or vasoactive therapy [2, 23].

The risk of toxicity was identified by many respondents as an important determinant for dosing of diuretics in AKI. Prior trials have generally administered diuretics by large IV bolus doses [10–14, 24]. Such a pattern of administration may predispose patients to toxicity that can persist indefinitely, especially deafness, tinnitus and vertigo [10, 13, 14, 24]. Recently, a small pilot RCT was performed that compared intermittent bolus with continuous infusion protocols of furosemide titrated to urine output in critically ill patients with AKI (Ostermann et al., personal commun.). This trial found that while both protocols were able to realize a similar diuresis and fluid balance, use of a continuous infusion needed approximately half the total daily dose, thus had greater effectiveness (milliliters urine per milligram furosemide) when compared with intermittent bolus dosing. This would suggest that there is a rationale for the use of a continuous infusion that is titrated to hourly urine output and fluid balance goals to improve efficacy while reducing the potential for toxicity.

The findings of this survey, taken with the fact that most respondents are willing to participate and believe an RCT is ethical, suggest that there is not only clinical agreement, but also a need for higher quality and more definitive evidence on early diuretic use in AKI. The limitations to the available trial data on diuretics in AKI may, perhaps, account for the significant proportion of clinicians who are uncertain of whether diuretics can influence outcomes, and which outcomes in particular, for critically ill patients with AKI. Recent insights have suggested that timely attention to early AKI is warranted and that intervention with diuretics may improve not just short-term physiology but have the potential to impact clinically important measures in terms of both kidney and non-kidney organ function. Importantly, many clinicians identified the presence of pulmonary edema as a potential barrier to enrollment, where allocation to placebo would be considered unethical unless, however, the study protocol clearly allowed for rescue therapy if deemed necessary. Otherwise, the majority reported a willingness to enroll AKI patients with an array of criteria signifying that a broad critically ill population could be represented in such a trial.

This study has several strengths and limitations. First, we used an unconventional method for sampling, however we believe this was the most efficient and expedient method to support the objective of capturing a broad range of practice across several centers in several countries. Second, the sampling frame of this survey, while comprised of nearly 30% from nephrologists, was largely focused on intensivists. However, we believe this is justified when considering that AKI now largely occurs as a complication of critically illness where not only AKI, but all accompanying care is performed in an intensive care setting. Similarly, sampling occurred largely in tertiary/academic centers. While this may not completely reflect practice patterns overall, our rationale was to capture responses from centers with potential interest in participation in a clinical trial.

Conclusions

We have conducted a multinational multicenter survey of intensive care specialists, nephrologists and other clinicians routinely involved in the management of critically ill patients with AKI to better understand the current viewpoints and practice patterns of diuretic use associated with AKI. We have shown that clinicians are most familiar and comfortable with use of furosemide given intravenously and titrated to the physiologic endpoint of urine output and fluid balance. While many clinicians either do not believe or are uncertain that diuretics can directly contribute to improved outcomes in AKI, most clinicians believe an RCT on diuretic use in AKI is justified and ethical. Moreover, most would be willing to participate and enroll patients with a variety of clinical presentations. This survey confirms both clinical agreement and the need for higher quality and more definitive evidence on diuretic use in AKI.

Acknowledgements

S.M.B. is supported by clinical fellowships from the Canadian Institutes for Health Research and the Alberta Heritage Foundation for Medical Research Clinical Fellowship.

References

1. Hoste EA, Clermont G, Kersten A, et al: RIFLE criteria for acute kidney injury are associated with hospital mortality in critically ill patients: a cohort analysis. Crit Care 2006;10:R73.
2. Uchino S, Kellum JA, Bellomo R, et al: Acute renal failure in critically ill patients: a multinational, multicenter study. JAMA 2005;294:813–818.
3. Abel RM, Beck CH Jr, Abbott WM, Ryan JA Jr, Barnett GO, Fischer JE: Improved survival from acute renal failure after treatment with intravenous essential L-amino acids and glucose. Results of a prospective, double-blind study. N Engl J Med 1973;288:695–699.
4. Ronco C, Bellomo R, Homel P, et al: Effects of different doses in continuous veno-venous haemofiltration on outcomes of acute renal failure: a prospective randomised trial. Lancet 2000;356:26–30.
5. Schiffl H, Lang SM, Fischer R: Daily hemodialysis and the outcome of acute renal failure. N Engl J Med 2002;346:305–310.
6. Uchino S, Doig GS, Bellomo R, et al: Diuretics and mortality in acute renal failure. Crit Care Med 2004;32:1669–1677.
7. Van Biesen W, Yegenaga I, Vanholder R, et al: Relationship between fluid status and its management on acute renal failure (ARF) in intensive care unit (ICU) patients with sepsis: a prospective analysis. J Nephrol 2005;18:54–60.
8. DeTorrente A, Miller PD, Cronin RE, Paulsin PE, Erickson AL, Schrier RW: Effects of furosemide and acetylcholine in norepinephrine-induced acute renal failure. Am J Physiol 1978;235:F131–F136.

9   Kramer HJ, Schuurmann J, Wassermann C, Dusing R: Prostaglandin-independent protection by furosemide from oliguric ischemic renal failure in conscious rats. Kidney Int 1980;17:455–464.
10   Cantarovich F, Fernandez JC, Locatelli A, Perez Loredo J: Frusemide in high doses in the treatment of acute renal failure. Postgrad Med J 1971;47(suppl):13–17.
11   Cantarovich F, Rangoonwala B, Lorenz H, Verho M, Esnault VL: High-dose furosemide for established ARF: a prospective, randomized, double-blind, placebo-controlled, multicenter trial. Am J Kidney Dis 2004;44:402–409.
12   Karayannopoulos S: High-dose frusemide in renal failure. Br Med J 1974;2:278–279.
13   Kleinknecht D, Ganeval D, Gonzalez-Duque LA, Fermanian J: Furosemide in acute oliguric renal failure. A controlled trial. Nephron 1976;17:51–58.
14   Shilliday IR, Quinn KJ, Allison ME: Loop diuretics in the management of acute renal failure: a prospective, double-blind, placebo-controlled, randomized study. Nephrol Dial Transplant 1997;12: 2592–2596.
15   Vargas Hein O, Staegemann M, Wagner D, et al: Torsemide versus furosemide after continuous renal replacement therapy due to acute renal failure in cardiac surgery patients. Ren Fail 2005;27: 385–392.
16   Mehta RL, Pascual MT, Zhuang S, McDonald BR, Gabbai FB, Pahl MV: Effect of diuretic use on outcomes from acute renal failure (ARF) in the intensive care unit (ICU) (abstract). J Am Soc Nephrol 2001;12:229A.
17   Kellum JA: The use of diuretics and dopamine in acute renal failure: a systematic review of the evidence. Crit Care 1997;1:53–59.
18   Lameire N, Vanholder R, Van Biesen W: Loop diuretics for patients with acute renal failure: helpful or harmful? JAMA 2002;288:2599–2601.
19   Noble DW: Acute renal failure and diuretics: propensity, equipoise, and the need for a clinical trial. Crit Care Med 2004;32:1794–1795.
20   Schetz M: Should we use diuretics in acute renal failure? Best Pract Res Clin Anaesthesiol 2004; 18:75–89.
21   Gettings LG, Reynolds HN, Scalea T: Outcome in post-traumatic acute renal failure when continuous renal replacement therapy is applied early vs. late. Intensive Care Med 1999;25:805–813.
22   Bagshaw SM, Mortis G, Godinez-Luna T, Doig CJ, Laupland KB: Renal recovery after severe acute renal failure. Int J Artif Organs 2006;29:1023–1030.
23   Bagshaw SM, Laupland KB, Doig CJ, et al: Prognosis for long-term survival and renal recovery in critically ill patients with severe acute renal failure: a population-based study. Crit Care 2005;9: R700–R709.
24   Brown CB, Ogg CS, Cameron JS: High dose frusemide in acute renal failure: a controlled trial. Clin Nephrol 1981;15:90–96.

Dr. Sean M. Bagshaw
Division of Critical Care Medicine
University of Alberta Hospital, University of Alberta
3C1.16 Walter C. Mackenzie Centre
8440-112 Street, Edmonton, Alta. T6G 2B7 (Canada)
Tel. 001 780 407 6755, Fax 001 780 407 1228
E-Mail bagshaw.sean@gmail.com

Stem Cells In Acute Kidney Injury

Benedetta Bussolati, Giovanni Camussi

Renal Vascular Pathophysiology Laboratory, Department of Internal Medicine, University of Turin, Turin, Italy

Abstract

The susceptibility of developing acute renal failure depends on the ability of the kidney to recover from acute injury and regain normal function. Recently, the possible contribution of stem cells (SCs) to the regeneration of acute tubular injury has been investigated. There is evidence indicating that, under pathophysiological conditions, SCs derived from bone marrow are able to migrate in the injured kidney but they seem to play a minor role in tubular regeneration in regard to the resident cells. However, the administration of ex vivo expanded bone marrow-derived mesenchymal SCs has proven to be beneficial in various experimental models of acute renal failure. The mechanism underlining this beneficial effect is still matter of debate. The transdifferentiation or fusion of SCs to repopulate tubules is considered to play a minor role. The administered SCs may, however, modify the microenvironment by inducing dedifferentiation and proliferation of tubular cells surviving to injury or by allowing expansion of resident SCs. The recent identification of resident progenitor/SC populations in the adult kidney supports the hypothesis that resident SCs may play a critical role in the repair of renal injury. Therefore, therapeutic strategies to exploit the regenerative potential of SCs may be based on the administration of ex vivo expanded SCs or on stimulation of expansion and differentiation of local progenitor/SC populations.

Copyright © 2007 S. Karger AG, Basel

The ability of the kidney to recover following episodes of acute injury has a critical impact on patient morbidity and mortality in the hospital setting. Renal tubular cells are particularly susceptible to injury when exposed to endogenous cytokines as in sepsis, to endogenous or exogenous toxins as myoglobin or aminoglycosides and radiocontrast agents, or to episodes of renal ischemia [1]. The susceptibility of developing acute renal failure (ARF) depends on the ability of renal tubules to regenerate and regain normal function. The age of the patient and the severity of injury may condition recovery. After severe or repeated episodes of renal injury, recovery can be impaired or even fail leading to the need for long-term dialysis and to an increase in patient mortality [2]. Necrosis and loss of tubular epithelial cells is the most common event in ARF [3] and the recovery of renal function following ARF is dependent on the replacement of necrotic tubular cells with functional tubular epithelium [4, 5]. The absence or reduction of epithelial and endothelial regeneration may predispose to tubulo-interstitial scarring and chronic renal disease [6]. Studies on the physiological response to renal injury indicate that, after the insult has occurred, tubular cells dedifferentiate and acquire a mesenchymal phenotype. Dedifferentiated cells then migrate into the regions where tubular cells undergo necrosis, apoptosis or detachment with denudation of the tubular basement membrane. This process is followed by cell proliferation and eventually by their subsequent differentiation into functional epithelial cells with restoration of tissue integrity. The process of dedifferentiation, migration, proliferation and eventual redifferentiation is thought to be orchestrated by the local release of growth factors such as hepatocyte growth factor, epidermal growth factor, and insulin-like growth factor-1 [for review see, 7, 8].

It has been also suggested that the interstitium of the kidney contains adult renal stem cells (SCs) capable of contributing to renal repair [9]. In addition, SCs from bone marrow have been implicated in renal repair either by secretion of factors that protect the tubular cells, or by migration to the injured tubule and transdifferentiation into tubular epithelial cells [9].

SCs are characterized by their ability to self-renew and differentiate into a variety of cell types. Several studies have established the plasticity of bone marrow-derived SCs as they have the ability to cross lineage boundaries forming components of different tissues [10]. In bone marrow the hematopoietic SCs (HSCs) are present that give rise to the majority of the cellular components of blood and mesenchymal SCs (MSCs) which are able to differentiate into several cell types including osteoblasts, chondrocytes, adipocytes and myocytes [11]. Recently, resident adult SCs have been isolated from several tissues, including the central nervous system [12], retina [13], skeletal muscle [14] and liver [15]. Tissue SCs preferentially generate differentiated cells of the same lineage as their tissue of origin; however, it has been shown that they can also generate cells of different embryonic lineages [11]. It has been suggested that resident SCs play a relevant role in the postnatal growth of organs and in the physiological turnover of tissues. This is particularly relevant for epithelial organs with a relatively high rate of cellular turnover, such as the intestine, skin and kidney. In addition, resident SCs could play a relevant role in the replacement of injured epithelial cells and in the tissue regeneration after injury.

Although the clinical management of ARF patients has significantly improved over recent years, we lack specific therapies to improve the rate or efficiency of the repair process. In the present review, we focused on the possible contribution and therapeutic implication of SCs both derived from the bone marrow or resident in the kidney in the repair of acute renal injury.

Contribution of Bone Marrow-Derived SCs in Renal Regeneration

The possibility that bone marrow-derived SCs might functionally contribute to renal tubule regeneration is still controversial. Several studies have demonstrated the presence of Y chromosome-bearing cells in female kidneys transplanted into male recipients, suggesting that SCs derived from male bone marrow migrate and differentiate into the transplanted organ [16–18]. However, only a small percentage of the total cells present in the kidney was derived from bone marrow. Using whole bone marrow transplantation in the mouse, Poulsom et al. [16] demonstrated that bone marrow-derived cells could contribute to the regeneration of the renal tubular epithelium. Bone marrow-derived cells were also shown to ameliorate renal disease in a mouse model of Alport syndrome [19, 20]. In particular, a partial restoration of the expression of type IV collagen α_3 chain was related to the recruitment of bone marrow-derived progenitor cells within the damaged glomeruli and to their differentiation in podocytes and mesangial cells.

Recent studies limited the role of bone marrow-derived SCs to the regeneration of tubular epithelial cells. Using accurate detection of Y-bearing cells, the percentage of SCs in the renal epithelium was considered extremely low, approximately 0.1% [21, 22]. On the other hand, Lin et al. [23] showed that renal tubular resident cells provide major contribution to renal repair after ischemia-reperfusion injury. However, the studies confirmed an effective improvement in renal function provided by the bone marrow-derived SCs, comparing irradiated animals and non-irradiated animals or irradiated animals with reconstituted bone marrow [22, 23]. The hypothesis supported by these data is that the role of bone marrow-derived SCs in renal repair appears mainly to be limited to the support of growth factors that may contribute to stimulation of resident mature or SCs to proliferate and differentiate [24]. In addition, it has been shown that MSCs possess an immunomodulatory function [25]. It is conceivable that MSCs recruited by the kidney and present in the renal vessels or interstitium may limit the proinflammatory reaction, thus favoring tissue survival [24].

Effect of Administration of in Vitro Expanded SCs in the Treatment of Acute Renal Failure

Several studies agree with the finding that the administration of in vitro expanded SCs may protect and reverse ARF [21–23, 26, 27]. The experiments from Morigi et al. [26] demonstrated that the beneficial effect of the administration of bone marrow-derived SCs has to be ascribed to MSCs rather than to HSCs. In addition, the experiments of bone marrow mobilization suggest that HSCs have a potential detrimental effect on kidney regeneration due to mobilization of cells that may enhance inflammation [28].

The infusion of in vitro expanded MSCs protected and improved the recovery from acute tubular injury induced by cis-platinum and glycerol [26, 27]. In these models, localization of MSCs within regenerating tubules was observed. Both these models were characterized by extensive necrosis of proximal and distal tubules that may favor migration of MSCs within regenerating tubules. We found that injection of transgenic GFP MSCs rapidly induced functional recovery and GFP+ cells were detectable in tubules after regeneration [27]. The observed increase in proliferating tubular cells after MSC administration suggest a trophic effect of these cells on resident tubular cells that survived the injury (fig. 1). The beneficial effect of MSC infusion in acute renal injury induced by ischemia-reperfusion was also reported by Duffield et al. [22]. However, they only found an interstitial localization of SCs without incorporation into tubules or endothelium. The discrepancy in the results of MSC integration into tubules may depend on the different severity of the models used. However, there is a general agreement on the beneficial effect of MSC infusion in ARF due either to the ability of these cells to determine a microenvironment favoring proliferation of dedifferentiated epithelial cells or to stimulate expansion of resident SCs [24].

Renal Resident SCs

The presence of organ-specific progenitor cells capable of differentiating into epithelia, myofibroblasts and smooth muscle cells has been described in the embryonic rat metanephric mesenchyme, indicating the presence of embryonic renal SCs [29].

In the adult rat kidney, Oliver et al. [30] identified slow cycling SCs, based on the assumption that, in tissues, SCs are characterized by a low proliferating rate that maintains the self-renewal of such a population. Cells with a slow cycling time could be identified by the retention of bromodeoxyuridine (BrdU), which is incorporated in cell DNA during synthesis. BrdU-retaining cells were mainly found to be present in renal papilla [30]. Based on these observations, the authors proposed that the renal papilla is a niche for the adult kidney SCs in rats.

Maeshima et al. [31] also described the presence of BrdU-labelled cells in the renal tubules of adult rats. These cells, termed renal progenitor-like tubular cells, were shown to be able to re-enter in mitosis in response to renal injury.

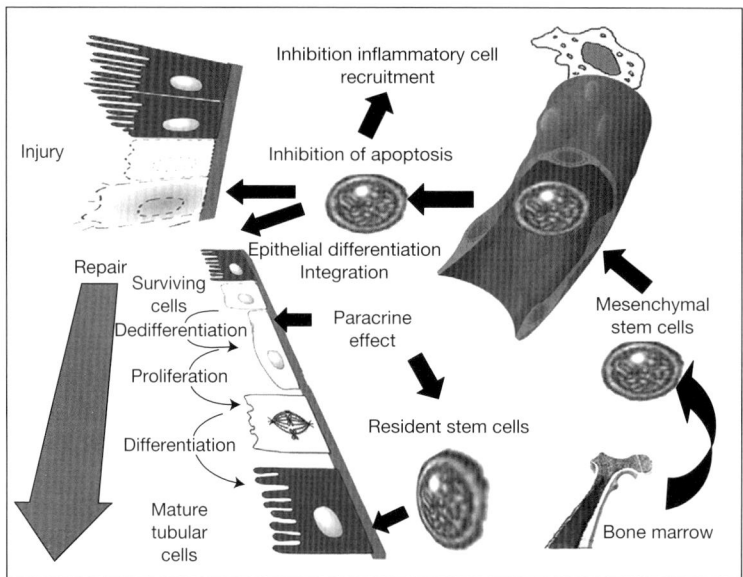

Fig. 1. Role of MSCs in the repair of acute tubular injury. MSCs derived from bone marrow may enter the circulation and reach the sites of tissue injury. MSCs can be recovered from bone marrow, expanded in vitro and administered for therapeutic purposes. The beneficial effects of MSCs on ARF may depend on their ability to inhibit apoptosis of tubular cells and recruitment of inflammatory cells. In several studies, the possibility has also been shown that some MSCs may transdifferentiate into mature epithelial cells. In addition, MSCs may have a paracrine effect on tubular cells surviving injury by stimulating their dedifferentiation, proliferation, migration and eventually redifferentiation into mature epithelial cells. In addition, the administered MSCs may modify the microenvironment by allowing expansion and differentiation of resident SCs that may contribute to the repopulation of injured tubules.

Kitamura et al. [32] identified a population of renal progenitor cells from the S3 segment of the nephron in the rat adult kidney. These cells, which express the renal embryonic markers Pax2, Wtn4 and Wtn1, were shown to be able to self-renew and to differentiate into mature epithelial cells expressing aquaporin 1 and 2, and responsive to parathyroid hormone and vasopressin. These cells contributed to tubular regeneration in rats with ischemia-reperfusion injury. Recently, Gupta et al. [33] demonstrated the presence in the adult rat kidney of a renal resident population expressing MSC markers, capable of self-renewal and multipotent differentiation. These cells expressed embryonic SC markers such as Octa-4 and Pax-2 and, when injected in injured kidneys after ischemia-reperfusion, contributed to tubular regeneration. Using a transgenic model of Octa-4-X-gal, it was found that Octa-4-positive cells were associated with the proximal tubules and were absent in the medulla. A MSC population has also been identified in the glomeruli of mice [34]. These cells were shown to be multipotent and to express smooth muscle myosin, like mesangial cells.

We recently demonstrated the presence of resident progenitors/SCs in the human adult kidney using CD133 as a SC marker [35]. CD133-positive cells were found in the interstitium in the proximity of proximal tubules and the glomerular capsule, and within tubular cells. These cells lacked the expression of hematopoietic markers (CD34 and CD45) and expressed some MSC markers, such as CD29, CD90, CD44 and CD73. Moreover, they expressed Pax-2, an embryonic renal marker, suggesting their renal origin. These cells were shown to undergo epithelial differentiation when cultured in vitro in the presence of hepatocyte growth factor and fibroblast growth factor-4. The differentiated epithelial cells expressed proximal and distal tubular markers and, when cultured in transwell filters, formed a polarized layer showing apical microvilli and junctional complexes. In vivo, undifferentiated CD133+ cells, injected subcutaneously in Matrigel into SCID mice, spontaneously differentiated in tubular structures expressing proximal and distal tubular epithelial markers. In addition, these cells were shown to undergo endothelial differentiation when cultured in vitro in the presence of VEGF. In vivo, when injected subcutaneously in Matrigel after the endothelial differentiation, CD133+ cells formed vessels connected with the mouse vasculature. Moreover, when injected into mice with ARF induced by glycerol, CD133+ renal progenitors homed to the kidney and integrated into proximal and distal tubules during repair, indicating the self-renewal ability of this population and the potential contribution to the repair of acute tubular injury.

Sagrinati et al. [36] recently reported the presence of a population of CD133+ CD24+ cells within Bowman's capsule of adult human kidneys which exhibited multipotent differentiation capabilities. Indeed these cells were able to generate not only mature tubular epithelial cells but also osteogenic cells, adipocytes and neuronal cells. CD133+ CD24+ SCs were shown to contribute to tubular regeneration in the model of glycerol-induced acute renal injury.

Recently, in mouse embryonic and adult kidneys, Challen et al. [37] described the presence of a SC population defined as a 'side' population. The side population is characterized by the unique ability to extrude Hoechst dye. This side population was found in particular in the tubular compartment and it was a heterogeneous population [37]. Some progenitor cells within this population displayed the characteristics of resident renal SCs and showed a multipotent differentiative ability and contributed to the repair of renal injury induced by adriamycin.

Conclusion

There is evidence that SCs may contribute to the regeneration of acute tubular injury. Under pathophysiological conditions, SCs derived from bone marrow, despite their ability to migrate in the injured kidney, play a minor role in tubular regeneration in regard to resident SCs. However, the administration of ex vivo expanded bone marrow-derived SCs was proved to be beneficial in various experimental models of ARF. The mechanism underlining this beneficial effect still remains unclear as transdifferentiation has been demonstrated only for a minimal portion of repopulated tubules. However, the administered SCs may modify the microenvironment by allowing expansion of resident SCs or by inducing dedifferentiation and proliferation of tubular cells surviving injury (fig. 1). The studies on the potential application of SCs may open new perspectives for a therapeutic approach to renal injury. This may be achieved either by administration of ex vivo expanded SCs or by strategies aimed to expand and differentiate local progenitor/SC populations.

References

1 Thadhani R, Pascual M, Bonventre JV: Acute renal failure. N Engl J Med 1996;334:1448–1460.
2 Star RA: Treatment of acute renal failure. Kidney Int 1998;54:1817–1831.
3 Kelly KJ, Molitoris BA: Acute renal failure in the new millennium: time to consider combination therapy. Semin Nephrol 2000;20:4–19.
4 Nash K, Hafeez A, Hou S: Hospital-acquired renal insufficiency. Am J Kidney Dis 2002;39: 930–936.
5 Sutton TA, Molitoris BA: Mechanisms of cellular injury in ischemic acute renal failure. Semin Nephrol 1998;18:490–497.
6 Tobrak FG: Regeneration after acute tubular necrosis. Kidney Int 1992;41:226–246.
7 Lameire N: The pathophysiology of acute renal failure. Crit Care Clin 2005;21:197–210.
8 Abouna GM, Al Adnani MS, Kremer GD, et al: Reversal of diabetic nephropathy in human cadaveric kidneys after transplantation into non diabetic recipients. Lancet 1983;2:1274–1276.
9 Cantley LG: Adult stem cells in the repair of the injured renal tubule. Nat Clin Pract Nephrol 2005;1:22–32.
10 Quesenberry PJ, Colvin GA, Abedi M, Dooner G, Dooner M, Aliotta J, Keaney P, Luo L, Demers D, Peterson A, Foster B, Greer D: The stem cell continuum. Ann NY Acad Sci 2005;1044:228–235.
11 Jiang Y, Jahagirdar BN, Reinhardt RL, Schwartz RE, Keene CD, Ortiz-Gonzalez XR, Reyes M, Lenvik T, Lund T, Blackstad M, Du J, Aldrich S, Lisberg A, Low WC, Largaespada DA, Verfaillie CM: Pluripotency of mesenchymal stem cells derived from adult marrow. Nature 2002;418:41–49.
12 Reynolds BA, Weiss S: Generation of neurons and astrocytes from isolated cells of the adult mammalian central nervous system. Science 1992;255:1707–1710.
13 Tropepe V, Coles BL, Chiasson BJ, Horsford DJ, Elia AJ, McInnes RR, van der Kooy D: Retinal stem cells in the adult mammalian eye. Science 2000;287:2032–2036.

14 Jackson KA, Mi T, Goodell MA: Hematopoietic potential of stem cells isolated from murine skeletal muscle. Proc Natl Acad Sci USA 1999;96:14482–14486.
15 Herrera MB, Bruno S, Buttiglieri S, Tetta C, Gatti S, Deregibus MC, Bussolati B, Camussi G: Isolation and characterization of a stem cell population from adult human liver. Stem Cells 2006;24:2840–2850.
16 Poulsom R, Forbes SJ, Hodivala-Dilke K, Ryan E, Wyles S, Navaratnarasah S, Jeffery R, Hunt T, Alison M, Cook T, Pusey C, Wright NA: Bone marrow contributes to renal parenchymal turnover and regeneration. J Pathol 2001;195:229–235.
17 Lin F, Cordes K, Li L, Hood L, Couser WG, Shankland SJ, Igarashi P: Hematopoietic stem cells contribute to the regeneration of renal tubules after renal ischemia-reperfusion injury in mice. J Am Soc Nephrol 2003;14:1188–1199.
18 Gupta S, Verfaillie C, Chmielewski D, Kim Y, Rosenberg ME: A role for extrarenal cells in the regeneration following acute renal failure. Kidney Int 2002;62:1285–1290.
19 Prodromidi EI, Poulsom R, Jeffery R, Roufosse CA, Pollard PJ, Pusey CD, Cook HT: Bone marrow derived-cells contribute to podocyte regeneration and amelioration of renal disease in a mouse model of Alport syndrome. Stem Cells 2006;24:2448–2455.
20 Sugimoto H, Mundel TM, Sund M, Xie L, Cosgrove D, Kalluri R: Bone-marrow-derived stem cells repair basement membrane collagen defects and reverse genetic kidney disease. Proc Natl Acad Sci USA 2006;103:7321–7326.
21 Kale S, Karihaloo A, Clark PR, Kashgarian M, Krause DS, Cantley LG: Bone marrow stem cells contribute to repair of the ischemically injured renal tubule. J Clin Invest 2005;112:42–49.
22 Duffield JS, Park KM, Hsiao LL, Kelley VR, Scadden DT, Ichimura T, Bonventre JV: Restoration of tubular epithelial cells during repair of the postischemic kidney occurs independently of bone marrow-derived stem cells. J Clin Invest 2005;115:1743–1755.
23 Lin F, Moran A, Igarashi P: Intrarenal cells, not bone marrow-derived cells, are the major source for regeneration in postischemic kidney. J Clin Invest 2005;115:1756–1764.
24 Krause D, Cantley LG: Bone marrow plasticity revisited: protection or differentiation in the kidney tubule? J Clin Invest 2005;115:1705–1708.
25 Le Blanc K: Immunomodulatory effects of fetal and adult mesenchymal stem cells. Cytotherapy 2003;5:485–489.
26 Morigi M, Imberti B, Zoja C, Corna D, Tomasoni S, Abbate M, Rottoli D, Angioletti S, Benigni A, Perico N, Alison M, Remuzzi G: Mesenchymal stem cells are renotropic, helping to repair the kidney and improve function in acute renal failure. J Am Soc Nephrol 2004;15:1794–1804.
27 Herrera MB, Bussolati B, Bruno S, Fonsato V, Mauriello-Romanazzi G, Camussi G: Mesenchymal stem cells contribute to the renal repair of acute tubular epithelial injury. Int J Mol Med 2004;14:1035–1041.
28 Togel F, Isaac J, Westenfelder C: Hematopoietic stem cell mobilization-associated granulocytosis severely worsens acute renal failure. J Am Soc Nephrol 2004;15:1261–1267.
29 Oliver JA, Barasch J, Yang J, Herzlinger D, Al-Awqati Q: Metanephric mesenchyme contains embryonic renal stem cells. Am J Physiol Renal Physiol 2002;283:799–809.
30 Oliver JA, Maarouf O, Martens TP, Cheema FH, Al-Awqati Q: The renal papilla is a niche for adult kidney stem cells. J Clin Invest 2004;114:795–804.
31 Maeshima A, Yamashita S, Nojima Y: Identification of renal progenitor-like tubular cells that participate in the regeneration processes of the kidney. J Am Soc Nephrol 2003;14:3138–3146.
32 Kitamura S, Yamasaki Y, Kinomura M, Sugaya T, Sugiyama H, Maeshima Y, Makino H: Establishment and characterization of renal progenitor like cells from S3 segment of nephron in rat adult kidney. FASEB J 2005;19:1789–1797.
33 Gupta S, Verfaillie C, Chmielewski D, Kren S, Eidman K, Connaire J, Heremans Y, Lund T, Blackstad M, Jiang Y, Luttun A, Rosenberg ME: Isolation and characterization of kidney-derived stem cells. J Am Soc Nephrol 2006;17:3028–3040.
34 da Silva Meirelles L, Chagastelles PC, Nardi NB: Mesenchymal stem cells reside in virtually all post-natal organs and tissues. J Cell Sci 2006;119:2204–2213.
35 Bussolati B, Bruno S, Grange C, Buttiglieri S, Deregibus MC, Cantino D, Camussi G: Isolation of renal progenitor cells from adult human kidney. Am J Pathol 2005;166:545–555.

36 Sagrinati C, Netti GS, Mazzinghi B, Lazzeri E, Liotta F, Frosali F, Ronconi E, Meini C, Gacci M, Squecco R, Carini M, Gesualdo L, Francini F, Maggi E, Annunziato F, Lasagni L, Serio M, Romagnani S, Romagnani P: Isolation and characterization of multipotent progenitor cells from the Bowman's capsule of adult human kidneys. J Am Soc Nephrol 2006;17:2443–2456.
37 Challen GA, Martinez G, Davis MJ, Taylor DF, Crowe M, Teasdale RD, Grimmond SM, Little MH: Identifying the molecular phenotype of renal progenitor cells. J Am Soc Nephrol 2004;15: 2344–2357.

Prof. G. Camussi
Cattedra di Nefrologia, Dipartimento di Medicina Interna
Corso Dogliotti 14
IT–10126 Torino (Italy)
Tel. +39 11 633 6708, Fax +39 11 663 1184, E-Mail giovanni.camussi@unito.it

Anticoagulation Options for Patients with Heparin-Induced Thrombocytopenia Requiring Renal Support in the Intensive Care Unit

Andrew Davenport

Centre for Nephrology, Division of Medicine, Department of Medicine, Royal Free and University College Medical School, London, UK

Abstract

World wide, heparins are the most commonly used anticoagulants for renal replacement therapy (RRT). In the intensive care unit (ICU) keeping the RRT circuit patent is more difficult than during routine outpatient hemodialysis, as ICU patients typically have sepsis and/or inflammation resulting in activation of the procoagulant pathways, with reduced antithrombin. One important cause of repeated RRT circuit clotting is heparin-induced thrombocytopenia (HIT), which should not be overlooked in patients with a reduced platelet count. If HIT is clinically suspected then all heparins should be withdrawn, and the patient systemically anticoagulated with either a direct thrombin inhibitor, such as argatroban and/or hirudin, or the heparinoid danaparoid. The availability and licensing of these alternative anticoagulants varies from country to country. Argatroban has to be continuously infused, which is an advantage for continuous RRT, but not for intermittent RRT, and can be monitored by activated partial thromboplastin time. Hirudin has a prolonged half life, which is extended by hirudin antibodies, and requires specialist monitoring to prevent over anticoagulation. Although the half life of danaparoid is increased in renal failure, it can be given as boluses for intermittent and continuous RRT, or by continuous infusion during continuous RRT, but requires factor Xa monitoring.

Copyright © 2007 S. Karger AG, Basel

Immune-mediated heparin-induced thrombocytopenia (HIT) is uncommon in general medical and/or surgical intensive care units (ICUs), with an incidence of <2% in those using low molecular weight heparins [1]. The incidence may be greater using unfractionated heparins, particularly bovine, and post-cardiac surgery [2]. In the ICU, critically ill patients have been reported to have up to 46% 'all cause' thrombocytopenia and many receive heparin (table 1) [1].

Table 1. Causes of thrombocytopenia in the ICU patient

Artefactual	Inadequate anticoagulation	Platelet clumping
	EDTA-activated agglutination	Platelet clumping
Dilutional	Massive blood transfusion	
Distributional	Hypersplenism	Sequestration
↓ Platelet production	Viral infections	Parvo virus
		Human immunodeficiency virus
		Epstein-Barr virus
	Drugs	Chemotherapy
		Radiotherapy
	Toxins	Alcohol
	Nutritional deficiencies	Folic acid
		Vitamin B_{12}
	Liver disease	Acute liver failure
	Bone marrow disorders	Severe sepsis
		Aplasia/hypoplasia
		Myelodysplastic syndromes
		Myeloproliferative syndromes
↑ Platelet destruction	Immune-mediated	Idiopathic ITP
	Drug	Heparin
		Sodium valproate
		Quinine
	Infection	Severe sepsis
		Human immunodeficiency virus
		Epstein-Barr virus
		Cytomegalovirus
	Allo-immune destruction	ABO incompatible transfusion
		ABO incompatible organ transplantation
	Nonimmune infection	Malaria
		Viral hemorrhagic fevers
		Disseminated intravascular coagulation
	Thrombotic thrombocytopenia	↓ Von Willebrand factor
		Cleaving protease
	Hemolytic uremic syndrome	Idiopathic
		Factor H deficiency/mutation
		Membrane complement-binding protein deficiency/mutation
		Drug induced

Table 1. (continued)

Antiphospholipid syndrome	
Pregnancy	HELLP syndrome
Drug-induced	Anti-platelet agents
	abciximab, eptifibatide
Mechanical destruction	Cardiopulmonary bypass

Table 2. The '4 Ts' scoring system to estimate the pretest probability of heparin-induced thrombocytopenia

	Score		
	2 points	1 point	zero
Thrombocytopenia ×10^9/l	20–100 or fall > 50%	10–19 or fall 30–50%	<10 or fall < 30%
Timing of onset in fall platelets	5–10 days heparin Rx	>10 days or timing not evident	<1 day heparin exposure
Thrombosis or acute systemic symptoms	Proven thrombosis Skin necrosis or acute systemic reaction	Progressive, recurrent, silent thrombosis or erythematous skin lesions	none
Other etiology for thrombocytopenia	None evident	possible	probable

Low probability ≤ 3; intermediate probability 4–6; high probability ≥ 6.

As there are many other causes of thrombocytopenia in the ICU patient other than HIT, a pre-test probability score, known as the '4 Ts' scoring system (table 2) has been developed based on both clinical and laboratory findings to help predict the likelihood that a patient will have HIT [1].

If HIT is clinically suspected, then all exposure to heparins (both unfractionated and low molecular weight heparin) must be stopped, as even heparin catheter locks and flushes can result in fatal reactions [3].

HIT develops due to the binding of heparin to platelet factor 4 (PF4), which is released from the α granules of activated platelets. When there is a key stoichimetric ratio of heparin molecules to PF4, binding leads to a conformational change in the PF4 molecule, allowing so-called neo-epitopes to be exposed, resulting in the formation of autoantibodies. Typically these are IgG antibodies against the PF4-heparin complex, although occasionally the autoantibodies are

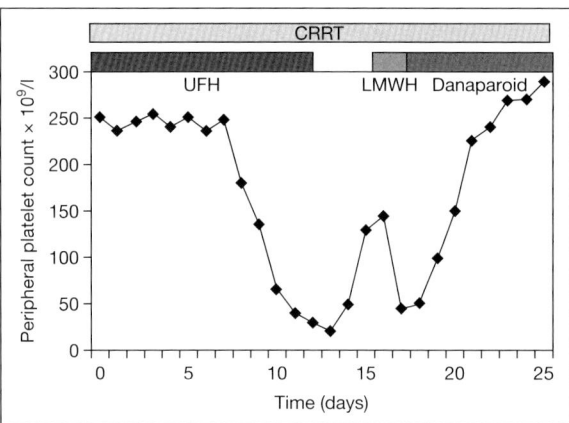

Fig. 1. The temporal change in peripheral platelet count in a patient who developed HIT during continuous renal replacement therapy (CRRT) and was anticoagulated with heparin, low molecular weight heparin (LMWH) and danaparoid. UFH = Unfractionated heparin.

of other isotypes and sometimes directed against other platelet-derived chemokines. Once formed the antibodies bind to platelets, causing platelet, monocyte and endothelial cell activation with increased potential for both arterial and venous thrombosis. The peripheral platelet count falls due to sequestration of the activated platelets (fig. 1). These antibodies can now be detected using commercially available ELISA kits and gel particle agglutination assays.

In patients with a high probability of HIT based on the '4 Ts' scoring system, then all heparin therapy should be withdrawn whilst awaiting confirmatory laboratory testing. If patients are subsequently found to be HIT positive, then despite heparin withdrawal there is an increased risk of thrombosis, even when the peripheral platelet count has normalized [4], and therefore there is a need for an alternative anticoagulant. Whereas in the chronic dialysis patient a regional anticoagulant may suffice to allow routine hemodialysis, in the ICU patient systemic anticoagulation is usually required. Currently two main groups of anticoagulants are available, the heparinoids and the direct thrombin inhibitors.

The Heparinoids

Danaparoid is a heparinoid composed of a mixture of 84% heparin sulfate, 12% dermatan sulfate and 4% chondroitin sulfate. Its mechanism of action remains to be fully elucidated, but it inhibits factor Xa and, to a lesser extent, thrombin [5]. Although in about 5–20% of cases danaparoid exhibits in vitro cross-reactivity with the HIT antibodies [2, 6], in vivo cross-reactivity is rare, with the occasional clinical case reported [7]. In non-renal patients, a bolus of 2,500 anti-Xa units is recommended followed by a continuous infusion reducing from 400 IU/h for 4 h, to 300 IU/h for a further 4 h, then to 200 IU/h as a maintenance dose [2]. However as danaparoid has a prolonged half life in renal failure, and is not cleared by renal replacement therapy (RRT), this dosing regimen has to be reduced in renal failure. Most centers use a bolus of 750 IU for continuous RRT, followed by a maintenance infusion starting at 1–2 U/kg · h, then adjusting the dose to maintain an anti-Xa of 0.2 to <0.35 IU/ml. For intermittent RRT, a bolus of 2,000–2,500 IU is used depending upon patient weight, and then adjusted according to the anti-Xa level. The recent development of bedside testing for anti-Xa will allow easier monitoring.

Dermatan sulfate has been used as a sole anticoagulant, with reports of a loading dose of 150 mg followed by an infusion of 15 mg/h. Dermatan sulfate acts through heparin cofactor II and inhibits thrombin and fibrin-bound thrombin. As such, the activated partial thromboplastin time (aPTT) has been used to monitor anticoagulation, aiming for a ratio of 1.0–1.4 over the laboratory baseline [6].

Fondaparinux is a synthetic analogue of the antithrombin-binding pentasaccharide found in heparins. Fondaparinux binds to antithrombin, predominantly inhibiting factor X, and so requires anti-Xa monitoring, but also has some action by inhibiting factor IXa [8]. It is renally excreted, so the plasma half life is markedly increased in renal failure [4], and single case reports have suggested an alternate day dose of 2.5 mg for hemodialysis. Although HIT antibodies have been reported with fondaparinux, these only poorly recognize fondaparinux, and are not thought to be clinically relevant [8].

Unlike unfractionated heparin, there is no specific antidote for the heparinoids in cases of hemorrhage, due to over anticoagulation. However recombinant factor VIIa has been used successfully to control hemorrhage in combination with tranexamic acid [9].

The Direct Thrombin Inhibitors

Argatroban directly inhibits free and clot bound thrombi. The half live is modestly prolonged in renal failure, to around 35 min, but it accumulates in liver failure [10]. Thus a bolus dose of argatroban (250 μg/kg) is followed by an infusion, starting at 0.5–2.0 μg/kg · min, then titrating the dose according to the aPTT, aiming for a ratio of 1.0–1.4. Currently, the clinical experience with argatroban is limited to North America.

Recombinant hirudin irreversibly inhibits free and clot-bound thrombin. The commercially available r-hirudin, lepirudin, has a markedly increased half life in renal failure, and accumulates during RRT. Although some lepirudin is removed by high flux membranes, in around 60% of patients antibodies develop and these prevent lepirudin removal during RRT. The relationship between aPPT and plasma hirudin is not linear, and as the aPTT ratio increases to 1.5 and above, the rate of rise in aPTT is much lower than the increase in the hirudin concentration, so increasing the risk of bleeding. Thus the risks of hemorrhage with lepirudin are associated with renal failure and increased aPPT ratios [11]. To overcome this problem, a direct thrombin test using viper venom, called the ecarin clotting time (ECT), has been introduced.

Bolus lepirudin doses of 0.2–0.5 mg/kg for intermittent hemodialysis have been suggested [2], but these are then often adjusted downwards in clinical practice, aiming for a hirudin concentration of 0.6–1.4 mg/l or an ECT of 80–100 s. For continuous RRT, hemorrhage has frequently been reported when a bolus was followed by a continuous infusion, thus an infusion starting at 0.005–0.01 mg/kg · h is recommended, and then adjusting the dose downwards. If the aPTT, ECT and or plasma hirudin start to increase, then it is probably best to stop the infusion and administer small boluses (0.007–0.05 mg/kg) [6], once the aPTT ratio has fallen to 1.0. As there is no specific antidote in cases of hemorrhage, activated factor VIIa has been used to control bleeding, and in cases without hirudin antibodies, hemodiafiltration can also be used to help clear lepirudin.

In addition with lepirudin, a small number of cases of anaphylaxis have been reported, with an estimated incidence of 0.015% on the first and 0.16% on the second exposure [2].

Conclusion

Extracorporeal anticoagulation for patients with acute kidney injury requiring RRT in the ICU is different from that of chronic kidney failure patients on routine outpatient hemodialysis. In the ICU, the majority of patients have some degree of sepsis and/or inflammation leading to activation of the procoagulant pathways shown by reduced levels of proteins S, C, antithrombin, and tissue factor pathway inhibitor, with increased plasminogen activator inhibitor-1 and tissue factor [12].

Although thrombocytopenia is relatively common in the ICU, an important cause not to be overlooked is HIT, as these patients are at risk of major thrombotic events, sudden cardiorespiratory collapse following bolus heparin administration, and persistent clotting of RRT circuits and vascular access catheters.

If HIT is clinically suspected then all heparin administration should cease, and patients systemically anticoagulated with a heparinoid or direct thrombin inhibitor, until laboratory testing for antibodies can be completed. Of the currently available anticoagulants argatroban has the shortest half life and has to be continued as a infusion, adjusted according to the aPPT ratio. However, argatroban is not universally available, and also accumulates in liver failure. Both the heparinoids, danaparoid and fondaparinux, and lepirudin have prolonged half lives in renal failure, and have no specific antidote in cases of over anticoagulation and hemorrhage. The heparinoids require anti-Xa monitoring, which has become easier with the development of bedside testing. Whereas hirudin should be monitored by either measuring the plasma hirudin concentration or ECT. After a few days of hirudin exposure, the majority of patients develop antihirudin antibodies, which further prolong the half life, and increase the risk of hemorrhage.

Argatroban is currently the systemic anticoagulant of choice for HIT in patients without severe liver disease but, if not available, then danaparoid is an effective alternative, and should be started if HIT is suspected and continued until the results of appropriate laboratory tests are available.

References

1. Wartentin TE, Cook DJ: Heparin, low molecular weight heparin, and heparin-induced thrombocytopenia in the ICU. Crit Care Clin 2005;21:513–529.
2. Keeling D, Davidson S, Watson H; Haemostasis and Thrombosis Task Force of the British Committee for Standards in Haematology: The management of heparin-induced thrombocytopenia. Br J Haematol 2006;133:259–269.
3. Davenport A: Sudden collapse during haemodialysis due to immune mediated heparin induced thrombocytopenia. Nephrol Dial Transplant 2006;21:1721–1724.
4. Arepally GM, Ortel TL: Heparin induced thrombocytopenia. N Engl J Med 2006;355:809–817.
5. Davenport A: Management of heparin induced thrombocytopenia during renal replacement therapy. Hemodial Int 2001;5:81–85.
6. Oudemans van straaten HM, Wester JPJ, de Pont ACJM, Schetz MRC: Anticoagulation strategies in continuous renal replacement therapy: can the choice be evidence based? Intensive Care Med 2006;32:188–202.
7. Keng TB, Chong BH: Heparin-induced thrombocytopenia and thrombosis syndrome: in vivo cross-reactivity with danaparoid and successful treatment with r-Hirudin. Br J Haematol 2001;114: 394–396.
8. Weitz JI: Emerging anticoagulants for the treatment of venous thromboembolism. Throm Haemostat 2006;96:274–284.
9. Huvers F, Slappendel R, Benraad B, van Hellemondt G, van Kraaij M: Treatment of postoperative bleeding after fondaparinux with rFVIIa and tranexamic acid. Neth J Med 2005;63:184–186.
10. Davenport A: Heparin induced thrombocytopenia during renal replacement therapy. Hemodial Int 2004;8:295–303.
11. Tardy B, Lecompte T, Boelhen F, Tardy-Poncet B, Elalamy I, Morange P, Gruel Y, Wolf M, Francois D, Racadot E, Camarasa P, Blouch MT, Nguyen F, Doubine S, Dutrillaux F, Alhenc-Gelas M, Martin-Toutain I, Bauters A, Ffrench P, de Maistre E, Grunebaum L, Mouton C, Huisse MG,

Gouault-Heilmann M, Lucke V; GEHT-HIT Study Group: Predictive factors for thrombosis and major bleeding in an observational study in 181 patients with heparin-induced thrombocytopenia treated with lepirudin. Blood 2006;108:1492–1496.
12 Russell JA: Management of sepsis. N Engl J Med 2006;19:1699–1713.

Dr. A. Davenport
UCL Centre for Nephrology, Royal Free and University College Medical School
Rowland Hill Street
London NW3 2PF (UK)
Tel. +44 20 783 02291, Fax +44 20 783 02125, E-Mail Andrew.Davenport@royalfree.nhs.uk

Nutritional Support during Renal Replacement Therapy

R. Chioléro, M.M. Berger

Department of Adult Intensive Care Medicine, CHUV, Lausanne, Switzerland

Abstract

Background/Aims: Malnutrition is common in critically ill patients with acute renal failure. The aim of this review is to describe the basis for nutritional support during renal replacement therapy. **Methods:** Review of the literature. **Results:** Techniques of nutritional support and nutritional requirements are described. **Conclusion:** Early aggressive enteral, parenteral or combine nutritional support is required in critically ill patients on replacement therapy.

Copyright © 2007 S. Karger AG, Basel

The prevalence of malnutrition is elevated in critically ill patients with acute renal failure (ARF), particularly in those requiring renal replacement therapy (RRT). Malnutrition is promoted by numerous factors related to the critical illness, renal failure, RRT and the nutritional support per se. Sepsis is commonly present in ARF and constitutes another risk factor of malnutrition in critically ill patients with ARF. The aim of this review is to describe and discuss the basis of nutritional support in such a condition.

Numerous data demonstrate the negative impact of malnutrition on clinical outcome in critically ill patients. There are few published data regarding ARF patients on RRT in the intensive care unit (ICU), but they also suggest that the occurrence of malnutrition throughout the stay is associated with poor outcome. In a prospective cohort of 309 patients requiring intermediate care, i.e. with moderately severe acute illness (mean APACHE II score 23.1), nutritional assessment was performed using the subjective global assessment, anthropometric, biochemical and immunology markers [1]. RRT was performed in 67% of the patients (intermittent hemodialysis in 56%, continuous in 11%). In-hospital mortality amounted to 39%. The overall prevalence of malnutrition was high: 42% of patients had a normal nutritional status, 16% were moderately malnourished and 42% had severe malnutrition. In-hospital mortality (OR 7.2, p 0.001) and the hospital length of stay (p < 0.001) were significantly increased in patients with malnutrition. Both septic and non-septic complications were significantly increased in severely malnourished patients, including the overall septic morbidity, septic shock, hemorrhage, intestinal complications, cardiogenic shock, arrhythmias, and acute respiratory failure. In multiple regression analysis, malnutrition was also an independent predictor of in-hospital mortality. These data confirm the high prevalence of malnutrition in ARF patients and its strong impact on clinical outcome and resource utilization.

Early and aggressive nutritional support is therefore a high therapeutic priority in such patients.

Metabolic Responses, Nutrient Losses

Extensive metabolic changes are induced by the critical illness: hypermetabolism, accelerated protein catabolism, and hyperglycemia related to insulin resistance constitute the hallmarks of the metabolic response to major stress. Published data suggest that ARF per se does not significantly influence the level of resting metabolism in critically ill patients, which is usually increased in surgical, trauma and septic patients. This contrasts with protein catabolism which is further enhanced by renal failure in comparison to critically ill patients without ARF [2].

RRT, particularly when continuous, also influences energy metabolism and substrate utilization. There is no direct effect of continuous RRT (CRRT) on energy metabolism [2]. A significant effect is only observed when the use of this supportive treatment is associated with thermal losses and temperature changes. Studies show that hypothermia is caused by CRRT in 5–50% of patients: this causes a proportional decrease in resting metabolic rate, except in patients with insufficient sedations, in whom shivering may induce a substantial elevation in the metabolic rate. The mean body temperature was decreased by 2.8°C in critically ill patients after CRRT initiation, while the mean VO_2 was decreased by 26% [3]. This will proportionally increase the caloric requirements.

Protein catabolism is markedly increased in most ICU patients requiring CRRT [2, 4]. In addition, substantial amino acid and protein losses, amounting to 10–20 g/day, are caused by CRRT according to the type of membrane and to the technique of replacement. Recent studies in ICU patients on CRRT suggest that protein and amino acid supply should be increased up to 2.0–2.5 kg/kg · day, to compensate for the accelerated protein catabolism and for the losses in the CCRT device [5, 6]. In one study, improved nitrogen balance and clinical outcome were associated with a high protein supply.

Hyperglycemia is present in 60–90% of fasted and nourished critically ill patients. In addition to the multiple factors which decrease insulin sensitivity in such a condition, glucose loss in CRRT may be substantial. Glucose loss is dependent on the glucose concentration of the replacement solutions: the use of low-glucose solutions (0.1–0.15%) is associated with small glucose loss (about 4% net glucose loss), whereas solutions containing 1% or more glucose induce net absorption of glucose [2]. In 8 patients receiving CRRT and replacement solution without glucose, we observed significant glucose loss in the effluent, amounting to 60 g/day [7]. Such loss obviously requires increased supply to insure sufficient carbohydrate administration.

Micronutrients

The critically ill are exposed to an increased oxidative stress, partly due to their illness, but also to the inflammatory response, and to the oxidative side effects of ICU treatments (high FiO_2 mechanical ventilation, transfusion, CRRT). A series of micronutrients are essential for the endogenous antioxidant defenses, mainly selenium, manganese, zinc, vitamins C and E, while copper and iron have both antioxidant and prooxidant properties [8]. A German randomized supplementation study in 42 infected ICU patients using moderate doses of selenium (500 μg/day) showed that 9 days of supplementation were associated with a reduction in severe ARF and a trend to better survival [9]: the higher plasma levels resulting from the supplements were associated with higher glutathione peroxidase activity – a major antioxidant selenoenzyme.

Trace element and vitamin metabolism is altered in both chronic renal failure and ARF, mostly due to losses through the membranes but also to insufficient nutritional intakes. Low plasma concentrations of selenium and zinc have repeatedly been shown, as well as vitamin C and vitamin E [10–12]. Balance studies have been difficult to realize due to the low detection limits of the analytical methods used until recently. Story et al. [11] analyzed micronutrient losses in the ultrafiltrate: they were small or undetectable except for vitamin C, chromium, and copper. Of note, the lipid-soluble vitamins were not detectable. Our group [12], using inductively coupled plasma mass spectrometry showed that CRRT causes significant losses and negative micronutrient balances whatever the buffer solution: the 4 micronutrients analyzed, selenium, copper, zinc and thiamine, were lost in large quantities (table 1) and contributed to low plasma concentrations. While zinc was lost in the effluent, the amounts provided by the replacement solutions counteracted the losses causing modestly

Table 1. Micronutrients at risk during CRRT: comparison of micronutrient losses in the effluent, recommended daily parenteral nutrition intakes, and quantities provided by industrial intravenous supplements

Micronutrient	Losses/24 h mean values	Daily PN, recommended intakes	Range of doses provided by industrial PN supplements[1]
Chromium	25 μmol	15 μg	10–15 μg
Copper	0.41 mg	1.0–1.2 mg	0.48–1.3 mg
Selenium	110 μg	60 μg	24–70 μg
Zinc	0.2 mg	6.5 mg	3.3–10 mg
Vitamin B1	4.1 mg	3 mg	3.0–3.51 mg
Vitamin C	10 mg	100 mg	100–125 mg
Vitamin E	ND	10 IU	10–10.2 IU

ND = Not determined; PN = parenteral nutrition.
Data from Story et al. [11] and Berger et al. [12].
[1]Trace elements: Tracutil®, BBraun; Addamel-N®/Tracitrans®, Fresenius-Kabi; Décan®, Aguettant; vitamins: Soluvit®/Vitalipid®, Fresenius-Kabi; Cernevit®, Baxter. Note: Decan® had the highest content for Cr-Se-Zn, lowest for Cu-Fe, and highest vitamin content was in Cernevit®.

positive balances. Considering these results, it is likely that all water-soluble micronutrients are lost in the effluent due to the membrane characteristics, resulting in rapid depletion of the patient's pools.

Nutritional Support in CRRT Patients

Prevention of malnutrition is a high priority goal in critically ill patients with ARF. This requires the administration of early nutritional support, i.e. within the first 24–48 h after ICU admission: such a strategy avoids the accumulation of markedly negative daily energy balances in patients with frequent pre-ICU malnutrition. Recent studies show that septic and nonseptic complications are increased in fasted critically ill patients or in patients receiving hypocaloric feeding: observations in critically ill patients show that the incidence of bacteremia was significantly increased after only 2 days of hypocaloric feeding [13].

The use of the enteral route should be promoted in all ICU patients, including those with ARF: numerous studies show that full or partial enteral feeding is possible in a large proportion of these patients. This strategy was recently supported by the implementation of the Canadian guidelines for nutritional support in critically ill patients on mechanical ventilation, showing that full enteral support or combined enteral and parenteral feeding were possible in 89% of critically ill patients [14].

In a multicenter cohort study including 17,126 ICU patients requiring RRT, the use of enteral feeding was associated with an improved probability of survival [15]. Combined enteral and parenteral or uncommonly exclusive parenteral nutrition should be administered to patients totally or partially intolerant to enteral nutrition to avoid prolonged hypocaloric feeding.

There are no fixed rules to determine the nutritional requirements of all critically ill patients with ARF, since they are not a homogenous group of patients: as previously stated, the type and severity of critical illness, severity of ARF and use of CRRT (technique and dose) are the main determinants of the nutrition requirements [16]. Keeping such limitations in mind, general recommendations can be provided for ARF patients on CRRT, which constitute a more homogenous group.

Provision of energy should be normal to high, amounting to 20–30 kcal/kg · day. It should be increased in patients with major burns, severe multiple injury or severe sepsis. It should be decreased in morbidly obese patients in whom energy supply should be based on ideal body weight. Monitoring of energy balance is mandatory in patients with prolonged and complicated evolution to avoid dangerous accumulation of energy deficits or inappropriate hypercaloric feeding. Indirect calorimetry, whenever available, is useful to provide the targeted caloric supply.

Protein supply should be increased during CRRT. In the recent ESPEN recommendations, they are set at 1.7 g/kg · day or below [16]. As explained previously, values up to 2.0–2.5 g/kg · day may improve both the nitrogen economy and the clinical outcome [5, 6, 17].

Fat supply should be limited to 20–30% of total energy supply, as in all critically ill patients. The literature does not furnish data to provide specific recommendation concerning the type of fat that should be administered. Recent data in critically ill patients without CRRT suggest that n–3 polyunsaturated fatty acids, derivate from fish oil, may be useful in patients with acute respiratory distress syndrome or sepsis/septic shock.

Carbohydrates should be the main source of caloric supply in CRRT patients, as in other critically ill patients.

Micronutrient Supply

Micronutrient requirements are increased in the vast majority of critically ill patients [8], and particularly in those with increased losses and acute deficiencies such as ARF patients. Therefore early supplementation/substitution is required to avoid aggravating the oxidative stress and altering macronutrient metabolism. There are no ready mixtures for this purpose on the market, as industrial intravenous supplements are indeed intended for stable parenteral nutrition patients, and not to match for additional losses (table 1). Considering that modestly positive zinc balances are not deleterious, the simplest option is to provide a double dose of one of the existing trace element preparations, even when the patients are on enteral nutrition. Selenium and thiamine appear from actual data to be the micronutrients at highest risk of depletion. Therefore additional 100 µg of selenium and 100 mg of thiamine should be delivered daily intravenously while on CRRT.

Glucose Control

Studies in critically ill patients during the last decade suggest that a tight control of the blood glucose concentration should be used in all critically ill patients. Recent studies show that such metabolic control decreases the occurrence of renal failure and improves clinical outcome in patients with or without ARF. In a first randomized controlled study performed by van den Berghe et al. [18] in 1,548 surgical critically ill patients, the probability of developing ARF and requiring RRT was decreased by 35 and 41%, respectively, by a strict control of glycemia (4.4–6.1 vs. <12.0 mmol/l). This was associated with a significant decrease in mortality, length of stay, prolonged mechanical ventilation and septic morbidity. Another study performed by the same authors in medical critically ill patients again found a decreased occurrence of ARF in patients receiving strict glycemic control but there was no effect on survival [19]. Despite such promising results, there is presently hot debate concerning the optimal target of blood glucose control in ICU patients. It may well be that different targets are required in different populations. Focusing on patients with ARF or requiring RRT there is no doubt that avoiding hyperglycemia is a high priority therapeutic goal, although the precise target remains yet controversial.

Conclusion

ARF patients have a high prevalence of malnutrition. Providing adequate nutritional support is a high priority goal in those critically ill patients requiring CRRT. Provision of early enteral nutrition, adequate amounts of energy, protein and micronutrients constitute the most important points with special accent on the elevated protein and micronutrient requirements. Metabolic monitoring is mandatory, particularly for blood glucose and energy balance control.

References

1 Fiaccadori E, Lombardi M, Leonardi S, Rotelli C, Tortorella G, Borghetti A: Prevalence and clinical outcome associated with preexisting malnutrition in acute renal failure: a prospective cohort study. J Am Soc Nephrol 1999;10:581–593.
2 Wooley JA, Btaiche IF, Good KL: Metabolic and nutritional aspects of acute renal failure in critically ill patients requiring continuous renal replacement therapy. Nutr Clin Pract 2005;20: 176–191.
3 Matamis D, Tsagourias M, Koletsos K, et al: Influence of continuous haemofiltration-related hypothermia on haemodynamic variables and gas exchange in septic patients. Intensive Care Med 1994;20:431–436.
4 Frankenfield D, Reynolds H: Nutritional effect of continuous hemodiafiltration. Nutrition 1995;11:388–393.
5 Scheinkestel CD, Adams F, Mahony L, et al: Impact of increasing parenteral protein loads on amino acid levels and balance in critically ill anuric patients on continuous renal replacement therapy. Nutrition 2003;19:733–740.
6 Scheinkestel CD, Kar L, Marshall K, et al: Prospective randomized trial to assess caloric and protein needs of critically Ill, anuric, ventilated patients requiring continuous renal replacement therapy. Nutrition 2003;19:909–916.
7 Bollmann MD, Revelly JP, Tappy L, et al: Effect of bicarbonate and lactate buffer on glucose and lactate metabolism during hemodiafiltration in patients with multiple organ failure. Intensive Care Med 2004;30:1103–1110.
8 Berger M, Shenkin A: Update on clinical micronutrient supplementation studies in the critically ill. Curr Opin Clin Nutr Metab Care 2006;109:711–716.
9 Angstwurm M, Schottdorf J, Schopohl J, Gaertner R: Selenium replacement in patients with severe systemic inflammatory response syndrome improves clinical outcome. Crit Care Med 1999;27:1807–1813.
10 Richard M, Ducros V, Foret M, et al: Reversal of selenium and zinc deficiencies in chronic hemodialysis patients by intravenous sodium selenite and zinc gluconate supplementation – time-course of glutathione peroxidase repletion and lipid peroxidation decrease. Biol Trace Elem Res 1993;39:149–159.
11 Story D, Ronco C, Bellomo R: Trace element and vitamin concentrations and losses in critically ill patients treated with continuous venovenous hemofiltration. Crit Care Med 1999;27: 220–223.
12 Berger M, Shenkin A, Revelly J, et al: Copper, selenium, zinc, and thiamine balances during continuous venovenous hemodiafiltration in critically ill patients. Am J Clin Nutr 2004;80: 410–416.
13 Rubinson L, Diette GB, Song X, Brower RG, Krishnan JA: Low caloric intake is associated with nosocomial bloodstream infections in patients in the medical intensive care unit. Crit Care Med 2004;32:350–357.
14 Heyland DK, Dhaliwal R, Day A, Jain M, Drover J: Validation of the Canadian clinical practice guidelines for nutrition support in mechanically ventilated, critically ill adult patients: results of a prospective observational study. Crit Care Med 2004;32:2260–2266.
15 Metnitz P, Krenn C, Steltzer H, et al: Effect of acute renal failure requiring renal replacement therapy on outcome in critically ill patients. Crit Care Med 2002;30:2051–2058.
16 Cano N, Fiaccadori E, Tesinsky P, et al: ESPEN Guidelines on Enteral Nutrition: adult renal failure. Clin Nutr 2006;25:295–310.
17 Bellomo R, Tan H, Bhonagiri S, et al: High protein intake during continuous hemodiafiltration: impact on amino acids and nitrogen balance. Int J Artif Organs 2002;25:261–268.

18 Van den Berghe G, Wouters P, Weekers F, et al: Intensive insulin therapy in critically ill patients. N Engl J Med 2001;345:1359–1367.
19 Van den Berghe G, Wilmer A, Hermans G, et al: Intensive insulin therapy in the medical ICU. N Engl J Med 2006;354:449–461.

René Chioléro, MD
Department of Intensive Care Medicine, CHUV, BH 08.610
CH–1011 Lausanne (Switzerland)
Tel. +41 21 314 20 02, Fax +41 21 314 30 45, E-Mail rene.chiolero@chuv.ch

Vascular Access for HD and CRRT

Miet Schetz

Department of Intensive Care Medicine, University Hospital Gasthuisberg, Leuven, Belgium

Abstract

A good functioning vascular access is an essential component for adequate renal replacement therapy (RRT) in acute kidney injury. Tunneled, cuffed catheters are preferred if the anticipated duration of RRT is more than 3 weeks. The right jugular vein is the preferred insertion site for the temporary dialysis catheter (TDC), with ultrasound-guided insertion reducing the risk of mechanical complications. The femoral vein is the second choice, whereas the subclavian vein should be avoided. The most important complications of a TDC are acute malfunction and infection. Intraluminal thrombosis, fibrin sleeve formation, malpositioning and kinking result in acute malfunction. Recirculation can be reduced by correct placement of the catheter and is more an issue for intermittent hemodialysis than for continuous RRT. Strict adherence to simple preventive strategies reduces catheter-related bloodstream infection. In selected patients more sophisticated strategies such as the use of antibiotic/antiseptic impregnated catheters and antibiotic/antiseptic lock solutions may be useful.

Copyright © 2007 S. Karger AG, Basel

The delivery of an adequate dialysis dose is important for survival both in acute and chronic renal failure [1–3]. An essential component to perform this treatment is a good functioning vascular access, the ideal access being able to deliver an adequate blood flow without associated complications. Whereas in chronic dialysis patients a native forearm arteriovenous fistula is the optimal access [4], renal replacement therapy (RRT) for acute kidney injury (AKI) is mostly performed with a temporary dialysis catheter (TDC), which is easily inserted at the bedside and immediately usable. The importance of an adequate access is illustrated by the fact that lower than prescribed blood flows contribute to inadequate dialysis doses during intermittent hemodialysis in patients with AKI [5]. In continuous RRT vascular access is a major determinant of filter life span, with access malfunction contributing to 3–25% of filter exchanges

[6–10]. Literature data on vascular access for acute dialysis are relatively scarce. Studies on TDCs are mostly performed in chronic dialysis patients, whereas studies in critically ill patients are mostly on central venous catheters (CVCs). TDCs have a different design than CVCs and are manipulated differently, and chronic dialysis patients differ in many respects from critically ill patients, raising the question whether results from these studies can be extrapolated to dialysis catheters in critically ill patients.

Catheter Characteristics

The material of TDCs should be sufficiently rigid to allow insertion and maintain lumen patency, it should be flexible to prevent kinking, thromboresistant and resistant to bacterial invasion. Modern TDCs are made of polyurethane or silicone which are less thrombogenic than older materials [11]. Polyurethane catheters are semi-rigid and easier to insert with the Seldinger technique but are associated with a higher risk of endothelial damage or venous perforation. Polyurethane has thermoplastic properties, being rigid during placement but softening when soaked at body temperature. Silicone is mostly used for tunneled catheters that are soft and flexible, but are more difficult to insert and are associated with more mechanical failure due to compression of the lumen. On the other hand the danger of perforation is lower. These catheters can therefore be placed in the right atrium, allowing better blood flow and a reducing the risk of recirculation and malfunction [12, 13].

Although some centers use two separate single lumen catheters inserted into two different large veins or into the same vein, most acute RRT is performed with a double-lumen catheter inserted into a large vein. The outer diameter of the double-lumen TDC varies between 11 and 14 french. The two lumens can be arranged side-by-side (double-D or double-O configuration) or can be concentric (coaxial). The arterial port ends about 3 cm proximal to the venous port. The optimal length depends on the location of insertion: 12–15 cm for the right internal jugular vein; 15–20 cm for the left internal jugular vein, and 19–24 cm for the femoral vein [4, 14]. Shorter catheters at the femoral site do not reach the inferior vena cava and are associated with more malfunction and recirculation and consequently reduced dialysis efficiency [15, 16]. Some jugular catheters have a curved extension allowing proper fixation and increasing the comfort of the patient.

For short-term RRT non-tunneled non-cuffed catheters can be used, but tunneled (5–10 cm subcutaneous course) and cuffed (fixed to the tissue by a Dacron or silver impregnated collagen cuff) catheters should be used if the duration of RRT is anticipated to be more than 3 weeks [4, 17].

Where to Insert a Dialysis Catheter?

The choice of the appropriate insertion site depends on the patient's characteristics (coagulopathy, morbid obesity, previous surgery, local infection, altered local anatomy, cardiopulmonary reserve capacity, prolonged immobilization), the availability of the insertion site in the often 'heavily catheterized' critically ill patients, the operator's skills and experience, and the risk of complications associated with the different insertion sites.

In order to prevent atrial perforation, semi-rigid catheters in the superior vena cava (SVC) should have their tip in the SVC or at the junction of the SVC and the right atrium. Prior to their use the correct position should be verified with a chest X-ray [4]. Silicone catheters function optimally with their tip in the right atrium [12, 13]. Insertion into the right internal jugular vein is preferred for both tunneled and non-tunneled catheters because it is a more direct route to the caval-atrial junction than a left-sided placement that results in a lower blood flow and more complications [4, 18]. With left-sided insertion, the access or removal lumen should be inside the catheter curve to prevent obstruction by sucking of the vessel wall. In critically ill patients the subclavian vein is the preferred insertion site for a CVC because it is associated with the lowest risk of infectious complications [17]. However, in a patient who may need permanent vascular access for dialysis, the use of the subclavian vein is discouraged for insertion of a TDC because it is associated with higher rates of central venous stenosis, precluding the ipsilateral arm for future dialysis access [19–21]. The subclavian vein should therefore be reserved for very short-term use or if there is no other alternative [4, 17]. The femoral vein is often preferred for patients with critical respiratory conditions or for urgent access because of the speed with which it can be performed. On the other hand, femoral catheters are associated with the highest risk of infection, both for CVC in critically ill patients [22] and for non-tunneled, non-cuffed TDC [23, 24]. In addition, femoral TDCs reduce patient mobilization and result in higher recirculation rates [15, 16].

Catheter Insertion and Care

Compared with a non-tunneled, non-cuffed catheter, the insertion of a tunneled cuffed catheter has a higher failure rate, requires more skill and time, and is associated with more tissue trauma. On the other hand, compared with non-cuffed catheters, cuffed catheters are less susceptible to infection (less extraluminal contamination). The mean incidence of catheter-related bloodstream infection (CRBSI) is ± 5 per 1,000 catheter days for non-cuffed non-tunneled dialysis catheters versus 1.6–3.5 for tunneled and cuffed catheters [25, 26]. According to the NF/KDOQI guidelines, non-cuffed dialysis catheters should therefore have a finite use-life, not exceeding 3 weeks for internal jugular or subclavian catheters and 5 days for femoral catheters [4]. A recent randomized comparison of tunneled and non-tunneled femoral catheters in AKI patients showed a lower incidence of vein thrombosis and catheter-related infection and a higher blood flow with greater delivered dialysis dose for the tunneled catheters [27].

A meta-analysis of 18 prospective randomized trials showed that, compared with the landmark method (blind insertion), the use of ultrasound guidance for insertion of a CVC was associated with a significantly lower failure rate for cannulation of the internal jugular vein, whereas the effect was limited for the subclavian and femoral vein [28]. In addition, a recent prospective randomized trial in 900 patients not only showed an increased overall success rate but also a reduction in hemothorax, pneumothorax and catheter-related infection with ultrasound guidance for internal jugular vein catheterization [29]. Use of ultrasound guidance is therefore advocated both for placement of CVCs [17] and for (tunneled and non-tunneled) TDCs [4].

Maximal sterile barrier precautions (surgical scrub, sterile gloves, long-sleeved sterile gowns, mask, cap and large sterile sheet drapes) reduce the rate of infection. Skin disinfection should preferably be performed with 2% chlorhexidine, but a tincture of iodine or 70% alcohol is also acceptable. For CVC the application of an antibiotic ointment is not advocated because it promotes fungal infections and antimicrobial resistance. However, in chronic dialysis patients this procedure has been shown to be effective [4, 17]. Whether it should be applied in critically ill patients with AKI is not clear. It should be borne in mind that some of these ointments may adversely affect the integrity of dialysis catheters [30]. The catheter exit site should be covered with a sterile dry gauze (changed every 2 days) or transparent dressing (changed every 7 days) [17].

Complications

Insertion-related complications include inadvertent arterial puncture, hemothorax, pneumothorax, pericardial tamponade, arrhythmias, air embolism and retroperitoneal hemorrhage [11, 14]. Early mechanical complications can be reduced by insertion under ultrasound guidance [29]. In addition, routine chest X-ray control allows early detection of hemo- and pneumothorax. Late complications include infection, central vein thrombosis/stenosis, and catheter dysfunction.

Infection

Intravascular catheters put patients at risk of infection. Exit site infection is defined as localized infection of the skin and soft tissue around the exit site, often without systemic signs of infection. In ICU patients CRBSIs are associated with increased morbidity, mortality and duration of hospitalization and additional medical costs [31]. Microorganisms that most commonly cause CRBSI include (in decreasing order of frequency): coagulase-negative staphylococci, *Staphylococcus aureus*, enterococci, gram-negative bacteria and yeasts. A subset of patients with CRBSI develop metastatic infection including endocarditis, thrombophlebitis, septic arthritis, osteomyelitis and epidural abcesses, especially when *S. aureus* or *Candida albicans* is the causative organism [32]. A recent systematic review showed an average rate for CVC-associated bloodstream infection of 2.7 per 1,000 catheter days for non-tunneled and 1.7 for tunneled catheters [25]. For dialysis catheters the incidence is ± 5 per 1,000 catheter days for untunneled TDCs and 1.6–3.5 for tunneled TDCs [25, 26]. In ICU patients with AKI colonization rates for TDCs do not appear to be significantly higher than for CVCs (9.1 vs. 5.9 per 1,000 catheter days) [33].

Different mechanisms may lead to catheter-related infections. Extraluminal contamination results from the migration of skin flora along the external surface of the catheter into the bloodstream, is the most common cause of CRBSI of short-term percutaneously inserted non-cuffed CVCs [34], and can be reduced by the use of cuffed and tunneled catheters. A less common cause of extraluminal colonization is hematogenous seeding from another focus of infection. Intraluminal colonization is the dominant mechanism in longer-dwelling catheters and mostly results from contamination of the catheter hub, whereas infusate contamination is rare. A bacterial biofilm forms rapidly in the lumen of most CVCs and is the major source of both catheter-related bacteremia and thrombosis [35]. In ICU patients risk factors for CRBSI include: catheter material (more thrombogenic catheters are associated with high infection rates), the number of infusion ports, the frequency of manipulation, urgent versus elective insertion, the operator's experience (traumatic insertion procedures increase the risk of infection), insertion site (see above), indwelling time, and the patient's illness severity [11, 36]. In chronic dialysis patients nasal carriage of *S. aureus*, previous bacteremia, peripheral atherosclerosis, diabetes, site of insertion, indwelling time, and the number of dialyses performed have been identified as risk factors [23, 37, 38].

Catheter colonization can be defined as the presence of ≥ 15 colony-forming units (CFU) on semiquantitative (roll plate) or ≥ 100 CFU on quantitative (vortex or sonication method) catheter culture in the absence of clinical signs of infection. CRBSI should be suspected when a patient with a catheter develops fever or chills and does not have clinical evidence for another source of infection. Several methods can be used to diagnose CRBSI. Methods requiring device removal include: (1) a qualitative catheter segment culture; (2) a semiquantitative catheter segment culture (growth of ≥ 15 CFU), or (3) quantitative catheter segment culture ($\geq 1,000$ CFU), all of them combined with a positive blood culture yielding the same microorganism. Methods not requiring catheter removal include: (1) qualitative blood culture through the device; (2) quantitative blood culture through the device (≥ 100 CFU/ml); (3) paired quantitative blood culture (3–5 times more organisms grown in blood cultures drawn through the catheter than in a blood culture drawn peripherally); (4) differential time to positivity (blood culture drawn through the catheter positive ≥ 2 h earlier than blood culture drawn peripherally), and (5) acridine orange leukocyte cytospin (visualization of any organism). Paired blood culture is the most accurate estimate. However most other methods show acceptable sensitivity and specificity [39]. In the absence of concurrent blood cultures from a peripheral vein, a clinical diagnosis of catheter-related bacteremia requires the exclusion of alternate sources of infection. If *S. aureus* or *C. albicans* are the causative organism or in the case of persistent bacteremia after catheter removal, aggressive evaluation for metastatic infections, including a transesophageal echocardiography, should be performed [32].

Guidelines for the prevention of intravascular CRBSIs were published in 2002 [17] and include: education and training of healthcare workers, surveillance of catheter-related infections, adequate hand hygiene, aseptic technique during catheter insertion and care and selection of the catheter, the insertion site and insertion technique with the lowest risk of complications for the anticipated time and duration of catheterization. CVCs should not be replaced routinely, and catheters should be removed when no longer needed. For hemodialysis, cuffed and tunneled catheters are indicated if the period of temporary access is anticipated to be prolonged (>3 weeks). TDCs should be used exclusively for RRT. Strict adherence to these evidence-based catheter insertion and maintenance policies reduces CRBSI [40–42].

More sophisticated preventive measures include the use of antibiotic-impregnated catheters and antibiotic lock solutions. Antimicrobial impregnated CVCs (chlorhexidine/silver sulfadiazine or minocycline/rifampin) are indicated if the anticipated duration is more than 5 days and if, after implementing a comprehensive strategy to reduce rates of CRBSI, the rate of infection remains too high [17]. The advantage of antibiotic impregnation has also been shown for TDCs [43]. On the other hand, allergic reactions and the emergence of resistance remain an important concern when antimicrobial-impregnated catheters are used. Together with the fear of systemic toxicity, the concern about antimicrobial resistance also applies to the prophylactic use of antibiotic/antiseptic locks (combinations of gentamicin, taurolidine, isopropyl alcohol, minocycline or a cephalosporin with heparin, citrate or EDTA), which have been shown to reduce the incidence of CRBSI in chronic dialysis patients [44]. A recent randomized trial demonstrated that, compared with heparin, the use of a citrate lock reduces catheter-related infections [45]. With other long-term central venous access devices, vancomycin locks have been shown to reduce the risk of CRBSI and may be useful in high-risk patients [46]. Another strategy consists of using hubs that contain an antiseptic chamber [47].

Empiric systemic antibiotics appropriate for the suspected organisms should be started on clinical suspicion of CRBSI after cultures have been taken. In centers with a high prevalence of methicillin-resistant staphylococci this empiric regimen should include vancomycin. In other centers a penicillinase-resistant penicillin should be used. Whether broader coverage (with e.g. a third- or fourth-generation cephalosporin) is required depends on the severity of sepsis, the patient's immune system and known previous infections, and on the spectrum of microorganisms in the unit. As soon as the culture results are available, antibiotic treatment should be tailored to the specific organism. The duration of antibiotic treatment depends on the infecting organism (shorter for coagulase-negative staphylococci), on the presence of an immunocompromised state, valvular heart disease, intravascular prosthetic devices or metastatic infections, and whether or not the catheter has been removed [32].

When infection of the catheter is suspected, the catheter can be exchanged over a guide-wire, with further insertion at a new site if the culture of the catheter tip is found to be positive. Exit-site infection, pocket infection of tunneled catheters and CRBSI with clinical signs of sepsis are an indication for immediate catheter removal. A new non-tunneled catheter may be inserted after antibiotic treatment has started, whereas reinsertion of a tunneled catheter should ideally be postponed until after completion of the antibiotic treatment [32]. When symptoms of CRBSI are mild without suspicion of metastatic complications, especially when *Staphylocccus epidermidis* is the causative organism, attempts can be undertaken to salvage the catheter in patients with tunneled catheters and limited access sites. If fever does not resolve over the first 24–48 h the catheter should still be removed. If fever resolves the suspected catheter may be exchanged over a guide-wire [48–50]. Another method allowing catheter salvage is the use of an 'antibiotic lock' or the instillation of a concentrated antibiotic solution (combined with an anticoagulant) into the catheter [44]. This method is mostly reserved for long-term catheters such as tunneled dialysis catheters or long-term devices used in oncology or chronic parenteral nutrition. It allows obtaining local antibiotic concentrations that are several orders of magnitude higher than the blood concentration and is a solution for the inability of therapeutic concentrations of most antibiotics to kill microorganisms growing in a biofilm.

Acute Catheter Malfunction

Acute catheter malfunction may result from complete or partial intraluminal thrombosis, from a fibrin sheet around the catheter or from malpositioning (sucking the vein wall) and kinking of the catheter, and is indicated by a decrease in the attainable blood flow and increased arterial and/or venous pressures during hemodialysis. For continuous RRT blood flows between 150 and 200 ml/min are able to deliver an adequate dialysis dose. However, intermittent hemodialysis mostly requires higher blood flows (at least 250–300 ml/min). These flows can only be delivered with an adequately functioning access.

In order to prevent intraluminal thrombosis the catheter is filled with an anticoagulant (heparin or citrate) during the interdialytic period. Endoluminal thrombosis can be treated with mechanical (brush) or chemical (thrombolytics) methods, but this is mostly reserved for tunneled catheters [51, 52].

Improper catheter tip placement is a common cause of access malfunction. Femoral catheters should be in the inferior vena cava, and jugular and subclavian catheters should be at the junction of the SVC and the right atrium. Malfunction of catheters in the SVC is further decreased when their tip is located in the right atrium [12, 13, 53], which, however, is only safe with silicone catheters. For non-tunneled, non-cuffed catheters malfunction is more frequent with femoral compared with jugular catheters [54].

Access recirculation (recirculation of blood from the outflow to the inflow part of the catheter) occurs whenever pumped flow through the extracorporeal circuit exceeds flow in the vein, reduces the effective clearance, and depends on the design, length and insertion site of the catheter and on the blood flow. Well-functioning and nonreversed internal jugular and subclavian venous catheters have, in general, recirculation rates of <5%. Because of the higher blood flows, recirculation is more important with intermittent than with continuous RRT. Femoral catheters result in higher recirculation rates than subclavian or jugular catheters, especially when they are too short [15, 16]. For catheters in the SVC recirculation can be minimized by placement close to the right atrium. Inversion of the connecting lines also increases recirculation from 3 to 12% [55]. Two observational trials report contradictory results with regard to the delivered dialysis dose when comparing femoral and internal jugular access [56, 57].

Central Vein Stenosis and Thrombosis

Catheter-related thrombosis can manifest itself as the formation of a fibrin sleeve around the catheter or a thrombus adherent to the vessel wall. The incidence may be as high as 33–67% and is largely dependent on the diagnostic method. Risk factors for catheter-related thrombosis are the insertion site (vein diameter, local hemodynamics), technical problems during insertion, the catheter material, the indwelling time, and the hypercoagulability or increased blood viscosity of the patient [11]. In chronic dialysis patients, catheter-related central venous stenosis and thrombosis are most frequent in the subclavian vein [19–21], although incidences as high as 25% have also been reported for the internal jugular vein [58] and the femoral vein [59]. In AKI patients treated with femoral catheters the incidence appears to be lower with a tunneled compared with a non-tunneled catheter [27]. In ICU patients the thrombotic risk associated with CVC is higher for femoral [60] and jugular [61] than subclavian catheters. The administration of anticoagulants [62], and also the use of less thrombogenic material or anticoagulant-bonded catheters [63, 64] has been shown to decrease the risk of catheter-related thrombosis and subsequently the risk of catheter-related infection.

References

1. NF/KDOQI clinical practice guidelines for hemodialysis adequacy: update 2000. Am J Kidney Dis 2001;37(suppl 1):S7–S164.
2. Ronco C, Bellomo R, Homel P, Brendolan A, Dan M, Piccinni P, La Greca G: Effects of different doses in continuous veno-venous haemofiltration on outcomes of acute renal failure: a prospective randomised trial: Lancet 2000;356:26–30.
3. Schiffl H, Lang SM, Fischer R: Daily hemodialysis and the outcome of acute renal failure. N Engl J Med 2002;346:305–310.
4. NF/KDOQI clinical practice guidelines for vascular access: update 2000. Am J Kidney Dis 2001;37(suppl 1):S137–S181.
5. Evanson JA, Himmelfarb J, Wingard R, Knights S, Shyr Y, Schulman G, Ikizler TA, Hakim RM: Prescribed versus delivered dialysis in acute renal failure patients. Am J Kidney Dis 1998;32: 731–738.
6. Monchi M, Berghmans D, Ledoux D, Canivet JL, Dubois B, Damas P: Citrate vs. heparin for anticoagulation in continuous venovenous hemofiltration: a prospective randomized study. Intensive Care Med 2004;30:260–265.
7. Kutsogiannis DJ, Gibney RT, Stollery D, Gao J: Regional citrate versus systemic heparin anticoagulation for continuous renal replacement in critically ill patients. Kidney Int 2005;67: 2361–2367.
8. Palsson R, Laliberte KA, Niles JL: Choice of replacement solution and anticoagulant in continuous venovenous hemofiltration. Clin Nephrol 2006;65:34–42.
9. Mehta RL, McDonald BR, Ward DM: Regional citrate anticoagulation for continuous arteriovenous hemodialysis. An update after 12 months; in Sieberth HG, Mann H, Stummvoll HK (eds): Continuous Hemofiltration. Contrib Nephrol. Basel, Karger, 1991, vol 93, pp 210–214.
10. Holt AW, Bierer P, Bersten AD, Bury LK, Vedig AE: Continuous renal replacement therapy in critically ill patients: monitoring circuit function. Anaesth Intensive Care 1996;24:423–429.
11. Polderman KH, Girbes AJ: Central venous catheter use. Part 1: mechanical complications. Intensive Care Med 2002;28:1–17.
12. Jean G, Chazot C, Vanel T, Charra B, Terrat JC, Calemard E, Laurent G: Central venous catheters for haemodialysis: looking for optimal blood flow. Nephrol Dial Transplant 1997;12:1689–1691.
13. Petersen J, Delaney JH, Brakstad MT, Rowbotham RK, Bagley CM Jr: Silicone venous access devices positioned with their tips high in the superior vena cava are more likely to malfunction. Am J Surg 1999;178:38–41.
14. Oliver MJ: Acute dialysis catheters. Semin Dial 2001;14:432–435.
15. Leblanc M, Fedak S, Mokris G, Paganini EP: Blood recirculation in temporary central catheters for acute hemodialysis. Clin Nephrol 1996;45:315–319.

16 Little MA, Conlon PJ, Walshe JJ: Access recirculation in temporary hemodialysis catheters as measured by the saline dilution technique. Am J Kidney Dis 2000;36:1135–1139.
17 O'Grady N, Alexander M, Dellinger P, Gerberding JL, Heard SO, Maki DG, Masur H, McCormick RD, Mermel LA, Pearson ML, Raad II, Randolph A, Weinstein RA; the Healthcare Infection Control Practices Advisory Committee: Guidelines for the prevention of intravascular catheter-related infections. Infect Control Hosp Epidemiol 2002;23:759–769.
18 Oliver MJ, Edwards LJ, Treleaven DJ, Lambert K, Margetts PJ: Randomized study of temporary hemodialysis catheters. Int J Artif Organs 2002;25:40–44.
19 Ciminowski GE, Worley E, Rutherford WE, Sartain J, Blondin J, Harter H: Superiority of the internal jugular over the subclavian access for temporary dialysis. Nephron 1990;54:154–161.
20 Schillinger F, Schillinger D, Montagnac R, Milcent T: Post catheterisation vein stenosis in haemodialysis: comparative angiographic study of 50 subclavian and 50 internal jugular accesses. Nephrol Dial Transpl 1991;6:722–724.
21 Trerotola SO, Kuhn-Fulton J, Johnson MS, Shah H, Ambrosius WT, Kneebone PH: Tunneled infusion catheters: increased incidence of symptomatic venous thrombosis after subclavian versus internal jugular venous access. Radiology 2000;217:89–93.
22 Lorente L, Henry C, Martin MM, Jiménez A, Mora ML: Central venous catheter-related infection in a prospective and observational study of 2595 catheters. Crit Care 2005;9:R631–R635.
23 Oliver MJ, Callery SM, Thorpe KE, Schwab SJ, Churchill DN: Risk of bacteremia from temporary hemodialysis catheters by site of insertion and duration of use: a prospective study. Kidney Int 2000;58:2543–2545.
24 Kairaitis LK, Gottlieb T: Outcome and complications of temporary dialysis catheters. Nephrol Dial Transplant 1999;14:1710–1714.
25 Maki DG, Kluger DM, Crinch C: The risk of bloodstream infection in adults with different intravascular devices: a systematic review of 200 published prospective studies. Mayo Clin Proc 2006;81:1159–1171.
26 Saxena AK, Panhotra BR: Haemodialysis catheter-related infections: current treatment options and strategies for prevention. Swiss Med Wkly 2005;135:127–138.
27 Klouche K, Amigues L, Deleuze S, Beraud JJ, Canaud B: Complications, effects on dialysis dose, and survival of tunneled femoral dialysis catheters in acute renal failure. Am J Kidney Dis 2007;49:99–108.
28 Hind D, Calvert N, Mc Williams R, Davidson A, Paisley S, Beverley C: Ultrasonic locating devices for central venous cannulation: meta-analysis. BMJ 2003;327:361–367.
29 Karakitsos D, Labropoulos N, De Groot E, Patrianakos AP, Kouraklis G, Poularas J, Samonis G, Tsoutsos DA, Konstadoulakis MM, Karabinis A: Real-time ultrasound-guided catheterisation of the internal jugular vein: a prospective comparison with the landmark technique in critical care patients. Crit Care 2006;10:R162–R169.
30 Riu S, Ruiz CG, Martinez-Vea A, Peralta C, Oliver CA: Spontaneous rupture of polyurethane peritoneal catheter: a possible deleterious effect of mupirocin ointment. Nephrol Dial Transplant 1998;13:1870–1871.
31 Pittet D, Tamara D, Wenzel RP: Nosocomial bloodstream infection in critically ill patients: excess length of stay, extra costs and attributable mortality. JAMA 1994;271:1598–1601.
32 Mermel LA, Farr BM, Sherertz RJ, raad II, O'Grady N, Harris JS, Craven DE: Guidelines for the management of intravascular catheter-related infections. Clin Infect Dis 2001;32:1249–1272.
33 Souweine B, Liotier J, Heng AE, Isnard M, Ackoundou-N'Guessan C, Deteix P, Traore O: Catheter colonization in acute renal failure patients: comparison of central venous and dialysis catheters. Am J Kidney Dis 2006;47:879–887.
34 Safdar N, Maki DG: The pathogenesis of catheter-related bloodstream infection with noncuffed short-term central venous catheters. Intensive Care Med 2004;30:62–67.
35 Costerton JW, Stewart PS, Greenberg EP: Bacterial biofilms: a common cause of persistent infections. Science 1999;284:1318–1322.
36 Polderman KH, Girbes AR: Central venous catheter use. Part 2: infectious complications. Intensive Care Med 2002;28:18–28.
37 Jean G, Charra B, Chazot C, Vanel T, Terrat JC, Hurot JM, Laurent G: Risk factor analysis for long-term tunneled dialysis catheter-related bacteremias. Nephron 2002;91:399–405.

38 Naumovic RT, Jovanovic DB, Djukanovic LJ: Temporary vascular catheters for hemodialysis: a 3-year prospective study. Int J Artif Organs 2004;27:848–854.
39 Safdar N, Fine JP, Maki DG: Meta-analysis: methods for diagnosing intravascular device-related bloodstream infection. Ann Intern Med 2005;142:451–466.
40 Pronovost P, Needham D, Berenholtz S, Sinopoli D, Chu H, Cosgrove S, Sexton B, Hyzy R, Welsh R, Roth G, Bander J, Kepros J, Goeschel C: An intervention to decrease catheter-related bloodstream infections in the ICU. N Engl J Med 2006;355:2725–2732.
41 Warren DK, Cosgrove SE, Diekema DJ, Zuccotti G, Climo MW, Bolon MK, Tokars JI, Noskin GA, Wong ES, Sepkowitz KA, Herwaldt LA, Perl TM, Solomon SL, Fraser VJ: Prevention Epicenter Program. A multicenter intervention to prevent catheter-associated bloodstream infections. Infect Control Hosp Epidemiol 2006;27:662–669.
42 Young EM, Commiskey ML, Wilson SJ: Translating evidence into practice to prevent central venous catheter-associated bloodstream infections: a systems-based intervention. Am J Infect Control 2006;34:503–506.
43 Chatzinikolaou I, Finkel K, Hanna H, Boktour M, Foringer J, Ho T, Raad I: Antibiotic-coated hemodialysis catheters for the prevention of vascular catheter-related infections: a prospective, randomized study. Am J Med 2003;115:352–357.
44 Manierski C, Besarab A: Antimicrobial locks: putting the lock on catheter infections. Adv Chronic Kidney Dis 2006;13:245–258.
45 Weijmer MC, van den Dorpel MA, Van de Ven PJ, ter Wee PM, van Geelen JA, Groeneveld JO, van Jaarsveld BC, Koopmans MG, le Poole CY, Schrander-Van der Meer AM, Siegert CE, Stas KJ; CITRATE Study Group: Randomized, clinical trial comparison of trisodium citrate 30% and heparin as catheter-locking solution in hemodialysis patients. J Am Soc Nephrol 2005;16: 2769–2777.
46 Safdar N, Maki DG: Use of vancomycin-containing lock or flush solutions for prevention of bloodstream infection associated with central venous access devices: a meta-analysis of prospective, randomized trials. Clin Infect Dis 2006;43:474–484.
47 Leon C, Alvarez-Lerma F, Ruiz-Santana S, Gonzalez V, de la Torre MV, Sierra R, Leon M, Rodrigo JJ: Antiseptic chamber-containing hub reduces central venous catheter-related infection: a prospective, randomized study. Crit Care Med 2003;31:1318–1324.
48 Robinson D, Suhocki P, Schwab SJ: Treatment of infected tunneled venous access hemodialysis catheters with guidewire exchange. Kidney Int 1998;53:1792–1794.
49 Beathard GA: Management of bacteremia associated with tunneled-cuffed hemodialysis catheters. J Am Soc Nephrol 1999;10:1045–1049.
50 Tanriover B, Carlton D, Saddekni S, Hamrick K, Oser R, Westfall AO, Allon M: Bacteremia associated with tunneled dialysis catheters: comparison of two treatment strategies. Kidney Int 2000;57: 2151–2155.
51 Clase CM, Crowther MA, Ingram AJ, Cina CS: Thrombolysis for restoration of patency to haemodialysis central venous catheters: a systematic review. J Thromb Thrombolysis 2001;11: 127–136.
52 Hilleman DE, Dunlay RW, Packard KA: Reteplase for dysfunctional hemodialysis catheter clearance. Pharmacotherapy 2003;23:137–141.
53 Abidi SM, Khan A, Fried LF, Chelluri L, Bowles S, Greenberg A: Factors influencing function of temporary dialysis catheters. Clin Nephrol 2000;53:199–205.
54 Hryszko T, Brzosko S, Mazerska M, Malyszko J, Mysliwiec M: Risk factors of nontunneled noncuffed hemodialysis catheter malfunction. A prospective study. Nephron Clin Pract 2004;96: c43–c47.
55 Level C, Lasseur C, Chauveau P, Bonarek H, Perrault L, Combe C: Performance of twin central venous catheters: influence of the inversion of inlet and outlet on recirculation. Blood Purif 2002; 20:182–188.
56 Liangos O, Sakiewicz PG, Kanagasundaram NS, Hammel J, Pajouh M, Seifert T, Paganini EP: Dialyzer fiber bundle volume and kinetics of solute removal in continuous venovenous hemodialysis. Am J Kidney Dis 2002;39:1047–1053.
57 du Cheyron D, Bouchet B, Bruel C, Daubin C, Ramakers M, Charbonneau P: Antithrombin supplementation for anticoagulation during continuous hemofiltration in critically ill patients with septic shock: a case-control study. Crit Care 2006;10:R45.

58 Wilkin TD, Kraus MA, Lane KA, Trerotola SO: Internal jugular vein thrombosis associated with hemodialysis catheters. Radiology 2003;228:697–700.
59 Maya ID, Allon M: Outcomes of tunneled femoral hemodialysis catheters: comparison with internal jugular vein catheters. Kidney Int 2005;68:2886–2889.
60 Merrer J, De Joughe B, Golliot F, lefrant JY, Raffy B, barre E, Rigaud JP, Casciani D, Misset B, Bosquet C, Outin H, Brun-Buisson C, Nitenberg G: Complications of femoral and subclavian venous catheterisation in critically ill patients. JAMA 2001;286:700–707.
61 Timsit JF, Farkas JC, Boyer JM, Martin JB, Misset B, Renaud B, Carlet J: Central vein catheter-related thrombosis in intensive care patients: incidence, risk factors and relationship with catheter-related sepsis. Chest 1998;114:207–213.
62 Randolph AG, Cook DJ, Gonzales CA, Andrew M: Benefit of heparin in central venous and pulmonary artery catheters: a meta-analysis of randomized controlled trials. Chest 1998;113:165–171.
63 Baumann M, Witzke O, Dietrich R, Haug U, Deppisch R, Lutz J, Philipp T, Heemann U: Prolonged catheter survival in intermittent hemodialysis using a less thrombogenic micropatterned polymer modification. ASAIO J 2003;49:708–712.
64 Long DA, Coulthard MG: Effect of heparin-bonded central venous catheters on the incidence of catheter-related thrombosis and infection in children and adults. Anaesth Intensive Care 2006;34: 481–484.

Miet Schetz, MD, PhD
Department of Intensive Care Medicine, University Hospital Gasthuisberg
Herestraat 49
BE–3000 Leuven (Belgium)
Tel. +32 16 344 021, Fax +32 16 344 015, E-Mail marie.schetz@uz.kuleuven.ac.be

Dialysate and Replacement Fluid Composition for CRRT

Filippo Aucella, Salvatore Di Paolo, Loreto Gesualdo

Department of Nephrology, Dialysis and Transplantation, University of Foggia, Foggia, Italy

Abstract

Continuous renal replacement therapies (CRRTs) are increasingly used in order to maintain normal or near-normal acid-base balance in intensive care unit (ICU) patients. Acid-base balance is greatly influenced by the type of dialysis employed and by the administration route of replacement fluids. In continuous veno-venous hemofiltration, buffer balance depends on losses with ultrafiltrate and gain with replacement fluid, while in techniques such as continuous veno-venous hemodiafiltration, clinicians should balance the role of the dialysate. The type of buffer greatly influences not only acid-base correction, but also clinical outcome. Lactate or bicarbonate fluids are currently used, but recent studies suggest that bicarbonate-buffered replacement fluids can improve acid-base status and reduce cardiovascular events better than lactate fluids. The buffer concentration should exert a buffer load that may compensate for deficits, for losses in the buffer process, and for extracorporeal losses and should therefore usually be supraphysiological. However, the dialysate buffer or electrolyte concentration need always to be balanced with that of the replacement fluids employed. Both fluids should contain electrolytes in concentrations aiming for a physiologic level and taking into account preexisting deficits or excess and all input and losses. Clinicians should be aware that in CRRTs the quality control for sterility, physical properties, individualized prescription and balance control are vitally important.

Copyright © 2007 S. Karger AG, Basel

Major changes have occurred in the medical management of intensive care unit (ICU) patients over the past two decades, and the role of extracorporeal blood treatments has broadened from conventional renal function replacement to a series of non-renal conditions.

Continuous renal replacement therapies (CRRTs) have found widespread use and acceptance because they incorporate several advantages, such as improved hemodynamic stability, gradual urea removal without fluctuations, and optimal fluid balance. Further, they obviate the need of nutritional restrictions in terms of both the volume and composition of the nutrients administered to critically ill patients [1].

In this setting, we need to keep in mind that patients with acute renal failure (ARF) depend on dialysis to maintain fluid and electrolyte balance. One crucial goal of CRRT in the treatment of ARF patients is to achieve and maintain normal or near-normal acid-base balance over time, thereby preventing the detrimental effects of acidemia on cardiovascular performance, hepatic metabolism and hormonal response [2]. Over the course of the procedure solutes are ultrafiltered or diffused between blood and dialysate such that the plasma composition is restored toward normal values. Obviously, the makeup of the dialysate and the replacement fluid, in combination with the features of the dialysis technique [3], are of paramount importance to accomplish this goal.

Choice of Buffer and Acid-Base Balance

Acid-base balance is greatly influenced by the type of dialysis technique employed, diffusion and convection being differently balanced. Moreover, the type of buffer in both replacement fluid and dialysate greatly influences not only acid-base correction, but also clinical outcome.

The dialysis techniques have a clear influence on buffer kinetics. Despite a comparable blood and dialysate/replacement fluid flow rate, different techniques are associated with varying effects on electrolyte and acid-base homeostasis.

In continuous veno-venous hemofiltration (CVVH) the buffer balance depends on buffer losses with the ultrafiltrate and buffer gain with the replacement fluid. If patients undergo isolated ultrafiltration, without infusing any replacement fluid, for example in chronic cardiac failure, bicarbonate losses are compensated by the reduction in the distribution volume of the buffer, such that bicarbonate serum levels do not change significantly [3]. When replacement fluids are infused along with fluid ultrafiltration, bicarbonate losses in the ultrafiltrate need to be balanced by equal amounts in the infusion solution. In the clinical setting, metabolic acidosis is the most frequent condition occurring in these patients and a positive buffer balance is commonly required. Moreover, clinicians need to be aware that the bicarbonate concentration in the ultrafiltrate is higher than the plasma level, because of a sieving coefficient of >1.

In continuous veno-venous hemodialysis (CVVHD), a pure diffusive technique, buffer balance is a function of the concentration gradient between the dialysate and blood, and a feedback between the base balance and blood bicarbonate also occurs. Thus, the correction of acidosis depends only on the buffer concentration in the dialysate [1, 3].

In techniques with both diffusive and convective processes, such as continuous veno-venous hemodiafiltration (CVVHDF), the buffer gain depends on both the dialysate and the replacement fluid, with the inherent risk of alkalemia. A comparison of CVVHDF and CVVH showed that they had a significantly different impact on bicarbonate and sodium control [4]. Specifically, CVVH was associated with a lower incidence of metabolic acidosis and a higher incidence of metabolic alkalosis [4], possibly due to small differences in the total amount of buffer infused during CVVH.

Which Buffer Should Be Preferred in CRRT?

Lactate, acetate, and bicarbonate have all been used as buffers during CRRT. Citrate has been used as a buffer and anticoagulant. Although bicarbonate is the natural buffer, bicarbonate-based solutions have not been available until recently, because of the higher risk of bacterial contamination and the instability of this buffer in the presence of calcium and magnesium ions. Moreover, the rate of administration of buffer solutions during CRRT is definitely lower than during discontinuous treatments, because the accumulation of acetate or lactate has rarely been reported, although this may be an important point in high-volume treatments [5]. It is now clear that lactate or bicarbonate fluids offer a better control of acid-base balance and improved cardiovascular stability compared to acetate fluids [5]. In clinical practice, both lactate and bicarbonate ions are used in replacement fluids and in dialysate for CRRT. These buffer have shown similar efficacy in correcting metabolic acidosis [6]. Obviously, lactate infusion increases serum lactate levels and this might lead to a misleading interpretation of the clinical situation. Solutions containing lactate are contraindicated in patients with concomitant lactic acidosis and in those with lactate intolerance, defined as a rise of 5 mmol/l during CRRT: these patients are at a high risk of worsening acidosis because of insufficient conversion of lactate into bicarbonate in the face of ongoing bicarbonate losses. Currently, lactate concentrations vary from 45 mmol/l (chloride 103 mmol/l) down to 35 mmol/l (chloride 110 mmol/l), and these differences in solute composition can lead to clinical consequences, namely to the development of hypochloremic metabolic alkalosis after several days of CRRT with high lactate/low chloride solutions, or hyperchloremic metabolic acidosis when a low lactate/high chloride fluid is used [5]. Moreover, several studies suggest that lactate-buffered replacement fluids can exert negative effects on different metabolic and hemodynamic parameters [2]. In ICU patients suffering from multiple organ dysfunction, the conversion of lactate to bicarbonate is frequently impaired and the resulting increase in blood lactate concentration may exert multiple negative effects. On the other hand, the administration of bicarbonate in patients with lactic acidosis has also been questioned, although CRRT makes it possible to overcome some of the unwanted effects of bicarbonate infusion, e.g. volume overload, hyperosmolarity, and a decrease in ionized calcium.

Bicarbonate-buffered replacement fluids are presently considered a valuable approach to improve the prognosis of critical ill patients with ARF. Several, but not all, studies support the view that bicarbonate-buffered replacement fluids can improve acid-base status and reduce the incidence of cardiovascular events in ICU patients, when compared with lactate-buffered fluids [7]. This suggestion should be viewed in light of the peculiar milieu of the critically ill patient, the specific features of which (systemic inflammation, increased energy expenditure and metabolic activation, cardiovascular stress, hyperdynamic cardiovascular response) cause a clinical picture characterized by an excess of lactate production, even in the absence of an increase in blood lactate levels, increased oxygen consumption and elevated protein catabolism. Blood lactate, if not further metabolized, acts as a strong anion, which has the same acidifying effect of chloride. Accordingly, iatrogenic hyperlactatemia can causes a state of metabolic acidosis. When oxidizable anions are used in replacement fluids, these anions, acetate, lactate or citrate, need to be completely oxidized to CO_2 and H_2O in order to generate bicarbonate and then the buffering capacity equals that of bicarbonate solutions. However, on the contrary, if metabolic conversion is not adequate, the increased blood concentration of the anions leads to an acidotic condition. The type and extent of these acid-base changes are governed by the intensity of plasma water exchange/dialysis, by the buffer content of the replacement fluids, and by the actual metabolic rate of the above-mentioned anions.

The advantage of using a buffer-free replacement solution is that the dose of bicarbonate can be titrated according to a given target value of base excess, which may vary according to the clinical picture of the ICU patient, as well as to specific treatment modalities, such as permissive hypercapnia. The application of a lung-protective strategy with reduced tidal volumes, effective lung recruitment, adequate positive end-expiratory pressure to minimize alveolar collapse during expiration, and permissive hypercapnia has been shown to be advantageous in adult patients who have acute respiratory distress syndrome [8]. Indeed, strategies such as permissive hypercapnia or permissive hypoxemia have been reported to favor the onset of ARF by compromising renal blood flow [9]. Therefore, the correction of acidosis, as suggested by the NIH protocol Acute Respiratory Distress Syndrome Network [10], has been questioned as experimental evidence and preliminary clinical data indicate that buffering hypercapnic acidosis abrogates its protective effects [11, 12]. Nevertheless, patients with permissive hypercapnia may require huge amounts of buffer to correct acidosis, the infusion of the buffer itself leading to a further increase in CO_2 production, thus leading to a vicious circle which, in the presence of limited CO_2 elimination, may worsen the acidotic state of the patient. The infusion of lactate in this condition does not appear appropriate because relatively more lactate than bicarbonate is required to achieve the same base excess target. In conclusion, the most reliable method to control respiratory acidosis of permissive hypercapnia, is the administration of lactate-free fluids and bicarbonate titration according to the requirements of individual patients [13].

The paradigmatic CRRT technique using bicarbonate-buffered replacement solutions is acetate-free CVVH (AF-CVVH) [14]. The technique is based on separate infusion of water and electrolytes administered pre-filtration, and isotonic sodium bicarbonate administered post-filtration. The setting of the technique is based on a model predicting the bicarbonate infusion rate for a targeted plasma bicarbonate level. Indeed, AF-CVVH allows fast control of acidosis and has the main advantage of separately controlling urea retention and metabolic acidosis in patients with severe ARF and cardiovascular instability. This method offers the possibility of titrating the amount of the buffer given more precisely, and has been used to treat severe metabolic acidosis.

A further particular point is the use of citrate for regional anticoagulation in CRRT. One absolute requirement for CRRT is anticoagulation which can expose patients to the risk of bleeding. Citrate anticoagulation may limit such risk, but the prevention of citrate side effects requires meticulous monitoring. Then, since citrate is metabolized to bicarbonate, each citrate ion producing three bicarbonate ions, no additional anionic base is required to control metabolic acidosis [5]. When adopting citrate anticoagulation, normal saline is infused as a predilution fluid, while a special hyponatremic dialysate is required because of the sodium load due to trisodium citrate and saline infusion. The amount of citrate to infuse is titrated on the basis of coagulation parameters, or on surrogate parameters, such as the total calcium-ionized calcium rate, and not on a target base excess. Consequently, the use of citrate may be associated with both metabolic alkalosis and acidosis [1]. Metabolic alkalosis is more common in patients with hepatic dysfunction or in those requiring support with large amounts of blood products containing acid citrate dextrose as anticoagulant. In this setting, alkalosis may be prevented using a high chloride load, i.e. large volumes of normal saline as predilution fluid, and infusing calcium chloride. When required, bicarbonate losses may be enhanced by increasing the dialysate flow or the predilution saline infusion. Metabolic acidosis may also occur in the presence of hypercitratemia and hypercalcemia, usually in the setting of ARF with rhabdomyolysis.

Fluid Composition

Electrolyte balance during ARF treated with CRRT is largely dependent on the electrolyte plasma concentration available for ultrafiltration, the ultrafiltration rate and the composition of the replacement solution. As CRRT works continuously, serious derangement in fluid and electrolyte homeostasis may occur in the absence of careful prescription and extremely vigilant monitoring [15].

Replacement fluids for CVVH should contain the following concentrations of electrolytes and glucose: sodium 140 mmol/l; chloride 108–112 mmol/l; potassium 0–4 mmol/l; calcium 1.5–1.75 mmol/l; magnesium 0.5–0.75 mmol/l, and glucose 0–15.00 mg/l [1].

Sodium

Sodium balance can be rather different according to the different CRRT techniques. In CVVH, a totally convective therapy, the value of ultrafilterable sodium must be taken into account to choose the correct sodium concentration in the substitution fluid [15]. It is worth remembering that the sodium concentration in the ultrafiltrate is around 7 mEq/l lower than the plasma water sodium concentration, because a certain amount of the total ion is complexed with anions, particularly with bicarbonate and proteins. When using techniques which combine diffusive and convective processes, such as CVVHD or CVVHDF, a dialysis solution is added to the system. Even with these techniques sodium transport is mainly due to the convective process; the diffusive flux is definitely lower and is strongly influenced by the sodium concentration of the dialysate. A retrospective study showed that CVVHDF was more likely to achieve serum sodium concentrations within the normal range than CVVH, thereby supporting the role of dialysate in the maintenance of an adequate sodium concentration [4]. It has been suggested that a supraphysiological sodium concentration in the dialysate or replacement solutions would improve the hemodynamic stability or prevent the increase in intracranial pressure, but there are no conclusive data concerning this issue [6].

Potassium

Hyperkalemia is a typical feature of ARF, particularly in the presence of tissue breakdown (crush syndrome, hemolysis, hypercatabolic syndromes). Unless important hyperkalemia is present, 3–4 mEq/l potassium needs to be added to the replacement fluid in CVVH in order to avoid hypokalemia. In techniques comprising a diffusive process, namely CVVHD or CVVHDF, the content of potassium in the dialysate must be titrated to about 2–4 mEq/l, to avoid the risk of hypokalemia [15]. The very slow flux of potassium from cells to the extracellular space would possibly contribute to the high hemodynamic tolerance of CRRT.

Divalent Ions

About 60% of total plasma calcium is ultrafilterable, then substitution fluid in CVVH must contain about 3 mEq/l of calcium in order to prevent hypocalcemia. When mixed techniques, such as CVVHD, are used, the dialysate concentration of calcium usually ranges from 3 to 3.5 mEq/l. Patients undergoing an exchange of very large volumes of ultrafiltrate during CRRT may be prone to the risk of a positive calcium balance, and therefore may require a lower content of calcium in replacement fluids. The same cautions are required to keep an appropriate magnesium balance.

Intensive schedules of CRRT can easily induce hypophosphatemia in the critically ill patient, the risk being magnified by prolonged parenteral nutrition, malnourishment or concomitant metabolic alkalosis. This condition may require oral or parenteral supplementation with phosphate salts, as well as supplementation of dialysate solutions. The addition of phosphate salts to the dialysate and/or the replacement solutions facilitates phosphate handling, while the risk of phosphate precipitation in the presence of calcium ions has recently been excluded [16].

Glucose

The recent publication of a randomized trial showing a significant survival advantage with strict glucose control in critically ill patients has important consequences for CRRT with glucose-containing dialysate or substitution fluids [17]. The use of physiologic concentrations of glucose in the dialysate and in replacement fluids is advised to prevent or compensate extracorporeal losses, while glucose-free solutions might be used when an adequate nutritional regimen has been established [6].

A special case is that of critically ill children weighing <10 kg who require blood as a priming solution for the extracorporeal circuit before initiating CRRT to prevent hemodilution and to maintain adequate oxygenation [18]. Blood preparations usually contain supraphysiological electrolyte concentrations and a nonphysiological acid-base balance that may exacerbate the critical condition of the small patient. In such cases, the pretreatment of blood bank-derived blood aimed to normalize pH and electrolyte concentration yields more physiological blood priming.

In conclusion, due to its continuous nature, CRRT need to be carefully monitored for the composition of dialysate and replacement fluids to avoid dangerous electrolyte imbalance. Moreover, it is possible that in the near future we may see more sophisticated CRRT solutions containing nutrients and antioxidants that may improve the outcome of ICU patients.

Physical Properties

All lactate-based or buffer-free solutions are acidic, while bicarbonate-based fluids have a physiological pH. Experimental findings suggest that acidic fluids cause intracellular acidosis and reduced activity of macrophages and blood mononuclear cells [5].

The use of replacement fluids or dialysate solutions at room temperature as well as continuous blood flow through the extracorporeal system cause an average 2°C reduction in body temperature, and an energy loss of about 1,000 kcal/day [1]. Although the clinical significance of this effect is not clear at this point, the energy loss measured during CVVHD has been suggested to be important in hemodynamic stability or even for patient prognosis [19]. Heat loss and consequent hypothermia may also affect immune functions and increase the risk of clotting of the CRRT circuit. The extent of these effects depends on the length of the circuit, on the flow rate of blood, replacement fluids and dialysate, on body weight and finally on the presence or absence of an intact autoregulatory mechanism to preserve core temperature. A heating device may avoid the above-cited side effects; on the other hand a lower body temperature may be desirable in patients with excessive oxygen consumption and low systemic vascular resistance. At present, it is not clearly known in which patients the net effect of CRRT-induced hypothermia is useful or harmful.

Obviously substitution fluids need to be sterile, and the same is true for the dialysate, not only in high-flux dialysis schedules when backfiltration is most likely to occur, but also in other continuous dialytic treatments. As in intermittent treatments, the feasibility of online production of substitution fluids in CRRT has recently been evaluated.

Administration Route

The CRRT prescription directly affects electrolyte and acid-base balance [5, 6, 20]. In diffusive or mixed dialysis schedules, the concentration of electrolytes in the blood is dependent on the corresponding concentration in the dialysate and in replacement fluids. In pure convective treatments most ions pass freely across the membrane, showing a sieving coefficient of around one. In this case the ionic concentration of the substitution fluids should be very similar to that of normal plasma, with an alkaline pH.

The different features of predilution (lower solute clearance, higher ultrafiltration rate and larger filter surface area) and of postdilution (higher filtration fraction) are well known [1, 5, 6, 20]. Predilution may be preferred to reduce the need for heavy anticoagulation or to enhance the ultrafiltration rate, especially relevant for HV-CVVH. Possibly, pre- and postdilution may be combined when extracorporeal clearance is limited by an inadequate blood flow, but this point deserves future studies.

Conclusions

The great complexity of current CRRT schedules, exchanging up to 50 l/day, has been made possible by the development of specialized dialysate and replacement fluid solutions. These strategies have been supported by studies reporting an improvement in the survival of the critically ill patient following the adoption of technical approaches providing the exchange of large amounts of fluids, and/or the use of high dialysate flows. However, with the increase in exchanged volumes, the control of fluid sterility, physical properties and composition, and the choice of individualized prescriptions are becoming increasingly crucial in the single ICU patient.

References

1 Manns M, Sigler MH, Teehan BP: Continuous renal replacement therapies. An update. Am J Kidney Dis 1998;32:185–207.
2 Macias WL: Choice of replacement fluid/dialysate anion composition in continuous renal replacement therapy. Am J Kidney Dis 1996;28(suppl):S15–S20.
3 Feriani M, Dell'Aquila R: Acid-base balance and replacement solutions in continuous renal replacement therapy. Kidney Int 1998;53(suppl 66):S156–S159.
4 Morimatsu H, Uchino S, Bellomo R, Ronco C: Continuous renal replacement therapy: does technique influence electrolyte and bicarbonate control? Int J Artif Organs 2003;26:289–296.
5 Davemport A: Dialysate and substitution fluids for patients treated by continuous forms of renal replacement therapy; in Ronco C, Bellomo R, La Greca G (eds): Blood Purification in Intensive Care. Contrib Nephrol. Basel, Karger, 2001, vol 132, pp 313–322.
6 Schetz M, Leblanc M, Murray PT: The Acute Dialysis Quality Initiative – Part VII: Fluid composition and management in CRRT. Adv Renal Replace Ther 2002;9:282–289.
7 Barenbrock M, Hausberg M, Matzkies F, De La Motte S, Shaefer RM: Effects of bicarbonate and lactate buffered replacement fluids on cardiovascular outcome in CVVH patients. Kidney Int 2000;58:1751–1757.
8 O'Croinin D, Ni Chonghaile M, Higgins B, Laffey JG: Bench-to-bedside review: permissive hypercapnia. Crit Care 2005;9:51–59.
9 Kuiper JW, Groeneveld AB, Slutsky AS, Plotz FB: Mechanical ventilation and acute renal failure. Crit Care Med 2005;33:1408–1415.
10 Acute Respiratory Distress Syndrome Network: Ventilation with lower tidal volumes as compared with traditional tidal volumes for acute lung injury and the acute respiratory distress syndrome. N Engl J Med 2000;342:1301–1308.

11 Gehlbach BK, Schmidt GA: Bench-to-bedside review: treating acid-base abnormalities in the intensive care unit – the role of buffers. Crit Care 2004;8:259–265.
12 Rotta AT, Steinhorn DM: Is permissive hypercapnia a beneficial strategy for pediatric acute lung injury? Respir Care Clin N Am 2006;12:371–387.
13 MacLean AG, Davemport A, Cox D, Sweny P: Effects of continuous haemodiafiltration against lactate-buffered and lactate-free dialysate in patients with and without liver dysfunction. Kidney Int 2000;58:1765–1772.
14 Wynckel A, Wuillai A, Bene B, Cornillet J, Chanard J: Assessment of acetate free continuous veno-venous hemofiltration in acute renal failure. ASAIO J 1998;44:M606–M609.
15 Locatelli F, Pontoriero G, Di Filippo S: Electrolyte disorders and substitution fluid in continuous renal replacement therapy. Kidney Int 1998;53(suppl 66):S151–S155.
16 Troyanov S, Geadah D, Ghannoum M, Cardinal J, Leblanc M: Phosphate addition to hemodiafiltration solutions during continuous renal replacement therapy. Intensive Care Med 2004;30: 1662–1665.
17 Van Der Berghe G, Wouters P, Weekers F, et al: Intensive insulin therapy in critically ill patients. N Engl J Med 2001;345:1359–1367.
18 Pasko DA, Mottes TA, Mueller BA: Pre dialysis of blood prime in continuous hemodialysis normalizes pH and electrolytes. Pediatr Nephrol 2003;18:1177–1183.
19 Yagi N, Leblanc M, Sakai K, Wright EJ, Paganini EP: Cooling effect of continuous renal replacement therapy in critically ill patients. Am J Kidney Dis 1998;32:1023–1030.
20 Leblanc M: Fluid composition for CRRT; in Ronco C, Bellomo R, Brendolan A (eds): Sepsis, Kidney and Multiple Organ Dysfunction. Contrib Nephrol. Basel, Karger, 2004, vol 144, pp 222–227.

Prof. Loreto Gesualdo
Department of Nephrology, Dialysis and Transplantation, University of Foggia
Viale Pinto, 1
IT–71100 Foggia (Italy)
Tel. +39 0881 732 054, Fax +39 0881 736 001, E-Mail l.gesualdo@unifg.it

Results from International Questionnaires

Zaccaria Ricci[a], Sergio Picardo[a], Claudio Ronco[b]

[a]Department of Pediatric Cardiology and Cardiac Surgery, Bambino Gesù Hospital, Rome, and [b]Department of Nephrology, Dialysis and Transplantation, St. Bortolo Hospital, Vicenza, Italy

Abstract

Background: The practice of renal replacement therapy (RRT) has reached an optimal standard of care worldwide. Nevertheless, some aspects of acute renal failure treatment and support still present wide variability between different centers. This is especially true for the mode and dose of RRT. This review describes the epidemiology of dialysis prescription and delivery around the world based on recent observational studies and international surveys. **Results:** Continuous RRT is delivered in 80% of intensive care units around the world. Since a certain consensus has been achieved on the adequacy of 35 ml/kg/h of clearance in continuous therapies, recent observations based on questionnaires and surveys demonstrated that such adequate therapy was only prescribed in the minority of patients. The number of centers prescribing adequate dialysis dose is increasing, but there are still many institutions where prescription is made with no specific adequacy targets and effective delivery is not measured. Several barriers to reaching adequacy targets have been identified including the lack of a high evidence multicentric trial, logistics, costs, personnel and technical difficulties. **Conclusion:** A trend to continuous therapies and increased RRT dosage over the last 10 years is shown by the surveys presented, even if scientific evidence is now very necessary as far as definitive RRT indications and prescriptions are concerned.

Copyright © 2007 S. Karger AG, Basel

The practice of renal replacement therapy (RRT) for acute renal failure (ARF) is extremely variable including different techniques from intermittent hemodialysis (IHD) to continuous renal replacement therapies (CRRT) [1]. There is a general consensus that optimal strategies to improve patient outcome in ARF should include delivery of an adequate treatment dose [2]. The Acute Dialysis Quality Initiative (ADQI) [3, 4] has identified several areas where consensus is lacking and recommendations for good clinical practice are very much needed. In this setting, the practice of CRRT seems to be a specific and important area where criteria for starting, the modality of therapy, and treatment prescription are still not carried out on solid bases of previous experience or evidence, but rather on local protocols [5]. However, ADQI has brought about a large consensus, and also encouraged possible studies and potential analyses that might be useful to generate not yet available evidence. One aspect that has been noted for urgent definition is the actual practice pattern in the choice and conduction of renal replacement techniques. Recently, several international surveys have tried to depict routine clinical practice in ARF management in order to provide knowledge on 'real-world' issues, on physicians' compliance to practice guidelines, on educational needs and research objectives. Here we review and discuss some results of these observations.

Beginning and Ending Supportive Therapy

With the intention of determining the association between outcome and different epidemiological parameters (period prevalence of ARF, etiology, illness severity, and clinical management of ARF), the Beginning and Ending Supportive Therapy for the Kidney (BEST Kidney) investigators conducted the largest multinational, multicenter, prospective, epidemiological survey of ARF in intensive care unit (ICU) patients [6] who either were treated with RRT or fulfilled at least 1 of the predefined criteria for ARF. ARF criteria were defined a priori in order to achieve a standardized definition that was easily reproducible for each center: oliguria, defined as a urine output of <200 ml in 12 h, and/or marked azotemia, defined as a blood urea nitrogen level of >30 mmol/l. The data were collected from September 2000 to December 2001 at 54 hospitals in 23 countries. Data about 29,269 critically ill patients were recorded during the 16-month study period. The median age of patients with ARF was 67 years, the median SAPS II score was 48, the median body weight was 74 kg. Approximately 30% of the patients had chronic renal dysfunction but were not receiving dialysis treatment. The estimated creatinine clearance on admission to the ICU was 35 (interquartile range 20–59) ml/min. 1,738 (5.7%) had ARF during their ICU stay, including 1,260 (4.3%) who were treated with RRT. Overall hospital mortality was 60.3%, significantly higher than the SAPS II predicted mortality. The most common contributing factor to ARF was septic shock (47.5%). Continuous RRT was the most common initial modality used (80.0%), followed by intermittent RRT (16.9%), peritoneal dialysis and slow continuous ultrafiltration (3.2%). 86.2% survivors were dialysis-independent at hospital discharge. Independent risk factors for hospital mortality included use of vasopressors, mechanical ventilation, septic shock, cardiogenic shock, and hepatorenal syndrome.

From the findings of this study we can reasonably finally conclude that: the prevalence of ARF among the ICU population is between 5 and 6%; more than 70% of these patients require RRT; hospital survival among ARF patients is still disappointing (about 40%), and renal recovery among survivors is very high. A systematic literature review [7] recently reported that no evidence of a substantial improvement in outcome from ARF over the last 50 years has been observed. The mortality rates remain superior to 50%, and it is likely that it will remain unchanged in the next decade or more: this mortality rate represents a level of adequate performance of the healthcare system but it should not be used to obtain comparisons with other periods. In other words, as therapeutic capability improves and the system continues to achieve a mortality of 50% for these very sick patients, the healthcare system will presumably admit and treat sicker and sicker patients with ARF. Future epidemiologic studies will need to take into account this confounder in order to appreciate the continuing change in illness severity [8].

CRRT through the New Millennium

Following such a practice-related approach, we have distributed a questionnaire on specific issues about practice patterns in the field of RRT during the 1st (1998) [5] and the 3rd (2004) International Course on Critical Care Nephrology [1]. The analysis comparing the answers from these 2 surveys covering 6 years presented many interesting results, especially with regard to the steep improvement in the standard of care for critical care nephrology. Responders were nephrologists (60%) and intensivists (40%) from all 5 continents who attended both meetings. 345 questionnaires in 1998 and 560 in 2004 were correctly completed. The two sets of completed questionnaires were collected into an access database and the results examined. Percentage values are reported in order to compare the two uneven populations. Continuous arterio-venous hemofiltration (HF), representing one of the available options for more than 70% of candidates in 1998, was abandoned by the responders in 2004. Continuous veno-venous RRTs were also considered by more than 90% candidates in both surveys. In 2004 intermittent techniques, available in only 20% centers in 1998, were routinely available in more than 80% institutions, administered as intermittent hemodialysis (two thirds) or as slow extended daily dialysis (one third). Peritoneal dialysis was a rare option (5%) in ICUs in both 1998 and 2004. HF was and remained the preferred RRT modality throughout the evaluated period: it was selected by a large majority of responders in 2004, whereas it had appeared to be less clearly predominant before: perhaps recent literature and new dedicated machines facilitated HF prescription and delivery [9]. Apparently, in the 6-year interval, ICU physicians prescribing RRT without further nephrological counseling increased from about 15% to almost 30%.

Anticoagulation management did not change significantly, and patient bleeding remained one of the most selected complains during both meetings, together with circuit clotting. Responders notably modified their RRT prescriptions from 1998 to 2004: urea clearance increased from 1–1.5 (range 0.5–2) to 2–3 (range 1–10) l/h. A more in-depth analysis of CRRT prescription in 2004 showed interesting results: surprisingly only about 50% of RRT are prescribed upon a standard protocol. Furthermore, a large part of our responders seemed to be uncertain about treatment prescription: this could mean that delivery is not personalized to the patient and the clinical setting. In 2004 participants mostly reported prescribing a dose of 35 ml/kg/h or 2–3 l/h as the urea efficiency target, with a range from 1 to more than 5 l/h which, in our opinion, is consistent with a trend to increased RRT dosage over the last 10 years. In 1998 CRRT dose prescription only ranged from 0.5 to 2 l/h. Different from that survey, in 2004 low treatment efficiency was not a matter of complaint anymore, whereas filter clotting and catheter dysfunction still represented a problem for operators in the field of RRT. As a matter of fact, from 1998 to 2004 heparin infusion remained the preferred anticoagulation technique and anticoagulation side effects (bleeding and hematoma) are still a problem. Less dangerous alternatives or more effective molecules are still being evaluated [10, 11]. Similarly, great technological improvements were evident between the two sets of questionnaires: more than 50% of the available equipment in 1998 consisted of continuous arterio-venous HF kits, or adapted machines from chronic therapies, whereas in 2004, 100% of the participants declared using dedicated integrated monitors. Exactly as in 1998, as far as non-renal indications are concerned (congestive heart failure, sepsis, anasarca, systemic inflammatory disease), 90% of responders stated that they agree with non-renal indications: two thirds in 1998 and only half in 2004 declared prescribing RRT for extended indications even in the absence of acute renal failure (fig. 1). The lack of scientific evidence is the primary reason for skepticism with regard to adopting extracorporeal treatment: presumably for this reason the number of the responders who in 1998 declared starting RRT in case of septic shock, even in the absence of ARF, decreased significantly in the last survey (fig. 2). This contradiction could indicate that current RRT practice might not completely apply to evidence-based medicine and that studies with a high level of evidence in the field of non-renal RTT indications are strongly needed. In the case of a non-renal indication, most of the meeting participants would prescribe routine treatment without changing the usual machines or settings. Nonetheless, our audit selected a number of alternative techniques as being feasible treatments during sepsis syndrome (namely, high-volume HF, continuous plasma filtration adsorption and hemoperfusion), showing that constant attention is paid to the most recent technical possibilities offered by extracorporeal treatments.

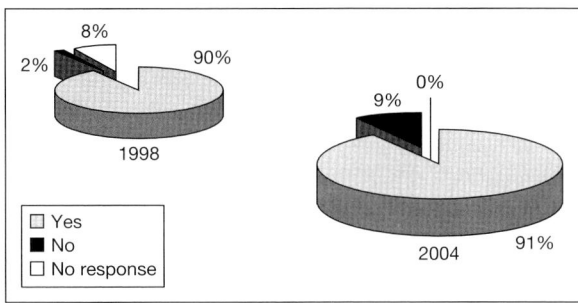

Fig. 1. Answers of participants to the 1st and 3rd International Course on Critical Care Nephrology to the question whether extracorporeal therapies are feasible treatments for 'non-renal' indications.

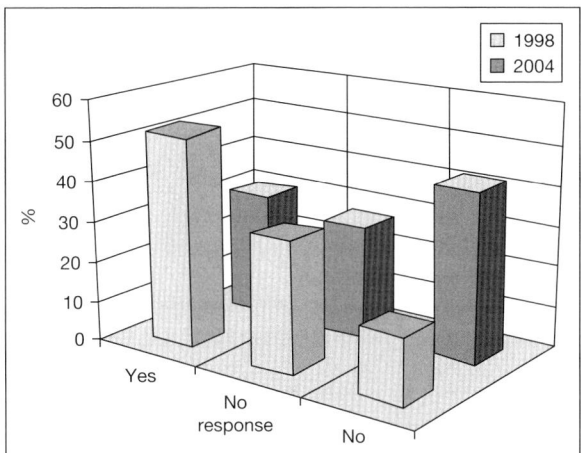

Fig. 2. Answers of participants to the 1st and 3rd International Course on Critical Care Nephrology to the question whether extracorporeal therapies can be used with 'non-renal' indications in the absence of acute renal failure.

DO-RE-MI

The Dose Response Multicenter International collaboration (DO-RE-MI) [12] is currently seeking to address the issue of how practice patterns are currently chosen and performed. DO-RE-MI is an observational, multicenter study conducted in ICUs. The primary aim is to study the dialysis dose delivered, which will then be compared with ICU mortality, 28-day mortality, hospital mortality, ICU length of stay, and the number of days of mechanical ventilation. It is hoped that this international collaboration will provide a clearer picture of how RRT is chosen, prescribed and delivered, and how such delivery may affect outcome. The preliminary results of this survey are presented in detail elsewhere in the present issue.

Conclusion

The syndrome known as 'acute renal failure' is common in the ICUs and may affect from 1 to 25% of the patients [6, 13]. This wide range might depend on the different patient populations present in different centers, and also on the different criteria used to define its presence. When severe ARF occurs in patients with severe systemic illness, septic shock and multiorgan dysfunction [14], it considerably complicates patient management, increases the cost of care, and is associated with a high level of morbidity and mortality [6]. Starting from the definition of ARF itself, many controversies surround its management [15]. Surveying routine clinical practice may provide precious knowledge on 'real-world' issues, on physicians' compliance to practice guidelines, on educational needs and research objectives. Participants who attend this kind of meeting are obviously a self-selected population and their answers cannot reasonably reflect the worldwide daily reality of patient care. Nonetheless, especially for the BEST study, the group of respondents was indeed quite large and a broad distribution of participants was evident.

Analysis of available techniques in different institutions showed a certain prevalence of continuous techniques. Nonetheless, in many institutions intermittent techniques are present together with continuous ones, thus showing the availability of different prescriptions and practices. Surprisingly, according to our survey, only in about 50% of cases is RRT managed as a standard protocol. Furthermore, a large number of responders seemed to be uncertain with regard to treatment prescription: this could mean that delivery is not personalized on a patient and clinical setting. In the late 1990s dose prescription was mostly described in terms of liters of effluent per day. Very seldom did the dose amount exceed 30 l/day leading to an average dose delivery in the range of 15–20 ml/kg/h. After 2000 a dose-related discrimination was made. Furthermore the respondents seemed not only to have an increased awareness of the importance of adequate doses in ARF, but also on the potential effect of higher doses in septic patients. Nonetheless, in the last survey, a large number of respondents still admitted to ignoring how prescription was made in their center.

A trend to continuous therapies and increased RRT dosage with respect to the last 10 years was shown by the presented surveys, even if scientific evidence is now more necessary than ever as far as definitive RRT indications and prescriptions are concerned [16].

References

1 Ricci Z, Ronco C, D'Amico G, et al: Practice patterns in the management of acute renal failure in the critically ill patient: an international survey. Nephrol Dial Transplant 2006;21:690–696.
2 Ricci Z, Bellomo R, Ronco C: Dose of dialysis in acute renal failure. Clin J Am Soc Nephrol 2006;1:380–388.
3 Bellomo R, Ronco C, Kellum A, Mehta RL, Palevsky P; ADQI Workgroup: Acute renal failure – definition, outcome measures, animal models, fluid therapy and information technology needs: the Second International Consensus Conference of Acute Dialysis Quality Initiative (ADQI) group. Crit Care 2004;8:R204–R212.
4 ADQI: Acute Dialysis Quality Initiative. http://www.adqi.net.
5 Ronco C, Zanella M, Brendolan A, Milan M, Canato G, Zamperetti N, Bellomo R: Management of severe acute renal failure in critically ill patients: an international survey in 345 centres. Nephrol Dial Transplant 2001;16:230–237.
6 Uchino S, Kellum JA, Bellomo R, Doig GS, Morimatsu H, Morgera S, Schetz M, Tan I, Bouman C, Macedo E, Gibney N, Tolwani A, Ronco C; Beginning and Ending Supportive Therapy for the Kidney (BEST Kidney) Investigators: Acute renal failure in critically ill patients: a multinational, multicenter study. JAMA 2005;294:813–818.
7 Ympa IP, Sakr Y, Reinhart K, Vincent JL: Has mortality from acute renal failure decreased? A systematic review of the literature. Am J Med 2005;118:827–832.
8 Bellomo R: The epidemiology of acute renal failure: 1975 versus 2005. Curr Opin Crit Care 2006; 12:557–560.
9 Ronco C, Bellomo R, Homel P, Brendolan A, Maurizio D, Piccini P, La Greca G: Effects of different doses in continuous veno-venous haemofiltration on outcomes of acute renal failure: a prospective randomised trial. Lancet 2000;355:26–30.
10 Fiaccadori E, Maggiore U, Rotelli C, Minari M, Melfa L, Cappe G, Cabassi A: Continuous haemofiltration in acute renal failure with prostacyclin as the sole anti-haemostatic agent. Intensive Care Med 2002;28:586–593.
11 Monchi M, Berghmans D, Ledoux D, Canivet JL, Dubois B, Damas P: Citrate vs. heparin for anticoagulation in continuous venovenous hemofiltration: a prospective randomized study. Intensive Care Med 2004;30:260–265.
12 Kindgen-Milles D, Journois D, Fumagalli R, et al: Study protocol: The Dose Response Multicentre International collaborative initiative (DO-RE-MI). Crit Care 2005;9:R396–R406.
13 Liano F, Junco E, Madero R, Pascual J, Verde E; Madrid Acute Renal Failure Study Group: The spectrum of acute failure in the intensive care unit compared with that seen in other settings. Kidney Int 1998;53:S16–S24.
14 Kleinknecht D: Risk factors for acute renal failure in critically ill patients; in Ronco C, Bellomo R (eds): Critical Care Nephrology. Dordrecht, Kluwer Academic, 1998, pp 143–152.
15 Kellum J, Palevsky P: Renal support in acute kidney injury. Lancet 2006;368:344–345.
16 Bellomo R: Do we know the optimal dose for renal replacement therapy in the intensive care unit? Kidney Int 2006;70:1202–1204.

Zaccaria Ricci
Department of Pediatric Cardiology and Cardiac Surgery
Bambino Gesù Hospital, Piazza S. Onofrio
IT–00100 Rome (Italy)
Tel. +39 06 6859 3333, E-Mail z.ricci@libero.it

Intermittent Hemodialysis for Renal Replacement Therapy in Intensive Care: New Evidence for Old Truths

W. Van Biesen, N. Veys, R. Vanholder

University Hospital Ghent, Ghent, Belgium

Abstract

Acute renal failure requiring dialysis is a frequent complication in critically ill patients with a high morbidity and mortality. Until recently, no evidence-based guidelines on the optimal treatment modality for renal replacement in the ICU could be issued because of a lack of well-performed randomized controlled trials (RCT). Over the last years however, some important new concepts and RCTs have been published on this topic. An important concept is the understanding that 'chronic dialysis strategies' are not suitable for acute renal failure patients in the ICU. From this understanding the necessity of daily dialysis followed, and later on, the need for flexible treatments related to the patients' need, using slow long extended daily dialysis (SLEDD). Several recent papers compared continuous renal replacement therapy and intermittent hemodialysis (IHD) in ICU patients, pointing to a lack of differences in outcome, but there were less practical problems using IHD, even in unstable patients. In conclusion, it can be stated that all patients can be treated with IHD when available, without jeopardizing their outcome. Slow extended daily dialysis emerged as a hybrid renal replacement therapeutic modality and has promising features because it combines the advantages of both continuous renal replacement therapy and IHD, but until now, no studies evaluating whether SLEDD is superior to 'regular IHD' are available.

Copyright © 2007 S. Karger AG, Basel

Some decades ago, continuous renal replacement therapy (CRRT) was developed in response to the frustration of intensivists caused by the fact that nephrologists at that time used intermittent hemodialysis (IHD) in acute renal failure patients in the same way as they did in their chronic patients. From the initially very simple and inefficient continuous arteriovenous hemofiltration, CRRT developed into a very complex treatment (fig. 1) [1], which at the end

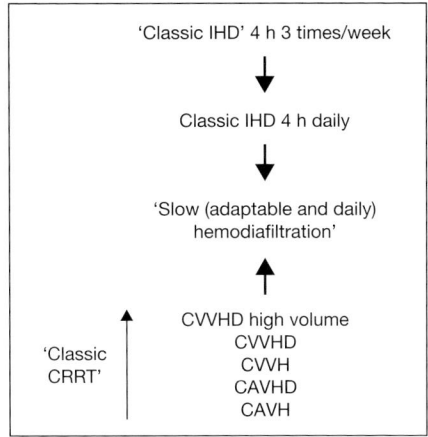

Fig. 1. Evolution of 'pure intermittent hemodialysis' (IHD) and 'pure continuous low efficient' dialysis towards the hybrid treatment of slow extended daily dialysis.

mimicked the highly efficient intermittent hemodiafiltration of regular dialysis, but without using dialysis monitors. On the other hand, nephrologists interested in intensive care pathology invested a lot of effort to get daily dialysis accepted, and to introduce slow low-efficient daily dialysis (SLEDD) [2]. The discussion whether CRRT or IHD was the preferred modality of choice for renal replacement therapy (RRT) in the ICU was for a long time fuelled by a lack of randomized controlled trials (RCT) and competing interests of nephrologists and intensivists [3]. Although many efforts were made to persuade physicians that CRRT was the preferred modality, it was also apparent that most nephrologists preferred IHD [4], whereas intensivists preferred CRRT [5]. From uncontrolled trials in centers using both CRRT and IHD, it was apparent that, if anything, the outcome of IHD was at least as good as of CRRT, and this independent of the severity of disease (fig. 2) [6]. The first RCT on this topic [7] 'unexpectedly' showed a superior outcome for the IHD group, an effect that was attributed to problems of randomization. A meta-analysis, however, confirmed the lack of difference in outcome concerning the hard endpoint 'mortality' [8], but the hard evidence was, until recently, lacking.

Recent Trials

A first RCT on the topic was performed by Uehlinger et al. [9]. In this single center study, 125 patients were randomized to either CRRT (hemodiafiltration, n = 70) or IHD (n = 55). The randomization procedure can be criticized

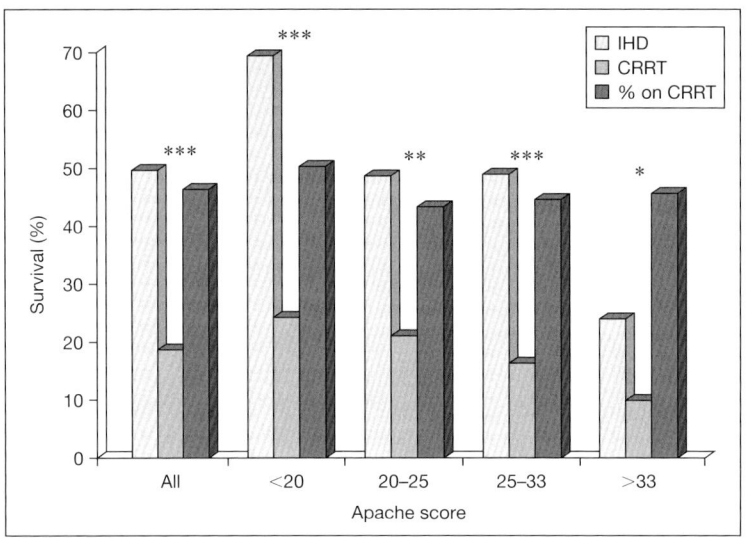

Fig. 2. CRRT vs. IHD. Ghent University Hospital 1995–1999 (N = 557) (*p < 0.05, **p < 0.01, ***p < 0.001).

because not all available patients (n = 191) could be randomized due to a lack of the availability of enough machines in each of the groups to perform the treatment the patient was randomized to at all time points of the study. However, this situation is quite usual in most hospitals were both modalities are offered, and should therefore be accepted as a real life condition, increasing the generalizability of the results to everyday clinical practice. Randomization of the included patients was, however, quite successful, as the two groups were comparable at the start of RRT with respect to age [62 ± 15 vs. 62 ± 15 years, continuous venovenous hemodiafiltration (CVVHDF) vs. IHD], gender (66 vs. 73% male sex), number of failed organ systems (2.4 ± 1.5 vs. 2.5 ± 1.6), Simplified Acute Physiology Scores (57 ± 17 vs. 58 ± 23), septicemia (43 vs. 51%), shock (59 vs. 58%) or previous surgery (53 vs. 45%). Both modalities were comparable with regard to the primary endpoint (mortality rates) in the hospital (47 vs. 51%, CVVHDF vs. IHD, p = 0.72) or in the ICU (34 vs. 38%, p = 0.71). Also hospital length of stay and duration of RRT required in the survivors was comparable in patients on CVVHDF [median (range) 20 days (6–71), n = 36] and in those on IHD [30 days (2–89), n = 27, p = 0.25].

A second large RCT was published by the Hemodiafe group [10]. In this large, multicentric trial, patients were randomized to either CVVHDF (n = 175) with ± 1 l/h dialysate flow and 1.3 l/h hemofiltration with predilution, with a blood flow of 150 ml/min or to IHD (n = 184), using a blood flow of 500 ml/min and a dialysate flow of 500 ml/min. IHD was performed daily for a duration of 5.2 h. Both IHD as CRRT were performed using the same type of membranes, excluding the fact that eventual differences in outcome might be attributable to differences in biocompatibility. Comorbidity scores were equally distributed among both groups. There was no difference between the two treatment modalities concerning the primary endpoint (patient mortality) at any moment of observation. In addition, there were no differences observed between the groups for occurrence of hypotension or of bleeding episodes. Also the length of ICU and hospital stay and the duration of the need for RRT were not different between the groups. Interestingly, although it was recommended to use a low dialysate temperature (35.5°C) in the IHD group, as it has indeed been suggested that the better hemodynamic tolerance in CRRT was mainly due to lowering the central body temperature, the number of cases suffering from hypothermia was significantly higher in the CRRT group ($p = 0.0005$). Despite the lack of differences in outcome, there was, however, a higher number of patients that was switched from CRRT to IHD than vice versa, the major reason for transfer being bleeding-related problems. In conclusion, the authors postulated that, if all efforts are made to optimize treatment, outcome of CRRT and IHD are comparable for all patient categories.

The results of these two trials are compatible with the recently published observational study published by the PICARD group [11]. Among 398 patients who required dialysis, the risk of death within 60 days was examined by assigned initial dialysis modality [CRRT (n = 206) vs. IHD (n = 192)]. Although the study was, by definition, not randomized, differences in comorbidity were accounted for by a propensity score approach. Crude survival rates were lower for patients who were treated with CRRT than IHD (survival at 30 days 45 vs. 58%; $p = 0.006$). Adjusted for age, hepatic failure, sepsis, thrombocytopenia, blood urea nitrogen, and serum creatinine and stratified by site, the relative risk of death associated with CRRT was 1.82 (95% confidence interval 1.26–2.62). Further adjustment for the propensity score did not materially alter the association (relative risk 1.92; 95% confidence interval 1.28–2.89). Although these results could still reflect residual confounding by the severity of illness, as it was not a randomized trial, the results at least do not support a survival benefit afforded by CRRT.

Conclusion

According to current knowledge, the outcome for patients needing RRT in the ICU is not influenced by the modality used. IHD is, however, much cheaper than CRRT, and therefore, it should be the first RRT of choice in centers where it is available. It is likely that newer hybrid techniques (SLEDD) offer an additional advantage but this remains to be proven.

References

1 Sever MS, Vanholder R, Lameire N: Management of crush-related injuries after disasters. N Engl J Med 2006;354:1052–1063.
2 Vanholder R, Van Biesen W, Lameire N: What is the renal replacement method of first choice for intensive care patients? J Am Soc Nephrol 2001;12(suppl 17):S40–S43.
3 Lameire N, Van Biesen W, Vanholder R, Colardijn F: The place of intermittent hemodialysis in the treatment of acute renal failure in the ICU patient. Kidney Int Suppl 1998;66:S110–S119.
4 Abdeen O, Mehta RL: Dialysis modalities in the intensive care unit. Crit Care Clin 2002;18: 223–247.
5 Uchino S, Kellum JA, Bellomo R, Doig GS, Morimatsu H, Morgera S, et al: Acute renal failure in critically ill patients: a multinational, multicenter study. JAMA 2005;294:813–818.
6 Swartz RD, Messana JM, Orzol S, Port FK: Comparing continuous hemofiltration with hemodialysis in patients with severe acute renal failure. Am J Kidney Dis 1999;34:424–432.
7 Mehta RL, McDonald B, Gabbai FB, Pahl M, Pascual MT, Farkas A, et al: A randomized clinical trial of continuous versus intermittent dialysis for acute renal failure. Kidney Int 2001;60: 1154–1163.
8 Tonelli M, Manns B, Feller-Kopman D: Acute renal failure in the intensive care unit: a systematic review of the impact of dialytic modality on mortality and renal recovery. Am J Kidney Dis 2002; 40:875–885.
9 Uehlinger DE, Jakob SM, Ferrari P, Eichelberger M, Huynh-Do U, Marti HP, et al: Comparison of continuous and intermittent renal replacement therapy for acute renal failure. Nephrol Dial Transplant 2005;20:1630–1637.
10 Vinsonneau C, Camus C, Combes A, Costa de Beauregard MA, Klouche K, Boulain T, et al: Continuous venovenous haemodiafiltration versus intermittent haemodialysis for acute renal failure in patients with multiple-organ dysfunction syndrome: a multicentre randomised trial. Lancet 2006;368:379–385.
11 Cho KC, Himmelfarb J, Paganini E, Ikizler TA, Soroko SH, Mehta RL, et al: Survival by dialysis modality in critically ill patients with acute kidney injury. J Am Soc Nephrol 2006;17:3132–3138.

W. Van Biesen
Renal Division, Department of Internal Medicine
University Hospital Ghent
BE–9000 Ghent (Belgium)
E-Mail wim.vanbiesen@ugent.be

Continuous Renal Replacement in Critical Illness

Claudio Ronco[a], Dinna Cruz[a], Rinaldo Bellomo[b]

[a]Department of Nephrology, San Bortolo Hospital, Vicenza, Italy;
[b]Department of Intensive Care, Austin Hospital, Melbourne, Vic., Australia

Abstract

Acute renal failure in the intensive care unit is usually part of the multiple organ dysfunction syndrome, and the complexity of illness in patients with this complication has risen in recent years. Continuous renal replacement therapy (CRRT) was introduced in the late 1970s and early 1980s to compensate for the inadequacies of conventional intermittent hemodialysis (IHD) in the treatment of these patients. IHD was considered aggressive and unphysiological, often resulting in hemodynamic intolerance and limited efficiency. Although CRRT has been shown to be physiologically superior with respect to IHD in both observational and randomized studies, it is not clear whether this physiological superiority translates into clinically important gains. A number of recent studies have tried to address this issue, and with these, there is a lack of evidence to suggest improved survival and major clinical outcomes with CRRT. However, these studies are generally underpowered and have certain aspects which may influence the interpretation of their results. In addition, the development of hybrid techniques, such as slow extended daily dialysis, makes this a dynamic area of study where the terms of comparison are constantly changing. This article reviews recent trials comparing CRRT and IHD, and discusses their results and limitations.

Copyright © 2007 S. Karger AG, Basel

Acute renal failure (ARF) affects 5–7% of all hospitalized patients [1]. Apparently, ARF continues to be associated with poor outcomes in spite of technological advances [2]. Deeper analysis suggests that pathophysiological mechanisms and clinical pictures are today more complex, including a large proportion of sepsis-related cases and significant involvement of other organs. ARF is today frequent in the intensive care unit (ICU) with a reported incidence of 1–25% [3] depending on the population studied and the criteria used for definition. Thus we observe two separate syndromes: uncomplicated ARF can be managed outside the ICU and usually carries good prognosis with mortality rates less than 10%. In contrast, complicated ARF in the ICU is part of multiple organ dysfunction syndromes and is associated with mortality rates of 50–70% [4]. The average outcome of the two syndromes has maintained mortality of ARF falsely constant over the last decades. In fact, the outcome of the less severe forms has improved while the outcome of most severe forms has remained stable or worse. The problem is then to analyze the different case mix, the differences in etiology of ARF and the different complexity of the syndrome when sepsis is involved. Thus, patients who would previously have died before they could develop ARF now survive long enough to develop ARF and complex derangements leading to multiple organ failure and high mortality.

In this setting, careful attention is paid to the type of renal replacement modality utilized for the treatment of ARF. Continuous (CRRT) and intermittent (IHD) renal replacement modalities have often been compared without reaching consensus on which form is optimal for the patients.

However, when comparing CRRT with IHD therapy, it is wise to remember that the very reason for the development and introduction of CRRT into clinical practice in the late 1970s and early 1980s was to compensate for the clear inadequacies of *conventional* IHD in the treatment of critically ill patients with multiorgan failure [5]. In the last 20 years, substantial clinical experience accumulated in the use of CRRT and its technology evolved a great deal from adaptive arteriovenous techniques to dedicated CRRT machines for venovenous therapy. During this period, multiple studies confirmed the many physiological advantages of CRRT over conventional (3–4 h/day every other day as typically delivered to end-stage renal failure patients) IHD [6–11]. This was especially true in centers which had developed appropriate and necessary physician and nursing expertise in its application. Importantly and predictably, not a single comparative study ever showed IHD to be physiologically better than CRRT.

In these years physicians have increasingly shifted from conventional IHD to CRRT especially for critically ill patients [12]. In fact, conventional IHD has essentially disappeared from ICUs in some countries [13]. However, the controversy as to which therapy is 'best' from the point of view of *mortality* or other *major clinical outcomes* (duration of ICU stay, hospital stay and rate and time to renal recovery) remains. This is because no sufficiently powered multicenter randomized controlled trials have yet been conducted to assess these outcomes and because there have been major and continuing changes in what experts consider optimal continuous therapy or optimal intermittent therapy.

Furthermore, all studies minimized any chance of detecting a differential effect on outcome by allowing crossover from continuous to intermittent therapy in ICU. On the other hand, looking at recent meta-analyses [14], one can clearly see that several of these studies were performed before the year 2000 and in most of them, patients with higher severity scores were allocated to the

CRRT arm. Data published in the year 2000 suggested a correlation between dose of therapy and outcome [15]. One could then speculate that populations treated with CRRT before the year 2000 were highly underdialyzed and thus comparisons with intermittent modalities in those times were flawed.

The concept and the objectives of RRT have evolved in these years together with the evolution of the ARF syndrome. As intensive care patients have become more complex, it has become clear that critically ill patients may require specific forms of RRT. Twenty-five years ago, several of these patients could not be dialyzed due to their hemodynamic instability or other technical and clinical problems. In order to facilitate renal support in this setting, CRRT techniques have been developed. Continuous arteriovenous hemofiltration was a safe and well-tolerated treatment but efficiency was not sufficient. New machines have subsequently made it possible to perform continuous venovenous hemofiltration (CVVH), hemodialysis (CVVHD) and hemodiafiltration (CVVHDF) routinely maintaining the good clinical tolerance but allowing adequate levels of blood purification [16–18]. Thus, when comparing CRRT and IHD, one should first analyze the two therapies on the ground of physiological endpoints.

The complexity of the patient with ARF associated with multiple organ failure suggests that continuous therapies should be utilized as a first choice treatment in intensive care settings. Furthermore, clinical conditions other than ARF, such as congestive heart failure, respiratory distress syndrome, cerebral edema and so on, may benefit from these forms of treatments when oliguria is present or there are early associated signs of renal insufficiency.

The patient with severe hemodynamic instability often cannot tolerate intermittent treatments such as hemodialysis or hemodiafiltration carried out for 3–4 h/day. The slow continuous fluid removal achieved with CRRT is generally well tolerated and an optimal hydration status can be reached within a relatively short period of time with adequate constancy of measured hemodynamic parameters. Direct measurements of blood volume during treatment have made it possible to demonstrate that even in the presence of small volumes of ultrafiltration, a significant drop in circulating blood volume can be observed in intermittent treatments. This phenomenon is not observed in continuous treatments. This aspect may be of tremendous importance in the phase of recovery from ARF. The recovering kidney is extremely sensitive to variations in perfusion pressures and blood flows. Accordingly, IHD may turn out to be unphysiological and may contribute to possible further damage to the renal parenchyma. In contrast, CRRT may be well tolerated and may contribute to a constant and progressive recovery of the kidney without major hemodynamic alterations. These considerations are supported by recent data from Uchino et al. [12], showing a higher chance of renal recovery in ARF patients treated with CRRT.

Continuous therapies can also effectively correct various forms of acidosis. In fact, while IHD produces a dramatic alkalinization during treatment but where a subsequent rebound of acidosis can be frequently observed, CRRT acts slowly but continuously and reaches a steady-state concentration both for uremic solutes and organic acids in blood.

In patients with cerebral edema, intermittent treatments may worsen the clinical condition because of a postdialytic influx of fluid into the brain tissue. After IHD brain density decreases due to osmotic fluid shifts. These alterations induced by intermittent treatments are not observed with continuous therapies that can therefore be utilized with maximal advantage in patients with or at risk of brain edema.

Several mechanisms have been proposed to explain the improvement of adult respiratory distress syndrome (ARDS) patients treated with CRRT. The continuous fluid withdrawal from the interstitium due to a progressive vascular refilling represents a major advantage. However, the modulation of the vascular inflammation thanks to the clearance or adsorption of specific proinflammatory substance onto the membrane has been recently hypothesized. This mechanism has also be invoked as an interesting possibility for patients with systemic inflammatory response syndrome or septic shock. Numerous ex vivo as well as animal and human studies have shown that synthetic filters can extract nearly every substance involved in sepsis to a certain degree [19]. Prominent examples are complement factors, TNF, IL-1, IL-6, IL-8 and PAF [17]. The progressive, continuous unselective removal of humoral mediators has been the theoretical base to generate the 'peak concentration hypothesis' of sepsis and to explain the beneficial immunomodulatory effects induced by CRRT.

Regarding plasma cytokine levels, their decrease appears minor in host defense to infection while high levels need to be modulated by anti-inflammatory feedback. As sepsis does not fit a one-hit model but shows the complex behavior of mediator levels that change over time, neither single-mediator-directed nor one-time interventions seem appropriate. Some studies showed no influence on cytokine plasma levels by CRRT. On the other hand, significant clinical benefits in terms of hemodynamic improvement have been achieved even without measurable decreases in cytokine plasma levels (the peak concentration hypothesis). The removal of substances other than the measured cytokines might have been responsible for the achieved effect. However, several mediators may act together to alter the functional responses of the circulating leukocytes. When the response to sepsis is viewed in a network perspective, absolute values seem to be less relevant than relative ones. Within an array of interdependent mediators, even small decreases could induce major balance changes. In spite of some encouraging results as already mentioned, the extent of achievable clinical benefit with conventional CRRT (using conventional filters and flow rates) in sepsis has generally been disappointing. Consequently, attempts have been made to improve the efficiency of soluble mediator removal in sepsis by increasing the amount of plasma water exchange, i.e. increasing ultrafiltration rates. Animal studies provide great support to this concept. These studies established that a convection-based treatment can remove substances which can induce hemodynamic effects resembling septic shock, when sufficiently high ultrafiltration rates are applied. More relevant to human sepsis was the finding that ultrafiltration dosage is correlated to outcome in critically ill patients with ARF. In a large randomized, controlled study including 425 patients, an ultrafiltration dosage of 35 ml/kg/h increased survival rate from 41 to 57% compared to a dosage of 20 ml/kg/h [15]. Eleven to 14% (per randomization group) of the patients had sepsis. In these subgroups there was a trend of a direct correlation between treatment dosage and survival even above 35 ml/kg/h in contrast to the whole group where a survival plateau was reached. Of note, there was no increase in adverse effects even with the highest ultrafiltration dosage. Impressive clinical results were obtained in an evaluation of short-term high-volume hemofiltration (HVHF) in patients with catecholamine-refractory septic shock [18] comprising a patient cohort with very poor expected survival. A control group was not defined. Only one 4-hour session of HVHF removing 35 l of ultrafiltrate replaced by bicarbonate-containing fluid was applied as soon as mean blood pressure could not be stabilized above 70 mm Hg with dopamine, norepinephrine and epinephrine after appropriate volume resuscitation. HVHF was followed by conventional CVVH.

With the extracorporeal fluxes used for CRRT, negative thermal balance of up to -100 kJ/h can be obtained depending on the length of the blood lines, the room temperature and the dialysate/replacement fluid temperature. This might contribute to modulating the inflammatory response as well as oxygen demand in several organs with the possibility of using such a mechanism for specific clinical targets.

When electrolyte disorders are life threatening or refractory to medical corrections, extracorporeal therapy is the treatment of choice. Because of the slow and gentle rate of fluid exchange, the treated blood operates in continuous equilibrium with peripheral tissues and organs, and the entire organism may benefit from a safe and effective restoration of water, sodium and electrolyte homeostasis. This restoration of homeostasis is particularly true for acid-base control (as administration of bicarbonate can be easily titrated to the necessary acid-base goals), intra-/extracellular potassium and phosphate equilibrium and water fluxes between the interstitium and the intracellular space.

After evaluating the physiological effects of continuous and intermittent treatment modalities, and considering the potential advantages of continuous therapies, one should still face the lack of evidence for improved survival and major clinical outcomes of one technique over another. In this setting the lack of evidence may induce physicians to draw false conclusions from a misleading interpretation of the available literature. In order to clarify the real content of the available trials, it is still useful to review some of their aspects even in the presence of the above-mentioned shortcomings. Mehta et al. [20] randomized 166 critically ill patients with severe acute kidney injury to either CRRT or IHD therapy. There was a significantly higher ICU mortality rate in subjects randomized to CRRT (60 vs. 42%, $p = 0.02$). After post hoc adjustment for severity of illness, the increased risk attributed to CRRT was no longer statistically significant (odds ratio 1.6). There was clear baseline imbalance with patients randomized to CRRT having greater illness severity (higher APACHE III and grater incidence of liver failure). The reasons for such imbalances remain unclear. However, several other aspects of this study are still of interest. First, patients were allowed to cross over making a true comparison impossible. Second, patients with hemodynamic instability (MAP < 70 mm Hg) were excluded. These are the very patients where the advantages of CRRT are most evident. This selection bias was the expression of an acknowledged lack of equipoise: in such patients intermittent therapy is physiologically inferior. Third, despite these limitations, one very relevant observation emerged: if patients received a sufficient trial of CRRT and survived, renal recovery was dramatically increased (92.3 vs. 59.4%, $p < 0.01$). In other words, intermittent therapy delayed or impeded renal recovery. Fourth, CRRT delivered superior control of uremia.

In another single-center randomized trial, 125 patients were randomized to CRRT or IHD. Inhospital mortality rates did not differ by treatment assignment (CRRT 47 vs. 51% for IHD, $p = 0.72$) [21]. However, this trial suffered from extraordinary logistic constraints in that patients could not be randomized on a 1:1 basis because of untrained staff in CRRT or a lack of CRRT machines. This is hardly the environment that would create a scientific 'level playing field' for the two therapies. More importantly, if the near 4% absolute decrease in mortality still seen with CRRT were true, it would have taken 5,000 patients to detect it at a beta of 0.2 and an alpha of 0.05. The investigators randomized only 125 patients! Such a 4% decrease might seem small but would be clinically relevant because it would imply that the number needed to treat to save one life is only 25. The number needed to treat to save one life with percutaneous coronary intervention in patients with myocardial infarction with ST segment elevation is 50 [22].

In a third single-center prospective randomized trial, 80 patients were randomized to treatment with CRRT or IHD [23]. Survival was 67.5% for CRRT and 70% for dialysis. Again the study was dramatically underpowered to detect differences in survival between the two therapies and crossover from CRRT to

IHD occurred in 22.5% of CRRT patients. Of interest, this trial confirmed the hemodynamic problems associated with IHD with 40% of patients requiring an increase in vasopressor therapy during initial treatment with IHD compared to only 12.5% for CRRT (p = 0.005). In this study, CRRT was associated with a 6-day (close to 15%) reduction in hospital stay. Nine patients were converted from CRRT to IHD because of frequent filter clotting, a concept that seems strange to practitioners who have only used CRRT in the well-established era of citrate and regional heparin/protamine anticoagulation.

Most recently, the HEMODIAF Study Group reported the results of a randomized multicenter study comparing the results of CRRT to IHD in 360 critically ill patients with acute kidney injury [24]. Overall, there was no difference in the primary endpoint of 60-day survival (33% with CRRT vs. 32% with IHD). However, the authors noted an unexpected and significant increase in survival rates within the IHD group over time (relative risk 0.67/year), an effect not seen in the CRRT group. This creates a major bias. How was IHD changed? Who changed it? Was advice offered to trial sites? This is of major concern because if it had not happened and the initial trend in outcome seen with IHD had continued, the conclusions would have been diametrically opposite. In this regard, it is of interest to see how the average duration of dialysis became 5.2 h (significantly longer than conventional IHD), while CRRT continued to deliver a calculated creatinine clearance of 25 ml/kg/h (not 29 ml/kg/h as reported by the authors who did not correct for the effects of predilution). Such clearance is much lower than what is considered an 'optimal' CRRT dose [15]. Finally, again, this trial reported a 10% crossover from CRRT to IHD.

Recently, the PICARD group compared the outcomes of different renal replacement therapy modalities [25]. This analysis incorporated five sites in the USA and used multivariable regression analysis, and a propensity score approach to address the effect of confounding variables. Within the PICARD cohort, using this methodology, the provision of CRRT in comparison to IHD was associated with a significantly higher relative risk of mortality. However, patients with CRRT were obviously sicker: they had more organ failure (CNS, liver, hematologic and respiratory), higher mechanical ventilation rate, more sepsis, lower blood pressure, higher total bilirubin and lower platelet count. Despite this, none of the respiratory and circulatory variables, like mechanical ventilation requirement, ARDS, heart rate or blood pressure, were included in the multiple regression analysis for mortality, although these variables had univariate p values of <0.0001. It is quite hard to understand why these variables were not selected as independent variables. These observations cast serious doubt on the accuracy of this analysis.

Thus, there is clear lack of suitably designed multicenter randomized controlled trials where all ICU patients with ARF are randomized to either CRRT or IHD (doses to be defined) from start to finish. Until such a trial is done, the question of *clinical* superiority cannot be answered.

Whatever future trials might wish or be able to compare, there is clear evidence that RRT is not like a 'tablet'. Its effects are modified by the expertise of those who prescribe it and guide it (physicians) and those who execute it (nurses). For example, it is clear from the HEMODIAF study that the quality of IHD can be improved and better outcomes follow. Another study has recently confirmed that priming the dialysis circuit with isotonic saline, setting dialysate sodium concentration above 145 mmol/l, discontinuing vasodilator therapy, and setting dialysate temperature below 37°C [26] improves IHD-related outcomes. Finally, the introduction of slow extended daily dialysis (SLEDD) introduces the final step in the rehabilitation of dialysis in ICU [27]. Indeed as IHD becomes more and more like CRRT through SLEDD, the protagonists of CRRT will be filled with delight: their battle was not with dialysis per se, but rather with the mindless application of it in conventional mode to critically ill patients. Once IHD in ICU is reformed, adjusted to take into account the needs of the critically ill, and extended to 6 h (or 8 or maybe even 12 h) so that fluid removal is performed safely and uremic control optimized, little of the controversy will remain.

As a continuous extracorporeal therapy, CRRT frequently requires continuous anticoagulation, which may increase the bleeding risk. Thus, it also needs 'reform'. Citrate and regional heparin/protamine anticoagulation are safe and effective, but underused. This might represent a chance of improvement for the future. Machine operation and safety should be given more attention especially providing more education and personnel training [28]. Finally and more importantly, the dose of CRRT might well require readjustment with two recent randomized controlled trials showing that an increasing dose of CRRT improves survival [29]. All studies comparing IHD to CRRT have used much lower CRRT doses than those shown to improve survival. The optimal weekly Kt/V for CRRT in an 80-kg man would appear to be close to 10. If multicenter trials currently under way confirm this observation, CRRT will also have to be reformed. More importantly, by implication, its twin (IHD) will have to be reformed further as well to be able to deliver such an optimal dose intermittently. Both techniques also have to look further than the dose. The issue of timing of intervention is likely very important and yet not well studied so far [30]. Timing of cessation may be equally important but has not yet been studied. Much work needs to be done in the field of acute RRT.

In conclusion, there is a reason for the human kidney working 24 h/day. The reason is to maintain homeostasis and to keep biological parameters within a tight steady-state control. Intermittent forms of renal replacement therapies are aggressive and unphysiological since they must accomplish their task in a short fraction of the day trying to correct in a few minutes derangements that have developed over hours or days. The result is a severe hemodynamic intolerance, a limited efficiency for waste product elimination, a high risk of brain edema, worsening of inflammatory conditions and a complete failure to keep acid-base and electrolytes in steady condition.

In contrast, continuous therapies perform a gentle and slow correction of derangements leading to a steady-state condition close to that provided by native kidneys. Most of the claimed drawbacks or complications can be prevented if adequate prescription and monitoring are carried out and the right treatment dose is delivered.

All these considerations apply to critically ill ICU patients with ARF. In these patients, intermittent dialysis can be problematic or even impossible to perform. The opposite may be true for patients in renal wards with uncomplicated ARF where daily (possibly long) hemodialysis sessions are perfectly capable to accomplish the required tasks. Even in these cases however, an intermediate form between fully intermittent and fully continuous dialysis regimes is strongly advisable since ARF patients are still very different from typical chronic hemodialysis patients.

The evidence that CRRT is physiologically superior to conventional IHD is clear. The evidence that this physiological superiority can be translated into clinically important gains is not. Appropriately powered, designed, conducted and analyzed studies have simply not yet been done. In addition the evolution of both therapies presents a fluid environment where the terms of comparison constantly change. The correct focus for clinicians might actually be not so much to compare the two, but to make sure that whatever therapy is applied, it is 'done right'. They need to ensure that patients receive the best therapy for their condition at a given time in the course of their illness in ICU, receive it safely, in a timely fashion, with the correct dose and for long enough.

References

1 DuBose TD Jr, Warnock DG, Mehta RL, Bonventre JV, Hammerman MR, Molitoris BA, Paller MS, Siegel NJ, Scherbenske J, Striker GE: Acute renal failure in the 21st century: recommendations for management and outcomes assessment. Am J Kidney Dis 1997;29:793–799.
2 Clermont G, Acker CG, Angus DC, Sirio CA, Pinsky MR, Johnson JP: Renal failure in the ICU: comparison of the impact of acute renal failure and end-stage renal disease on ICU outcomes. Kidney Int 2002;62:986–996.
3 Liano F, Junco E, Pascual J, Madero R, Verde E: The spectrum of acute renal failure in the intensive care unit compared with that seen in other settings. The Madrid Acute Renal Failure Study Group. Kidney Int 1998;66:S16–S24.
4 De Mendonca A, Vincent J-L, Suter PM, et al: Acute renal failure in the ICU: risk factors and outcome evaluated by the SOFA score. Intensive Care Med 2000;26:915–921.
5 Kramer P, Wigger W, Rieger J, Matthaei D, Scheler F: Arteriovenous haemofiltration: a new and simple method for treatment of over-hydrated patients resistant to diuretics. Klin Wochenschr 1977;55:1121–1122.

6 Bellomo R, Boyce N: Continuous veno-venous hemodiafiltration compared with conventional dialysis in critically ill patients with acute renal failure. ASAIO J 1993;39:M794–M797.
7 Bellomo R, Martin H, Parkin G, Love J, Boyce N: Continuous arteriovenous haemodiafiltration in the critically ill. Influence on major nutrient balances. Intensive Care Med 1991;17:399–402.
8 Bellomo R, Farmer M, Bhonagiri S, Porceddu S, Ariens M, M'pisi D, Ronco C: Changing acute renal failure treatment from intermittent hemodialysis to continuous hemofiltration: impact on azotemic control. Int J Artif Organs 1999;22:145–150.
9 Ronco C, Bellomo R, Brendolan A, Pinna V, La Greca G: Brain density changes during renal replacement therapy in critically ill patients with acute renal failure. Continuous hemofiltration versus intermittent hemodialysis. J Nephrol 1999;12:173–178.
10 Tan HK, Bellomo R, M'pisi DA, Ronco C: Phosphatemic control during acute renal failure: intermittent hemodialysis vs. continuous hemodiafiltration. Int J Artif Organs 2001;24:186–191.
11 Manns M, Sigler MH, Teehan BP: Intradialytic renal haemodynamics – potential consequences for the management of the patient with acute renal failure. Nephrol Dial Transplant 1997;12: 870–872.
12 Uchino S, Kellum JA, Bellomo R, Doig GS, Morimatsu H, Morgera S, Schetz M, Tan I, Bouman C, Macedo E, Gibney N, Tolwani A, Ronco C: Acute renal failure in critically ill patients: a multinational, multicenter study. JAMA 2005;294:813–818.
13 Silvester W, Bellomo R, Cole L: The epidemiology, management and outcome of severe acute renal failure of critical illness in Australia. Crit Care Med 2001;29:1910–1915.
14 Kellum JA, Angus DC, Johnson JP, Leblanc M, Griffin M, Ramakrishnan N, Linde-Zwirble WT: Continuous versus intermittent renal replacement therapy: a meta-analysis. Intensive Care Med 2002;28:29–37.
15 Ronco C, Bellomo R, Homel P, et al: Effects of different doses in continuous veno-venous haemofiltration on outcomes of acute renal failure: a prospective randomised trial. Lancet 2000;356: 26–30.
16 Ronco C, Bellomo R: Continuous versus intermittent renal replacement therapy in the treatment of acute renal failure. Nephrol Dial Transplant 1998;13/6:79–85.
17 Ronco C, Tetta C, Lupi A, et al: Removal of platelet-activating actor in experimental continuous arteriovenous hemofiltration. Crit Care Med 1995;23:99–107.
18 Honore PM, Jamez J, Wauthier M, et al: Prospective evaluation of short-term, high-volume isovolemic hemofiltration on the hemodynamic course and outcome in patients with intractable circulatory failure resulting from septic shock. Crit Care Med 2000;28:3581–3587.
19 Bellomo R, Tipping P, Boyce N: Continuous veno-venous hemofiltration with dialysis removes cytokines from the circulation of septic patients. Crit Care Med 1993;21:522–526.
20 Mehta RL, McDonald B, Gabbai FB, et al: A randomized clinical trial of continuous versus intermittent dialysis for acute renal failure. Kidney Int 2001;60:1154–1163.
21 Uehlinger DE, Jakob SM, Ferrair P, et al: Comparison of continuous and intermittent renal replacement therapy for acute renal failure. Nephrol Dial Transplant 2005;20:1630–1637.
22 Keeley EC, Hillis LD: Primary PCI for myocardial infarction with ST-segment elevation. N Engl J Med 2007;356:47–54.
23 Augustine JJ, Sandy D, Seifert TH, Paganini EP: A randomized controlled trial comparing intermittent with continuous dialysis in patients with ARF. Am J Kidney Dis 2004;44:1000–1007.
24 Vinsonneau C, Camus C, Combes A, et al: Continuous venovenous haemodiafiltration versus intermittent haemodialysis for acute renal failure in patients with multiple-organ dysfunction syndrome: a multicentre randomised trial. Lancet 2006;368:379–385.
25 Cho KC, Himmelfarb J, Paganini E, et al: Survival by dialysis modality in critically ill patients with acute kidney injury. J Am Soc Nephrol 2006;17:3132–3138.
26 Schortgen F, Soubrier N, Delclaux C, et al: Hemodynamic tolerance of intermittent hemodialysis in critically ill patients: usefulness of practice guidelines. Am J Respir Crit Care Med 2000;162: 197–202.
27 Marshall MR, Ma T, Galler D, et al: Sustained low-efficiency daily diafiltration (SLEDD-f) for critically ill patients requiring renal replacement therapy towards an adequate therapy. Nephrol Dial Transplant 2004;19:877–884.

28 Schultz DG: FDA Updated Public Health Notification: Gambro Prisma® Continuous Renal Replacement System. September 22, 2006. www.fda.gov/cdrh/safety/022706-gambro.html
29 Bellomo R: Do we know the optimal dose for renal replacement therapy in the intensive care unit? Kindey Int 2006;70:1202–1204.
30 Bent P, Tan HK, Bellomo R, Buckmaster J, Doolan L, Hart G, Silvester W, Gutteridge G, et al: Early and intensive continuous veno-venous hemofiltration for severe acute renal failure after cardiac surgery. Ann Thorac Surg 2001;71:832–837.

Claudio Ronco
Department of Nephrology, San Bortolo Hospital
Viale Rodolfi 37
IT–36100 Vicenza (Italy)
Tel. +39 0444 753650, Fax +39 0444 753973, E-Mail cronco@goldnet.it

Sustained Low-Efficiency Dialysis

Ashita J. Tolwani, Thomas S. Wheeler, Keith M. Wille

University of Alabama at Birmingham, Birmingham, Ala., USA

Abstract

Sustained low-efficiency dialysis (SLED) is an increasingly popular form of renal replacement therapy for patients with renal failure in the intensive care unit. Advantages of SLED are efficient clearance of small solutes, good hemodynamic tolerability, flexible treatment schedules, and reduced costs. Studies comparing outcomes of SLED with those of other dialysis modalities are being performed.

Copyright © 2007 S. Karger AG, Basel

Hybrid hemodialysis, also known as sustained low-efficiency dialysis (SLED), extended daily dialysis (EDD), or slow continuous dialysis, has emerged as an alternative to intermittent hemodialysis (IHD) and continuous renal replacement therapy (CRRT) in the treatment of renal failure in the intensive care unit (ICU) setting. Hybrid hemodialysis is utilized in an increasing number of hospitals in the US, Europe, South America, Asia, New Zealand and Australia [1]. Pesacreta et al. [2] interviewed 131 physicians in 27 medical centers and found SLED to be the primary therapy for ICU renal replacement therapy in 7% of patients compared to 56% for IHD and 35% for CRRT.

Hybrid hemodialysis, which was first described in 1988, utilizes an IHD machine to perform dialysis for extended periods of time using low blood flow and dialysate rates of 100–300 ml/min. In contrast to CRRT, hybrid modalities are not continuous and usually run for 6–12 h/day. Hybrid hemodialysis employs characteristics of both CRRT and IHD, hence, the use of the term 'hybrid'. The rationale for its use is its ability to provide improved hemodynamic stability as in CRRT and high solute clearances and flexible scheduling as in IHD without the need for expensive CRRT machines, costly customized solutions, and trained staff.

Technical Issues

Machinery

In the US, the Fresenius 2008H/K® model hemodialysis machines are capable of delivering the low dialysate and blood flow rates required of hybrid modalities [3, 4]. Outside of the US, the Fresenius 4008S ArRT Plus®, Fresenius Genius®, and Gambro 200S Ultra® are used to deliver hybrid therapy [3, 4]. The machinery can be categorized as single pass machines or batch machines. Single pass machines use dialysate generated on-line from reverse osmosis purified water and a bicarbonate proportioning system. The dialysate in batch machines is generated from prepackaged salts and sterile water and is stored within the machine.

Prescription

Duration and Timing of Treatment

The duration of SLED therapy can be individualized according to the needs of patients. Treatment durations may range from 6 to 18 h/day. In a study of 20 patients on SLED vs. 19 patients on CRRT, Kielstein et al. [5] found that the urea reduction ratio was similar between the two groups [continuous venovenous hemofiltration (CVVH) therapy $53 \pm 2\%$; EDD therapy $52 \pm 3\%$]. Although urea reduction ratio was equivalent, EDD patients were dialyzed for 11.7 ± 0.1 h compared with 23.3 ± 0.2 h for CRRT patients. This suggests that the effect of 12 h of SLED is comparable to 23 h of CRRT.

Since SLED is not a continuous modality, the timing of therapy can be tailored to the benefit of the patient. Many centers now perform nocturnal SLED so that patients may be available during the day for various diagnostic and therapeutic procedures, thereby avoiding interruptions of therapy [4].

Dialysate and Ultrafiltration Rates

Published studies use dialysate flow rates ranging from 100–300 ml/min, depending on the dialysis machine specifications, treatment duration, and tolerance of ultrafiltration. Higher dialysate flows of 300 ml/min are generally used with treatments of less than 8 h, while lower dialysate flow rates are used with longer treatments. The ultrafiltration rate is varied according to the patient's clinical need and hemodynamic stability.

Dialysate Composition

The composition of the dialysate varies according to clinical needs. Typical dialysate baths consist of 3.0–4.0 mEq/l potassium, 1.5–2.5 mEq/l calcium, and 24–35 mmol/l bicarbonate. Phosphate removal with hybrid therapies can be extensive, and phosphorus can be repleted intravenously or by adding 45 ml of

Fleets Phosphasoda to 9.5 l of bicarbonate concentrate, giving a final concentration of 0.81 mmol/l [6].

A potential concern with on-line dialysate generation as used in the single pass machine is the possibility of backfiltration of endotoxin from the dialysate compartment into the patient. However, insufficient data exist as to the recommended purity of the dialysate.

Anticoagulation

The reported incidence of SLED circuit clotting without anticoagulation is 26–46% using single pass machines and much less using batch systems [7–9]. If anticoagulation is required, unfractionated heparin is most commonly given as a 1,000- to 2,000-unit bolus followed by a maintenance dose of 500–1,000 units/h with a goal APTT of 1.5 times baseline. With heparin, the reported incidence of clotting is 17–26% [10].

Several regional citrate anticoagulation protocols have been published. These utilize a zero calcium dialysate with intravenous calcium replacement or a low calcium dialysate without intravenous calcium replacement [11–13]. Case reports using prostacyclin and argatroban for anticoagulation have also been published [10, 14].

Clinical Outcomes

Solute Control

SLED, compared to conventional IHD, offers greater small solute clearance (Kt/V 1.3–1.5), less small solute disequilibrium (single pool urea kinetic modeling), and, with high-flux dialyzers, greater large solute clearance [5]. Kinetic models indicate that both SLED and CRRT provide effective control of azotemia in hypercatabolic acute kidney injury (AKI) patients [15]. SLED, however, is less effective than CRRT for larger solute control [5]. Sustained low efficiency diafiltration (SLEDD-f), a variation of SLED, combines diffusive and convective solute transport to improve clearance of larger molecular weight solutes [9].

Of note, while albumin removal is minimal with SLED, amino acid losses are significant, and protein supplementation of 0.2 g/kg/day should be administered on treatment days.

Hemodynamic Tolerance

Reports of hybrid therapy have consistently shown that ultrafiltration is generally well tolerated. Only a minority (0–7%) of patients in all published reports had to discontinue hybrid treatment because of refractory hypotension. Kumar et al. [6] prospectively compared SLED (n = 25) with CVVH (n = 17) and found no difference in mean arterial pressure or net daily ultrafiltration rates but significant differences in treatment duration (7.5 h with SLED vs. 19.5 h with CVVH) and anticoagulation requirement (median heparin dose 4,000 U/day with SLED vs. 21,000 U/day with CVVH). Similarly, a prospective controlled trial, in which 39 ICU patients were randomized to either hybrid therapy or CRRT, found no significant difference in hemodynamic parameters, inotrope dose, or outcome [5].

Mortality

Marshall et al. [8] performed SLED using a Fresenius 2008H® IHD machine at a dialysate flow rate of 100 ml/min. Sufficient solute removal and ultrafiltration were achieved in most patients, and hospital mortality (62%) did not exceed the predicted mortality rate based on the acute physiology and chronic health evaluation (APACHE II) score.

Currently, the VA/NIH Acute Renal Failure Trial Network (ATN) Study, a prospective, multicenter study, is recruiting ICU patients with AKI and randomizing them to high- vs. low-dose dialysis [16]. Since patients may undergo intermittent hemodialysis, CRRT, or hybrid therapy, this study may allow for outcome comparisons using the different modalities of therapy. The Stuivenberg Hospital Acute Renal Failure (SHARF) study, which is ongoing, is a prospective, randomized multicenter clinical trial comparing hybrid therapy with CRRT for patients with AKI. No significant outcome differences were observed following an interim analysis of 996 patients, but continued patient recruitment is underway [17].

Cost

Hybrid therapies are less expensive than CRRT. This is due in part due to avoidance of expenses associated with CRRT machinery and preparation of specialized fluids. Studies have found that the daily cost of SLED may be up to 8 times less expensive than CRRT [18–20].

Conclusion

Hybrid hemodialysis has proven to be a viable option in the treatment of renal failure since it combines the relatively low cost and complexity of IHD with the advantages of gradual fluid and solute removal of CRRT. In addition, its intermittent nature allows for scheduling of other diagnostic and therapeutic procedures between treatments.

References

1   Fliser D, Kielstein JT: Technology insight: treatment of renal failure in the intensive care unit with extended dialysis. Nat Clin Pract Nephrol 2006;2:32–39.

2 Pesacreta M, Overberger P, Palevsky PM, the VA/NIH Acute Renal Failure Trial Network: Management of renal replacement therapy in acute renal failure: a survey of practitioner prescribing practices. J Am Soc Nephrol 2004;15:350A.
3 Fliser D, Kielstein JT: A single-pass batch dialysis system: an ideal dialysis method for the patient in intensive care with acute renal failure. Curr Opin Crit Care 2004;10:483–488.
4 Lonnemann G, Floege J, Kliem V, Buckhurst R, Koch K: Extended daily veno-venous high-flux haemodialysis in patients with acute renal failure and multiple organ dysfunction syndrome using a single path batch dialysis system. Nephrol Dial Transplant 2000;15:1189–1193.
5 Kielstein J, Kretschmer U, Ernst T, Hafer C, Bahr M, Haller H, Fliser D: Efficacy and cardiovascular tolerability of extended dialysis in critically ill patients: a randomized controlled study. Am J Kidney Dis 2004;43:342–349.
6 Kumar VA, Yeun JY, Depner TA, Don BR: Extended daily dialysis vs. continuous hemodialysis for ICU patients with acute renal failure: a two-year single center report. Int J Artif Organs 2004;27: 371–379.
7 Kumar V, Craig M, Depner T, Yeun J: Extended daily dialysis: a new approach to renal replacement for acute renal failure in the intensive care unit. Am J Kidney Dis 2000;36:294–300.
8 Marshall M, Golper T, Shaver M, Alam M, Chatoth D: Sustained low-efficiency dialysis for critically ill patients requiring renal replacement therapy. Kidney Int 2001;60:777–785.
9 Marshall M, Tianmin M, Galler D, Rankin A, Williams A: Sustained low-efficiency daily diafiltration (SLEDD-f) for critically ill patients requiring renal replacement therapy: towards an adequate therapy. Nephrol Dial Transplant 2004;19:877–884.
10 Marshall M, Golper T: Sustained low efficiency or extended daily dialysis. Up-to-date 2006, online version 13.3. http://www.uptodate.com
11 Finkel K, Foringer J: Safety of regional citrate anticoagulation for continuous sustained low efficiency dialysis (C-SLED) in critically ill patients. Ren Fail 2005;27:541–545.
12 Marshall MR, Ma TM, Eggleton K, Ferencz A: Regional citrate anticoagulation during simulated treatments of sustained low efficiency diafiltration. Nephrology 2003;8:302–310.
13 Morgera S, Scholle C, Melzer C, Slowinski T, Liefeld L, Baumann G, Peters H, Neumayer H: A simple, safe, and effective citrate anticoagulation protocol for the Genius dialysis system in acute renal failure. Nephron Clin Pract 2004;98:35–40.
14 Fiaccadori E, Maggiore U, Parenti E, Giacosa R, Picetti E, Rotelli C, Tagliavini D, Cabassi A: Sustained low-efficiency dialysis (SLED) with prostacyclin in critically ill patients with acute renal failure. Nephrol Dial Transplant 2006;22:529–537.
15 Liao Z, Zhang W, Hardy PA, et al: Kinetic comparison of different acute dialysis therapies. Artif Organs 2003;27:802.
16 Palevsky PM, O'Connor T, Zhang JH, Star RA, Smith MW: Design of the VA/NIH Acute Renal Failure Trial Network (ATN) Study: intensive versus conventional renal support in acute renal failure. Clin Trials 2005;2:423.
17 Malbrain M, Elseviers M, Van der Niepen P, et al: Interim results of the SHARF4 Study: outcome of acute renal failure with different modalities (abstract). Crit Care 2004;8(suppl 1):153.
18 Alam M, Marshall M, Shaver M, Chatoth D: Cost comparison between sustained low efficiency hemodialysis (SLED) and continuous venovenous hemofiltration (CVVH) for ICU patients with ARF (abstract). Am J Kidney Dis 2000;35:A9.
19 Ma T, Walker R, Eggleton K, Marshall M: Cost comparison between sustained low efficiency dialysis/diafiltration (SLEDD) and continuous renal replacement therapy for ICU patients with ARF (abstract). Nephrology 2002;7:A54.
20 Berbece A, Richardson R: Sustained low-efficiency dialysis in the ICU: cost, anticoagulation, and solute removal. Kidney Int 2006;70:963–968.

Ashita J. Tolwani, MD
ZRB 604
1530 3rd Ave. S.
Birmingham, AL 35294-0007 (USA)
Tel. +1 205 975 2021, Fax +1 205 996 2156, E-Mail atolwani@uab.edu

The Role of the International Society of Nephrology/Renal Disaster Relief Task Force in the Rescue of Renal Disaster Victims

R. Vanholder[a], W. Van Biesen[a], N. Lameire[a], M.S. Sever[b]

[a]Renal Division, Department of Internal Medicine, University Hospital, Ghent, Belgium; [b]Departments of Nephrology and Internal Medicine, Istanbul School of Medicine, Istanbul, Turkey

Abstract

Disasters are a major cause of distress and material as well as corporal damage. Next to direct trauma, the crush syndrome inducing multiorgan problems as a consequence of muscle compression and the release of muscular contents into the bloodstream is the most important cause of death; this is to a large extent related to the induction of severe acute kidney injury, for which dialysis is a life-saving therapy. The practical means (both hardware and personnel) to do so are, however, often lacking in disaster conditions. The Renal Disaster Relief Task Force (RDRTF) offered support for renal problems in the aftermath of several disasters, e.g. the Marmara earthquake (1999) in Turkey, the Bam earthquake (2003) in Iran, and the Kashmir earthquake (2005) in Pakistan. A preconceived intervention plan is followed with adaptations according to local conditions. Material and personnel are dispatched to the disaster areas. These interventions have been life-saving for a substantial number of victims. The current article describes the structure and approach of the RDRTF.

Copyright © 2007 S. Karger AG, Basel

The Concept of Disaster

Disasters are sudden calamities provoking substantial damage, loss and distress, and most often strike extended areas. At first instance, natural disasters, e.g. earthquakes, come to mind, but they can be man-made as well. If a natural or man-made disaster results in the collapse of buildings or other solid structures, this might result in the entrapment of numerous victims under the rubble, which in turn may be the origin of the crush syndrome (CS; see below). In some disasters, a substantial number of victims suffering from the CS will subsequently develop acute kidney injury (AKI). This sequence of events has been defined as renal disaster [1]. During the past few years, several earthquakes could be classified as renal disasters: the Armenian earthquake in 1988 [2], the Marmara earthquake in Turkey (1999) [3], the Bam earthquake in Iran (2003) [4], and the Kashmir earthquake in Pakistan and surrounding countries (2005) [5].

In this review, we define the CS and its pathophysiology; we then describe the concept of renal disaster and of the Renal Disaster Relief Task Force (RDRTF), and finally the measures to be taken to potentially limit the number of renal victims.

The CS and Rhabdomyolysis

Crush injury develops due to entrapment under the rubble and the ensuing compression of the muscular system. If these events give rise to systemic manifestations, such as the systemic inflammatory reaction syndrome, the adult respiratory distress syndrome, diffuse intravascular coagulation or shock, the term CS is justified. One of the most devastating epiphenomena of the CS is AKI. The latter is to a large extent the consequence of rhabdomyolysis, a term which refers to the disintegration of striated muscle as a consequence of muscle compression.

It might seem as if the CS affects only a minor fraction of the total potential number of disaster victims, e.g. if compared to the overall number of fatalities. However, no support can be given to those who die immediately and no substantial help is needed for those with minor injuries; this leaves the proportionally small group of subjects with CS as the one in whom intervention, especially by dialysis, might increase survival chances. Actually, CS is the second most frequent cause of death after direct trauma, and the only potentially fatal complication for which life-saving therapy is available.

Pathophysiology of AKI in Rhabdomyolysis

One of the central pathophysiological events in rhabdomyolysis is the sequestration of large quantities of body water in the damaged muscles [6], which can easily amount to a total volume of 5 l or more. This process induces intravascular dehydration and is one of the main elements at the origin of AKI.

Together with water, calcium is also attracted into the damaged muscles. This may sometimes give rise to metastatic calcification [7]. The subsequent hypocalcemia is a potential cause of heart failure and/or cardiac arrhythmias which provoke further ischemic kidney damage. Simultaneously, several other compounds with toxic impact are released from the damaged muscles. The most important one here is myoglobin, which induces tubular obstruction, especially together with dehydration. At the same time, hepatic metabolism of myoglobin generates bilirubin which also has a nephrotoxic potential.

Release of potassium out of the damaged muscle cells induces hyperkalemia. Its deleterious effect is potentialized by the hypocalcemia. Other compounds released from the muscles are phosphate aggravating hypocalcemia, acids enhancing hyperkalemia, and nucleic acids, which are metabolized to uric acid, another obstructive nephrotoxin.

The Concept of Renal Disaster

In many of the areas recently struck by major earthquakes, the potential to perform dialysis was present, but the facilities to deal appropriately with a large number of crush victims were restricted for socioeconomic and/or logistic reasons.

The first awareness of a major kidney problem subsequent to an earthquake occurred after the 1988 Armenian earthquake in Spitak. Although a vast quantity of dialysis material and personnel was transported into the damaged area, the initiative to start support was taken without appropriate preparation several days after the disaster, so that adequate help became possible only when most renal victims had either died or recovered.

Together with the development of the concept of 'renal disaster' [1] grew the awareness that if such support were needed in the future, a preconceived plan would be necessary, with volunteer lists and stocks of material.

The Renal Disaster Relief Task Force

The International Society of Nephrology (ISN) installed the RDRTF in 1989, planning to organize rescue structures for three areas (Northern, Central and South America, South-East Asia and Europe). For the time being, the European branch is the most operational one. Lists of volunteers (physicians, nurses, technicians) available to leave at short notice are registered in the headquarters of the RDRTF. Materials, such as drugs or dialysis devices, are available in the pharmacy of the hospital where the actions are coordinated (University Hospital Ghent) or in the warehouses of Médecins Sans Frontières (MSF – Doctors without Borders), the nongovernmental relief organization under which the RDRTF operates. Further needs are supplemented by contacts with pharmaceutical and dialysis companies.

Table 1. Main interventions of the ISN/RDRTF, European branch

- Iran, March 1997: material support
- Moldova, March 1999: material support
- Macedonia, May 1999: evacuation of chronic patients
- Macedonia/Kosova, July 1999: material support
- Turkey, August 1999: major intervention
- Kosova, February 2000: educational support
- India, January 2001: scouting
- Turkey, May 2003: material support
- Algeria, May 2003: scouting
- Iran, December 2004: major intervention
- Luisiana, August 2005: advisory role
- Pakistan, October 2005: major intervention
- Poland, January 2006: advisory role
- Indonesia, May 2006: scouting
- Lebanon, July 2006: material support

A summary of the main activities of the RDRTF over the last few years is given in table 1.

Organizational Aspects

Severity Assessment
One of the first tasks in renal rescue is to make a severity assessment so as to know how many renal victims, especially those potentially in need of dialysis, should be anticipated. The number of AKI patients in need of dialysis and their proportion to the number of fatalities might be largely different from disaster to disaster (table 2), and depends on the intensity of the earthquake, its location, the time of occurrence, the quality of the buildings, the body constitution of the potential victims, and the quality of rescue activities.

The general perception is that disasters in densely populated areas with good rescue facilities, medium quality buildings (not good enough to avoid collapse but solid enough to cause a lot of crushing after collapse), efficient primary care immediately after extrication, sufficient transport possibilities and adequate hospital infrastructure in safe areas at a reasonable distance of the affected zone result in a high prevalence of CS cases with kidney damage.

Fluid Administration
One of the mainstays of therapy and prevention of dialysis is a timely and appropriate administration of fluids. In the Bingöl earthquake in Turkey 2003,

Table 2. Ratio dialyzed/deaths (×1,000)

Location	Country	Year	Ratio
Spitak	Armenia	1988	9.0–15.4
Northern Iran	Iran	1990	3.9
Kobe	Japan	1995	24.6
Marmara	Turkey	1999	28.1
Chi-Chi	Taiwan	1999	13.3
Gujarat	India	2001	1.7
Boumerdès	Algeria	2003	6.6
Bam	Iran	2003	3.7
Kashmir	Pakistan	2005	2.4
Yogyakarta	Indonesia	2006	<0.1

Table 3. Fluid administration

Early fluid resuscitation (first 6 h, preferably starting before extrication)
While still under the rubble
   1 l/h isotonic saline
After rescue
   1 l/h half isotonic saline
   50 mEq Na bicarbonate to be added to each second or third liter of half isotonic saline
   5 g/h mannitol to be added if urinary flow >20 ml/h

After hospitalization
Under well-controlled conditions
   Target an urinary flow >300 ml/h
   Due to muscular sequestration daily infusion may be needed to exceed urinary output by
     ≥5 l/day during the first 3 days
   Same combination of half isotonic saline complemented with Na bicarbonate and
     mannitol in the same proportions as above
   To be continued until myoglobinuria disappears
   Central venous pressure measurements may offer additional guidance
In chaotic mass disasters (follow-up inadequate)
   Restrict to 6 l/day, especially in the elderly and the presumed anuric

which was a disaster of limited extent, 12 victims with CS who ultimately needed no dialysis treatment, received twice as much fluids (on average 50.6 vs. 24.0 l over the first 3 days following their extrication) as compared to 4 other victims needing dialysis [8]. This indicates that appropriate fluid administration might obviate the need for dialysis. In table 3, we describe a schematic approach to fluid administration as previously summarized [9].

Hospital Infrastructure

Hospitals in the affected area are often damaged to an extent that treatment of severely wounded victims becomes difficult. In the case of aftershocks, the damage may be extended further, leading to the total collapse of hospital infrastructure. It is therefore preferable to transport severe casualties with CS with a short delay to a safer area which is not damaged. Transportation at a later stage when patients have become dependent on ventilation or other intensive care therapies may be fatal. Hence, early transport should be organized. Dialyzing patients in local field hospitals near the damaged areas might be considered, but because these can seldom be embedded into appropriate intensive care or other hospital infrastructures, this option is less preferable.

Transport

As roads are often damaged, alternative means such as boats, air bridges or helicopters should be taken into account [10]. Since hyperkalemia is a major and potentially fatal problem, all severely affected victims with crush injury should receive kayexalate salt before transportation.

Dialysis Treatment

Dialysis treatment may become necessary in an overwhelming number of crush victims.

Intermittent Standard Hemodialysis

This option offers the possibility to treat several patients at the same position, is relatively efficient in correcting the concentration of small water soluble compounds such as potassium, and can be applied with minimal or no anticoagulation. Due to heavy electrolyte disturbances, it may sometimes be necessary to repeat hemodialysis several times per day.

Continuous Renal Replacement Therapy

Continuous renal replacement therapy (CRRT) offers the theoretical advantage of maintaining better hemodynamic stability, although up to now no controlled studies have shown superiority as to the clinical outcome [11, 12]. Only one patient can be treated per position and the need for continuous anticoagulation may be a handicap in patients with bleeding or at risk of bleeding. The strategy may be useful for areas where no traditional hemodialysis infrastructure is available, but necessitates the availability of bulky amounts of substitution fluid.

Peritoneal Dialysis

Also peritoneal dialysis offers the theoretical advantage of maintaining better hemodynamic stability than intermittent standard hemodialysis. Removal of small molecules is less efficient than with intermittent standard hemodialysis which may be a handicap in case of hyperkalemia. Peritoneal dialysis is less appropriate in subjects with abdominal or thoracic surgery or trauma, and in subjects with respiratory distress. Like CRRT, this strategy may be useful for the rare areas where no traditional hemodialysis infrastructure is available but similar to CRRT, it also necessitates the availability of bulky amounts of fluid.

Conclusions

The experience with the RDRTF has shown that preplanned structures for renal support in severe disaster conditions may be life saving for a substantial number of victims affected by the CS. This approach necessitates the availability of volunteer lists, a well-defined action plan, the possibility to deploy logistic support, and easy access to stocks of medication and dialysis material.

References

1 Solez K, Bihari D, Collins AJ, et al: International dialysis aid in earthquakes and other disasters. Kidney Int 1993;44:479–483.
2 Collins AJ: Kidney dialysis treatment for victims of the Armenian earthquake. N Engl J Med 1989;320:1291–1292.
3 Vanholder R, Sever MS, De Smet M, Erek E, Lameire N: Intervention of the Renal Disaster Relief Task Force in the 1999 Marmara, Turkey earthquake. Kidney Int 2001;59:783–791.
4 Hatamizadeh P, Najafi I, Vanholder R, et al: Epidemiologic aspects of the Bam earthquake in Iran: the nephrologic perspective. Am J Kidney Dis 2006;47:428–438.
5 Vanholder R, van der Tol A, De Smet M, et al: Earthquakes and crush syndrome casualties: lessons learned from the Kashmir disaster. Kidney Int 2007;71:17–23.
6 Vanholder R, Sever MS, Erek E, Lameire N: Rhabdomyolysis. J Am Soc Nephrol 2000;11:1553–1561.
7 Thyssen EP, Hou SH, Alverdy JC, Spiegel DM: Temporary loss of limb function secondary to soft tissue calcification in a patient with rhabdomyolysis-induced acute renal failure. Am J Kidney Dis 1990;16:491–494.
8 Gunal AI, Celiker H, Dogukan A, et al: Early and vigorous fluid resuscitation prevents acute renal failure in the crush victims of catastrophic earthquakes. J Am Soc Nephrol 2004;15:1862–1867.
9 Sever MS, Vanholder R, Lameire N: Management of crush-related injuries after disasters. N Engl J Med 2006;354:1052–1063.
10 Sever MS, Erek E, Vanholder R, et al: The Marmara earthquake: epidemiological analysis of the victims with nephrological problems. Kidney Int 2001;60:1114–1123.
11 Tonelli M, Manns B, Feller-Kopman D: Acute renal failure in the intensive care unit: a systematic review of the impact of dialytic modality on mortality and renal recovery. Am J Kidney Dis 2002;40:875–885.

12 Vinsonneau C, Camus C, Combes A, et al: Continuous venovenous haemodiafiltration versus intermittent haemodialysis for acute renal failure in patients with multiple-organ dysfunction syndrome: a multicentre randomised trial. Lancet 2006;368:379–385.

R. Vanholder
Renal Section, 0K12, University Hospital
De Pintelaan, 185
BE–9000 Ghent (Belgium)
Tel. +32 9240 4525, Fax +32 9240 4599, E-Mail raymond.vanholder@ugent.be

Renal Replacement Therapy for the Patient with Acute Traumatic Brain Injury and Severe Acute Kidney Injury

Andrew Davenport

Centre for Nephrology, Division of Medicine, Department of Medicine, Royal Free and University College Medical School, London, UK

Abstract

Fortunately with improvements in initial medical resuscitation, such as the avoidance of nephrotoxins, the incidence of acute kidney injury requiring renal support in patients with acute traumatic brain injury remains low. However the incidence of cerebral hemorrhage in patients on chronic dialysis programs appears to be increasing. By carefully adapting renal replacement to minimize cardiovascular instability and reduce the rate of change of serum osmolality, patient survival in this group of critically ill patients is increasing and starting to approach that of patients with traumatic brain injury without kidney injury.

Copyright © 2007 S. Karger AG, Basel

Patients with acute traumatic brain injury (TBI) may develop acute kidney injury (AKI) requiring renal support [1], due to ischemic AKI following hypoperfusion as a complication of the original injury, or due to subsequent sepsis and/or administration of nephrotoxins. Fortunately the number of such patients requiring renal replacement therapy (RRT) each year is small, as they are typically young with previous good health and receive high quality resuscitation at the accident scene. However, TBI can also occur in patients on dialysis for established chronic kidney disease. Coumarins are prescribed for atrial fibrillation, severe cardiac failure, previous ischemic stroke, and also in dialysis units to prevent clotting of dialysis catheters, and in patients with central venous thrombosis. The risk of bleeding in hemodialysis (HD) patients prescribed coumarins is greater than that of the general population [2], resulting in an increased risk of subdural/intracranial hemorrhage.

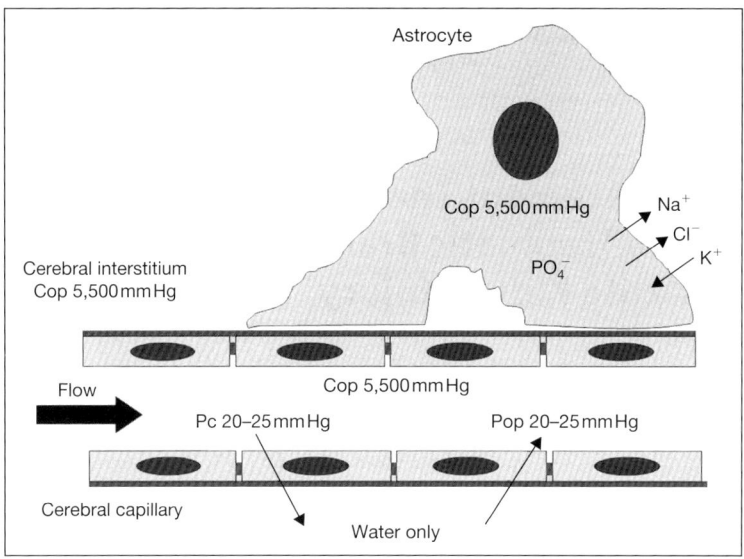

Fig. 1. Drawing of the intact blood brain barrier, which prevents free movement of proteins and solutes into the brain when transcapillary hydrostatic pressure (Pc) and plasma oncotic pressure (Pop) are balanced so that there is no net water. Crystalloid osmotic pressure (Cop) is equal in all three compartments. Active solute transport by carrier transport systems is vital for brain nutrition.

Standard Medical Management of Patients with TBI

Over the last decade the optimal management of patients with TBI has moved to a strategy designed to maintain cerebral perfusion pressure (CPP), with controlled ventilation and limited use of hyperventilation to treat surges in intracranial pressure (ICP) [3–5]. However, some centers continue to focus their management on attempts to control cerebral volume [6]. The brain contains more glial cells than neurons, and in particular the astrocytes are involved in maintaining the integrity of the cerebral capillary endothelial blood brain barrier (BBB) and homeostasis of the cerebral extracellular matrix (fig. 1). The brain lies in a fixed volume cranium, so any volume expansion due to acute hemorrhage/hematoma will result in an increase in ICP unless the expansion can be compensated by a reduction in cerebrospinal fluid or cerebral blood volume.

If cerebral autoregulation is preserved, sustaining the mean arterial blood pressure will reduce the intracranial blood volume by autoregulatory cerebral vasoconstriction and improve cerebral perfusion [3]. However if autoregulation fails, higher pressure will lead to increased cerebral capillary hydrostatic pressure with increased transcapillary leak causing further cerebral volume expansion and higher ICP. Thus control of cerebral volume, advocated by the Lund group [6], by attempting to reduce cerebral capillary hydrostatic pressure, whilst maintaining cerebral capillary plasma osmotic pressure, may improve clinical outcome [7]. Although there may appear to be some marked differences in the management strategy of the Brain Trauma Foundation compared to the Lund group, the key similarity is to determine the optimum CPP and ICP for any individual patient to maximize potential recovery [8].

Effect of Renal Replacement Therapies on ICP and CPP

In AKI, urea and other solutes increase and patients may develop metabolic acidosis. The BBB breaks down in cases of TBI allowing solute passage into the brain. This influx is initially compensated by astrocytes taking up additional ions and water, and generating idiogenic osmolarity.

The goals of any RRT are not only to clear urea and other solutes, but also to correct metabolic acidosis and restore sodium and water balance. Urea and water do not simply diffuse across the BBB, but pass through urea transporters and aquaporin channels, respectively. As the relative speed of passage through aquaporin channels is faster than that of urea transporters, this leads to the development of an osmotic gradient. RRT can impact on the brain in patients with TBI [9] by rapidly altering plasma solute concentrations. In addition, the rapid infusion of bicarbonate leads to a rapid correction of plasma pH. Bicarbonate cannot readily cross cell membranes, but reacts with hydrogen ions to form H_2CO_3 which can dissociate into water and carbon dioxide which can readily cross cell membranes. Thus the fluxes of HCO_3^-, H_2CO_3, and carbon dioxide lead to changes in pH, according to the Henderson-Hasselbach equation, resulting in intracellular acidosis and the compensatory production of intracellular osmolarity, with consequent water movement into the brain. Hypotension due to excessive ultrafiltration may reduce cerebral perfusion and CPP, with a consequent increase in ICP. Anticoagulation may be required for some RRT modalities, and this might increase the risk of intracerebral hemorrhage, depending upon the nature of the TBI and the ICP monitor used [10].

Peritoneal Dialysis
Although the rate of change in serum urea and osmolality is slower during peritoneal dialysis (PD) compared to intermittent HD, the dialysis disequilibrium syndrome has been reported during PD [11]. PD solutions are hyponatremic and this leads to a fall in serum sodium and water retention. This is contrary to current neurosurgical intensive care practice designed to maintain

plasma osmolality as advocated by the Lund hypothesis, and many centers now infuse hypertonic saline to maintain a high or supraphysiological serum sodium. In addition performing large volume exchanges of hypertonic glucose can adversely impact cerebral perfusion [12] and CPP by reducing right atrial filling [13]. Thus PD prescriptions should use the lowest glucose concentrations possible and avoid major swings in intraperitoneal volumes.

PD may well provide adequate clearances in patients with TBI and AKI alone, but may not be so effective in cases complicated by systemic sepsis. Intraperitoneal infection remains the commonest complication of PD, although in cases of TBI nosocomial infection, in particular ventilator-associated pneumonia, is a well-recognized complication and may be increased by PD.

Intermittent Hemodialysis/Hybrid Therapies

Brain edema occurs during routine outpatient thrice weekly HD. Thus intermittent HD increases brain swelling in the patient with TBI [14]. ICP increases not only due to changes in osmotic gradients, but also due to abrupt falls in mean arterial pressure and thus cerebral perfusion and perfusion pressure [15]. Intermittent RRT should be prescribed to maximize cardiovascular stability using a high sodium dialysate concentration, cooled dialysate, and minimizing changes in effective blood volume. As there are now HD machines with blood volume monitoring and/or blood volume control, these should be used preferentially. In addition the rate of change in serum osmolality should be minimized by increasing the frequency of RRT to a daily schedule with an extended session time, utilizing slower blood pump flows, small surface area dialyzer membranes and even reducing dialysate flow. There are no randomized prospective trials which have investigated the optimum predialysis urea to minimize changes in ICP during dialysis, however clinical practice suggests that a predialysis urea of <15 mmol/l (preferably <12 mmol/l or BUN <30 mg/dl) reduces the risk of ICP increasing during treatment.

Continuous Dialysis/Hemofiltration Therapies

Continuous arteriovenous hemofiltration was shown to have a much lesser effect on ICP than intermittent RRT [16]. However the introduction of pumped venous RRT and, in particular, high volume exchanges in patients with critical cerebral perfusion is associated with changes in ICP [17] (fig. 2). Thus when performing continuous RRT in critically ill patients with AKI, circuit design is important in that filtration is preferable to dialysis as this provides a slower rate of change in serum urea and other small solutes. Sodium balance during hemofiltration is positive as sodium sieving is <1.0, but even so a replacement fluid with a sodium concentration of >140 mmol/l should be used initially. At the start of treatment, small volume exchanges of 1.0 l/h should be used, and

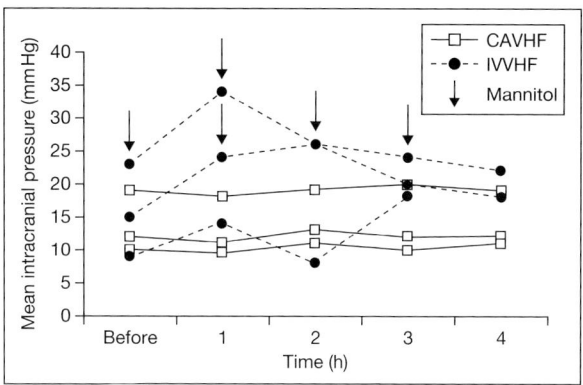

Fig. 2. Change in intracerebral pressure in a patient with cerebral edema who was treated by both continuous arteriovenous hemofiltration (CAVHF) and intermittent high volume veno-venous hemofiltration (4.0-liter exchange/h for 4 h; IVVHF). Mannitol boluses: 100 ml × 20% mannitol.

only when the patient has been shown to be stable should the exchange volume be increased.

Anticoagulation Strategies

Patients with AKI are at a high risk of hemorrhage, due either to hemorrhage associated with the initial TBI, or immediately post-neurosurgery to remove hemorrhage, or due to the presence of ICP-monitoring devices [10]. There is a greater risk of hemorrhage associated with intraventricular compared to brain parenchymal and subdural monitors/catheters. Thus ideally RRT should be anticoagulant free or a regional anticoagulant [17] such as citrate, nafamostat, and vasodilatory prostanoids used. In patients with compromised CPP, the vasodilatory effect of the potent prostanoids, prostacyclin and epoprostenol, may reduce CPP and thereby increase ICP [18]. Prior to using these agents, it is important to ensure that hypovolemia is corrected and that CPP is adequate and, if necessary, pressor support is temporarily increased.

Management of Sustained Surges in ICP during RRT

If sustained surges in ICP occur during RRT then standard medical management should be instituted, first checking that patients are adequately oxygenated with controlled hyperventilation (PaO_2 >11 kPa (82.5 mm Hg) with a $PaCO_2$ of 4.5–5 kPa (49.5–55 mm Hg)) [4, 5]. Then if the CPP is high, slow boluses of propofol and/or thiopentone could be administered, whereas if the CPP is normal or low, then hypertonic saline and/or mannitol should be given.

Individual centers differ in concentrations of hypertonic saline used (ranging from 3 to 10%), and these can be infused during RRT aiming for a serum sodium of 145, up to a maximum of 155 mmol/l, depending upon the ICP response. 100 ml of a 20% mannitol solution can also be given over 10–15 min, although most units now favor hypertonic saline. There is debate as to whether colloid solutions are deleterious because, when the BBB has broken down, colloids could enter the cerebral extracellular space and, due to their oncotic pressure, attract water and increase cerebral edema. The Lund group advocate albumin to maintain a normal serum albumin [6], whereas the SAFE study reported an adverse effect of colloid resuscitation in neurotrauma [19].

In refractory cases, other medical strategies include a short period of hyperventilation, deliberate hypothermia, CSF removal if a ventricular drain is in place [20], and surgical craniotomy as a last resort.

Conclusion

In patients with TBI requiring RRT, the RRT should be tailored to minimize abrupt changes in serum osmolality and to maintain cardiovascular stability. Continuous therapies are therefore advantageous for short intermittent RRT. Similarly anticoagulation free or regional anticoagulants are to be preferred to reduce the risk of further cerebral hemorrhage and bleeding around ICP-measuring devices.

References

1 Mehta RL, Kellum JA, Shah SV, Molitoris BA, Ronco C, Warnock DG, Levin A: Acute Kidney Injury Network (AKIN): report of an initiative to improve outcomes in acute kidney injury. Crit Care 2007;11:R31.
2 Lo DS, Rabbat CG, Clase CM: Thromboembolism and anticoagulant management in haemodialysis patients: a practical guide to clinical management. Thromb Res 2006;118:385–395.
3 Rosner MJ, Rosner SD, Johnson AH: Cerebral perfusion pressure. Management protocol and clinical results. J Neurosurg 1995;83:949–962.
4 Brain Trauma Foundation. American Association of Neurological Surgeons. Joint Section on Neurotrauma and Critical Care. Methodology. J Neurotrauma 2000;17:561–562.
5 Brain Trauma Foundation. American Association of Neurological Surgeons. Joint Section on Neurotrauma and Critical Care. Guidelines for the management of severe head injury. J Neurotrauma 1996;13:641–734.
6 Grande PO: The 'Lund Concept' for the treatment of severe head trauma – physiological principles and clinical applications. Intensive Care Med 2006;32:1475–1484.
7 Howells T, Elf K, Jones PA, Ronne E, Piper I, Nilsson P, Andrews PJD, Enblad P: Pressure reactivity as a guide in the treatment of cerebral perfusion pressure in patients with brain trauma. J Neurosurg 2005;102:311–317.
8 Andrews PJD, Citerio G: Lund Therapy – pathophysiology based therapy or contrived over interpretation of limited data? Intensive Care Med 2006;32:1461–1463.

9 Gondo GK, Fujitsu T, Kuwabara Y, Mochimatsu Y, Ishiwata H, Oda N, Takagi T, Yamashita H, Fujino I, Kim I, et al: Comparison of five modes of dialysis in neurosurgical patients with renal failure (in Japanese). Neurol Med Chir (Tokyo) 1989;29:1125–1131.
10 Blei AT, Olafsson S, Webster S, Levy R: Complications of intracranial pressure monitoring in fulminant hepatic failure. Lancet 1993;341:157–158.
11 Pai MF, Hsu SP, Peng YS, Hung KY, Tsai TJ: Hemorrhagic stroke in chronic dialysis patients. Ren Fail 2004;26:165–170.
12 Davenport A: Is there a role for continuous renal replacement therapies in patients with liver and renal failure? Kidney Int Suppl 1999;72:S62–S66.
13 Selby NM, Fonseca S, Hulme L, Fluck RJ, Taal MW, McIntyre CW: Automated peritoneal dialysis has significant effects on systemic hemodynamics. Perit Dial Int 2006;26:328–335.
14 Davenport A: Renal replacement therapy in the patient with acute brain injury. Am J Kidney Dis 2001;37:457–466.
15 Davenport A, Will EJ, Davison AM: Early changes in intracranial pressure during haemofiltration treatment in patients with grade 4 hepatic encephalopathy and acute oliguric renal failure. Nephrol Dial Transplant 1990;5:192–198.
16 Davenport A, Will EJ, Losowsky MS, Swindells S: Continuous arteriovenous haemofiltration in patients with hepatic encephalopathy and renal failure. Br Med J (Clin Res Ed) 1987;295:1028.
17 Davenport A: Is there a role for continuous renal replacement therapies in patients with liver and renal failure? Kidney Int Suppl 1999;72:S62–S66.
18 Davenport A, Will EJ, Davison AM: The effect of prostacyclin on intracranial pressure in patients with acute hepatic and renal failure. Clin Nephrol 1991;35:151–157.
19 Finfer S, Bellomo R, Boyce N, French J, Myburgh J, Norton R; SAFE Study Investigators: A comparison of albumin and saline for fluid resuscitation in the intensive care unit. N Engl J Med 2004;350:2247–2256.
20 Morris KP, Forsyth KP, Parslow RC, Tasker RC, Hawley CA; UK Paediatric Traumatic Brain Injury Study, and Paediatric Intensive Care Society Study Group: Intracranial pressure complicating severe traumatic brain injury in children: monitoring and management. Intensive Care Med 2006;32:1606–1612.

Andrew Davenport, MD
Centre for Nephrology, Division of Medicine, Department of Medicine
Royal Free and University College Medical School
Rowland Hill Street
London NW3 2PF (UK)
Tel. +44 20 783 022 91, Fax +44 20 783 021 25, E-Mail andrew.davenport@royalfree.nhs.uk

Cardiopulmonary Bypass-Associated Acute Kidney Injury: A Pigment Nephropathy?

Michael Haase[a,b], Anja Haase-Fielitz[a,b], Sean M. Bagshaw[a], Claudio Ronco[c], Rinaldo Bellomo[a]

[a]Department of Intensive Care, Austin Health, Melbourne, Vic., Australia; [b]Department of Nephrology, Charité University Medicine, Berlin, Germany; [c]Department of Nephrology, St. Bortolo Hospital, Vicenza, Italy

Abstract

Acute kidney injury (AKI) is a common and serious postoperative complication following exposure to cardiopulmonary bypass (CPB). Several mechanisms have been proposed by which the kidney can be damaged and interventional studies addressing known targets of renal injury have been undertaken in an attempt to prevent or attenuate CPB-associated AKI. However, no definitive strategy appears to protect a broad heterogeneous population of cardiac surgery patients from CPB-associated AKI. Although the association between hemoglobinuria and the development of AKI was recognized many years ago, this idea has not been sufficiently acknowledged in past and current clinical research in the context of cardiac surgery-related AKI. Hemoglobin-induced renal injury may be a major contributor to CPB-associated AKI. Accordingly, we now describe in detail the mechanisms by which hemoglobinuria may induce renal injury and raise the question as to whether CPB-associated AKI may actually be, in a significant part, a form of pigment nephropathy where hemoglobin is the pigment responsible for renal injury. If CPB-associated AKI is a pigment nephropathy, alkalinization of urine with sodium bicarbonate might protect from: (1) tubular cast formation from met-hemoglobin; (2) proximal tubular cell necrosis by reduced endocytotic hemoglobin uptake, and (3) free iron-mediated radical oxygen species production and related injury. Sodium bicarbonate is safe, simple to administer and inexpensive. If part of AKI after CPB is truly secondary to hemoglobin-induced pigment nephropathy, prophylactic sodium bicarbonate infusion might help attenuate it. A trial of such treatment might be a reasonable future investigation in higher risk patients receiving CPB.

Copyright © 2007 S. Karger AG, Basel

Dr. Haase holds a postdoctoral Feodor-Lynen Research Fellowship from the Alexander von Humboldt-Foundation, Germany.

Epidemiology and Risk Factors

With over one million operations a year worldwide, cardiac surgery is one of the most common major surgical procedures [1]. Acute kidney injury (AKI) is a common postoperative complication following exposure to cardiopulmonary bypass (CPB) [2–5]. AKI requiring dialysis occurs in up to 5% of patients undergoing elective cardiac surgery. An additional 8–15% of patients have moderate AKI with an increase in serum creatinine level of >1.0 mg/dl (88.4 μmol/l). A lesser degree of AKI with a greater than 25% increase in serum creatinine from baseline to a postoperative peak level may affect more than 50% of patients [5, 6].

Some patients are at particular risk of developing CPB-related acute renal failure (ARF) such as those with an increased duration of CPB, a preoperative serum creatinine level of >1.2 mg/dl, insulin-dependent diabetes mellitus, age >70 years, reduced left ventricular function, valve surgery, preoperative atrial fibrillation, and vascular disease [7]. Interestingly, there is evidence that a longer duration of CPB is associated with an increased likelihood of more severe AKI [4, 8, 9].

In these patients, reducing the incidence and severity of post-CPB AKI might prevent expenditure and morbidity and improve other outcomes.

Costs and Outcomes

AKI carries a significant cost and is a serious postoperative complication [10]. Also AKI leads to a significant increase in hospital expenditure especially if complicated by the need for dialysis [11].

Adverse outcomes of AKI after cardiac surgery include prolonged intensive care unit and hospital stay and discharge to extended-care facilities [12, 13]. After adjustment for comorbidities and intraoperative variables, all degrees of AKI are associated with increased mortality [2, 4, 5]. Even minimal increments in serum creatinine are associated with an independent increase in mortality [14, 15].

Pathophysiological Mechanisms and Prevention

Multiple causes of AKI following cardiac surgery have been proposed including perioperative hemodynamic instability and impaired renal blood flow, ischemia-reperfusion injury, and CPB-induced activation of inflammatory pathways and the generation of reactive oxygen species (ROS) [5, 16–19]. Other less common sources of renal injury include atheroembolism into the renal arteries and exogenous nephrotoxins such as nephrotoxic antibiotics, nonsteroidal anti-inflammatory drugs and anesthetics, all of which may contribute to AKI in selected patients [5].

However, ischemia-reperfusion injury and the generation of oxido-inflammatory stress represent two conventionally accepted major mechanisms in the pathogenesis of CPB-related AKI. The evidence supporting such mechanisms, however, is indirect and weak. In particular no randomized controlled trials (RCTs) seeking to prophylactically affect such pathways has yet to be shown effective.

Several RCTs have attempted to prevent or attenuate AKI, but most of these interventions have been found to be ineffective [6, 20–29] or inconclusive, and/or have only been studied in specific cardiac surgery subpopulations [30, 31].

Renal Hypoperfusion and Hypoxia/Ischemia-Reperfusion Injury

Although the kidneys receive more blood flow per gram of tissue than other major organs, they are also the most susceptible to ischemic injury. Metabolic demands from active tubular reabsorption and the oxygen diffusion shunt characteristic of the renal circulation contribute to the vulnerable physiology of renal perfusion including low medullary oxygen tension (10–20 mm Hg) [32]. Therefore, maintenance of physiological blood flow and near normal mean arterial pressure (MAP) before, during and after CPB is believed to be important for the prevention of postoperative AKI [33, 34].

Perioperative hemodynamic instability, which exceeds the autoregulatory reserve of the renal circulation may contribute to renal hypoperfusion and hypoxia in the renal medulla. Renal hypoxia may be increased by anemia resulting from hemodilution during and after CPB. Any oxygen supply-demand imbalances may play a crucial role in the development of AKI. However, at present, it remains unclear whether hypoxia per se or rather re-oxygenation (possibly through ROS) or [35] an impaired blood flow regulation (possibly through nitric oxide) cause AKI [36].

Ischemia-reperfusion injury of the kidney frequently occurring during cardiac surgery, for example due to cross clamping and reopening of the aorta, may be another important factor contributing to postoperative AKI [37, 38]. Decreased tissue oxygen tension promotes mitochondrial generation of ROS [39].

Unfortunately, there is a shortage of clinical trials randomizing patients according to hemodynamic targets (e.g. high versus low systemic MAP/CI/blood flow or renal blood flow) or high versus low hemoglobin levels to investigate renal outcomes. The only RCT available studied 21 cardiac surgery patients and found no influence of intraoperative systemic MAP on postoperative renal function [20]. In the absence of any RCT, a retrospective study of

1,760 patients evaluating the effect of low on-pump hematocrit on postoperative renal outcome found that CPB hemodilution to hematocrit <24% was associated with an increased likelihood of renal injury and worse operative outcomes [40].

Several RCTs have investigated pharmacological interventions, which were believed to improve renal perfusion, to increase renal blood flow or to decrease cortical oxygen consumption. None of these medications has been consistently found to prevent or attenuate AKI including fenoldopam, dopamine, clonidine, diltiazem, nesiritide [21–23, 31], pentoxifylline [24] and angiotensin-converting enzyme inhibitors and diuretics [25–28]. Alternatively, these interventions have only been studied in specific cardiac surgery subpopulations [30, 31].

Oxido-Inflammatory Stress

CPB has been shown to stimulate neutrophils and to induce the generation of ROS and inflammatory mediators [41–44]. Increased levels of serum lipid peroxidation products and an intra- and postoperatively decreased total serum antioxidative capacity have also been found [16, 17]. There is evidence indicating that the generation of ROS may contribute to the initiation and maintenance of acute tubular necrosis [45]. Oxidative stress is considered to be an important cause of renal injury in patients exposed to CPB [16–19].

CPB is proinflammatory and activates components of the nonspecific immune system. The inflammatory response to CPB generates cytokines (e.g. TNF and IL-6), both systemically, locally and in the kidney, that have major effects on the renal microcirculation and may lead to tubular injury [46–48].

Several RCTs aimed to achieve a reduction in inflammation and ROS, and to prevent AKI following CPB. However, N-acetylcysteine, which directly scavenges ROS, regenerates the glutathione pool, and reduces oxidative stress during CPB [49], did not provide renal protection [29, 50, 51]. Also, dexamethasone and enoximone [52–54] failed to demonstrate renal protection despite a reduction in the inflammatory response.

In summary, although diverse mechanisms exist by which the kidney can be damaged during cardiac surgery, no key mechanism of CPB-associated AKI and no corresponding preventive strategies have yet been identified.

A Pigment Nephropathy?

Pigment nephropathy is known to result from hemoglobinuria and myoglobinuria [55–58]. A wide range of causative factors are involved in the release of free hemoglobin or free myoglobin into the serum including hemolysis from extracorporeal circulation (e.g. CPB), rhabdomyolysis, chemical agents and also various venoms, malaria infection, mechanical destruction occasioned by valvular prosthesis or a transfusion reaction, heat stroke, burns as well as some genetic defects predisposing to reduced erythrocyte membrane stability [55].

Hemoglobin and myoglobin have a similar chemical core structure called heme protein. At the center of the heme group is the iron metal ion (Fe^{2+}). Hemoglobin consists of four protein chains and four heme groups whereas myoglobin only consists of a single protein chain and one heme group. Given the ability of both molecules to release free iron, which can act as nephrotoxin, similar pathogenetic mechanisms in the development of hemoglobinuric and myoglobinuric AKI can be assumed [55, 56]. However, due to the involvement of CPB in the release of free hemoglobin and not myoglobin, in the following we will focus on hemoglobinuria.

Although the association between hemoglobinuria and the development of AKI was recognized many years ago, this idea has not been sufficiently acknowledged in past and current clinical research focusing on cardiac surgery-related AKI [56, 57, 59–64]. Hemoglobin-induced AKI may be a clinically relevant cause for CPB-associated renal injury. Here we refer to hemoglobin-induced AKI as pigment nephropathy with free serum hemoglobin as the toxic pigment (fig. 1).

Hemolysis induced by mechanical destruction of erythrocytes through the CPB circuit releases free hemoglobin into the plasma, where it combines with haptoglobin to form a complex, which is carried to the liver, bypassing the kidney, and metabolized [65]. Thus, under normal circumstances hemoglobin does not exist in the serum in its free form. However, when the quantity of free serum hemoglobin exceeds the binding capacity of haptoglobin, free serum hemoglobin is able to scavenge endothelium-derived nitric oxide [36], but it will also pass through the glomerulus, appear in urine, release iron which is involved in the generation of ROS, and cause occlusion of renal tubules with hemoglobin casts and necrosis of tubular cells [65, 66].

In summary, hemolysis by the extracorporeal circulation may be an important contributor to AKI after cardiac surgery [67]. Accordingly, we now describe the mechanisms of injury proposed for hemoglobinuria in detail and we raise the question of whether CPB-associated AKI may eventually be, in part, a form of pigment nephropathy [57, 61, 68].

Sources and Magnitude of Hemolysis during CPB

CPB is involved in causing hemolysis by mechanical destruction of the erythrocytes thus generating free hemoglobin [59, 60, 67]. Many sources of hemolysis contribute to increased plasma levels of free serum hemoglobin during the use of CPB and in the early postoperative period [69].

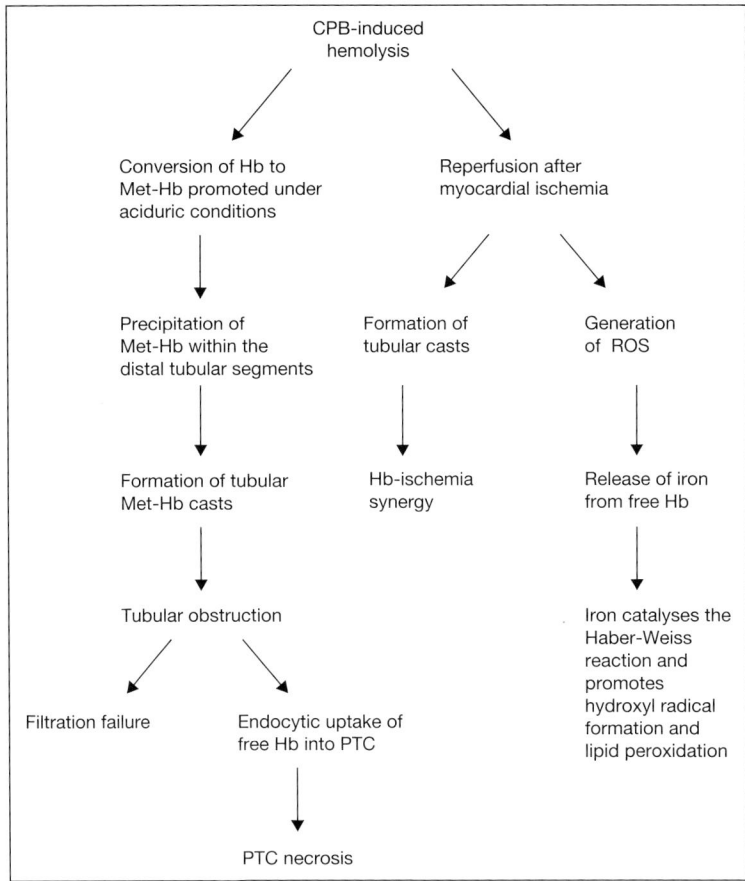

Fig. 1. Possible mechanisms of injury in pigment nephropathy induced by cardiopulmonary bypass (CPB). Hb = Hemoglobin; PTC = proximal tubular cell; ROS = reactive oxygen species.

Shear stress on erythrocytes resulting from contact with foreign surfaces of the bypass circuit (boundary layer of oxygenator, filters, tubing), cross-sectional area, the number of circuit connectors, blood aspiration by cardiotomy suction, and the roughness of surface of the pump, all of which are aggravated by high flow and high pressure conditions, are important determinants of hemolysis [70–72].

Free hemoglobin levels of >150 mg/dl, which is about 10-fold the upper physiological range, have been observed during the use of CPB until several hours postoperatively [69] despite the short duration of CPB (85 min). However, the detrimental effect of CPB on red cell destruction is accentuated by prolongation of CPB time [73, 74]. Thus, the longer the duration of CPB the more hemolysis occurs and the more free hemoglobin is generated. This may be of importance to the current clinical situation where complex surgery of the aortic arch and aortic valve is performed, and an increasing number of cardiac surgical centers have implemented time-consuming arterial coronary revascularization aiming to improve long-term results. Interestingly, there is evidence that a longer duration of CPB is associated with an increased likelihood of a more severe AKI [8, 9]. In addition, the use of CPB appears to have a close relation to hemolysis-induced gallstone formation after open cardiac surgery, which can be prevented by ursodeoxycholic acid [75, 76].

Relationship of Free Hemoglobin and CPB-Associated Pigment Nephropathy

It is well recognized that free serum hemoglobin may be a major contributor to AKI [57, 60, 61]. Several mechanisms for the development of pigment nephropathy have been suggested including the formation of tubular hemoglobin cast, free iron-induced generation of ROS, and scavenging of endothelium-derived nitric oxide [36, 56, 57, 63, 64].

In animal studies, infusion of free hemoglobin causes ARF [56–58]. The conversion of hemoglobin to met-hemoglobin is thought to be an important pathophysiological step in the induction of pigment nephropathy. An acid environment typical of tubular urine facilitates this conversion. Met-hemoglobin precipitates within the distal tubular segments forming casts, producing tubular obstruction and hence, filtration failure [77]. Such heme pigment casts have been described in virtually every study of pigment nephropathy [56–58, 77]. Also, under aciduric conditions, tubular obstruction may allow greater time for endocytic uptake of free hemoglobin into proximal tubular cells, which is associated with proximal tubular cell necrosis [57].

In addition, pigment nephropathy is drastically accentuated by additional ischemia-reperfusion (renal artery occlusion) resulting in widespread met-hemoglobin cast formation and proximal tubular cell necrosis [56, 57]. One mechanism identified that may help to explain a degree of hemoglobin-ischemia synergy is that ischemia-triggered cast formation enhances tubular obstruction, which facilitates proximal tubular cell hemoglobin uptake [57].

Infusion of hemoglobin under alkalinuric conditions causes virtually no renal injury and urine alkalinization attenuates renal failure in animal models [57, 64]. There is significantly reduced conversion of hemoglobin to met-hemoglobin, less met-hemoglobin cast formation, hence, less tubular obstruction. Proximal tubular cell necrosis appears to be extremely rare under these circumstances [57].

$$O^{2-} + 2Fe^{3+} \longrightarrow 2Fe^{2+} + O_2$$
$$2O^{2-} + 2H^+ \longrightarrow H_2O_2 + O_2$$
$$H_2O_2 + Fe^{2+} \longrightarrow OH^\bullet + OH^- + Fe^{3+}$$

Fig. 2. The superoxide-driven Haber-Weiss describes one possible mechanism in the generation of hydroxyl radicals that is catalyzed by free iron ions and most active at acid pH.

Another cast formation-independent pathogenetic mechanism of pigment nephropathy has been suggested: reperfusion following myocardial ischemia generates oxygen-derived free radicals (e.g. in cardiac surgery patients), which may release iron from free hemoglobin [78]. Iron promotes hydroxyl radical formation and lipid peroxidation [56]. In fact, several recent studies have demonstrated the potential for iron [78, 79] to catalyze the Haber-Weiss reaction (fig. 2) whereby superoxide radical (O_2^-) and hydrogenperoxide (H_2O_2) yield hydroxyl radical ($OH^\bullet$) [80].

At neutral or alkaline pH, free ferric ions precipitate as insoluble ferric hydroxide, reducing the production of injurious hydroxyl radicals [81].

It is also known that an acid environment typical of tubular urine enhances the formation of reactive hydroxyl radicals as the Haber-Weiss reaction is pH-dependent with a right shift when pH decreases. There is little argument that hydroxyl radicals are injurious in a wide variety of settings [80]. Accordingly, the beneficial effect of higher proximal tubular pH by urinary alkalinization, achieved for example with the use of sodium bicarbonate infusion, was protective in a rat model of ARF [64].

Finally, free serum hemoglobin is able to scavenge endothelium-derived nitric oxide 600-fold faster than erythrocytic hemoglobin [36]. This may lead to vasoconstriction, decreased blood flow, platelet activation, increased endothelin-1 expression and AKI [36].

Potential Strategies for the Prevention of Pigment Nephropathy following CPB

Several potential strategies for the prevention of pigment nephropathy have been proposed [56, 57, 60, 64, 82–85]. Administration of haptoglobin has been shown to have prophylactic and therapeutic effects on renal injury secondary to hemolysis [60, 82–84].

Also, iron chelation with deferoxamine has been found to be protective against pigment nephropathy in some animal models [56, 57, 85].

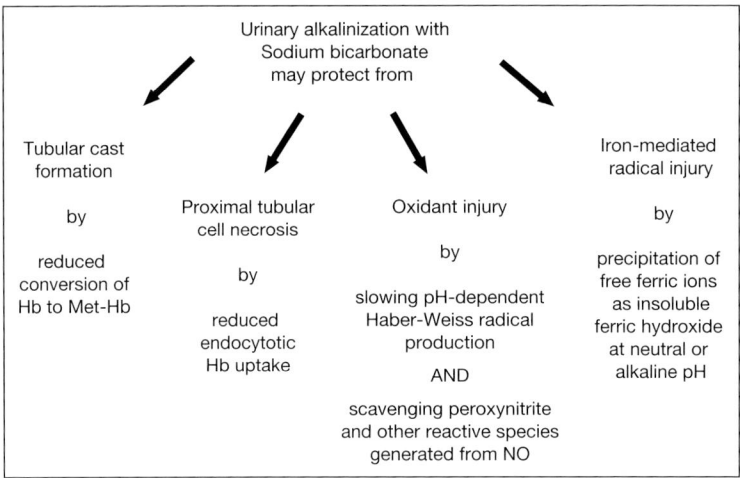

Fig. 3. Potential targets for sodium bicarbonate to protect the kidney from pigment-induced injury. Hb = Hemoglobin; NO = nitric oxide.

Taking into account the pathogenetic mechanisms proposed above for the notion of pigment nephropathy following exposure to CPB, there is sufficient biological rationale to target the urine pH as a means of attenuating AKI after CPB. An acid urine pH appears to be harmful in this setting by promoting increased formation of distal met-hemoglobin casts and tubulo-toxic ROS including hydroxyl radicals. An alkaline urine should be protective.

Sodium bicarbonate might potentially be able to target several injurious mechanisms of pigment nephropathy following CPB and might therefore represent a promising approach (fig. 3).

Urinary alkalinization with sodium bicarbonate might protect from: (1) tubular cast formation by reduced conversion of hemoglobin to met-hemoglobin; (2) proximal tubular cell necrosis by reduced endocytotic hemoglobin uptake; (3) oxidant injury by slowing pH-dependent Haber-Weiss radical production; (4) oxidant injury by direct scavenging of peroxynitrite and other reactive species generated from nitric oxide [86], and (5) free iron-mediated radical injury by precipitation of free ferric ions as insoluble ferric hydroxide at neutral or alkaline pH [81].

In addition, there is already evidence from a double-blind RCT that bicarbonate might attenuate AKI in patients undergoing infusion of contrast media [87]. The reduction of contrast-induced AKI by bicarbonate infusion would also be consistent with the hypothesis that, in part, AKI following exposure to contrast media is from ROS generated within the acid environment of the renal medulla [88, 89].

Conclusion

It is estimated that up to 50% of patients develop AKI following cardiac surgery. CPB-associated AKI may, at least in part, be a pigment nephropathy. Urine alkalinization might protect the kidney from such pigment nephropathy. Sodium bicarbonate is known to be a safe, simple to administer, and inexpensive intervention, which can alkalinize urine. It might, therefore, effectively protect the kidney from injury in this setting. Given these theoretical and experimental premises, it would seem reasonable to consider conducting a pilot RCT to test whether prophylactic sodium bicarbonate infusion is indeed able to prevent or attenuate AKI in patients receiving CPB.

References

1 Albert MA, Antman EM: Preoperative evaluation for cardiac surgery; in Cohn LH, Edmunds LH Jr (eds): Cardiac Surgery in the Adult. New York, McGraw-Hill, 2003, pp 235–248.
2 Chertow GM, Levy EM, Hammermeister KE, et al: Independent association between acute renal failure and mortality following cardiac surgery. Am J Med 1998;104:343–348.
3 Loef BG, Epema AH, Smilde TD, et al: Immediate postoperative renal function deterioration in cardiac surgical patients predicts in-hospital mortality and long-term survival. J Am Soc Nephrol 2005;16:195–200.
4 Conlon PJ, Stafford-Smith M, White WD, et al: Acute renal failure following cardiac surgery. Nephrol Dial Transplant 1999;14:1158–1162.
5 Stafford-Smith M, Podgoreanu M, Swaminathan M, et al: Association of genetic polymorphisms with risk of renal injury after coronary bypass graft surgery. Am J Kidney Dis 2005;45:519–530.
6 Bove T, Landoni G, Calabro MG, et al: Renoprotective action of fenoldopam in high-risk patients undergoing cardiac surgery: a prospective, double-blind, randomized clinical trial. Circulation 2005;111:3230–3235.
7 Thakar CV, Arrigain S, Worley S, et al: A clinical score to predict acute renal failure after cardiac surgery. J Am Soc Nephrol 2005;16:162–168.
8 Boldt J, Brenner T, Lehmann A, et al: Is kidney function altered by the duration of cardiopulmonary bypass? Ann Thorac Surg 2003;75:906–912.
9 Fischer UM, Weissenberger WK, Warters RD, et al: Impact of cardiopulmonary bypass management on postcardiac surgery renal function. Perfusion 2002;17:401–406.
10 Landoni G, Zangrillo A, Franco A, et al: Long-term outcome of patients who require renal replacement therapy after cardiac surgery. Eur J Anaesthesiol 2006;23:17–22.
11 Fischer MJ, Brimhall BB, Lezotte DC, et al: Uncomplicated acute renal failure and hospital resource utilization: a retrospective multicenter analysis. Am J Kidney Dis 2005;46: 1049–1057.
12 Mangano CM, Diamondstone LS, Ramsay JG, et al: Renal dysfunction after myocardial revascularization: risk factors, adverse outcomes, and hospital resource utilization. The Multicenter Study of Perioperative Ischemia Research Group. Ann Intern Med 1998;128:194–203.

13 Antunes PE, Prieto D, Ferrao de Oliveira J, Antunes MJ: Renal dysfunction after myocardial revascularization. Eur J Cardiothorac Surg 2004;25:597–604.
14 Lassnigg A, Schmidlin D, Mouhieddine M, et al: Minimal changes of serum creatinine predict prognosis in patients after cardiothoracic surgery: a prospective cohort study. J Am Soc Nephrol 2004;15:1597–1605.
15 Zanardo G, Michielon P, Paccagnella A, et al: Acute renal failure in the patient undergoing cardiac operation. Prevalence, mortality rate, and main risk factors. J Thorac Cardiovasc Surg 1994;107: 1489–1495.
16 Starkopf J, Zilmer K, Vihalemm T, et al: Time course of oxidative stress during open-heart surgery. Scand J Thorac Cardiovasc Surg 1995;29:181–186.
17 McColl AJ, Keeble T, Hadjinikolaou L, et al: Plasma antioxidants: evidence for a protective role against reactive oxygen species following cardiac surgery. Ann Clin Biochem 1998;35:616–623.
18 Doi K, Suzuki Y, Nakao A, et al: Radical scavenger edaravone developed for clinical use ameliorates ischemia/reperfusion injury in rat kidney. Kidney Int 2004;65:1714–1723.
19 McCord JM: Oxygen-derived free radicals in postischemic tissue injury. N Engl J Med 1985;312: 159–163.
20 Urzua J, Troncoso S, Bugedo G, et al: Renal function and cardiopulmonary bypass: effect of perfusion pressure. J Cardiothorac Vasc Anesth 1992;6:299–303.
21 Butler J, Emerman C, Peacock WF, et al: The efficacy and safety of B-type natriuretic peptide (nesiritide) in patients with renal insufficiency and acutely decompensated congestive heart failure. Nephrol Dial Transplant 2004;19:391–399.
22 Wang DJ, Dowling TC, Meadows D, et al: Nesiritide does not improve renal function in patients with chronic heart failure and worsening serum creatinine. Circulation 2004;110:1620–1625.
23 Sackner-Bernstein JD, Skopicki HA, Aaronson KD: Risk of worsening renal function with nesiritide in patients with acutely decompensated heart failure. Circulation 2005;111:1487–1491.
24 Boldt J, Brosch C, Piper SN, et al: Influence of prophylactic use of pentoxifylline on postoperative organ function in elderly cardiac surgery patients. Crit Care Med 2001;29:952–958.
25 Colson P, Ribstein J, Mimran A, et al: Effect of angiotensin converting enzyme inhibition on blood pressure and renal function during open heart surgery. Anesthesiology 1990;72:23–27.
26 Ryckwaert F, Colson P, Ribstein J, et al: Haemodynamic and renal effects of intravenous enalaprilat during coronary artery bypass graft surgery in patients with ischaemic heart dysfunction. Br J Anaesth 2001;86:169–175.
27 Lassnigg A, Donner E, Grubhofer G, et al: Lack of renoprotective effects of dopamine and furosemide during cardiac surgery. J Am Soc Nephrol 2000;11:97–104.
28 Sirivella S, Gielchinsky I, Parsonnet V: Mannitol, furosemide, and dopamine infusion in postoperative renal failure complicating cardiac surgery. Ann Thorac Surg 2000;69:501–506.
29 Burns KE, Chu MW, Novick RJ, et al: Perioperative N-acetylcysteine to prevent renal dysfunction in high-risk patients undergoing cabg surgery: a randomized controlled trial. JAMA 2005;294: 342–350.
30 Sward K, Valsson F, Odencrants P, et al: Recombinant human atrial natriuretic peptide in ischemic acute renal failure: a randomized placebo-controlled trial. Crit Care Med 2004;32:1310–1315.
31 Mentzer RM Jr, Oz MC, Sladen RN, et al: Effects of perioperative nesiritide in patients with left ventricular dysfunction undergoing cardiac surgery. J Am Coll Cardiol 2007;49:716–726.
32 Brezis M, Rosen S: Hypoxia of the renal medulla – its implications for disease. N Engl J Med 1995;332:647–655.
33 Bhat JG, Gluck MC, Lowenstein J, Baldwin DS: Renal failure after open heart surgery. Ann Intern Med 1976;84:677–682.
34 Abel RM, Buckley MJ, Austen WG, et al: Acute postoperative renal failure in cardiac surgical patients. J Surg Res 1976;20:341–348.
35 Rosenberger C, Rosen S, Heyman SN: Renal parenchymal oxygenation and hypoxia adaptation in acute kidney injury. Clin Exp Pharmacol Physiol 2006;33:980–988.
36 Gladwin MT, Crawford JH, Patel RP: The biochemistry of nitric oxide, nitrite, and hemoglobin: role in blood flow regulation. Free Radic Biol Med 2004;36:707–717.
37 Andersson LG, Bratteby LE, Ekroth R, et al: Renal function during cardiopulmonary bypass: influence of pump flow and systemic blood pressure. Eur J Cardiothorac Surg 1994;8:597–602.

38　Abbott WM, Austen WG: The reversal of renal cortical ischemia during aortic occlusion by mannitol. J Surg Res 1974;16:482–489.
39　Dada LA, Chandel NS, Ridge KM, et al: Hypoxia-induced endocytosis of Na,K-ATPase in alveolar epithelial cells is mediated by mitochondrial reactive oxygen species and PKC-zeta. J Clin Invest 2003;111:1057–1064.
40　Habib RH, Zacharias A, Schwann TA, et al: Role of hemodilutional anemia and transfusion during cardiopulmonary bypass in renal injury after coronary revascularization: implications on operative outcome. Crit Care Med 2005;33:1749–1756.
41　Chello M, Mastroroberto P, Patti G, et al: Simvastatin attenuates leucocyte-endothelial interactions after coronary revascularisation with cardiopulmonary bypass. Heart 2003;89:538–543.
42　Boyle EM Jr, Lille ST, Allaire E, et al: Endothelial cell injury in cardiovascular surgery: atherosclerosis. Ann Thorac Surg 1997;63:885–894.
43　Paparella D, Yau TM, Young E: Cardiopulmonary bypass induced inflammation: pathophysiology and treatment. An update. Eur J Cardiothorac Surg 2002;21:232–244.
44　Partrick DA, Moore EE, Fullerton DA, et al: Cardiopulmonary bypass renders patients at risk for multiple organ failure via early neutrophil priming and late neutrophil disability. J Surg Res 1999;86: 42–49.
45　Nath KA, Norby SM: Reactive oxygen species and acute renal failure. Am J Med 2000;109: 665–678.
46　Cunningham PN, Dyanov HM, Park P, et al: Acute renal failure in endotoxemia is caused by TNF acting directly on TNF receptor-1 in kidney. J Immunol 2002;168:5817–5823.
47　Segerer S, Nelson PJ, Schlondorff D: Chemokines, chemokine receptors, and renal disease: from basic science to pathophysiologic and therapeutic studies. J Am Soc Nephrol 2000;11: 152–176.
48　Heyman SN, Rosen S, Darmon D, et al: Endotoxin-induced renal failure. II. A role for tubular hypoxic damage. Exp Nephrol 2000;8:275–282.
49　Sucu N, Cinel I, Unlu A, et al: N-Acetylcysteine for preventing pump-induced oxidoinflammatory response during cardiopulmonary bypass. Surg Today 2004;34:237–242.
50　Ristikankare A, Kuitunen T, Kuitunen A, et al: Lack of renoprotective effect of i.v. N-acetylcysteine in patients with chronic renal failure undergoing cardiac surgery. Br J Anaesth 2006;97:611–616.
51　Hynninen MS, Niemi TT, Poyhia R, et al: N-Acetylcysteine for the prevention of kidney injury in abdominal aortic surgery: a randomized, double-blind, placebo-controlled trial. Anesth Analg 2006;102:1638–1645.
52　Morariu AM, Loef BG, Aarts LP, et al: Dexamethasone: benefit and prejudice for patients undergoing on-pump coronary artery bypass grafting: a study on myocardial, pulmonary, renal, intestinal, and hepatic injury. Chest 2005;128:2677–2687.
53　Boldt J, Brosch C, Lehmann A, et al: The prophylactic use of the beta-blocker esmolol in combination with phosphodiesterase III inhibitor enoximone in elderly cardiac surgery patients. Anesth Analg 2004;99:1009–1017, table of contents.
54　Boldt J, Brosch C, Suttner S, et al: Prophylactic use of the phosphodiesterase III inhibitor enoximone in elderly cardiac surgery patients: effect on hemodynamics, inflammation, and markers of organ function. Intensive Care Med 2002;28:1462–1469.
55　Evenepoel P: Acute toxic renal failure. Best Pract Res Clin Anaesthesiol 2004;18:37–52.
56　Paller MS: Hemoglobin- and myoglobin-induced acute renal failure in rats: role of iron in nephrotoxicity. Am J Physiol 1988;255:F539–F544.
57　Zager RA, Gamelin LM: Pathogenetic mechanisms in experimental hemoglobinuric acute renal failure. Am J Physiol 1989;256:F446–F455.
58　Tam SC, Wong JT: Impairment of renal function by stroma-free hemoglobin in rats. J Lab Clin Med 1988;111:189–193.
59　Takami Y, Makinouchi K, Nakazawa T, et al: Effect of surface roughness on hemolysis in a pivot bearing supported Gyro centrifugal pump (C1E3). Artif Organs 1996;20:1155–1161.
60　Tanaka K, Kanamori Y, Sato T, et al: Administration of haptoglobin during cardiopulmonary bypass surgery. ASAIO Trans 1991;37:M482–M483.
61　Feola M, Simoni J, Tran R, Canizaro PC: Nephrotoxicity of hemoglobin solutions. Biomater Artif Cells Artif Organs 1990;18:233–249.

62 Lehotsky J, Kaplan P, Matejovicova M, et al: Ion transport systems as targets of free radicals during ischemia reperfusion injury. Gen Physiol Biophys 2002;21:31–37.
63 Halliwell B, Gutteridge JM: Role of free radicals and catalytic metal ions in human disease: an overview. Methods Enzymol 1990;186:1–85.
64 Atkins JL: Effect of sodium bicarbonate preloading on ischemic renal failure. Nephron 1986;44: 70–74.
65 Keene WR, Jandl JH: The sites of hemoglobin catabolism. Blood 1965;26:705–719.
66 Loebl EC, Baxter CR, Curreri PW: The mechanism of erythrocyte destruction in the early postburn period. Ann Surg 1973;178:681–686.
67 Kanamori Y, Tanabe H, Shimono T, et al: The effects of administration of haptoglobin for hemolysis by extracorporeal circulation (in Japanese). Rinsho Kyobu Geka 1989;9:463–467.
68 Clyne DH, Kant KS, Pesce AJ, Pollak VE: Nephrotoxicity of low molecular weight serum proteins: physicochemical interactions between myoglobin, hemoglobin, bence-jones proteins and tamm-horsfall mucoprotein. Curr Probl Clin Biochem 1979;9:299–308.
69 Loef BG, Epema AH, Navis G, et al: Off-pump coronary revascularization attenuates transient renal damage compared with on-pump coronary revascularization. Chest 2002;121:1190–1194.
70 Tamari Y, Lee-Sensiba K, Leonard EF, et al: The effects of pressure and flow on hemolysis caused by Bio-Medicus centrifugal pumps and roller pumps. Guidelines for choosing a blood pump. J Thorac Cardiovasc Surg 1993;106:997–1007.
71 Crane KA, Brown D, Anderson R, Begelman KM: Further decrease in subclinical hemolysis utilizing 12.7 mm tubing in the arterial roller head. Ann Thorac Surg 1983;35:463–465.
72 Skrabal CA, Khosravi A, Westphal B, et al: Effects of poly-2-methoxyethylacrylate (PMEA)-coating on CPB circuits. Scand Cardiovasc J 2006;40:224–229.
73 Yamaguchi H, Shimizu T, Akutsu H, et al: Hemolysis and red cell deformability during cardiopulmonary bypass – the effect of prostaglandin E1 for prevention of hemolysis (in Japanese). Nippon Kyobu Geka Gakkai Zasshi 1990;38:625–629.
74 Hirayama T, Herlitz H, Jonsson O, Roberts D: Deformability and electrolyte changes of erythrocytes in connection with open heart surgery. Scand J Thorac Cardiovasc Surg 1986;20:253–259.
75 Azemoto R, Tsuchiya Y, Ai T, et al: Does gallstone formation after open cardiac surgery result only from latent hemolysis by replaced valves? Am J Gastroenterol 1996;91:2185–2189.
76 Ai T, Azemoto R, Saisho H: Prevention of gallstones by ursodeoxycholic acid after cardiac surgery. J Gastroenterol 2003;38:1071–1076.
77 Jaenike JR: The renal lesion associated with hemoglobinemia: a study of the pathogenesis of the excretory defect in the rat. J Clin Invest 1967;46:378–387.
78 Das DK, Engelman RM, Liu X, et al: Oxygen-derived free radicals and hemolysis during open heart surgery. Mol Cell Biochem 1992;111:77–86.
79 Gutteridge JM: Iron promoters of the Fenton reaction and lipid peroxidation can be released from haemoglobin by peroxides. FEBS Lett 1986;201:291–295.
80 Halliwell B, Gutteridge JM: Oxygen toxicity, oxygen radicals, transition metals and disease. Biochem J 1984;219:1–14.
81 Cohen G: The Fenton reaction; in Greenwald RA (ed): CRC Handbook of Methods for Oxygen Radical Research. Boca Raton, CRC Press, 1985, pp 55–64.
82 Ohshiro T, Kosaki G, Funakoshi S: Haptoglobin therapy: effect on prevention and treatment of hemoglobinuria. Med J Osaka Univ 1978;29:269–279.
83 Ohshiro TU, Mukai K, Kosaki G: Prevention of hemoglobinuria by administration of haptoglobin. Res Exp Med (Berl) 1980;177:1–12.
84 Yoshioka T, Sugimoto T, Ukai T, Oshiro T: Haptoglobin therapy for possible prevention of renal failure following thermal injury: a clinical study. J Trauma 1985;25:281–287.
85 Shah SV, Walker PD: Evidence suggesting a role for hydroxyl radical in glycerol-induced acute renal failure. Am J Physiol 1988;255:F438–F443.
86 Caulfield JL, Singh SP, Wishnok JS, et al: Bicarbonate inhibits N-nitrosation in oxygenated nitric oxide solutions. J Biol Chem 1996;271:25859–25863.
87 Merten GJ, Burgess WP, Gray LV, et al: Prevention of contrast-induced nephropathy with sodium bicarbonate: a randomized controlled trial. JAMA 2004;291:2328–2334.

88 Bakris GL, Lass N, Gaber AO, et al: Radiocontrast medium-induced declines in renal function: a role for oxygen free radicals. Am J Physiol 1990;258:F115–F120.
89 Katholi RE, Woods WT Jr, Taylor GJ: Oxygen free radicals and contrast nephropathy. Am J Kidney Dis 1998;32:64–71.

Prof. Rinaldo Bellomo
Department of Intensive Care, Austin Hospital
Melbourne, Vic. 3084 (Australia)
Tel. +61 3 9496 5992, Fax +61 3 9496 3932, E-Mail rinaldo.bellomo@austin.org.au

CRRT Technology and Logistics: Is There a Role for a Medical Emergency Team in CRRT?

Patrick M. Honoré[a], Olivier Joannes-Boyau[b], Benjamin Gressens[a]

[a]St-Pierre Para-University Hospital, Ottignies-Louvain-la-Neuve, Belgium, and
[b]Haut Leveque University Hospital, University of Bordeaux, Pessac, France

Abstract

Implementing continuous renal replacement therapy (CRRT) in a intensive care unit (ICU) is a somewhat difficult issue and quiet different from starting a new ventilation mode or a new hemodynamic device. It may indeed require an on-call medical emergency CRRT team as expertise in this field is really a key issue to success. Education for the nursing team is another key point, especially as ongoing or continuous education is changing very quickly. Uniformity of the type of device used is another crucial part in the organization process with regard to CRRT implementation in the ICU. Involvement of both the ICU and nephrology teams is another key to success especially when different modes and higher exchange rates are used. Also, a nursing group devoted to the ongoing implementation and education of the ICU team is very useful in order to attain the goals that have been set. Already in 1984 acute renal failure was described as one of the remaining and challenging problems in the ICU. Hemodialysis was not always feasible then because of the hemodynamic instability of critically ill patients. Under those circumstances continuous arteriovenous hemofiltration (CAVH) was advocated as an efficient alternative method with less detrimental hemodynamic effects. At the time it was thought that CAVH would be found to be an effective 'artificial kidney' (control of body fluid, electrolyte and acid-base homeostasis and uremia) and this without serious side effects. But already nearly 25 years ago, it was found that continuous anticoagulation was a major problem that could cause life-threatening complications in posttraumatic and surgical patients. At the time, it was thought that running a protamine infusion on the venous line would help to diminish these complications. CRRT has been carried out in our ICU since 1985, first with CAVH and since 1989 with some early forms of continuous veno-venous hemofiltration (CVVH). The unit has used BSM 22, BM 25 and Prisma for nearly 10 years, and Aquarius since the end of 2001. The educational process started at the beginning of 1990 with the implementation of CVVH using BSM 22 and BM 25. Very soon it was realized that a new strategy implementing pulse high-volume hemofiltration (pulse-HVHF) was really needed. Therefore, a nursing group composed of 5–8 nurses who would be taught beforehand was started, and this dedicated group would then teach the rest of the staff nurses. This group exists today and has at least 6–8 meetings/year in which all the problems that must be faced in the implementation of CRRT are dealt with. Here all the steps made by our and other units in this field will be discussed, including an overview of the various protocols implemented and a description of our dedicated nursing group with regard to CRRT.

Copyright © 2007 S. Karger AG, Basel

Global Approach to the Problem in the ICU

Implementing continuous renal replacement therapy (CRRT) in an intensive care unit (ICU) is a somewhat difficult issue and quiet different from starting a new ventilation mode or a new hemodynamic device. It may indeed require an on-call medical emergency CRRT team (MECT) as expertise in this field is a key issue for success. Education for the nursing team is another key point, especially as is it rapidly changing. Uniformity of the type of device used is another crucial part in the organization process with regard to implementing CRRT in the ICU. The involvement of both the ICU and nephrology teams is another key to success especially when different modes and higher exchange rates are being used. Also a nursing group devoted to the ongoing implementation and education of the ICU team is very useful in order to achieve the goals set as we have done at Saint-Pierre Hospital [1].

Already in 1984 (nearly a quarter of a century ago), Schetz et al. [2] described acute renal failure as one of the remaining and challenging problems in the ICU because hemodialysis was not always feasible due to the hemodynamic instability of critically ill patients. Under these circumstances continuous arteriovenous hemofiltration (CAVH) was advocated as an efficient alternative method with less detrimental hemodynamic effects. At the time Schetz et al. [2] were also convinced that CAVH would be found to be an effective 'artificial kidney' (control of body fluid, electrolyte and acid-base homeostasis and uremia) and this without serious side effects. But already nearly 25 years ago, they emphasized that a major problem was continuous anticoagulation, which could cause life-threatening complications in posttraumatic and surgical patients. At the time, they were in favor of running a protamine infusion on the venous line in order to diminish these complications [2].

As many experts have explained, the key to developing a successful CRRT course is the involvement of clinical experts – the staff nurses. It is an ongoing process that requires continual improvement. A devoted course on CRRT can greatly help to assure the competence of the ICU nurse in caring for the critically ill patient while on CRRT as Clevenger [3] has elegantly shown by his own experience.

As Giuliano and Pysznik [4] have highlighted, implementing a program as complex as continuous veno-venous hemofiltration (CVVH) without the involvement of nephrology nurses is a real challenge. However, with proper planning, appropriate staff support, and the ability to make changes as implementation proceeds, a successful program can be developed. Indeed, our reward will be that we will be able to offer therapy that is important and potentially life-saving to those critically ill patients with renal failure who are unable to tolerate intermittent hemodialysis [4].

More and more units are becoming convinced that a clinical educator is the right answer to this problem, and we tend to agree. Harvey et al. [5] described the new role of a renal critical care educator based in a regional pediatric renal unit with an 'outreach' to three pediatric ICUs (PICUs) within the region. Harvey et al. [5] demonstrated that after 18 months the training objectives for the PICU staff progressed in all aspects of hemofiltration and all units used flexible training programs which, at the same time, were under constant evaluation using questionnaires. This type of renal critical care educator also worked alongside staff whilst CVVH was being performed to further instill confidence. During the same period of time, equipment and protocols have largely been standardized throughout the region, and an ongoing survey of CVVH use was initiated which could help to inform audit standards such as complication rates. Harvey et al. [5] concluded that the renal critical care educator was the catalyst for the formation of a regional hemofiltration group which shared in the development of guidelines and protocols and discussion of clinical data. Since CRRT is used infrequently in many PICUs the development of a renal critical care educator could serve as a model for the development and maintenance of skills in other regions [5]. This latest part could be seen in some ways as a nurse emergency team for CRRT in ICU.

Policies concerning the choice of vascular access are also a crucial. As outlined by East and Jacoby [6] complications related to central venous line use are known to increase patient morbidity and mortality and increase costs and length of hospital stay. Education programs to promote best central line practice have been shown to reduce central line complications. This was especially demonstrated with regard to central line care policy in the pediatric cardiovascular ICU [6].

In our unit, we were able to demonstrate that, when reaching 35 ml/kg/h of exchange, a blood flow of 300 ml/min was required and this could not be sustained for more than 24 h other than using the right internal jugular approach [7] as the femoral access is not reliable in more than 50% of the patients due to the very high incidence of abdominal compartment syndrome in the ICU [8].

Guidelines and written instructions as protocols are extremely important in this setting. This has been very well illustrated by the work of Kingston et al. [9]

who introduced the concept of patient group direction (PGD) which is a specifically written instruction for supplying or administrating named medicines in an identified clinical situation. The introduction of a PGD must demonstrate a benefit for patients. They were able to easily show that hemofiltration was a very good candidate for this type of specifically written instruction. Using a hemofiltration PGD, patient care was improved by providing standardization in the administration of fluids and electrolytes and as well enabling nurses to respond rapidly to changes in biochemistry during the procedure [10]. In our unit we use similar types of written instructions [11], and in other units similar guidelines also exist, as shown in the literature [12].

The Saint-Pierre Way

CRRT has been performed in our ICU since 1985, first with CAVH and after 1989 with some early forms of CVVH. The unit has used BSM 22, BM 25 and Prisma for nearly 10 years and Aquarius since the end of 2001. The education process really started from the beginning of 1990 with the implementation of CVVH using BSM 22 and BM 25. At that time, a medical expert was teaching the nursing staff directly. Very soon after, it was found that with the implementation of pulse high-volume hemofiltration (pulse-HVHF) [13, 14], a new strategy was needed. Therefore, a nursing group, composed of 5–8 nurses, was formed in order that they be taught beforehand, and this dedicated group would then teach the remainder of the nursing staff. This group still exists and has at least 6–8 meetings/year in which all problems regarding the implementation of CRRT are dealt with.

In the early phase, pulse-HVHF was performed using Gambro AK-10 Ultra. This technique was initially developed by our nephrologist team who, like us, were on call 24 h/day throughout the year. Indeed, they have been doing HVHF in almost 50% of their chronic renal failure patients for more than 25 years [15]. Later this technique was taken over by us but still we work in close relationship with the nephrology team. Indeed, the both the ICU and nephrology units are very close allowing easy access to every box in the ICU that is equipped with an internal circuit of ultrapure water as part of the big nephrology circuit.

The biggest challenge that we faced was the change from BSM 22 and BM 25 to Prisma at the end of 1993, and further changes from Prisma to Aquarius at the end of 2001. The latest change occurred at the end of 2001 when the 35-ml/kg/h rule was implemented in our ICU. In 2001 we also took over the pulse-HVHF technology. The next challenge will be the implementation of citrate in our unit [1, 7].

These frequent and important changes require a MECT to be on call, sometimes just for technical problems. This team allows very rapid implementation of a new technique/machine in the unit. Undoubtedly, this type of team enables the use of various approaches with regard to CRRT in the unit including pulse-HVHF. Thereby filter lifespan has also greatly increased.

It appears that the integration of a dedicated medical team plus a nursing group is really a condition for success with this kind of therapy. A close relationship with the nephrology team is also a key importance. Continuous education is also a crucial point regarding this technique which needs continuous review of all existing protocols [1, 7]. So, the world 'continuous' is not only for the technique but also for 'educational' process itself as shown by other groups as well [16, 17].

Our Strategy for Implementing the 35-ml/kg/h Rule

All the rigorous education processes performed inside the nursing and the medical teams in order to enable the change from 20 to 35 ml/kg/h [18] will now be described.

At the beginning of 2002, we decided to implement 35 ml/kg/h as a standard CRRT dose for all our patients according to the study of Ronco et al. [19]. For the past 15 years, we have been using CRRT with BSM 22, BM 25, Prisma and finally Aquarius.

A nursing group specially dedicated to CRRT was created in our unit more than 10 years ago. This group allows us to have nurses dedicated to CRRT in every shift. This is also very helpful when new machines or techniques are implemented. Such a change needs a rigorous education process, first in the nursing group and then in the whole nursing team as well physicians and colleagues.

Our hospital is a regional general hospital with 455 beds, serving a population of 150,000 inhabitants; and it has a medical-surgical general ICU with 15 beds and 1,000 admissions a year. The unit performs CRRT on between 50 and 70 patients per year.

All the changes made in daily practice are summarized below and need seven crucial changes (fig. 1).

(1) Vascular access. A 14-french coaxial catheter is needed. The catheter site must exclusively be by a right internal jugular approach (posterior), 20 cm length, with the tip of the catheter placed in the right atrium (fig. 2). In order to perform 35 ml/kg/h in each patient regardless of the body weight (from 50 up to 150 kg), a blood flow of 300 ml/min is required in order to keep the filtration fraction around and below 30% (fig. 3).

For a 50-kg patient = $35 \times 50 = 1,750$ l/h which gives an filtration fraction of less than 15% but for a 150-kg patient = $35 \times 150 = 5,250$ l/h, this gives an

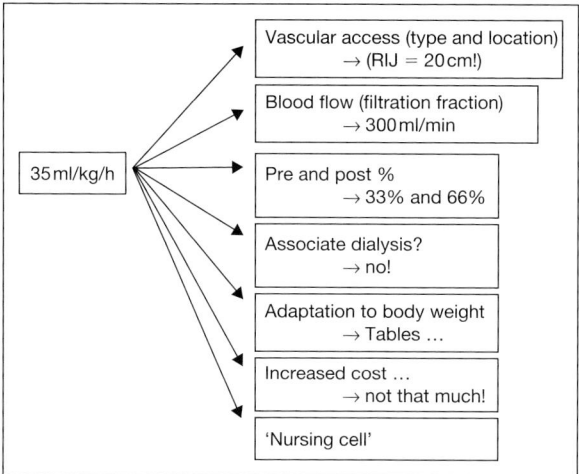

Fig. 1. The seven arms of the hemofiltration tree.

filtration fraction of 29%. Obviously the blood flow cannot keep changing dependent on the body weight of the patient. Very strict rules must be applied.

The choice of the right internal jugular approach needs a strong consensus among all the ICU consultants, and again very strict rules have to be applied.

(2) Pre- and post-dilution policy. According to the latest literature, we chose 33% pre-dilution and 66% post-dilution. This policy allows us to prevent some clotting and some clogging by using a certain amount of pre-dilution. This policy also permits a good convection rate. Indeed, the loss of convection by the concomitant use of pre-dilution reduces the risk of fouling and protein cake formation and finally preserves further loss of convection.

(3) Pre-determination of the exchange rate according to the body weight. In order to reduce the workload at the beginning of the CRRT, we introduced full tables which automatically give the amount of pre- and post-dilution for a given body weight, eliminating the need for sophisticated calculation that can lead to further mistakes (fig. 4). Additional tables exist also for the adaptation of antibiotics to 35 ml/kg/h and other drugs as well. Control of delivery is also very Important and therefore, we use special dedicated forms to monitor nursing (fig. 5).

(4) No further need of associated dialysis. During the last 3 years, even in very severe rhabdomyolysis cases, no need of associated dialysis was required no matter what the release of potassium was. This enabled us to use the

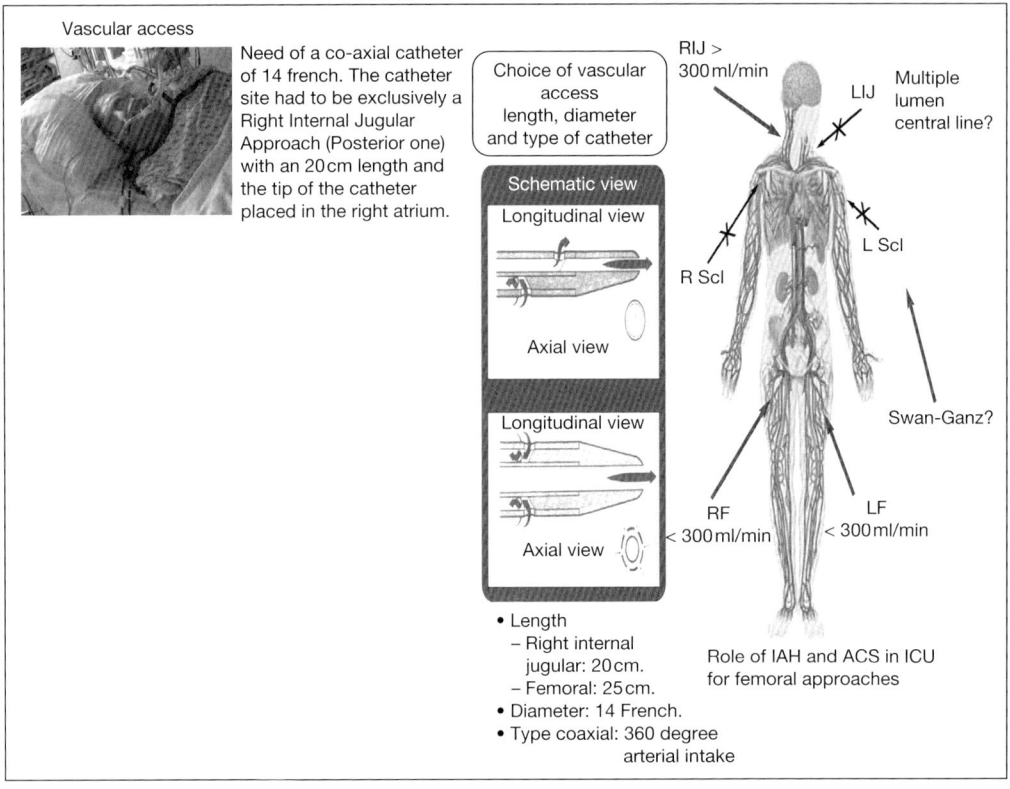

Fig. 2. Vascular access for CRRT at 35 ml/kg/h.

Aquarius with combined pre- and post-dilution and also to save money regarding additional cost of fluids.

In conclusions, the implementation of new guidelines concerning the 35-ml/kg/h rule cannot be decided in one day. This will need a rigorous long-standing educational process for the nursing staff as well as the medical team. A nursing group can be of help in order to achieved this. We will soon embark upon the implementation of citrate using the same implementation and educational process.

Medical Emergency CRRT and Nursing Emergency CRRT Teams

As said earlier, very big challenges occurred with the change from BSM 22 and BM 25 to Prisma at the end of 1993, and further changes from Prisma to Aquarius at the end of 2001. The latest change was at the end of 2001 when the

Blood flow and filtration fraction

Barriers to Achieve Prescribed Treatment Dose
Adequate blood access and flow

FF
- 35 ml/kg/h Policy need higher blood flow = 300 ml/min.
- When applied at a patient of 50 kg = 1.750 l/h = FF < 10%.
- When applied at a patient of 150 kg = 5.250 l/h = FF = 29%.
- You need a fixed blood flow.

FR

$$FR = \frac{QUF}{QBF + \rho QUF - Q\delta P}$$

ρ = predilution ratio
$Q\delta P$ = weight loss

Fig. 3. Ideal blood flow for hemofiltration at 35 ml/kg/h.

Pre-determination of the exchange rate according the body weight

Pre-determination of the exchange rate according the body weight: in order to reduce the workload at the beginning of the CRRT, we did introduce full tables

Use of tables for 'Adaptation' to body weight

Weight	Dose T 35 ml/kg/h	Pre-dilution 1/3 dose T	Post-dilution 2/3 dose T
50	1,800	600	1,200
55	1,900	600	1,300
60	2,100	700	1,400
65	2,300	800	1,500
70	2,400	800	1,600
75	2,600	900	1,700
80	2,700	900	1,800
85	3,000	1,000	2,000
90	3,200	1,100	2,100
95	3,300	1,100	2,200
100	3,500	1,200	2,300
115	4,000	1,300	2,700
120	4,200	1,400	2,800

Fig. 4. Predetermined tables for pre- and post-dilution ratio at 35 ml/kg/h.

35-ml/kg/h rule was implemented in our ICU. Pulse-HVHF technology was also taken over in 2001. The next challenge will be the implementation of citrate in our unit [1, 7].

Monitoring														

Control of treatment dose and dose delivery
In practice: The St-Pierre way

Vignette D'identification	Surveillance hemofiltration continue Aquarius Edwards Clinique St Pierre Ottignies	Pressions de références Débit de sang : 300 ml/min Pression artérielle : −170 à − 190 Pression veinéuse : +200 à + 220 Pres. trans. men : +30 à − 400
n° du set utilisé :	Filtre placé : le .... à .... changé : le .... à	Date : ....

Thérapie : CWHF Nb J d' E.E.R :	Débits Sang .... ml/min Substitution totale .... ml/h 2/3 Post-dilution .... ml/h 1/3 Pré-dilution .... ml/h Déplétion Horaire .... ml/h	Anticoagulation Bolus OUI/NON Continu OUI/NON

Heures	Bicaflac/ physio	Anticoagulation			Data				Pressions			Fraction de filtration <30%	Perte de charge
		TCA	Dilution	Doses	Substitution	Modification déplétion	Déplet/3h	Déplet/tot	Art	Vein	PTM		

Fig. 5. Nursing monitoring data form.

These frequent and important changes require a MECT, on call sometimes just for technical problems. This team allows a very quick implementation of a new technique/machine inside the unit. Undoubtedly, this type of team makes possible the use of various approaches regarding CRRT in the unit including pulse-HVHF. Filter lifespan has also greatly increased using this MECT approach.

Also other similar approaches have occurred in the pediatric world this time using nursing emergency CRRT teams (NECT). In the PICU, Harvey et al. [5] have described the new role of a renal critical care educator based in a regional pediatric renal unit reaching out to three PICUs within the region. They demonstrated that, after 18 months, the training objectives for the PICU staff in all aspects of hemofiltration had progressed in all units using flexible training programs which were simultaneously under constant evaluation using questionnaires. This type of renal critical care educator also worked alongside staff whilst CVVH was being performed to further instill confidence. During the same time period, equipment and protocols have been largely standardized throughout the region, and an ongoing survey of CVVH use was initiated which could help to inform audit standards such as complication rates. It was concluded that the renal critical care educator was the catalyst for the formation of a regional hemofiltration group which shared in the development of guidelines and protocols and discussion of clinical data. Indeed, since CRRT is used infrequently in many PICUs the development of the renal critical care educator could serve as a model for the development and maintenance of skills in other regions [5]. This could be seen in some ways by the nurse emergency team for CRRT in ICU.

Conclusions and Perspectives

The implementation of guidelines, MECT, NECT, and educational processes is really the key issue for the future of CRRT in the ICU. The implementation of new guidelines with regard to the 35-ml/kg/h rule cannot be decided in one day. This will need a rigorous longstanding educational process for the nursing staff as well the medical team. A nursing group can be of help in order to achieved this. We will soon embark upon the use of citrate using the same implementation and educational processes.

First education processes and ongoing education are the keys to success in the use of CRRT. Local experts are needed in the nursing team, the medical ICU team and the nephrology team as well.

References

1 Renard D, Douny N, Verly Ch, Lourtie A-M, Honoré PM: Nursing monitoring of continuous haemofiltration; in Robert R, Honoré PM, Bastien O (eds): Extra-Corporeal Circulation in ICU. The French SRLF Collection Europe, Collection Director: Prof F Saulnier. Paris, Elsevier, 2006, pp 179–194.
2 Shetz M, Lauwers P, Ferdinande P, Van de Walle J: The use of continuous arteriovenous haemofiltration in intensive care medicine. Acta Anaesthesiol Belg 1984;35:67–78.
3 Clevenger K: Setting up a continuous venovenous hemofiltration educational program. A case study in program development. Crit Care Nurs Clin North Am 1998;10:235–244.
4 Giuliano KK, Pysznik EE: Renal replacement therapy in critical care: implementation of a unit-based continuous venovenous hemodialysis program. Crit Care Nurs 1998;18:40–51.
5 Harvey B, Watson AR, Jepson S: A renal critical care educator: the interface between paediatric intensive care and nephrology. Intensive Crit Care Nurs 2002;18:250–254.
6 East D, Jacoby K: The effect of a nursing staff education program on compliance with central line care policy in the cardiac intensive care unit. Pediatric Nurs 2005;31:182–184.

7 Honoré PM, Piette V, Galloy A-C, Almpanis C, Pelgrim J-P, Dugernier Th: High volume haemofiltration: general review; in Robert R, Honoré PM, Bastien O (eds): Extra-Corporeal Circulation in ICU. The French SRLF Collection Europe, Collection Director: Prof F Saulnier. Paris, Elsevier, 2006, pp 195–220.
8 Malbrain MNG, Chiumello D, Pelosi P, Wilmer A, Brienza N, Malcangi V, et al: Prevalence of intra-abdominal hypertension in critically ill patients: a multicentre epidemiological study. Intensive Care Med 2004;30–822–829.
9 Kingston D, Sykes S, Raper S: Protocol for the administration of haemofiltration fluids and using patients group electrolytes direction. Nurs Crit Care 2002;7:193–197.
10 Honoré PM, Joannes-Boyau O, Merson L, Boer W, Piette V, Galloy AC: The big bang of haemofiltration: the beginning of a new era in the third millennium for extra-corporeal blood purification! Int J Artif Organs 2006;29:649–659.
11 Rahman TM, Treacher D: Management of acute renal failure on the intensive care unit. Clin Med 2002;2:108–113.
12 Winkelman C: Haemofiltration: a new technique in critical care nursing. Heart Lung 1985;14: 265–271.
13 Honoré PM, Jamez J, Wauthier M, Dugernier Th: Prospective evaluation of short time high volume isovolemic haemofiltration on the haemodynamic course and outcome of patients with refractory septic shock. Crit Care Nephrol 1998;90:87–99.
14 Honoré PM, Jamez J, Wauthier M, Lee PA, Dugernier Th, Pirenne B, Hanique G, Matson JR: Prospective evaluation of short term high volume isovolemic haemofiltration on the haemodynamic course and outcome in patients with intractable circulatory failure resulting from septic shock. Crit Care Med 2000;28:3581–3587.
15 Troch R, Van Ypersele de Strihou C: Home haemofiltration/dialysis: experience in Belgium. Acta Clin Belg 1974;29:218–224.
16 Craig M: Continuous venous to venous haemofiltration. Implementing and maintaining a program: examples and alternatives. Crit Care Nurs Clin North Am 1998;10:219–233.
17 Hameleers P: 24 hour CVVH treatment on the intensive care unit: a reality. EDTNA ERCA J 1998;24:21–22.
18 Honoré PM, Joannes-Boyau O, Kotulak T, Boer W, Renard D, Verly C: Implementation of 35 ml/kg/h of CRRT dose in ICU: a combined medical and nursing approach. Blood Purif 2006;24:261–262.
19 Ronco C, Bellomo R, Homel P, et al: Effects of different doses in continuous veno-venous haemofiltration in outcomes of acute renal failure: a prospective randomised trial. Lancet 2000;356:26–30.

Patrick M. Honoré, MD
St-Pierre Para-University Hospital
Avenue Reine-Fabiola, 9
BE–1340 Ottignies-Louvain-la-Neuve (Belgium)
Tel. +32 10 437 346, Fax +32 10 437 123, E-Mail Pa.honore@clinique-saint-pierre.be

Continuous Hemodiafiltration with Cytokine-Adsorbing Hemofilter in the Treatment of Severe Sepsis and Septic Shock

Hiroyuki Hirasawa[a], Shigeto Oda[a], Kenichi Matsuda[b]

[a]Department of Emergency and Critical Care Medicine, Chiba University Graduate School of Medicine, Chiba, and [b]Department of Emergency and Critical Care Medicine, Yamanashi University School of Medicine, Yamanashi, Japan

Abstract

Continuous hemodiafiltration (CHDF) using a polymethymethacrylate (PMMA) membrane hemofilter (PMMA-CHDF) can effectively and continuously remove various cytokines from the circulating blood. PMMA-CHDF can decrease the blood levels of various cytokines when the blood levels of cytokines are high prior to the initiation of CHDF. The main mechanism of cytokine removal with PMMA-CHDF is the adsorption of cytokines to the hemofilter membrane and this characteristic was not observed in the other membrane material. PMMA-CHDF could improve blood pressure, the depressed monocytic HLA-DR expression, and recover the delayed neutrophil apoptosis in septic patients. Thus, cytokine removal with PMMA-CHDF would be effective for the treatment of severe sepsis and septic shock.

Copyright © 2007 S. Karger AG, Basel

It is widely accepted that cytokines play a pivotal role in the pathophysiology of severe sepsis and septic shock [1–3]. However, despite recent progress in understanding the pathophysiology of sepsis under the wide application of molecular biology, no anticytokine therapy has been effectively applied in clinical settings in the management of severe sepsis and septic shock [4]. On the other hand, there is ongoing controversy regarding the efficacy of cytokine removal in septic patients with various types of continuous blood purifications such as

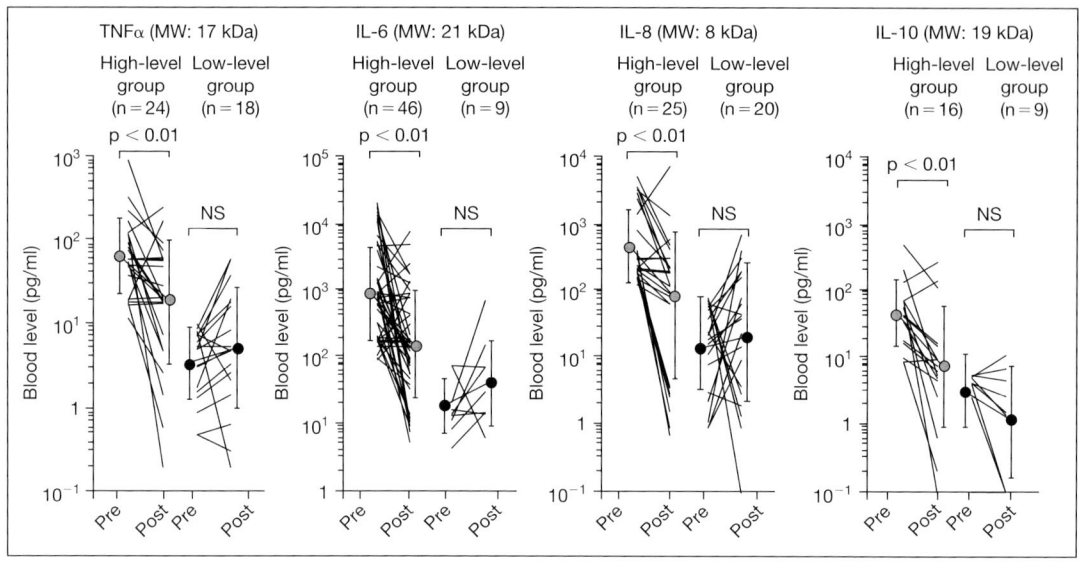

Fig. 1. Changes in blood levels of cytokines with 3 days of CHDF (mean ± SD).

continuous hemofiltration, continuous hemodialysis and continuous hemodiafiltration (CHDF) [5, 6]. Ronco et al. [7] proposed the 'peak concentration hypothesis' indicating that the removal of cytokines with continuous renal replacement therapy or continuous blood purification is effective to modulate the inflammatory response in sepsis syndrome. Some investigators indicated that hemofiltration is effective in the treatment of sepsis if it is performed with high-volume filtration [8, 9], especially as salvage therapy in patients with severe hyperdynamic septic shock [9].

We reported that CHDF using a polymethylmethacrylate (PMMA) membrane hemofilter (PMMA-CHDF) could effectively and continuously remove various cytokines from the circulating blood of a patient mainly through the adsorption of cytokines to the hemofilter membrane, and that PMMA-CHDF decreases the blood levels of not only proinflammatory cytokines but also anti-inflammatory cytokines when the blood levels of those cytokines are high prior to the initiation of CHDF [10, 11] (fig. 1). Furthermore, we also reported that such cytokine removal with PMMA-CHDF was effective for the treatment of hypercytokinemia-related pathophysiology [12] such as severe sepsis, septic shock [13], septic acute respiratory distress syndrome, septic multiple organ failure [14, 15], and severe acute pancreatitis [16].

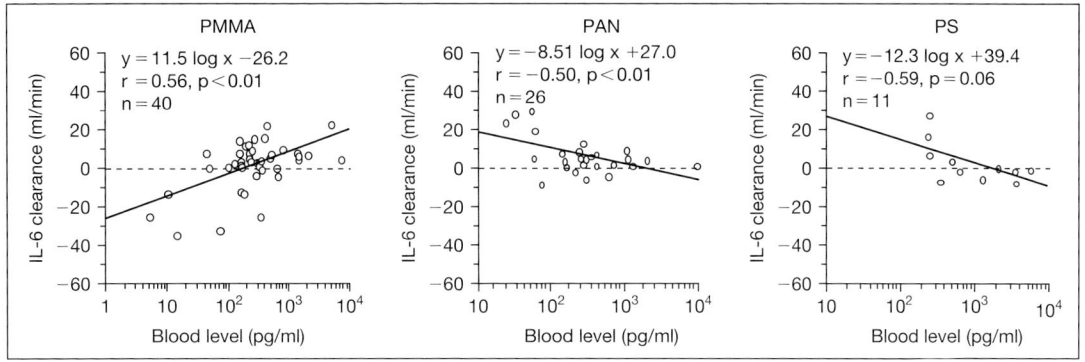

Fig. 2. Correlation between clearance and IL-6 blood level among various hemofilters SFI. Hemofilter = $1.0 \, m^2$; Q_B = 60 ml/min; Q_D = 500 ml/h; Q_F = 300 ml/h. PAN = Polyacrylonitrile; PMMA = polymethylmethacrylate; PS = polysulfone.

There are three different mechanisms for the removal of cytokines with CHDF, namely convection, diffusion and adsorption of cytokines to the hemofilter. As shown in figure 2, there is a significant positive correlation between the blood level of interleukin (IL)-6 and the clearance of IL-6 with PMMA-CHDF: the higher the blood level, the greater the clearance. These results indicate that the main mechanism of cytokine removal with PMMA-CHDF is through the adsorption of cytokines to the hemofilter. Furthermore, these features of PMMA-CHDF on cytokine clearance are very much clinically beneficial when we apply PMMA-CHDF as a cytokine remover because PMMA-CHDF is especially effective when a patient shows a severe degree of hypercytokinemia [10, 11, 13]. On the other hand, as shown in figure 2, there is no significant positive correlation or even a significant negative correlation between the IL-6 blood level and the clearance of IL-6 with CHDF when CHDF was performed with a hemofilter made from membrane materials other than PMMA, such as polyacrylonitrile (PAN) and polysulfone (PS). Thus the membrane material of the hemofilter is crucial when we apply CHDF as a cytokine modulator. We tried CHDF with various hemofilters made from many kinds of membrane materials and found that the PMMA hemofilter is superior to hemofilters made from other membrane materials when CHDF is applied as cytokine modulator [10, 11].

Since the main mechanism of cytokine removal with PMMA-CHDF is the adsorption of cytokine to the hemofilter membrane, we need not to apply troublesome high-volume CHDF [8, 9] to remove cytokines as proposed by others who use hemofilters made of materials other than PMMA. We also do not need large blood flows to perform PMMA-CHDF as a cytokine modulator. Actually we perform PMMA-CHDF with a filtration rate of 300–500 ml/h and dialysate rate of 500–1,000 ml/h and blood flow of 60–100 ml/h [10, 11].

In 2000, we introduced a rapid measurement system for the IL-6 blood level using an automated chemiluminescent enzyme immunoassay which can measure the IL-6 blood level within 30 min in the clinical laboratory [13]. Since then, we scientifically determine the indication and the timing of initiation and termination of PMMA-CHDF for cytokine removal [13]. Taking the results from previous studies in our laboratory [12, 14, 15], we now apply PMMA-CHDF when a septic patient shows IL-6 blood levels of >1,000 pg/ml [13]. We reported that PMMA-CHDF could improve blood pressure and could recover decreased urinary output within 2 h following the initiation [13]. One of the important aspects of severe sepsis and septic shock is immunoparalysis due to an excessive anti-inflammatory response caused by anti-inflammatory cytokines, such as IL-10. A decrease in monocytic HLA-DR expression has been implicated as an indicator of immunoparalysis. We found that PMMA-CHDF was able to remove IL-10 and that such a removal of anti-inflammatory cytokines resulted in an improvement in the depressed monocytic HLA-DR expression in septic patients. We also reported that delayed apoptosis of neutrophils, which is an important pathogenesis of organ failure in sepsis, is recovered through the removal of cytokines with PMMA-CHDF [17]. These are the mechanisms of the efficacy of PMMA-CHDF in the treatment of severe sepsis and septic shock.

Recently it was reported that cytokine-related gene polymorphism plays an important role in determining the degree of inflammatory response to an insult such as infection [18]. Therefore, we investigated the effect of cytokine-related genetic polymorphism on septic patients and reported that septic patients with a cytokine-related genetic polymorphism have more severe hypercytokinemia compared to septic patients without a cytokine-related genetic polymorphism [19–21]. More importantly we also reported that such cytokine-related genetic polymorphisms affected the efficacy of PMMA-CHDF on septic patients. In septic patients with a cytokine-related genetic polymorphism, PMMA-CHDF is less effective in controlling the cytokine storm and the survival rate is worse in those patients with genetic polymorphisms compared to those without cytokine-related genetic polymorphisms, even though patients in both groups received same therapeutic approaches including PMMA-CHDF [19, 20]. Taking these results, we recently started a prospective study to investigate the efficacy of personalized application of PMMA-CHDF based on the examination of cytokine-related genetic polymorphism in our intensive care unit population. We applied two PMMA-CHDF lines at the same time to a patient to enhance the removal rate of cytokine with PMMA-CHDF when the patient was diagnosed as having a cytokine-related genetic polymorphism. The results of this prospective study will hopefully be published in the near future.

References

1 Tetta C, Bellomo R, D'Intini V, et al: Do circulating cytokines really matter in sepsis? Kidney Int 2003;63(suppl):S69–S71.
2 Wang H, Czura CJ, Tracey KJ: Lipid unites disparate syndromes of sepsis. Nat Med 2004;10: 124–125.
3 Lotze MT, Tracey KJ: High-mobility group box 1 protein (HMGB1): nuclear weapon in the immune arsenal. Nat Rev Immunol 2005;5:331–342.
4 Kox WJ, Volk T, Kox SN, et al: Immunomodulatory therapies in sepsis. Intensive Care Med 2000;26:S124–S128.
5 Schetz M, Ferdinande P, van den Berghe G, et al: Removal of pro-inflammatory cytokines with renal replacement therapy: sense or nonsense? Intensive Care Med 1995;21:169–176.
6 Heering P, Grabensee B, Brause M: Cytokine removal in septic patients with continuous venovenous hemofiltration. Kidney Blood Press Res 2003;26:128–134.
7 Ronco C, Tetta C, Mariano F, et al: Interpreting the mechanisms of continuous renal replacement therapy in sepsis: the peak concentration hypothesis. Artif Organs 2003;27:792–801.
8 Cole L, Bellomo R, Journois D, et al: High-volume haemofiltration in human septic shock. Intensive Care Med 2001;27:978–986.
9 Cornejo R, Downey P, Castro R, et al: High-volume hemofiltration as salvage therapy in severe hyperdynamic septic shock. Intensive Care Med 2006;32:713–722.
10 Matsuda K, Hirasawa H, Oda S, et al: Current topics on cytokine removal technologies. Ther Apher 2001;5:306–314.
11 Nakada T, Hirasawa H, Oda S, et al: Blood purification for hypercytokinemia. Transfusion Apher Sci 2006;35:253–264.
12 Hirasawa H, Oda S, Matsuda K, et al: Clinical aspect of hypercytokinemia-induced pathophysiology in critical care; in Ogawa M, Yamamoto T, Hirota M (eds): The Biological Response to Planned and Unplanned Injuries: Cellular, Molecular and Genetic Aspects. Excerpta Medica. Int Congr Ser 1255. Amsterdam, Elsevier, 2003, pp 39–40.
13 Oda S, Hirasawa H, Shiga H, et al: Sequential measurement of IL-6 blood levels in patients with systemic inflammatory response syndrome/sepsis. Cytokine 2005;29:169–175.
14 Hirasawa H, Baue AE: Blood purification therapy to prevent or treat MOF; in Baue AE, Faist E, Fry DE (eds): Multiple Organ Failure: Pathophysiology, Prevention, and Therapy. New York, Springer, 2000, pp 501–504.
15 Hirasawa H, Oda S, Shiga H, et al: Endotoxin adsorption or hemodiafiltration in the treatment of multiple organ failure. Curr Opin Crit Care 2000;6:421–425.
16 Oda S, Hirasawa H, Shiga H, et al: Management of intra-abdominal hypertension in patients with severe acute pancreatitis with continuous hemodiafiltration using a polymethyl methacrylate membrane hemofilter (PMMA-CHDF). Ther Apher Dial 2005;9:355–361.
17 Hirano T, Hirasawa H, Oda S, et al: Modulation of polymorphonuclear leukocyte apoptosis in the critically ill by removal of cytokines with continuous hemodiafiltration. Blood Purif 2004;22: 188–197.
18 Ma P, Chen D, Pan J, et al: Genomic polymorphism within interleukin-1 family cytokines influences the outcome of septic patients. Crit Care Med 2002;30:1046–1050.
19 Watanabe E, Hirasawa H, Oda S, et al: Extremely high interleukin-6 blood levels and outcome in the critically ill are associated with tumor necrosis factor- and interleukin-1-related gene polymorphisms. Crit Care Med 2005;33:89–97.

20 Nakada K, Hirasawa H, Oda S, et al: Influence of toll-like receptor 4, CD14, tumor necrosis factor, and interleukin-10 gene polymorphism on clinical outcome in Japanese critically ill patients. J Surg Res 2005;129:322–328.
21 Watanabe E, Hirasawa H, Oda S, et al: Cytokine-related genotypic differences in peak interleukin-6 blood levels of patients with SIRS and septic complications. J Trauma 2005;59:1181–1190.

Dr. Shigeto Oda
1-8-1 Inohana, Chuo
Chiba City
Chiba 260–8677 (Japan)
Tel. +81 43 226 2341, Fax +81 43 226 2371, E-Mail odas@faculty.chiba-u.jp

Blood and Plasma Treatments: High-Volume Hemofiltration – A Global View

Patrick M. Honoré[a], Olivier Joannes-Boyau[b], Benjamin Gressens[a]

[a]St-Pierre Para-University Hospital, Ottignies-Louvain-la-Neuve, Belgium;
[b]Haut Leveque University Hospital, University of Bordeaux, Pessac, France

Abstract

From the recent past, hemofiltration, particularly high-volume hemofiltration, has rapidly evolved from a somewhat experimental treatment to a potentially effective 'adjunctive' therapy in severe septic shock and especially refractory or catecholamine-resistant hypodynamic septic shock. Nonetheless, this approach lacks prospective randomized studies (PRTs) evaluating the critical role of early hemofiltration in sepsis. An important milestone, which could be called the 'big bang' in terms of hemofiltration, was the publication of a PRT in patients with acute renal failure (ARF). Before this study, nobody believed that hemofiltration could change the survival rate in intensive care. Since that big bang, many physicians consider that hemofiltration at a certain dose can change the survival rate in intensive care. We now must try to define what the exact dose in septic ARF should be. As suggested by many studies this dose might well be higher than 35 ml/kg/h in the septic ARF group. The issue of the dosage of continuous high-volume hemofiltration must be tested in future randomized studies. Since the Vicenza study has shown that 35 ml/kg/h is the best dose in terms of survival when dealing with nonseptic ARF in the intensive care unit (ICU), several studies from different groups have shown that a higher dose might be correlated with better survival in septic ARF. This has also been shown in some way by the Vicenza group but not with a statistically significant value. New PRTs have just started in Europe such as the IVOIRE (hIgh VOlume in Intensive Care) study. The RENAL study is another large study looking more basically at dose in nonseptic ARF in Australasia. The ATN study in the USA is also testing the importance of dose in the treatment for ARF. Nevertheless, 'early goal-directed hemofiltration therapy' has to be studied in our critically ill patients. Regarding this issue, fewer studies, mainly retrospective, exist; but again the IVOIRE study will address this issue by studying septic patients with acute renal injury according to the RIFLE classification. This chapter will focus on the early application and adequate dose of continuous high-volume hemofiltration in septic shock in order to improve not only the hemodynamics but also survival in this very severely ill cohort of patients. This could be called the big bang of hemofiltration as one could have never anticipated that an adequate dose of hemofiltration could markedly influence the survival rate of septic ARF patients in the ICU. Apart from the use of an early and adequate dose of hemofiltration in sepsis, a higher dose could also provide a better renal recovery rate and reduce the risk of associate chronic dialysis in these patients. Furthermore, this presentation will also review brand-new papers regarding the use of hemofiltration in systemic inflammatory response syndrome and out-of-hospital cardiac arrest.

Copyright © 2007 S. Karger AG, Basel

Global Approach to the Problem in the ICU

Over the last two decades, hemofiltration and especially high-volume hemofiltration (HVHF) have rapidly evolved from a somewhat experimental treatment to a potentially effective 'adjunctive' therapy in severe septic shock and refractory or catecholamine-resistant (hypodynamic) septic shock (CRSS). Nevertheless, this approach lacks prospective randomized studies (PRTs) evaluating the critical role of early hemofiltration in sepsis.

An important step forward, which could be called the 'big bang' in terms of hemofiltration [1], was the publication of a PRT in patients with acute renal failure (ARF) [2]. Before this study [2] nobody believed that hemofiltration could change the survival rate in intensive care. Since that 'big bang', many physicians think that 'correctly dosed' hemofiltration has the potential to change the survival rate in intensive care. So the world of hemofiltration in the intensive care unit (ICU) is not a definite world; it is still in expansion. Right now, we have to try to define what will be the exact dose needed in septic ARF. This dose might well be higher than 35 ml/kg/h in the septic ARF group as suggested by many studies [2–5].

In order to challenge this hypothesis, a continuous dose rather than a pulsed dose of HVHF will be tested in ongoing or future randomized studies. As the Vicenza study [2] has shown that 35 ml/kg/h is the best dose in terms of survival, while dealing with nonseptic ARF in ICU, several studies from different groups have shown that, in septic ARF, a higher dose might correlate with better survival [3–5]. In some way this has also been shown by the study of Ronco et al. [2], but unfortunately not with statistically significance.

New PRTs have just started in Europe, such as the IVOIRE (hIgh VOlume in Intensive Care) study [6] and the RENAL study; another large study looking more basically at dose in nonseptic ARF is ongoing in Australasia [7], as well as the ATN study in the USA which also tests the importance of dose in the treatment of ARF. Nevertheless, early goal-directed hemofiltration therapy [8] has to be studied in critically ill patients. In this regard, fewer mainly retrospective studies do exist, but again the IVOIRE study [6] will address this issue by studying septic patients with acute renal injury according to the RIFLE classification [9] rather than ARF, which is already late in the illness process.

This review will focus on the early application and adequate dosing of continuous HVHF in septic shock in order to improve not only hemodynamics but also survival in this very severely ill cohort of patients.

This could be called the 'big bang of hemofiltration' as one could never have anticipated the fact that an adequate dose of hemofiltration could markedly influence the survival rate in the ICU while treating septic ARF patients.

As well as the use of early and adequate dose of hemofiltration in sepsis, a higher dose could also provide better renal recovery rate and reduce the risk of chronic dialysis dependency in these patients.

Big Bang of Hemofiltration and Hemofiltration-Derived Therapies

The 'big bang' of clinical hemofiltration occurred when the study of Ronco et al. [2] was published in the *Lancet* in 2000. This study comprised 425 patients and prospectively assessed three dosages, 20, 35 and 45 ml/kg/h, respectively. Exchange was exclusively realized in post-dilution in order to maximize convection. The membrane was also changed every 24 h. This study demonstrated that in nonseptic ARF a dose of 35 ml/kg/h was correlated with the best survival rate. This difference was statistically significant. This can be expressed as a big bang in terms of hemofiltration [10]. Before that study, nobody was really thinking realistically that hemofiltration could ever change survival in intensive care. Since that big bang, it is now believed by many that the use of a correctly dosed hemofiltration has the potential to change the survival rate in intensive care patients. Now the exact dose needed in septic ARF must be defined. This dose will probably be higher than 35 ml/kg/h in the septic ARF group, as has been suggested by many studies [4, 5, 11–16].

This has been put into perspective by an interesting recently published review paper [17] highlighting all the potentials of extracorporeal therapy with regard to the wide spectrum of molecules that could be removed, ranging in theory from 500 up to 900,000 kDa when using plasmafiltration for instance. Other outcome-dose studies also participate in the expansion of this initial big bang, among which the IVOIRE [6], RENAL [7] and ATN take the lead. The IVOIRE study [6] is trying to expand the findings of the initial study by Ronco et al. [2] to the septic ICU. It will include more than 480 patients with septic shock plus acute renal injury as defined by the RIFLE classification in ICU [9]. Allocation into the two arms will be determined by computerized randomization. One group will receive 35 vs. 70 ml/kg/h in the other group. This study will try to demonstrate that the higher dose (70 ml/kg/h) will further improve the survival rate of septic ARF patients in the ICU at 28, 60 and 90 days, respectively, after ICU admission.

Animal Models with Their Strengths and Pitfalls

Animal models have shown benefits in term of survival when an early and strong hemofiltration dose was applied in septic animal models. The early use of hemofiltration has been well applied in many animal models. The earlier experiments by Grootendorst et al. [18–21] most of the time used hemofiltration before or just after the injection of a bolus or even infusion of endotoxin.

Only in the late 1990s, with the studies of Rogiers et al. [22, 23], did the investigators start waiting for about 6–12 h before using HVHF, thereby allowing the animals to become extremely ill, hemodynamically unstable with early multiple organ dysfunction syndrome. In this way, the animal model was able to mimic the clinical situation in some way. Only animal models that were submitted to the early application of HVHF were shown to be very beneficial (and in some ways with very impressive results) mainly by the fact that in addition to early use, the investigators applied a much stronger dose of HVHF. By aggregating 12 studies over the last 10 years with regard to animal models, Honoré et al. [24] have shown that the mean dose used in those experiments was about 100 ml/kg/h whereas for humans (last 13 human studies or so) only 40 ml/kg/h was effectively given. The most beneficial effects of HVHF have been shown by these animal models, although the maximum delay between the septic insult and intervention was less than 12 h. This is totally different from the clinical situation in which the delay is rarely below 24 h and/or even below 48 h. The literature shows that in animal models not only a stronger application of dose was very important but also that early application was the second most important condition to make the use of hemofiltration in sepsis beneficial in terms of hemodynamics and survival [25] (table 1). It has also been advocated that the best response seen in the animal model was obtained when sepsis was intravascular as opposed to extravascular or when sepsis was restricted in another more confined compartment, for instance peritonitis. This might explain why the use of high-permeability hemofiltration (HPHF) in a sheep model of peritonitis, described by Rogiers [26], was not able to show any beneficial effect. If we believe the 'mediator delivery hypothesis' [27], we can see that the absence of large intakes of fluids in HPHF has major consequences: no increase in the lymphatic flow. This increased lymphatic flow is in charge of retrieving massive amounts of cytokines and mediators from the interstitium and the tissue level back to the blood compartment level, making them available for extra-renal removal.

It makes more sense to think that HVHF could be efficient in an acute model of peritonitis as demonstrated by Rogiers et al. [23], but HPHF remains ineffective in that specific setting of this particular study [27]. It is also known that in this kind of animal model the cytokine pattern is very different from the

Table 1. HVHF in animal studies: survival

Reference	Material	Membrane + surface	Hemofiltration technique	UF/h	Timing	Animal weight kg	LD %	UF/h indexed to body size ml/kg/h	Survival %	Survival Length	Effect
Freeman et al. [59]	Dog Peritonitis	PS S: NA	HVHF (CAVH)	600	60 min after peritoneal clot	10	100	60	142 (T) 14.2 (C)	7 days (T) 7 days (C)	N
Rogiers et al. [60]	Sheep Peritonitis Cecostomy	PS S: NA	VHVHF (CVVH)	NA000	4 h after surgery	NA	100	100	100 (T) 0 (C)	14 h (T) 14 h (C)	P
Yekebas et al. [30]	Pig pancreatitis + sepsis	CA-SMS S: 0.6 m²	VHVHF (CVVH)	NA	Immediately after induction of pancreatitis	NA	100	100 20	67 (T) 0 (C)	60 h (T) 60 h (C)	P
Lee et al. [61]	Pig S. aureus infusion	PS S: NA	VHVHF (CAVH)	1,000	Immediately after S. aureus infusion	7.5	100	133	38.5 (T) 0 (C)	7 days (T) 7 days (C)	P
Grootendorst et al. [21]	SMA clamping	C.P S: 2.3 m²	VHVHF (CVVH)	6,000	Before clamping	37	100	150	66 (T) 0 (C)	24 h (T) 24 h (C)	P

C = Control; CAVH = continuous ambulatory venous hemofiltration; CP = cuprophane; CVVH = continuous veno-venous hemofiltration; HVHF = high-volume hemofiltration; N = negative; NA = not available; P = positive; PS = polysulfone; S = surface; T = treated; VHVHF = very high-volume hemofiltration; Y = yes.

human situation because the duration of the proinflammatory phase is much longer in the animal setting and is not always followed by an secondary immunoparalysis phase. What the animal model has brought to light is the use of HVHF as a prophylactic measure in the second phase of sepsis, the so-called immunoparalysis phase or the compensatory anti-inflammatory response syndrome phase as explained by Bone [28]. Yekebas et al. [29–31] and Wang et al. [32] have worked to modernize this immunoparalysis post- systemic inflammatory response syndrome (SIRS). Therefore, they induced traumatic pancreatitis in healthy pigs. Hemofiltration started 12 h after pancreatic trauma but before sepsis and shock occurred. They waited about 12 h and, after this time, bacterial translocation occurred and very fulminant peritonitis and intravascular sepsis occurred inducing a shock state in these pigs. By comparing different settings of hemofiltration, especially low dose (20 ml/kg/h) plus adsorption and HVHF at the rate of 100 ml/kg/h, the authors were able to demonstrate that the prophylactic use of HVHF (100 ml/kg/h) was able to reduce the immunoparalysis level and the subsequent risk of secondary infection and, ultimately, the death rate. So, for the first time, HVHF was able to show that it could work not only in the proinflammatory phase but also in the secondary immunoparalysis phase per se as a prophylactic measure.

Transposition of these findings to the human setting is even more difficult because most of the animal models used a so-called hypodynamic septic shock [33]. It is only in the last few years that researchers are really challenging the so-called hyperdynamic septic shock concept, which is much closer to the human situation. Researchers like Sun et al. [34] or other groups [35] have nicely shown that this option is really feasible.

Human Studies and Translations or Transpositions from Animal Models

One of the greatest remaining problems with human studies (especially the mechanistic studies) is the fact that the number of patients is very limited resulting from the high cost of the technique. What is important in these human studies is to understand that HVHF applied at a continuous dose for 96 h can be compared in some ways to activated protein C (APC) [36, 37]. Obviously, we cannot rely on the same level of evidence as we can for APC. HVHF like APC can have a pleiotropic action on sepsis in the human setting. Indeed, it can interfere with the proinflammatory phase and, by decreasing the so-called proinflammatory phase, it can potentially reduce the unbound part of cytokines and reducing the corresponding remote organ-associated damages [38, 39].

The second point is that it can also alter and reduce some cardiovascular compounds (in the blood compartment) that are responsible for the shock state in the human situation. Indeed, endothelin-I can be removed and is held responsible for early pulmonary hypertension in sepsis, whereas endocannabinoids are responsible for the vasoplegia and myodepressant factor responsible for the cardiodepression seen in sepsis [40–42]. All these factors can be easily removed by HVHF.

Thirdly, HVHF can also alter the clotting system similar to the way it decreases platelet-activating inhibitor (PAI) factor 1, thus eventually reducing the level of diffuse intravascular coagulopathy [43]. It is well known that the level of PAI-1 is correlated with sepsis with a increased APACHE II score and a higher mortality rate [44].

Fourthly, it has been shown many times in animals that HVHF can reduce the risk of immunoparalysis after sepsis and the subsequent risk of nosocomial or secondary infection [29–32].

It has also been shown that HVHF can reduce the level of inflammatory cell apoptosis occurring during sepsis as it can extract caspase-3 products with a molecular weight of about 35,000 kDa, as well as some products of the caspase-8 pathways which are heavily involved in the setting of inflammatory cell apoptosis, especially in macrophages and neutrophils [45].

We know that clinical studies cannot reproduce the mean 100-ml/kg/h exchange that has been realized in animal models (only 40 ml/kg/h in human studies vs. 100 ml/kg/h) [24]. As a consequence, many anticipated effects seen in animal models can never been reproduced in human settings related to the use of inadequately low doses of HVHF.

What we do know is that there is huge variability between clinical trials concerning the range of doses applied. It can vary from 1 to 15 in terms of dose [24] when we aggregate all the recent studies.

If we decide to show that hemofiltration can be considered as a treatment in the ICU, it must be adapted to the body weight and it must also to be adapted to the severity of illness of the ICU patients. If we are dealing with nonseptic ARF, perhaps a lower dose will be optimal. On the contrary, if we are dealing with septic ARF than we might need a higher dose, close to 50 or 70 ml/kg/h. From the data presently available, we can say that in CRSS (or refractory hypodynamic septic shock), the use of pulse HVHF running at about 100 ml/kg/h during 4 consecutive hours (and then back at 35 ml/kg/h) is an important adjunctive treatment that can dramatically increase the survival (table 2) of these severely ill patients as compared with classical treatment [4, 5, 14, 15]. The monocentric study of Oudemans-van Straaten et al. [3], which was realized in a cohort of mainly cardiac surgery patients with oliguria at the time of inclusion, showed that the patient subgroup with the best improvement (in term of observed versus expected

Table 2. HVHF in human studies: survival

Reference	Patients n	Diagnosis severity	Design	CRRT technique	Mb + S m²	Ultrafiltrate ml/h	Observed effect	Survival	Timing	Weight kg	UF volume indexed to body ml/kg/h
Sander et al. [62], 1997	26	SIRS	R	13 CVVH 13 Co,st LV-CVVH LVHF	CA-SMS 0.6	1,000	No effect	NA	NA	NA 75 (E)	13 (E)
Matamis et al. [63], 1994	20	Sepsis MODS	P; UNC	LV-CVVH LVHF	PS 7	1,500	↗ MAP ↗ SVR	NA	NA	NA 75 (E)	20
Ronco et al. [2], 2000	425	USI-ARF	R	LV-CVVH MV-CVVH LVHF	PS 0.7–1.3	–1,500 –2,700 –3,400	ND	↗ survival with ↗ doses	NA	NA 75 (E)	20 35 45 (sepsis)
John et al. [64], 1968	30	Sepsis	R	20 CVVH 10 IHD	Mb + S =NA	NA	↗ SVR	NA	NA	NA	NA
Grootendorst et al. [65], 1996	26	Sepsis MODS	P; UNC	HV-CVVH HFHV	PA S: NA	4,500	↗ MAP CI?	↗ survival with no relation to hemodynamics	NA	NA	60 (E)
Oudemans-van straaten et al. [3], 1999	306	USI-ARF cohort	P HV-CVVH	Intermittent HFHV	Cell tri S: 1.9	5,000	NA	↗ survival	NA	NA 75 (E)	65 (E)

| Cole et al. [66], 2001 | 11 | Septic shock + MODS | R Cross | HV-CVVH HFHV | CA-SMS S: 1.6 | 6,000 | ↗ vaso | NA | NA | NA 75 | 80 (E) |
| Honoré et al. [4], 2000 | 20 | Refractory septic shock | P, inter, UNC | Short-term HV-CVVH VHVHF | PS S: 1.6 | 9,000 | ↗ MAP + SVR ↗ IC ↗ LVSWI | ↗ survival related to hemo-dynamics | 6.5 h (S) 14 h (NS) | 66.2 (S) 82.5 (NS) | 140 110 |

ARF = Acute renal failure; C = controlled; CA-SMS = copolymer acrylonitrile and sodium methal sulfate (AN 69); Cell Tri = cellulose triacetate; Cross = cross-over study; E = estimated; HVHF = high-volume hemofiltration; Inter = interventional; LVHF = low-volume hemofiltration; NA = not available; NS = non-survivors; P = prospective; PA = polyamide; PS = polysulfone; R = randomized; S = survivors; UNC = uncontrolled; Vaso = vasopressor; VHVHF = very high-volume hemofiltration; VLVHF = very low-volume hemofiltration.

mortality) was the septic subgroup of patients in this specific study. The technique used was intermittent hemofiltration at a dose of 60 ml/kg/h. Since the publication of several positive trials dealing with CRSS [4, 5, 14, 15], it is at present almost accepted for hemofiltration in the ICU that, when dealing with CRSS, a short-term procedure applying a very high dose should be the preferred procedure, whereas for classical hyperdynamic septic shock with acute renal injury, a continuous moderate high dose (during 96 h) might be the ideal choice in order to achieve a dose of 50–70 ml/kg/h (for 96 h). In this setting work on the pleiotropic background is needed, mainly on immunoparalysis post-septic insult, as shown especially by Yekebas [29–31] and Wang et al. [32].

Pulsed high volume has been shown to be still very effective in septic shock as recently outlined in the literature [46, 47]. Those studies confirmed the initial findings of Honoré et al. [4, 14, 15] and Joannes-Boyau et al. [5].

They also showed that the threshold dose needed to improve patients with septic ARF was about 45 ml/kg/h as suggested by Ronco et al. [2] and Piccinni et al. [47]. Recent studies have also shown that in CRSS, but this time hyperdynamic, HVHF might be a salvage therapy if a protocol-guided approach is used [48].

Recommendations for Clinical Practice, Future Research, HVHF Evolving Technology and Drug Adaptation during Hemofiltration and Hemofiltration-Derived Therapies

Regarding the recommendations for clinical practice: CRSS, either hypodynamic [4, 5, 14, 15] or hyperdynamic [48], could be seen as an indication (level V evidence and grade E recommendation) for experienced clinicians in the field of HVHF.

A patient with septic shock and ARF should receive a renal replacement dose of at least 35 ml/kg/h (level II evidence and grade C recommendation). Despite the numerous studies published and the ongoing IVOIRE study [6], there are no sufficient hard data yet to support a higher extended dose in this condition. This is true for other potential indications, such as septic shock with or without renal failure or injury or even sepsis and SIRS (with or without failure or injury). In the case of SIRS induced by out-of-hospital cardiac arrests [49], the existing data are too scarce to allow guidelines yet.

Regarding recommendations for future research concerning CRSS, it will be very difficult in this case to apply a PRT and we should stick to the available data or perform small bi-centric randomized studies.

Evaluating hyperdynamic septic shock patients, more numerous and larger PRTs are needed to detect potential interference with APC. Indeed, this potential interference deserves more attention as the molecular weight of APC (55,000 kDa) creates the theoretical possibility that the membrane during HVHF can adsorb the drug. Yet the risk is really minimal.

Nevertheless we have to think also about a possible synergy between HVHF and APC as shown by experimental work completed by Heylen et al. [50]. Regarding sepsis and SIRS, the aim is still to reduce immunoparalysis shown by various animal studies (level II evidence). For this type of patient, more mechanistic studies evaluating the potential risk of the technique are needed that must be balanced with the anticipated beneficial effects. In many cases, a short-term procedure will not be the ideal technique in these patients because to reduce immunoparalysis a long technique, for instance 96 h, will be required.

What is the best environment for future research?

First a safer and more efficient technique must be developed in order to increase the clinical operability, safer application and better clinical effectiveness.

Secondly a better understanding of the mechanisms of sepsis and SIRS is needed in order to identify the molecular as well the proteinomic targets for HVHF.

In order to meet the specific needs of control and restoration of immune homeostasis in sepsis and SIRS, better designed new HVHF technology must be ensured rather than modification of already existing technology.

Thirdly an appropriately designed (and with suitable power), trial of HVHF is necessary to test the clinical effectiveness of this therapy in patients with SIRS and/or sepsis. Lastly this trial should be conducted by indexing the dose to body size and paying more attention to controlling (and analyzing) the effects of time delay between the onset of SIRS or sepsis and HVHF initiation in order to better define the duration of treatment as well the appropriate initiation window.

Conclusions and Perspectives

The so-called big bang of hemofiltration has taught us that hemofiltration is not a definitive world. As demonstrated by Ronco et al. [2] in 2000 hemofiltration is still in expansion. As we are aware of this phenomenon concerning nonseptic ARF in ICU, we are now trying to define what should be the exact dose for continuous hemofiltration in septic ARF.

HVHF can still be seen as a potent immunomodulatory treatment in sepsis or SIRS. Since the mediator delivery hypothesis has been unravel [27], we know that not only is the extraction important but also the amount of fluids exchanged, and so the intake of fluids per se can increase the lymphatic flow dramatically up to 20- to 40-fold. As a consequence, circulatory cytokines are

Table 3. Renal recovery and dose of RRT

Reference	Study design	Patients n	Interventions	Recovery groups	Main outcome	Survival	Level of evidence
Jacka et al. [53], 2005	Retrospective	116	Renal recovery	CRRT = 87% IHD = 37.5%	Better recovery in CRRT	Not affected	IV
Manns et al. [54], 2003	Retrospective cohort	261	Cost of acute renal failure	CRRT = less dialysis dependency	Better renal recovery CRRT	Not affected	IV
Best III study[1] (2004)	Data collection epidemiological	1,260	Type of CRRT	CRRT >75% IHD = 60%	Better renal recovery CRRT	Not affected	III
Schiffl et al. [55], 2002	Prospective randomized	74 DHD 72 AHD	DHD AHD	DHD = 9 days renal recovery	Faster recovery of renal function	DHD = 72% AHD = 54%	II

[1]Best III study: Unpublished data 2004.

no longer valuable players [51], with the exception perhaps of very severe CRSS [52], and now what is important is the crucial relationship between immunological changes at the tissue level (where mediators do harm), hemodynamic modifications and survival. A last point is obviously the possible synergism in terms of therapy between APC and HVHF as both treatments have a lot of similarities in terms of pleiotropic effects, and recent research work has shown that synergy is possible between these two therapies. Nevertheless, many more studies are needed to precisely define what the exact role (and the exact impact on survival) of HVHF is, especially in hyperdynamic septic shock without acute renal injury.

Higher doses of treatment may also be important whatever the choice of the initial therapy, as it is able to influence the rate of secondary chronic dialysis dependency or conversely the rate of renal recovery [53, 54] as shown also by the work of Schiffl et al. [55] and Ronco et al. [56] (table 3).

In other words, the cost-effectiveness of continuous hemofiltration therapy when compared to intermittent techniques may be changing very quickly with time.

Recent publications have shown that we have to be very careful with regard to studies as the conclusions may not be fully supported by the data and therefore may be in some ways misleading [57, 58].

The expansion and the odyssey of the hemofiltration universe continues.

References

1 Lemaitre GH: The Big Bang Theory (Eddington's translation). London, Royal Astronomical Society, Monthly Notices, March 1931.
2 Ronco C, Bellomo R, Homel P, Brendolan A, Dan M, Piccini P, et al: Effects of different doses in continuous veno-venous haemofiltration. Lancet 2000;356:26–30.
3 Oudemans-van Straaten HM, Bosman RJ, Van der Spoel JL, Zanstra DF: Outcome of critically ill patients treated with intermittent high-volume haemofiltration: a prospective cohort analysis. Intensive Care Med 1999;25:814–821.
4 Honoré PM, Jamez J, Wauthier M, Lee P, Dugernier T, Pirenne B, Hanique G, Matson JR: Prospective evaluation of short-term, high volume isovolemic hemofiltration on the hemodynamic course and outcome in patients with intractable circulatory failure resulting from septic shock. Crit Care Med 2000;28:3581–3587.
5 Joannes-Boyau O, Rapaport S, Bazin R, Fleureau C, Janvier G: Impact of high volume hemofiltration on hemodynamic disturbance and outcome during septic shock. ASAIO J 2004;50: 102–109.
6 Honoré PM, Joannes-Boyau O: The IVOIRE (hIgh VOlume in Intensive caRE) study: impact of high volume haemofiltration in early septic shock with acute renal injury. A prospective multicentric randomized study. Presented for the Stoutenbeek Award of the 18th Annual Congress of ESICM Society, Berlin, 2004.
7 Bellomo R: http//clinical trials.gov/show/NCT221013
8 Rivers E, Nguyen B, Havstad S, Ressler J, Muzzin A, Knoblich B, Peterson E, Tomlanovitch M: Early goal-directed therapy in the treatment of severe sepsis and septic shock. N Engl J Med. 2001;345:1368–1377.
9 Bellomo R, Kellum JA, Mehta R, Palevsky PM, Ronco C: The acute dialysis II: The Vicenza Conference. Adv Ren Replace Ther 2002;9:290–293.
10 Lemaitre GH: La théorie cosmologique du 'Big Bang'. Cosmologie dynamique et théorie de l'atome primitif. Ann Soc Sci Bruxelles, 1927.
11 Bellomo R, Honoré PM, Matson JR, Ronco C, Winchester J: Extracorporeal blood treatment (EBT) methods in SIRS/Sepsis. Int J Artif Organs 2005;28:450–458.
12 Honoré PM, Joannes-Boyau O: High volume hemofiltration (HVHF) in sepsis: a comprehensive review of rationale, clinical applicability, potential indications and recommendations for future research. Int J Artif Organs 2004;27:1077–1082.
13 Honoré PM, Matson JR: Extracorporeal removal for sepsis: acting at the tissue level – the beginning of a new era for this treatment modality in septic shock. Crit Care Med 2004;32:896–897.
14 Honoré PM, Jamez J, Wauthier M, Dugernier T: Prospective evaluation of short-time high volume isovolemic hemofiltration on the haemodynamic course and outcome of patients with refractory septic shock. Crit Care Nephrol 1998;90:87–99.
15 Honoré PM, Jamez J, Wittebole X, Wauthier M: Influence of high volume haemofiltration on the haemodynamic course and outcome of patients with refractory septic shock. Retrospective study of 15 consecutives cases. Blood Purif 1997;15:135–136.
16 Klouche K, Cavadore P, Portales P, Clot J, Canaud B, Beraud JJ: Continuous veno-venous hemofiltration improves hemodynamic in septic shock with acute renal failure without modifying TNF-α and IL-6 plasma concentrations. J Nephrol 2002;15:150–157.
17 Matson JR, Zydney RL, Honoré PM: Blood filtration: new opportunities and the implications on system biology. Crit Care Resusc 2004;6:209–218.
18 Grootendorst AF, van Bommel AF, van der Hoeven B, van Leengoed LA, van Oosta AL: High volume haemofiltration improves right ventricular function in endotoxin induced shock in the pig. Intensive Care Med 1992;18:235–240.
19 Grootendorst AF, van Bommel AF, van der Hoeven B: High volume haemofiltration improves haemodynamics in endotoxin induced shock in the pigs. J Crit Care 1992;7:67–75.
20 Grootendorst AF, van Bommel AF, van Leengoed LA, van Zande AR, Huygens HJ, Groenefeld ABJ: Infusion of ultrafiltrate from endotoxemic pigs depressed myocardial performance in normal pigs. J Crit Care 1993;8:61–69.

21 Grootendorst AF, van Bommel AF, van Leengoed LA, Nabuurs M, Boumans CS, Groenefeld ABJ: High volume haemofiltration improves haemodynamics and survival of pigs exposed to gut ischemia reperfusion. Shock 1994;2:72–78.
22 Rogiers P, Zhang H, Smail N, Pauwels D, Vincent JL: Continuous veno-venous haemofiltration improves right cardiac performance by mechanisms other than tumor necrosis factor alpha attenuation during endotoxin shock. Crit Care Med 1999;27:1848–1855.
23 Rogiers P, Zhang H, Pauwels D, Vincent JL: Comparison of polyacrylonitrile (AN69) and polysulfone membrane doing haemofiltration in canine endotoxic shock. Crit Care Med 2003;31: 1219–1225.
24 Honoré PM, Zydney AL, Matson JR: High volume and high permeability haemofiltration in sepsis. The evidences and the key issues. Care Crit Ill 2003;3:69–76.
25 Matson JA, Lee PA: Evolving concepts of therapy for sepsis and septic shock and the use of hyperpermeable membranes. Curr Opin Crit Care 2000;6:431–436.
26 Rogiers P: Hemofiltration treatment for sepsis: is it time for controlled trials? Kidney Int Suppl 1999;72:99–103.
27 Di Carlo JV, Alexander SR: Hemofiltration for cytokine-driven illness: the mediator delivery hypothesis. Int J Artif Organs 2005;28:777–786.
28 Bone RC: Sir Isaac Newton, sepsis, SIRS and CARS. Crit Care Med 1996;24:1125–1128.
29 Yekebas EF, Treede H, Knoefel WT, Bloechle C, Fink E, Isbicki JR: Influence of zero balanced haemofiltration on the course of severe experimental pancreatitis in pigs. Ann Surg 1999;229: 514–522.
30 Yekebas EF, Eisenberger CF, Ohnnesorge H, Saalmuller, Elsmer HA, Engelhardt M, et al: Attenuation of sepsis-immunoparalysis with continuous veno-venous haemofiltration in experimental porcine pancreatitis. Crit Care Med 2001;29:1423–1430.
31 Yekebas EF, Strate T, Zolmajd S, Eisenberger CE, Erbesdobler A, Saalmuller A, et al: Impact of different modalities of continuous veno-venous hemofiltration on sepsis-induced alterations in experimental pancreatitis. Kidney Int 2002;62:1806–1818.
32 Wang H, Zhang ZH, Yan XW, Li WQ, Ji DX, Quan ZF, Gong DH, Li N, Li JS: Amelioration of haemodynamics an oxygen metabolism by continuous veno-venous hemofiltration in experimental pancreatitis. World J Gastroenterol 2005;11:127–131.
33 Piper RD, Cook DJ, Bone RC, Sibbald WJ: Introducing critical appraisal to studies of animal models investigating novel therapies in sepsis. Crit Care Med 1996;24:2059–2070.
34 Sun Q, Rogiers P, Pauwels D, Vincent JL: Comparison of continuous thermodilution and bolus cardiac output measurements in septic shock. Intensive Care Med 2002;28:1276–1280.
35 Natanson C, Hoffman WD, Suffredini AF, Eichacker PQ, Danner RL: Selected treatment strategies for septic shock based on proposed mechanisms of pathogenesis. Ann Intern Med 1994;120: 771–783.
36 Bernard GR, Vincent JL, Laterre PF, LaRosa SP, Dhainaut JF, Lopez-Rodriguez A, Steingrub JS, Garber GE, Helterbrand JD, Ely EW, Fischer CJ Jr; Recombinant Human Protein C Worldwide Evaluation in Severe Sepsis (PROWESS) Study Group: Efficacy and safety of recombinant human activated protein C for severe sepsis. N Engl J Med 2001;344:699–709.
37 Bernard GR, Margolis BD, Shanies HM, Ely EW, Wheeler AP, Levy H, Wong K, Wright TJ; for the US Investigators: Extended evaluation of recombinant human activated protein C United States Trial (ENHANCE US): a single-arm, phase 3, multicenter study of drotrecogin alfa (activated) in severe sepsis. Chest 2004;125:2206–2216.
38 Tetta C, Bellomo R, Kellum J, Ricci Z, Pohlmeiere R, Passlick-Deetjen J, Ronco C: High volume hemofiltration in critically ill patients: why, when and how?; in Ronco C, Bellomo R, Brendolan A (eds): Sepsis, Kidney and Multiorgan Dysfunction. Contrib Nephrol. Basel, Karger, 2004, vol 144, pp 362–375.
39 Brendolan A, D'Intini V, Ricci Z, Bonello M, Ratanarat R, Salvatori G, Bordoni V, De Cal M, Andrikos E, Ronco C: Pulse high volume hemofiltration. Int J Artif Organs 2004;27:398–403.
40 Bellomo R, Kellum JA, Gandhi CR, Pinsky MR, Ondulik B: The effect of intensive plasma water exchange by hemofiltration on hemodynamics and soluble mediators in canine endotoxemia. Am J Respir Crit Care Med 2000;161:1429–1436.

41 Kohro S, Imaizumi H, Yamakage M, Masuda Y, Namiki A, Asai Y: Reductions in levels of bacterial superantigens/cannabinoids by plasma exchange in a patient with severe toxic shock syndrome. Anaesth Intensive Care 2004;32:588–591.
42 Court O, Kumar A, Parrillo JE, Kumar A: Clinical review: myocardial depression in sepsis and septic shock. Crit Care 2002;6:500–508.
43 Garcia Fernandez N, Lavilla FJ, Rocha E, Purroy A: Haemostatic changes in systemic inflammatory response syndrome during continuous renal replacement therapy. J Nephrol 2000;13:282–289.
44 Zouaoui Boudjeltia KZ, Piagnerilli M, Brohée D, Guillaume M, Cauchie P, Vincent JL, Remacle C, Bouchaert Y, Vanhaverbeck M: Relationship between CRP and hypofibrinolysis: is this a possible mechanism to explain the association between CRP and outcome in critically ill patients. Thromb J 2004;2:7–12.
45 Bordoni V, Balgon I, Brendolan A, Crepaldi C, Gastoldon F, D'intini V, et al: Caspase 3 and 8 activation and cytokine removal with a novel cellulose tracetate super-permeable membrane in vitro sepsis model. Int J Artif Organs 2003;26:897–905.
46 Ratanarat R, Brendolan A, Piccinni P, Dan M, Salvatory G, Ricci Z, Ronco C: Pulse high-volume haemofiltration for treatment of severe sepsis: effects on hemodynamics and survival. Crit Care 2005;9:294–302.
47 Piccinni P, Dan M, Barbacini S, Carraro R, Lieta E, Marafon S, et al: Early isovolaemic haemofiltration in oliguric patients with septic shock. Intensive Care Med 2006;32:80–86.
48 Cornejo R, Downey P, Castro R, et al: High-volume hemofiltration as salvage therapy in severe hyperdynamic septic shock. Intensive Care Med 2006;42:713–722.
49 Laurent I, Adrie C, Vinsonneau C, et al: High-volume hemofiltration after out-of-hospital cardiac arrest: a prospective randomized study. J Am Coll Cardiol 2005;46:432–437.
50 Heylen A, Bervoets K, Smet M, Alexander JP, Rogiers P: Combination of Drotecogin α and early hemofiltration can improve outcome in severe septic patients with acute renal failure (abstract). Spring Meeting of the Belgium Society of ICM, June 2005.
51 Tetta C, Bellomo R, D'Intini V, et al: Do circulating cytokines really matter in Sepsis? Kidney Int 2003;63:69–71.
52 Honoré PM, Ernst Y, Jamez J, Lemaire M, Pirenne B, Lebaupin C, et al: VHVHF is accompanied by a significant drop in cytokines only in survivors during early refractory septic shock. Crit Care 2003;7:106–107.
53 Jacka MY, Ivancinova X, Gibney RT: Continuous renal replacement therapy improves renal recovery from acute renal failure. Can J Anesth 2005;52:327–332.
54 Manns B, Doig CJ, Lee H, Dean S, Tonelli M, Johnson D, Donalson C: Cost of acute renal failure requiring dialysis in the intensive care unit: clinical and resource implications of renal recovery. Crit Care Med 2003;31:644–646.
55 Schiffl H, Lang SM, Fischer R: Daily hemodialysis and the outcome of acute renal failure. N Engl J Med 2002;34:305–310.
56 Ronco C, Belomo R, Homel P, Brendolan A, Dan M, Piccinni P, La Greca G: Effects of different doses in continuous veno-venous haemofiltration on outcomes of acute renal failure: a prospective randomised trial. EDTNA ERCA J 2002;2:7–12.
57 Vinsonneau C, Camus C, Combes A, et al: Continuous venovenous haemofiltration versus intermittent haemodialysis for acute renal failure in patients with multiple-organ dysfunction syndrome: a multicentre randomised trial. Lancet 2006;368:379–345.
58 Honoré PM, Joannes-Boyau O: The 'French Hemodiafe Trial': this study is neither decisive nor definitive in resolving the controversy on renal replacement therapy in ICU. Int J Artif Organs 2006;29:1190–1192.
59 Freeman BD, Yatsiv I, Natanson C, Solomon MA, Quezado ZM, Danner RL, Banks SM, Hoffman WD: Continuous arteriovenous hemofiltration does not improve survival in a canine model of septic shock. J Am Coll Surg 1995;180:286–292.
60 Rogiers P, et al: 21st ISICEM, Brussels, 2001.
61 Lee PA, Matson JR, Pryor RW, Hinshaw LB: Continuous arteriovenous hemofiltration therapy for *Staphylococcus aureus*-induced septicemia in immature swine. Crit Care Med 1993;21:914–924.

62 Sander A, Armbruster W, Sander B, Daul AE, Lange R, Peters J: Hemofiltration increases IL-6 clearance in early systemic inflammatory response syndrome but does not alter IL-6 and TNF alpha plasma concentrations. Intensive Care Med 1997;23:878–884.
63 Matamis D, Tsagourias M, Koletsos K, Riggos D, Mavromatidis K, Sombolos K, Bursztein S: Influence of continuous haemofiltration-related hypothermia on haemodynamic variables and gas exchange in septic patients. Intensive Care Med 1994;20:431–436.
64 John et al: Abstract, Intensive Care Med 1998;S75.
65 Grootendorst AF, Bouman CS, Hoeben KH: The role of continuous replacement therapy in sepsis and multiorgan failure. Am J Kidney Dis 1996;28:50–57.
66 Cole L, Bellomo R, Journois D, Davenport P, Baldwin I, Tipping P: High-volume haemofiltration in human septic shock. Intensive Care Med 2001;27:978–986.

Patrick M. Honoré, MD
ST-Pierre Para-University Hospital
Avenue Reine Fabiola, 9
BE–1340 Ottignies-Louvain-La-Neuve (Belgium)
Tel. +32 10 437 346, Fax +32 10 437 123, E-Mail Pa.honore@clinique-saint-pierre.be

Blood and Plasma Treatments: The Rationale of High-Volume Hemofiltration

Patrick M. Honoré[a], *Olivier Joannes-Boyau*[b], *Benjamin Gressens*[a]

[a]St-Pierre Para-University Hospital, Ottignies-Louvain-La-Neuve, Belgium, and
[b]Haut Leveque University Hospital, University of Bordeaux, Pessac, France

Abstract

Since the early 1990s, experts in the field have thought that a reduction in cytokines in the blood compartment could, in theory, reduce mortality, but this is perhaps too naïve as the pharmacodynamics and pharmacokinetics of cytokines throughout the body are not well known and are probably much more complicated than previously thought. This ha now led to three leading theories and concepts. Ronco and Bellomo conceived the peak concentration hypothesis in which clinicians concentrate their efforts to remove mediators and cytokines from the blood compartment at the proinflammatory phase of sepsis. By reducing the amount of free cytokines, it is hoped that the level of remote organ (associated) damages can be dramatically decreased and, as a consequence, the overall death rate. In this regard, it is still not known what will happen at the interstitial and tissue level with regard to mediators and cytokines which are obviously the most important part in terms of consequences at the tissue level. In this setting, techniques that can more rapidly and substantially remove great amounts of cytokines or mediators are privileged. Among these, there is high-volume and very high-volume hemofiltration and a number of hybrid therapies encompassing high-permeability hemofiltration, super high-flux hemofiltration, hemo-adsorption or coupled filtration and adsorption and other types of adsorption using physical or chemical forces rather than driving forces as used normally in hemofiltration-derived techniques. The second concept is called the threshold immunomodulation hypothesis, also called the Honoré concept. In this concept the view of the system is much more dynamic. In experiments when removal is occurring on the blood compartment side, the level on the interstitial side and the tissue side is also changing and, because not only mediators but also pro-mediators are being removed, some pathways have really stopped when enough pro-mediators have been removed by this technique. At this point, the cascade is blocked and this point is called the threshold point. At this level, the cascade is lost and no further harm can be done to the tissue of the organism. Obviously, it is difficult to know when this point has been reached once high-volume hemofiltration is applied. But what is known, is that hemodynamics and survival can be improved in some patients as shown by various studies using high-volume hemofiltration without any significant drop in mediators inside the blood compartment itself. This effect is obtained without a dramatic fall in the plasma cytokine level because the cytokine or mediator levels should fall at the tissue level and not specifically at the blood compartment level. Nevertheless, the exact mechanism by which high-volume hemofiltration increases the flow of mediators and cytokines between the interstitial compartment and the blood compartment (and back to the blood side) is not known. Before the end of 2005, it was found that this missing step is perhaps well explained by the last theory and/or concept. The third theory and concept is called the mediator delivery hypothesis and has also been called the Alexander concept. In this theory, the use of high-volume hemofiltration and especially high intakes of incoming fluids (3–5 l/h) is able to increase the lymphatic flow 20- to 40-fold, even more so for mediators and cytokine lymphatic flow (drag). This has been demonstrated by several reports and is obviously extremely important. Perhaps this can explain why some very recent studies using high-permeability hemofiltration in sepsis have not been effective in improving hemodynamics and survival in septic acute animal models. In summary various brand new theories will be reviewed here in depth.

Copyright © 2007 S. Karger AG, Basel

Rationale Revisited Based upon Recent Human and Animal Model Publications

Since the early 1990s, it has been advocated that a reduction in cytokines in the blood compartment could, in theory, lead to a reduction in mortality [1, 2], but this is too naïve as the pharmacodynamics and pharmacokinetics of cytokines throughout the body are not well know and are probably much more complicated than previously thought.

This has led to three leading theories and concepts. The Ronco and Bellomo concept of the peak concentration hypothesis [3–5] (fig. 1), in which efforts are made to remove mediators and cytokines from the blood compartment at the proinflammatory phase of sepsis. It is hoped that by reducing the amount of free cytokines the level of remote organ (associated) damages can be decreased and as a consequence, the overall associated death rate.

In this regard, it is still not know what will happen at the interstitial and tissue level with regard to mediators and cytokines, which are obviously the most important part in terms of consequences at the tissue level.

In this setting, techniques that can more rapidly and more substantially remove great amounts of cytokines or mediators are privileged. Among these, a large place has been given to high-volume hemofiltration (HVHF) and very HVHF and quite a lot of hybrid therapies encompassing high-permeability hemofiltration (HPHF) [6], super high-flux hemofiltration (SHFHF) [7], hemoadsorption [8] or coupled filtration and adsorption (CPFA) [9] and any other type of adsorption using physical or chemical forces rather than driving forces as normally used in hemofiltration-derived techniques.

Also with regard to this issue, semantics is very crucial. Indeed, it can be argued that the term 'aDsorption' is probably not the right term because blood

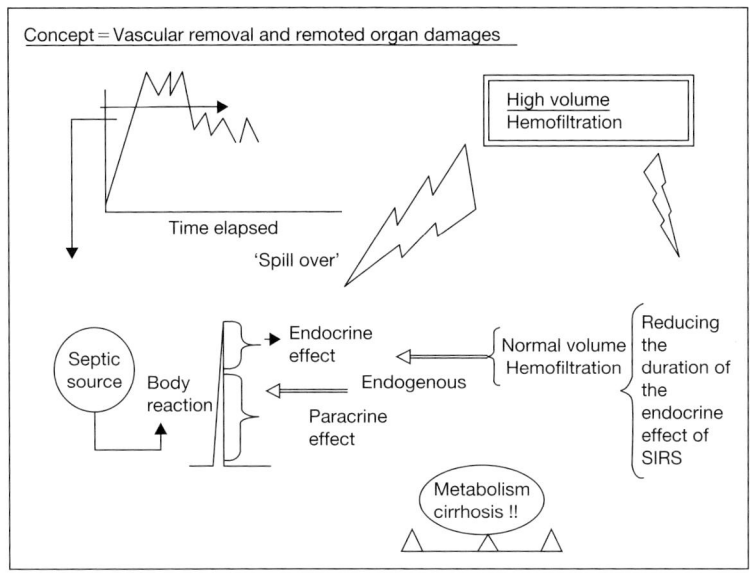

Fig. 1. Rationale for extracorporeal removal. The peak concentration hypothesis.

is not flooding through a semi-permeable membrane and it is not the net effect of convection forces plus oncotic forces that results in the passage of mediators through this kind of device.

In this type of device, it is more appropriated to use the term 'aBsorption' as chemical and physical forces are really engaged in that setting. So, we should be very careful about the use of appropriate terms when describing this kind of technique [10]. Indeed, membrane separation only occurs with 'aDsorption'.

The second concept is called the threshold immunomodulation hypothesis (fig. 2) and has been called the Honoré concept [11, 12]. In this concept the view of the system is much more dynamic. In some experiments when removal occurs on the blood compartment side, the level on the interstitial side (and also on the tissue side) also changes and, because not only mediators but also pro-mediators are being removed, some pathways are really stopped when enough pro-mediators have been removed by this technique. At this point, the cascade is blocked and when reached, is called the threshold point. At this level, the cascade is lost and no further harm can be done to the tissue of the organism. Obviously, it is difficult to know when this point is reached once HVHF is applied at the clinical level. But hemodynamics and survival can be improved in some patients as shown by various studies using HVHF without any significant

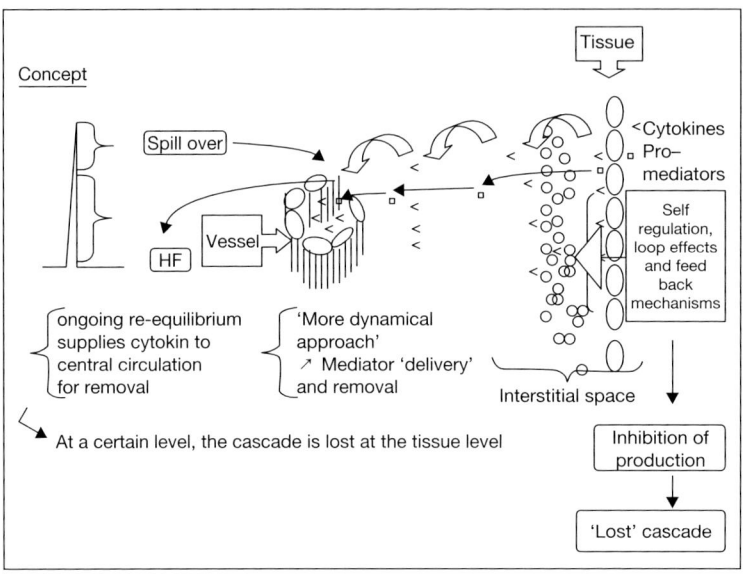

Fig. 2. Rationale for extracorporeal removal. The threshold modulation hypothesis.

drop in mediators in the blood compartment itself [13–15]. This effect is obtained without any dramatic fall in plasma cytokine level because where the cytokine or mediator level should fall is at the tissue level (where they do harm) and not specifically at the blood compartment level.

Nevertheless, the exact mechanism by which HVHF can increase the flow of mediators and cytokines between the interstitial compartment and the blood compartment (and back to the blood side) is not known. Before the end of 2005, it was not known if this missing step was perhaps well explained by the last theory and/or concept.

The third theory and concept is called the mediator delivery hypothesis (fig. 3) [16] and has been called the Alexander concept. In this theory, the use of HVHF and especially high intakes of incoming fluids (3–5 l/h) is able to increase the lymphatic flow by 20- to 40-fold, especially for mediator and cytokine lymphatic flow (drag) [17–19], and is obviously extremely important. Thus, the use of exchange fluid might be very important (and not only extraction) in order to increase the flow of lymphatic transport between the interstitial tissue and the blood compartment.

We can now understand why high-flow hemofiltration is able to dramatically increase the lymphatic transport from tissue and the interstitial space including cytokines and mediators back to the blood compartment in order to be potentially removed afterwards.

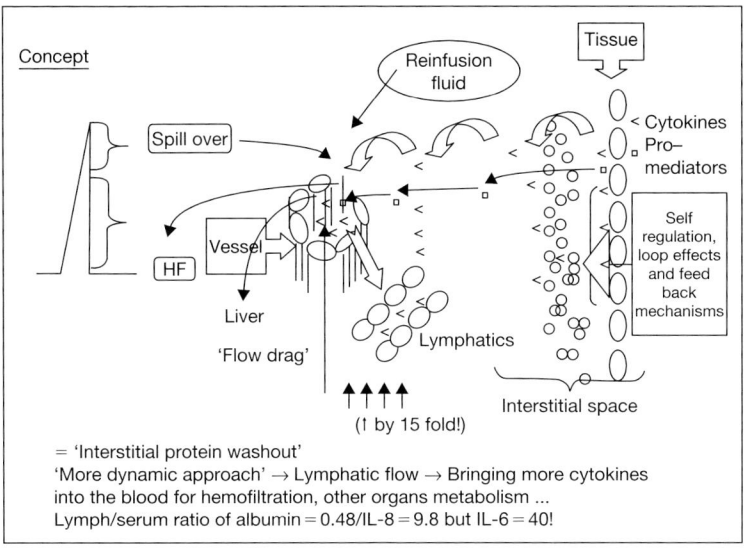

Fig. 3. Rationale for extracorporeal removal. The mediator delivery hypothesis.

In comparison, HPHF is able to remove larger amounts of mediators and cytokines in the blood compartment but is not able to increase lymphatic flow and, as a consequence, is not able to remove some very crucial cytokines and mediators at the interstitial and tissue level (where they do harm).

Therefore, this can explain why some very recent studies have shown that using HPHF in sepsis is not effective for improving hemodynamics and survival in septic acute animal models [20].

Therefore, clinicians should be aware of these new insights regarding the rationale for extracorporeal removal in severe septic shock in order to choose the best option with regard to adjunctive treatment for severe septic shock at the bedside.

Future of HPHF and Increased Filter Porosity

With regard to mediators and despite the increased complexity of the rationale, one should think that increasing filter porosity could be a good option [21]. Many mediators have a greater molecular weight and could be eliminated by using more sophisticated techniques as HPHF, SHFHF and hemo-adsorption. These techniques are able to remove more substantial amounts and perhaps more mediators, but the question remains about removal in the right compartment.

There is also the risk of losing many important nutrients, hormones, drugs, especially antibiotics, and many unknown metabolites.

Hybrid techniques have been attempted, taking advantage simultaneously of different techniques without having to support their drawbacks. CPFA and cascade hemofiltration (CCHF) [22] are able to retrieve large amounts and large molecules without the risk of losing important nutrients because part of the so-called purified blood is going back to the patient.

With regard to filter porosity, if we stick to hybrid techniques such as CCHF and CPFA and if a significant part of this so-called purified blood is going back to the patients, there would be no real theoretical limits as nothing important would be lost and only target molecules would be adsorbed.

Along these lines, it can be seen that a complete neglected domain is HVHF, and related techniques are searching for molecules below 45 kDa and plasma filtration looks for molecules around 900 kDa [23]. As a consequence, molecules between these two limits are really neglected and clinicians (and as well investigators) should pay much more attention to them. If we want to guarantee that all the purified blood will go back to the patient, the limits must be widened eluding to the potential risk of losing many important blood components.

Clinical Implications for the Intensivist in 2006

Regarding the use of extracorporeal treatment in sepsis as an adjunctive therapy in intensive care, it can be said that the 35-ml/kg/h rule should be applied widely in intensive care units (ICUs) as recent unpublished surveys have shown that less than 20% of the units (at least in continental Europe) are using them.

A recent position paper published by an Acute Dialysis Quality Initiative (ADQI) group has stressed that HVHF could be used in catecholamine-resistant septic shock (level V evidence and grade E recommendation) [24]. The same ADQI position paper also widely supported the extended use of the 35-ml/kg/h rule with level II evidence and grade C recommendation [24].

In classical hyperdynamic septic shock, especially ICU acute septic renal failure or ICU acute septic renal injury (according to the RIFLE classification), we are eagerly waiting the results of several outcome dose studies in the ICU.

Amongst the ongoing studies regarding the appropriate dose of hemofiltration in critically ill septic acute renal failure, the hIgh VOlume in Intensive caRE (IVOIRE) study must be mentioned [25]. This ongoing study will potentially give us very important insights for the future regarding the exact dose to use in subgroups of septic patients with acute renal injury. The IVOIRE study will include more than 480 patients with septic shock plus acute renal injury defined by the RIFLE classification in the ICU. Allocation into the two groups will be determined by computerized randomization. One group will receive 35 versus 70 ml/kg/h in the other group. The exact and complex methodology will be available soon [26].

This study will try to demonstrate that a higher dose (70 ml/kg/h) will further improve the survival rate of septic acute renal failure patients in the ICU. Ronco et al. [5] have already alluded to this with the 45-ml/kg/h subgroup, whereas the septic subpopulation had better survival although the nonseptic one did not improve further.

Conclusions and Perspectives

Clinicians should be aware of the new insights regarding the rationale for extracorporeal removal in severe septic shock in order to choose the best option for adjunctive treatment at the clinical level. The exchange volume is not only important for removal of mediators but also for displacement of mediators throughout the body [27–30]. Membrane porosity or system complexity can never replace systems that just use high-volume exchange rates with simple membrane separation technology.

As a final note, the world of hemofiltration and associated hybrid therapies is still evolving rapidly. Not only the investigator but also the clinician should be aware of the recent advances as several ongoing dose outcome studies may profoundly change daily practice. The expansion and odyssey of the hemofiltration universe continues.

References

1 Damas P, Canivet JL, de Groote D, Vrindts Y, Albert A, Franchimont P, et al: Sepsis and serum cytokines concentrations. Crit Care Med 1997;25:405–412.
2 Casey LC, Balk RA, Bone RC: Plasma cytokine and endotoxin levels correlate with survival in patients with the sepsis syndrome. Ann Intern 1993;119:771–778.
3 Ronco C, Tetta C, Mariano F, Wratten ML, Bonello M, Bellomo R: Interpreting the mechanism of continuous renal replacement therapy in sepsis. The peak concentration hypothesis. Artif Organs 2003;27:792–801.
4 Ronco C, Bellomo R: Acute renal failure and multiple organ dysfunction in the ICU: from renal replacement therapy (RRT) to multiple organ support therapy (MOST). Int J Artif Organs 2002;25:733–747.
5 Ronco C, Ricci Z, Bellomo R: Importance of increased ultrafiltration volume and impact on mortality: sepsis and cytokine story and the role for CVVH. EDTRA ERCA J 2002;2:13–18.
6 Lee PA, Weger G, Pryor RW, Matson JR: Effects of filter pore size on efficacy of continuous arterio-venous haemofiltration therapy for *Staphylococcus aureus*-induced septicaemia in immature swine. Crit Care Med 1998;26:730–737.

7   Lee WC, Uchino S, Fealy N, Baldwin I, Panagiotopoulos S, Goehl H, Morgera S, Neumayer HH, Bellomo R: Super high flux hemodialysis at high dialysate flows: an ex vivo assessment. Int J Artif Organs 2004;27:24–28.
8   Honoré PM, Matson JR: Hemofiltration, adsorption, sieving and the challenge of sepsis therapy design. Crit Care 2002;6:394–396.
9   Bellomo R, Tetta C, Ronco C: Coupled plasma filtration adsorption. Intensive Care Med 2003;29: 1222–1228.
10 Bellomo R, Honoré PM, Matson JR, Ronco C, Winchester J: Extracorporeal blood treatment (EBT) methods in SIRS/Sepsis. Consensus statement. ADQI III Conference. Electronic Supplement Material.www.adqi.net (2005).
11 Honoré PM, Joannes-Boyau O: High volume hemofiltration (HVHF) in sepsis: a comprehensive review of rationale, clinical applicability, potential indications and recommendations for future research. Int J Artif Organs 2004;27:1077–1082.
12 Honoré PM, Matson JR: Extracorporeal removal for sepsis: acting at the tissue level – the beginning of a new era for this treatment modality in septic shock. Crit Care Med 2004;32:896–897.
13 Honoré PM, Jamez J, Wauthier M, Dugernier T: Prospective evaluation of short-time high volume isovolemic hemofiltration on the hemodynamic course and outcome of patients with refractory septic shock. Crit Care Nephrol 1998;90:87–99.
14 Honoré PM, Jamez J, Wittebole X, Wauthier M: Influence of high volume haemofiltration on the haemodynamic course and outcome of patients with refractory septic shock. Retrospective study of 15 consecutives cases. Blood Purif 1997;15:135–136.
15 Klouche K, Cavadore P, Portales P, Clot J, Canaud B, Beraud JJ: Continuous veno-venous hemofiltration improves hemodynamic in septic shock with acute renal failure without modifying TNF-α and IL-6 plasma concentrations. J Nephrol 2002;15:150–157.
16 Di Carlo JV, Alexander SR: Hemofiltration for cytokine-driven illness: the mediator delivery hypothesis. Int J Artif Organs 2005;28:777–786.
17 Olszewski WL: The lymphatic system in body homeostasis: physiological condition lymph fat rest. Lymphat Res Biol 2003;1:11–21.
18 Onarherim H, Missavage E, Gunther RA, Kramer GC, Reed RK, Laurent TC: Marked increase of plasma hyaluronan after major thermal injury and infusion therapy J Surg Res 1991;50:259–265.
19 Wasserman K, Mayerson HS: Dynamics of lymph and plasma protein and exchange. Cardiologia 1952;21:296–307.
20 Rogiers P: High volume haemofiltration: high volume, high permeability: which target (abstract). 4th ERTIC Meeting, Nice, 2005.
21 Honoré PM, Zydney AL, Matson JR: High volume and high permeability haemofiltration in sepsis. The evidence and the key issues. Care Crit Ill 2003;3:69–76.
22 Valbonesi M, Carlier P, Icone A, Accorsi P, Borberg H, Schreiner T, et al: Cascade filtration: a new filter for secondary filtration – a multicentric study. Int J Artif Organs 2004;27:513–515.
23 Matson JR, Zydney RL, Honoré PM: Blood filtration: new opportunities and the implications on system biology. Crit Care Resusc 2004;6:209–218.
24 Bellomo R, Honoré PM, Matson JR, Ronco C, Winchester J: Extracorporeal blood treatment (EBT) methods in SIRS/sepsis. Int J Artif Organs 2005;28:450–458.
25 Honoré PM, Joannes-Boyau O: The IVOIRE study: impact of high volume haemofiltration in early septic shock with acute renal injury: a prospective multicentric randomized study. Presented for the Stoutenbeek Award, 18th Ann Congr ESICM Soc, Berlin, 2004.
26 Joannes-Boyau O, Honoré PM, Boer W, et al: The IVOIRE Study. Description of the methodology and the design used. Submitted to Crit Care 2007.
27 Joannes-Boyau O, Honoré PM, Boer W: Hemofiltration: the case for removal of sepsis mediators from where they do harm. Crit Care Med 2006;34:2244–2246.
28 Honoré PM, Jamez J, Wauthier M, Lee PA, Dugernier Th, Pirenne B, Hanique G, Matson JR: Prospective evaluation of short-term high-volume isovolemic hemofiltration on the hemodynamic course and outcome in patients with intractable circulatory failure resulting from septic shock. Crit Care Med 2000;28:3581–3587.

29 Honoré PM, Joannes-Boyau O, Meurson L, Boer W, Piette V, Galloy AC, Janvier G: The big bang of hemofiltration: the beginning of a new era in the third millennium for extra-corporeal blood purification! Int J Artif Organs 2006;29:649–659.
30 Honoré PM, Joannes-Boyau O: The 'French Hemodiafe Trial': this study is neither decisive nor definitive in resolving the controversy on renal replacement therapy in ICU. Int J Artif Organs 2006;29:1190–1192.

Patrick M. Honoré, MD
St-Pierre Para-University Hospital
Avenue Reine Fabiola, 9
BE–1340 Ottignies-Louvain-La-Neuve (Belgium)
Tel. +32 10 437 346, Fax +32 10 437 123, E-Mail Pa.honore@clinique-saint-pierre.be

Liver Support Systems

Antonio Santoro[a], *Elena Mancini*[a], *Emiliana Ferramosca*[a], *Stefano Faenza*[b]

[a]Unità Operativa di Nefrologia, Dialisi e Ipertensione, Dipartimento di Medicina Interna, Scienze Nefrologiche ed Invecchiamento, [b]Dipartimento di Discipline Chirurgiche, Rianimatorie e dei Trapianti, Policlinico Sant'Orsola-Malpighi, Bologna, Italia

Abstract

Liver insufficiency is a dramatic syndrome with multiple organ involvement. A multiplicity of toxic substances (hydrophilic like ammonia and lipophilic like bilirubin or bile acids or mercaptans) are released into the systemic circulation, thus altering many enzymatic cellular processes. Patients frequently die while on the transplantation waiting list because of organ scarcity. Systems supporting liver function may be useful to avoid further complications due to the typical toxic state, 'bridging' the patients to the transplantation, or, in the event of an acute decompensation of a chronic liver disease, sustain liver function long enough to permit the organ's regeneration and functional recovery. An ideal liver support system should substitute the main functions of the liver (detoxification, synthesis and regulation). Extracorporeal systems now available may be totally artificial or bioartificial. While the first are only able to perform detoxification, the second may add the functions of synthesis (plasma proteins, coagulation factors) and regulation (neurotransmitters). Bioartificial liver working with isolated hepatocytes and a synthetic membrane in an extracorporeal system are however still far from being ready for clinical use. At present, liver insufficiency may be treated with an extracorporeal support technology aimed either at detoxification alone or at a real purification. Charcoal hemoperfusion or exchange/absorption resins may be used for blood detoxification. Blood or plasma exchange, from a theoretical point of view, could be suitable for a polyvalent intoxication, such as liver failure; however, the multicompartmental distribution of some solutes largely endangers the efficacy of these procedures. Selective plasmapheresis techniques are now available for some solutes (e.g. styrene for bilirubin) and may progressively reduce the plasma levels and presumably the deposits of the solute. Novel treatments introduced to improve detoxification, mainly of the protein-bound substances, are the molecular adsorbent recirculation system (MARS) and Prometheus™ systems. MARS performs an albumin dialysis, where albumin is the exogenous carrier for the toxic substances, and different experiences have proved its efficacy mainly in the treatment of hepatic encephalopathy, while data on survival are still limited to small case series. With Prometheus, the most recent system developed for a wide detoxification, albumin-bound toxins are directly removed in two separate cartridges with different solute affinity, without the need for exogenous albumin; plasmadsorption is then coupled with a real dialysis process. After promising initial results, the efficacy of Prometheus in the patients' hard endpoints will be evaluated in a large international trial. On the whole, liver support systems may offer, in many cases, a survival benefit. Stem cells are however, even in this filed, the real great hope for the future of patients with end-stage liver disease.

Copyright © 2007 S. Karger AG, Basel

In presence of liver insufficiency a multiplicity of toxic substances, both lipophilic and hydrophilic, are released into the systemic circulation, thus altering many enzymatic cellular processes. The lipophilic substances interfere with structural cellular processes, such as the reconstruction of cellular membranes, while the hydrophilic ones alter and block functional processes both enzymatic and nonenzymatic in nature. Still, the different forms of liver failure (acute, subacute, acute-on-chronic, chronic) are characterized by a high patient mortality [1].

The only proven successfully therapeutic solution for such an extremely critical situation is orthotopic liver transplantation. Unfortunately, organ scarcity cannot fulfill all the needs for transplantation arising during acute and chronic liver diseases. Furthermore, as in the case of decompensation of a chronic liver disease, it is necessary to support the liver function long enough to permit the regeneration and functional recovery of the organ.

An ideal liver support system has to support or substitute the main functions of the liver, providing detoxification, synthesis and regulation. For detoxification (e.g. removal of bilirubin, bile acids, and toxins), artificial detoxification systems have been shown to reach varying degrees of efficiency. The complex tasks of regulation (e.g. central nervous system transmitter precursors) and synthesis (e.g. coagulation factors) remain to be addressed by the use of live hepatocytes in bioartificial livers. Yet in this field, the current research is still quite far from accomplishing the ideal bioartificial liver (BAL) for clinical use, even though several studies have been performed on limited patient series.

In this review, we describe the most promising extracorporeal artificial systems for liver support, while we do not take into consideration the biological system such as BAL and even more the stem cells that are the great hope for the future of patients with liver disease (table 1). The initial concept of BAL was to simply fill a hemodialyzer with cells. BAL basically integrates isolated hepatocytes which come into contact with patient's blood or plasma with the interposition of a synthetic membrane. Excellent reviews on the history of BAL, and the various systems so far used in experimental models and clinical trials are available in the literature [2–4].

Table 1. Systems for extracorporeal liver support

Blood detoxification
- Sorbent-based therapy (charcoal, exchange or adsorption resins)

Blood purification techniques
- Plasma exchange
- Selective plasma adsorption (cartridge of styrene)
- High-performance hemodiafiltration/hemofiltration

Provision of whole liver function
- BAL
    Porcine hepatocytes
    Human hepatocytes
- Extracorporeal hepatic perfusion

Table 2. Therapeutic beneficial effects of artificial support systems

- Bridge to recovery from liver failure before liver transplantation or during the course of severe acute or acute-on-chronic liver diseases
- Physiological support of liver function after delayed graft function
- Physiological support after liver surgery (for trauma or neoplasia)

A recent consideration focuses on stem cells beyond primary hepatocytes and cell lines. Successful generation of hepatocyte-like cells was obtained from human embryonic stem cells [2]. Despite the richness of publications, especially in the USA, related to the huge potential of embryonic stem cells to generate different types of tissues, there is a scarcity of research focused on liver. Recently, the European Parliament banned the use of embryonic stem cells for research purposes in Europe [5].

Artificial Liver Support Systems

Clinical indications for an extracorporeal liver support have been steadily evolving since the beginning of orthotopic liver transplantation in an increasing number of centers. Table 2 provides a schematic summary of the most frequent clinical situations which may benefit from the use of extracoporeal systems for liver support.

Table 3. Biological actions of bilirubin and biliary salts

- Inhibit hydrolytic enzymes, dehydrogenases, and enzymes involved in the electron transport
- Act as an uncoupler of oxidative phosphorylation
- Decrease the Na-K-adenosine triphosphatase activity
- Inhibit the tyrosine uptake
- Reduce the gluthatione-8-transferase activity of ligandine
- Inhibit protein kinase C

Conventional Systems

In 1956 Kiley et al. [3] reported the use of hemodialysis in the treatment of 5 patients with hepatic encephalopathy: in 4 patients the state of consciousness improved, but with no improvement in long-term survival.

During the last 20 years, there has been a movement from hemodialysis to hemodiafiltration due to the availability of highly permeable membranes capable of removing molecules with a molecular weight of over 15,000 Da. Currently, hemofiltration has proven particularly useful in resolving the intracranial/intracerebral hypertension which accompanies hepatic encephalopathy [6].

Hemoperfusion on charcoal was introduced for the therapy of liver failure by Yatzidis [4] in 1964, and since then has received a broad consensus, even if the rather few controlled studies have failed to demonstrate a significant clinical advantage as compared with conventional medical therapies. Combined hemoperfusion, charcoal plus resins [7], could theoretically offer advantages even if no controlled prospective studies have been performed yet.

Total blood exchange is definitely, at least on a theoretical level, a treatment suitable for a polyvalent intoxication, such as hepatic failure. Currently, since many toxic substances generated during liver failure, bilirubin first and foremost, are distributed into several pools, such as the interstitial and the cellular compartments beyond plasma, one treatment cannot provide a complete therapeutic solution.

Novel Treatments

Bilirubin and the biliary salts are toxic in many respects as they can induce a series of chain-linked toxic effects leading to multiorgan dysfunctions (table 3). Treatments that reduce bilirubin levels and thus the toxicity of hyperbilirubinemia may be relevant. Modern technologies now exist and include selective cascade

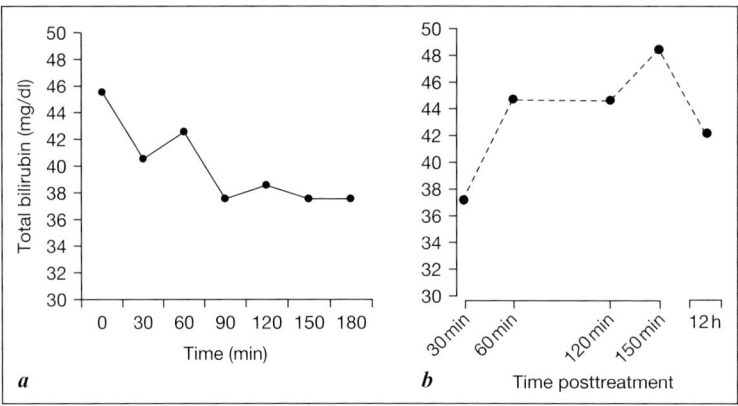

Fig. 1. Bilirubin behavior (cascade plasmapheresis). Progressive reduction throughout the treatment (*a*) and posttreatment rebound (*b*).

plasmapheresis, the molecular adsorbent recirculating system (MARS) and the Prometheus™ system.

Although plasma exchange was introduced in the 1960s after some positive results [8], it has never attained popularity due to the risk of viral infections, high costs and depression of specific immunity. Recently, the use of high exchange rate plasmapheresis has provided promising results, above all in cases of drug-induced liver intoxication. In order to overcome these disadvantages some variations have been suggested, as in the cascade plasmapheresis system [9].

Whatever the treatment, bilirubin reduction occurs according to an exponential pattern with a rapid decline during the first 30 min followed by a slower decline at different velocities depending on the initial bilirubin value and with patient-to-patient pattern variability (fig. 1). There is a rebound phase in bilirubin plasmatic values following each treatment session, appearing during the first 30 min after treatment end. Rebound velocities in the immediate posttreatment phase are extremely high, while the rise of bilirubin values during the interval between the two treatments is much slower. The initial rebound surge is the expression of bilirubin release from the tissue-interstitial and cellular pools towards the vascular one in order to reequilibrate the two sectors. As mentioned for bilirubin reduction, also the rebound, at the same time intervals after treatment end, has a patient-to-patient pattern variability. This variability is probably correlated with the pattern of bilirubin pool and its dynamics of accumulation. It is known that the cellular pool is much slower than the interstitial one. Furthermore, the transfer constraints of the extravasal pool, with a high case to case variability, also depend on the accumulation and production pattern.

MARS is a purification system aimed at removing albumin-bound toxic molecules. The principle is that dialysis membranes are impermeable to albumin but able to clear toxic substances bound to albumin, such as endogenous benzodiazepines, mercaptanes, and biliary acids when an albumin-rich dialysate is used. The dialysate also contains electrolytes and bicarbonate as buffer, and is regenerated by passage through an anionic-exchange resin charcoal absorption and sequential hemodialysis.

Worldwide, various nonrandomized trials and clinical applications in more than twenty clinical indications [10] have demonstrated some efficacy of the system in the treatment of hepatic encephalopathy. A recently published prospective, controlled and randomized trial, but on small numbers (8 patients in the study and 5 in the group treated with conventional standard medical therapy alone), demonstrated a greater survival for the study group (25 days as compared with 4 in the control group), an enhanced prothrombin activity and lower levels of bilirubin [11]. Other randomized controlled trials are, however, needed in order to prove the efficacy of this liver support in increasing survival over 3 months of the liver failure patients [12].

In 2001 the first Prometheus treatment was performed in a patient. This fractionated plasma separation, adsorption and dialysis system uses dialysis and adsorption for detoxification of water-soluble and albumin-bound toxins like other extracorporeal liver support systems, too. However, in contrast to precursor systems like MARS and fractionated plasma separation and adsorption system (FPSA) [13, 14] Prometheus [15] separates both procedures (fig. 2).

First of all, blood flows through a special albumin-permeable filter AlbuFlow™ with a membrane cutoff close to 300,000 Da. Here the patient's *own* albumin is separated from the blood, while cells and higher molecular weight solutes like fibrinogen are retained in the blood. The albumin is then perfused through the prometh™ 01 and prometh™ 02 adsorber, whereby the bound toxins are captured by direct contact with the high-affinity adsorbing material. The purified albumin then reenters the bloodstream. Afterwards the blood passes through a high-flux dialyzer.

Due to this two-stage procedure the efficiency is considerably increased. Albumin-bound toxins are directly transported, by convection, to the site where removal takes place. In this way, no dissociation from albumin and diffusion is necessary, known to be the limiting factors in removal kinetics when high-flux membranes are used for mass transfer from blood. In addition, there is no need for exogenous albumin.

Afterwards the blood undergoes high-flux hemodialysis, one of the most effective treatments for removal of water-soluble toxins. The use of Fresenius Polysulfone™ membranes and ultrapure dialysis fluid prepared with the 4008H dialysis machine ensures maximum biocompatibility. Maintenance and monitoring

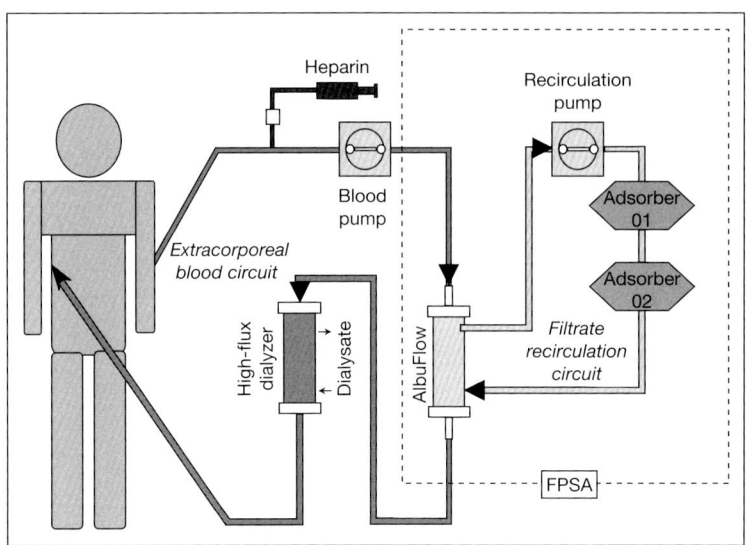

Fig. 2. Scheme of Prometheus system circuit. At first blood flows through a special albumin-permeable filter AlbuFlow with a membrane cutoff close to 300,000 Da. Here the patient's *own* albumin is separated from the blood, while cells and higher molecular weight solutes like fibrinogen are retained in the blood. The albumin is then perfused through the prometh 01 and prometh 02 adsorber. The prometh 01 adsorber cartridge contains a highly porous neutral resin for the removal of albumin-bound toxins such as bile acids, aromatic amino acids, and phenolic substances while the prometh 02 adsorber cartridge contains a high-performance anion exchange resin. The purified albumin then reenters the bloodstream. Afterwards the blood passes through a high-flux dialyzer.

of the extracorporeal circuit is performed by the Prometheus unit. It is a comprehensive 'all-in-one' extracorporeal liver support system based on the Fresenius 4008H dialysis unit with FPSA and hemodialysis integrated in one unit. Actually the Prometheus unit may not only be used for liver support therapy but also for conventional dialysis alone [16].

Clinical safety and adsorber efficiency as expressed by laboratory findings were confirmed in 11 patients treated with Prometheus on 2 consecutive days [14]. Total bilirubin, bile acids and ammonia were reduced during 5-hour treatments by 21, 43 and 40%, respectively. This was confirmed in a study with 14 patients with acute or acute-on-chronic liver failure where citrate was used as anticoagulant [17]. Here bilirubin decreased by 39.9%, total bile acids by 28.1% and ammonia by 39.0% during treatments of 6 h. Successful treatment of fulminant hepatic failure was also reported [18].

We used [19] the Prometheus system in 12 patients with acute or acute-on-chronic liver insufficiency: 8 cirrhosis, 1 posttransplant dysfunction, and

3 secondary liver insult (2 cardiogenic shocks and 1 rhabdomyolysis). All the patients were severely hyperbilirubinemic, hypercholemic and hyperammoniemic. Twenty-eight sessions were performed, 2.5/patient, each lasting 340 ± 40 min.

Mean total bilirubin decreased from 33.6 ± 20 to 22.2 ± 13.6 mg/dl ($p < 0.001$); the reduction ratios for cholic acid and ammonia were 48.6 and 51.6%, respectively. The pre- to postsession reduction of urea was $57.6 \pm 9.5\%$ and of creatinine $42.7 \pm 10\%$. A significant reduction was observed in the circulating level of the soluble receptor for interleukin 2 (before $2,687.2 \pm 1,434.7$ IU/ml, after $1,977.1 \pm 602$ IU/ml, $p < 0.001$) and in IL-6 (before 56.1 ± 11.1 pg/ml, after 35.9 ± 10.3 pg/ml, $p = 0.05$). Intratreatment hemodynamics was stable.

Two patients received liver transplant. Secondary liver insult was completely overcome in all 3 patients. The overall survival at 30 days was 41.6% (5/12 patients). Prometheus, based on FPSA, produced high clearance for protein-bound and water-soluble markers, which resulted in a high treatment dose. The efficacy in the patients' outcome of this highly efficient system is expected and we are taking part in an European multicenter trial in a larger population with acute-on-chronic liver insufficiency.

Conclusions

For patients with acute-on-chronic liver failure and perhaps also for patients with acute liver insufficiency the artificial liver support devices may offer a survival benefit. However, the evidence from both the artificial systems and the bioartificial ones is anything but conclusive and, as is always the case with pioneering methods, there is a definite need for randomized clinical trials that can shed light on the real survival advantages over standard medical therapy for liver failure.

References

1 Santoro A, Mancini E: The kidney in hepatorenal syndrome. I Part I. Int J Artif Organs 2004;27: 95–103.
2 Rambhatla L, Chiu CP, Kundu P, Peng Y, Carpenter MK: Generation of hepatocyte-like cells from human embryonic stem cells. Cell Transplant 2003;12:1–11.
3 Kiley J, Welch HF, Pender JC: Removal of blood ammonia by hemodialysis. Proc Soc Exp Biol Med 1956;91:489–490.
4 Yatzidis H: A convenient hemoperfusion microapparatus over charcoal for the treatment of endogenous and exogenous intoxification: its use as an effective artificial kidney. Proc Eur Dial Transplant Assoc 1964;1:83–87.
5 Commission of the European Community, March 2002.

6 Davenport A, Will EJ, Losowsky MS: Rebound surges of intracranial pressure as a consequence of forced ultrafiltration used to control intracranial pressure in patients with severe hepatorenal failure. Am J Kidney Dis 1989;14:516–519.
7 Tetta C, Cavaillon JM, Camussi G, et al: Continuous plasma filtration coupled with sorbents. Kidney Int 1998;66(suppl):S186–S189.
8 Trey MB, Bums DG, Saunders SJ: Treatment of hepatic coma by exchange blood transfusion. N Engl J Med 1966;42:394–398.
9 Santoro A, Mancini E, Buttiglieri S, Krause A, Yakubovich M, Tetta C: Extracorporeal support of liver function. Part II. Int J Artif Organs 2004;27:176–185.
10 Stange J, Hassanein TI, Mehta R, Mitzner SR, Bartlett RH: The molecular adsorbents recycling system as a live support system based on albumin dialysis: a summary of preclinical investigations, prospective randomised, controlled clinical trial and clinical experience from 19 centers. Artif Organs 2002;26:103.
11 Stange J, Mitzer S, Klammt S, et al: Liver support by extracorporeal blood purification: a clinical observation. Liver Transpl 2000;6:603–613.
12 Ichai P, Samuel D: Extracorporeal liver support with MARS in liver failure: has it a role in the treatment of severe alcoholic hepatitis? J Hepatol 2003;38:104–106.
13 Strobl W, Vogt G, Mitteregger R, et al: Das 'Fractionated Plasma Separation and Adsorption System' (FPSA), ein neues membran-adsorptionsgestütztes adjunktives extrakorporales Blutreinigungssystem für das Leberversagen. Biomed Technik Ergänzungsband 1998;43/1: 168–169.
14 Falkenhagen D, Strobl W, Vogt G, et al: Fractionated plasma separation and adsorption system: a novel system for blood purification to remove albumin bound substances. Artif Organs 1999;23: 81–86.
15 Krause A: Prometheus – a new extracorporeal liver support therapy; in Arroyo V, Forns X, Garcia-Pagán JC, Rodés J (eds): Progress in the Treatment of Liver Diseases. Toronto, Ars Medica, 2003, pp 437–443.
16 Rifai K, Manns PM: Clinical experience with Prometheus. Ther Apheresis Dial 2006;10:132–137.
17 Herget-Rosenthal S, Treichel U, Saner F: Citrate anticoagulated modified fractionated plasma separation and adsorption: first clinical efficacy and safety data in liver failure (poster). American Society of Nephrology (ASN), San Diego, 2003.
18 Kramer L, Bauer E, Schenk P, et al: Successful treatment of refractory cerebral oedema in ecstasy/cocaine-induced fulminant hepatic failure using a new high-efficacy liver detoxification device (FPSA-Prometheus). Wien Klin Wochenschr 2003;115:599–603.
19 Santoro A, Faenza S, Mancini E, Ferramosca E, Grammatico F, Zucchelli A, Facchini MG, Pinna AD: Prometheus system: a technological support in liver failure. Transplant Proc 2006;38: 1078–1082.

Antonio Santoro
Malpighi Nephrology, Dialysis and Hypertension Unit
Policlinico S.Orsola-Malpighi
Via P. Palagi 9
IT–40138 Bologna (Italy)
Tel. +39 051 6362430, Fax +39 051 6362511, E-Mail santoro@aosp.bo.it

Coupled Plasma Filtration Adsorption

Marco Formica[a], *Paola Inguaggiato*[a], *Serena Bainotti*[a], *Mary Lou Wratten*[b]

[a]Nephrology and Dialysis Unit, S. Croce and Carle Hospital, Cuneo, and
[b]Sorin Group Italia (Bellco), Scientific Research Department, Mirandola, Italy

Abstract

Sepsis is one of the main causes of death in critically ill patients worldwide, and in many cases it is associated with renal and/or other organ failure. However, we do not have a unique efficient therapy to reduce this extremely high mortality rate. In the last years interest around the use of extracorporeal blood purification techniques has increased. One of the emerging treatments in patients with severe sepsis and septic shock is coupled plasma filtration adsorption (CPFA), a novel extracorporeal blood purification therapy aimed at a nonselective reduction of the circulating levels and activities of both pro- and anti-inflammatory mediators. Early experimental studies and the following clinical trials have demonstrated impressive results regarding hemodynamics and respiratory parameters, even in patients without concomitant acute renal injury, paralleled by a quick tapering of vasoactive drugs. Considering the still high morbidity and mortality rates in septic shock patients, this new blood purification technique seems to have benefits when applied early in the course of sepsis, also without renal indications, suggesting that it might be performed to prevent rather than to treat acute kidney injury.

Copyright © 2007 S. Karger AG, Basel

Epidemiology of Sepsis

Sepsis represents a crucial source of morbidity and mortality in critically ill patients: in the United States it is the tenth most frequent cause of death in intensive care units (ICU) [1]. As reported in the literature, patients diagnosed with sepsis and severe sepsis or septic shock have a mortality rate ranging from 20 to 80%. This extreme variability is due to several factors: the huge disparity of the clinical picture, the number of involved organs, the study design and the timing and quality of treatment. The kidney is often injured during sepsis; then renal failure alone or in association with other organ failure, in the so-called

M.L. Wratten is a full-time employee of the Sorin Group Italia (Bellco).

multiorgan dysfunction syndrome, may be the clinical expression of the exaggerated host response to infection [2]. Indeed, sepsis is the leading cause of acute renal injury (AKI) and its prevalence ranges from 19% in cases of sepsis, to 23% in severe sepsis and up to 51% in septic shock [3].

Pathophysiology of Sepsis

Sepsis is considered an exaggerated immune response to infection, leading to an overproduction and release of a wide array of pro- and anti-inflammatory molecules. To summarize, there are two important pathways: the proinflammatory response and the opposite immunosuppressive (or immunodysfunctional) response. The first pathway determines the release of tumor necrosis factor-α (TNF-α), interleukin-1 (IL-1), and IL-6 that have a predominantly proinflammatory role, whereas the second pathway leads to the release of IL-10 and IL-4, cytokines with a predominantly anti-inflammatory activity. These two pathways might be present at the same time, and not necessarily in sequence as previously thought. Cytokines seem to be the principal factors causing diffuse endothelial injury, because they are able to induce vasoparalysis and drive selective permeability with important consequences on systemic hemodynamics. Moreover, during sepsis monocytes lose their ability to synthesize and deliver cytokines, determining the so-called 'immunoparalytic' state [4].

Treatment of Sepsis

The prognosis of patients admitted to ICU with septic shock and multiorgan dysfunction syndrome today still has high mortality and all the attempts to find a 'magic bullet' to restore the immune derangements have failed, due to the complex interactions that take place between the pro- and anti-inflammatory responses along with the clinical course of sepsis. At the present time no effective therapy is available for sepsis. Several attempts have been made in recent years, but clinical trials did not show any good therapeutic agent targeting the specific components of this pathological cascade and having a positive effect on outcome [5]. Then, the attention moved to techniques able to remove different circulating cytokines, that is extracorporeal blood purification [6, 7]. Standard continuous renal replacement therapies do not show a great ability to remove sepsis-related molecules because of the small volumes employed and low membrane sieving coefficient [8], whereas the use of a large-pore membrane seems to have a higher clearance of cytokines because of an increased convective transfer, but at the expense of added albumin loss [9, 10].

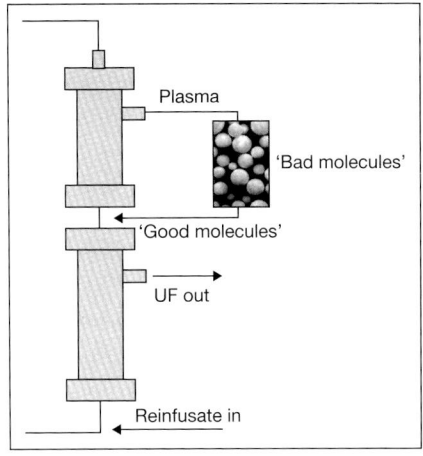

Fig. 1. Schematic diagram of CPFA. UF = Ultrafiltrate.

Coupled Plasma Filtration Adsorption: Technical Characteristics

Thus, a new extracorporeal blood purification technique was developed, coupled plasma filtration adsorption (CPFA), which couples plasma filtration and adsorption using a resin cartridge, along with a second hemofiltration system that allows convective exchange. The nonselective removal of inflammatory mediators is due to a hydrophobic styrenic resin, which has high affinity and a large capacity for many cytokines and mediators [11]. The sorbent adsorption allows reinfusion of the endogenous plasma after nonselective, simultaneous removal of different sepsis-associated mediators through a specific cartridge. CPFA is performed using a four-pump modular treatment (Lynda, Bellco®, Mirandola, Italy) that consists of a plasma filter (0.45 m² polyethersulfone with an approximate cutoff of 800 kDa), a nonselective hydrophobic resin cartridge (140 ml) with a surface of about 700 m²/g, and a synthetic, high-permeability, 1.4-m² polyethersulfone hemofilter where convective exchanges may be applied to the reconstituted blood in a postdilutional mode (fig. 1). The innovative aspect of this technique is the application of the sorbent to plasma instead of blood [12]. This aspect has important advantages: the lower flow of plasma allows a longer contact time with the sorbent, and biocompatibility problems are avoided. The postdilution reinfusion rate can be set for up to a maximum of 4 liters/h. The blood flow is usually 150–180 ml/min while the plasma filtration rate is maintained at a fractional filtration of the blood flow (approximately 15–20%). The treatment is usually run for approximately 10 h, after which the cartridge begins to show saturation of the mediators.

Clinical Results

The treatment goal of CPFA is to target the excess of circulating pro- and anti-inflammatory mediators to restore the normal immune function. Ronco et al. [13] performed some of the first clinical CPFA studies with a prototype machine, and showed that a 10-hour session of CPFA had a better impact on hemodynamics, expressed as increased mean arterial pressure (MAP) and decreased norepinephrine requirement, if compared to standard continuous venovenous hemodiafiltration (CVVHDF). Moreover, they demonstrated that monocytes obtained from blood treated with CPFA recovered their ability in responding to the lipopolysaccharide challenge with TNF-α production to a much greater extent than blood treated by CVVHDF. Subsequently, it was hypothesized that CPFA might have a role in the treatment of septic patients, not only for renal function substitution but also for immunomodulation activity. Still employing the prototype machine, Formica et al. [14] evaluated the hemodynamic performance to CPFA taking into account two unique features: (1) repeated application of the technique (for a mean of 10 h/day) along the course of the septic shock and (2) the use in patients without concomitant AKI. These authors demonstrated that hemodynamic and respiratory parameters, such as MAP, cardiac index, peripheral vascular resistances and oxygen arterial pressure/inspired oxygen fraction ratio, were improved, so that norepinephrine administration could be stopped after a mean of 5 CPFA sessions (fig. 2). CPFA seems to be feasible and safe, without clinical side effects and risks, and might be reliable even in the absence of AKI; in patients with AKI, more special anticoagulation schedules may be customized for different needs [14, 15]. The improvement in splanchnic perfusion, evaluated by means of tonometry of gastric-mucosal PCO_2 diffusion, is also of interest, which could further underline the resolution of the hyperdynamic-vasoparalytic state displayed in septic shock. Thus, CPFA is a treatment targeted to the nonselective removal of soluble mediators involved in the septic shock scenario. We can furthermore speculate that the association of different removal mechanisms (diffusion/convection/ adsorption) may play a role in reestablishing a new immune balance (immune modulation) with a significant reduction of acute phase reactants by hampering their peaks [16, 17]. The results may be related to its ability to restore leukocyte responsiveness to immunoactive stimuli and this may be clinically beneficial because of the link to hemodynamic improvement [13]. Despite the fact that some clinical results appear to be quite good, it is mandatory to interpret them with caution because of the small sample size considered, but this underlines that these good outcome results, where achieved, fueled the relationship between the nephrologists and the intensivists in managing these very complex, critically ill patients.

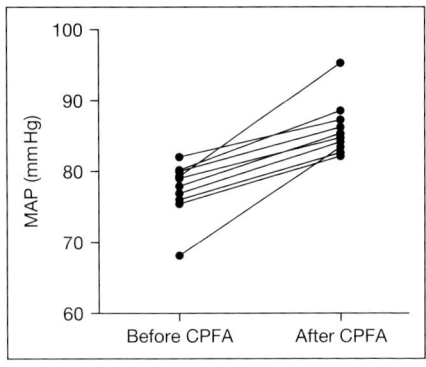

Fig. 2. MAP behavior before and after CPFA treatments. Before vs. after: 77.2 ± 12.5 vs. 83.3 ± 14.1 mm Hg ($p < 0.001$).

Future Developments

It is reasonable to propose to extend this technique also to former stages of septic shock, such as severe sepsis or systemic inflammatory response syndrome along with pancreatitis. Further ongoing studies may provide evidence about the potential efficacy of blood purification systems such as CPFA in critically ill septic patients. Among these, the target of one prospective randomized trial is to compare the clinical outcome of septic patients with AKI treated either with CPFA or pulse high volume hemofiltration. Furthermore, a large Italian multicenter study has been initiated, registered on Clinical Trial.gov with the identifier NCT00332371 and on ISRCTN with the code ISRCTN24534559, identified by the acronym COMPACT (Combined Plasmafiltration and Adsorption Clinical Trial), regarding treatment of septic shock patients with early initiation (within 6 h of diagnosis) of CPFA. The main target of this study is to compare hospital and ICU mortality and morbidity rates between patients treated with standard medical care alone and those treated with standard medical care and CPFA. In this trial, started in December 2006, 330 patients will be recruited, in order to show a mortality difference of 25% between the two arms.

References

1 Martin GS, Mannino DM, Eaton S, Moss M: The epidemiology of sepsis in the United States from 1979 through 2000. N Engl J Med 2003;348:1546–1554.
2 Schrier RW, Wang W: Acute renal failure and sepsis. N Engl J Med 2004;351:159–169.
3 Rangel-Frausto MS, Pittet D, Costigan M: The natural history of the systemic inflammatory response syndrome (SIRS). A prospective study. JAMA 1995;273:117–123.

4　Cavaillon JM, Adib-Conquy M, Cloez-Tayarani I, Fitting C: Immunodepression in sepsis and SIRS assessed by ex vivo cytokine production is not a generalized phenomenon: a review. J Endotoxin Res 2001;7:85–93.
5　Zeni D, Freeman B, Nathanson C: Anti-inflammatory therapies to treat sepsis and septic shock. Crit Care Med 1997;25:1095–1100.
6　Schetz M: Nonrenal indications for continuous renal replacement therapy. Kidney Int 1999;56 (suppl 72):S88–94.
7　Cole L, Bellomo R, Journois D, Davenport P, Baldwin I, Tipping P: High-volume hemofiltration in human septic shock. Intensive Care Med 2001;27:978–986.
8　De Vriese AS, Vanholder RC, De Sutter JH, Colardyn FA, Lameire NH: Continuous renal replacement therapies in sepsis: where are the data? Nephrol Dial Transplant 1998;13:1362–1364.
9　Tetta C, Cavaillon JM, Schulze M, Ronco C, Ghezzi PM, Camussi G, Serra AM, Curti F, Lonnemann G: Removal of cytokines and activated complement components in an experimental model of continuous plasma filtration coupled with sorbent adsorption. Nephrol Dial Transplant 1998;13:1458–1464.
10　Tetta C, Gianotti L, Cavaillon JM, Wratten ML, Fini M, Braga M, Bisagni P, Giavaresi GL, Bolzani R, Giardino R: Continuous plasma filtration coupled with sorbent adsorption in a rabbit model of endotoxic shock. Crit Care Med 2000;28:1526–1533.
11　Winchester JF, Kellum JA, Ronco C, Brady JA, Quartararo PJ, Salsberg JA, Levin NW: Sorbents in acute renal failure and the systemic inflammatory response syndrome. Blood Purif 2003;21: 79–84.
12　Reeves JH, Butt WW, Shann F, Layton JE, Stewart A, Waring PM, Presneill JJ: Continuous plasmafiltration in sepsis syndrome. Crit Care Med 1999;27:2096–2104.
13　Ronco C, Brendolan A, Lonnemann G, Bellomo R, Piccinni P, Digito A, Dan M, Irone M, La Greca G, Inguaggiato P, Maggiore U, De Nitti C, Wratten ML, Ricci Z, Tetta C: A pilot study of coupled plasma filtration with adsorption in septic shock. Crit Care Med 2002;30:1250–1255.
14　Formica M, Olivieri C, Livigni S, Cesano G, Vallero A, Maio M, Tetta C: Hemodynamic response to coupled plasmafiltration-adsorption in human septic shock. Intensive Care Med 2003;29: 703–708.
15　Mariano F, Tetta C, Stella M, Biolino P, Miletto A, Triolo G: Regional citrate anticoagulation in critically ill patients treated with plasmafiltration and adsorption. Blood Purif 2004;22:313–319.
16　Opal S: Hemofiltration-adsorption systems for the treatment of experimental sepsis: it is possible to remove the 'evil humors' responsible for septic shock? Crit Care Med 2000;28:1681–1682.
17　Ronco C, Tetta C, Mariano F, Wratten ML, Bonello M, Bordoni V, Cardona X, Inguaggiato P, Pilotto L, D'Intini V, Bellomo R: Interpreting the mechanisms of continuous renal replacement therapy in sepsis: the peak concentration hypothesis. Artif Organs 2003;27:792–801.

Dr. Marco Formica
Nephrology and Dialysis Unit, S. Croce and Carle Hospital
Via Carle, 25
IT–12100 Cuneo (Italy)
Tel. +39 0171 616241, Fax +39 0171 616120, E-Mail formica.m@ospedale.cuneo.it

Albumin Dialysis and Plasma Filtration Adsorption Dialysis System

Federico Nalesso[a], Alessandra Brendolan[a], Carlo Crepaldi[a], Dinna Cruz[a], Massimo de Cal[a], Rinaldo Bellomo[b], Claudio Ronco[a]

[a]Department of Nephrology, Dialysis and Transplantation, San Bortolo Hospital, Vicenza, Italy; [b]Department of Intensive Care, Austin Hospital, Melbourne, Vic., Australia

Abstract

Albumin-bound toxins are important in the pathophysiology of liver failure, systemic inflammatory response syndrome, and poisoning. Due to its intrinsic ability to bind molecules, albumin has been used in blood purification techniques, such as single pass albumin dialysis, the molecular adsorbent recirculating system and the Prometheus systems. Plasma filtration adsorption dialysis is the latest technology that can combine the best processes of blood/plasma purification in order to determine a selective and effective purification of molecules implicated in liver failure.

Copyright © 2007 S. Karger AG, Basel

Molecules are present in the plasma, in the plasmatic water or bound to specific or unspecific carriers. The most important plasmatic carrier is albumin. According to their characteristics of solubility, plasmatic molecules are present as solution in the plasmatic water, if water soluble, or bound to carriers, if hydrophobic.

Techniques such as hemodialysis, hemofiltration, and their combination are able to remove small molecules and medium molecules acting on plasmatic water and its solutes. According to the membrane cutoff and high volume of infusion, they can improve the total removal of molecules with high molecular weight compared with small molecules. In order to improve the efficiency of the removal of the molecules, it is possible to combine the convective and/or diffusive processes with adsorption on specific materials.

The adsorption allows removing a wide range of hydrophobic and higher molecular weight substances such as bilirubin, salt acids, cytokines, myoglobin and others. The possibility of using specific physical interaction in some molecule absorbers (ion exchange, chemical affinity, Van der Waals forces) allows for the removal of specific molecule targets such as bile acids and bilirubin during liver failure. The adsorption process acts on protein-bound substances and high molecular weight toxins present in the plasma. Convection and diffusion are not able to obtain good clinical clearances of high molecular weight or hydrophobic molecules due to their theoretical and practical limitations (volume of infusion and cutoff membrane).

Summarizing all these concepts, we can understand and highlight the central role of plasma as a *transporter* of toxic molecules and its potential function in the purification of blood. Therefore, thanks to its intrinsic capacity to bind and transport molecules, plasma is the best fluid to perform a purification process. It seems useful to combine the physical and chemical principles of purification (diffusion, convection and adsorption) in order to improve and obtain the best removal of substances. According to this view, there is the possibility to use the albumin as a medium of purification thanks to its capacity to bind toxins.

A critical issue of the clinical syndrome in liver failure is the accumulation of toxins not cleared by the failing liver. Based on this hypothesis, the removal of lipophilic, albumin-bound substances such as bilirubin, bile acids, metabolites of aromatic amino acids, medium-chain fatty acids and cytokines should be beneficial to the clinical course of a patient in liver failure. This led to the development of artificial filtration and adsorption devices.

Hemodialysis, hemofiltration and their combination are used in renal failure and primarily remove water-soluble toxins; however, they do not remove toxins bound to albumin that accumulate in liver failure because of the technical limitations to use high cutoff membrane and shift and removal of albumin-bound toxins.

Artificial detoxification devices currently under clinical evaluation include the molecular adsorbent recirculating system (MARS), single pass albumin dialysis and the Prometheus system. A new system is going to be developed in the Dialysis Unit of San Bortolo Hospital, Vicenza thanks to its innovative conception of patient's plasma as substrate and medium of purification; the system is the plasma filtration adsorption dialysis (PFAD) technology.

Molecular Adsorbent Recirculation System

The MARS, developed by Teraklin of Germany, is the best-known extracorporeal liver dialysis system. It consists of two separate dialysis circuits. The first circuit consists of human albumin, is in contact with the patient's blood through a semipermeable membrane and has two special devices to clean the albumin after it has absorbed toxins from the patient's blood. The second circuit consists of a hemodialysis machine and is used to purify the albumin in the first circuit, before it is recirculated to the semipermeable membrane in contact with the patient's blood. The MARS system [1] can remove a number of toxins, including ammonia, bile acids, bilirubin, copper, iron and phenols.

Single Pass Albumin Dialysis

Single pass albumin dialysis is a simple method of albumin dialysis using standard renal replacement therapy machines without an additional perfusion pump system [2]: the patient's blood flows through a circuit with a high-flux hollow fiber hemodiafilter. The other side of this membrane is cleansed with an albumin solution in counterdirectional flow (such as a standard dialysate during a bicarbonate dialysis), which is discarded after passing the filter. The albumin can be used in single pass or regenerated by adsorber and reused in a closed system.

Prometheus

The Prometheus system (Fresenius Medical Care, Bad Homburg, Germany) is a new device based on the combination of albumin adsorption with high-flux hemodialysis after selective filtration of the albumin fraction through a specific polysulfone filter (AlbuFlow) [3].

Plasma Filtration Adsorption Dialysis

The PFAD technology [4] is based on a new principle of purification that utilizes a tricompartmental dialyzer (TD) to purify the patient's blood by a combination of three sequential techniques: convection and adsorption both on plasma followed by a process of 'whole blood dialysis' provided by the regenerated patient's own plasma (fig. 1).

The TD is the core of this new technology (fig. 2). It is composed of hollow fibers like a regular dialyzer for hemodialysis. The compartments are located in different areas, and each has its own particular function. The hollow fibers form three compartments along the extension of the dialyzer (fig. 2): the first compartment is formed by the inner space of hollow fibers in which the blood goes through the length of the whole fibers thanks to a roller pump

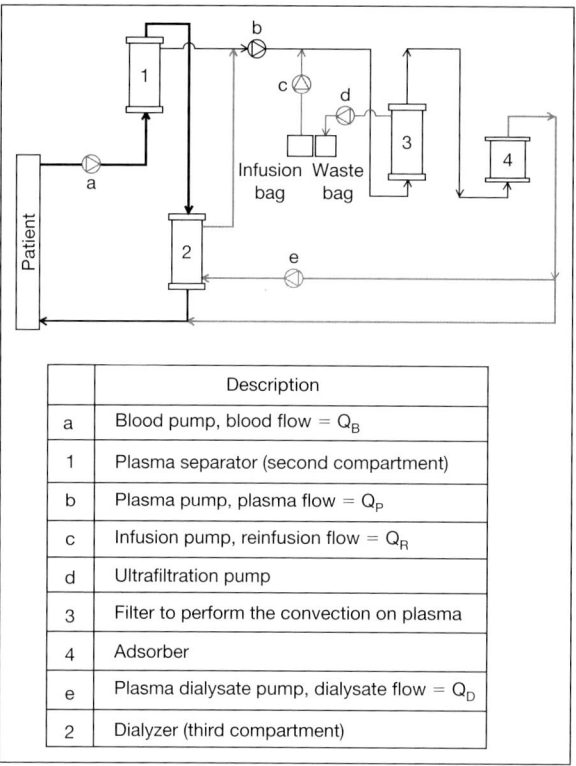

	Description
a	Blood pump, blood flow = Q_B
1	Plasma separator (second compartment)
b	Plasma pump, plasma flow = Q_P
c	Infusion pump, reinfusion flow = Q_R
d	Ultrafiltration pump
3	Filter to perform the convection on plasma
4	Adsorber
e	Plasma dialysate pump, dialysate flow = Q_D
2	Dialyzer (third compartment)

Fig. 1. Schematic representation of PFAD circuits. The second (1) and third (2) compartment are described as separate devices in order to simplify the explanation of single processes.

(blood pump). The internal compartment of the dialyzer is divided into two more compartments separated by a wall along the extension of the hollow fibers: the second compartment forms a stage for filtering plasma, and the third compartment forms a stage for dialysis. The second compartment is the delineated space where the patient's plasma can be filtered from the whole blood across the hollow fiber membrane (fig. 1, number 1). The third compartment is the space where the patient's regenerated plasma performs a process of purification based on a 'diffusive and binding process'; in this way the regenerated patient's plasma is used as a dialysate in countercurrent to purify the blood flowing in the first compartment (fig. 1, number 2). The second and third compartment have a specific cutoff of hollow fiber membranes according to their specific function and their area is able to assure the processes that occur (filtration and dialysis). The second and the third compartment communicate through

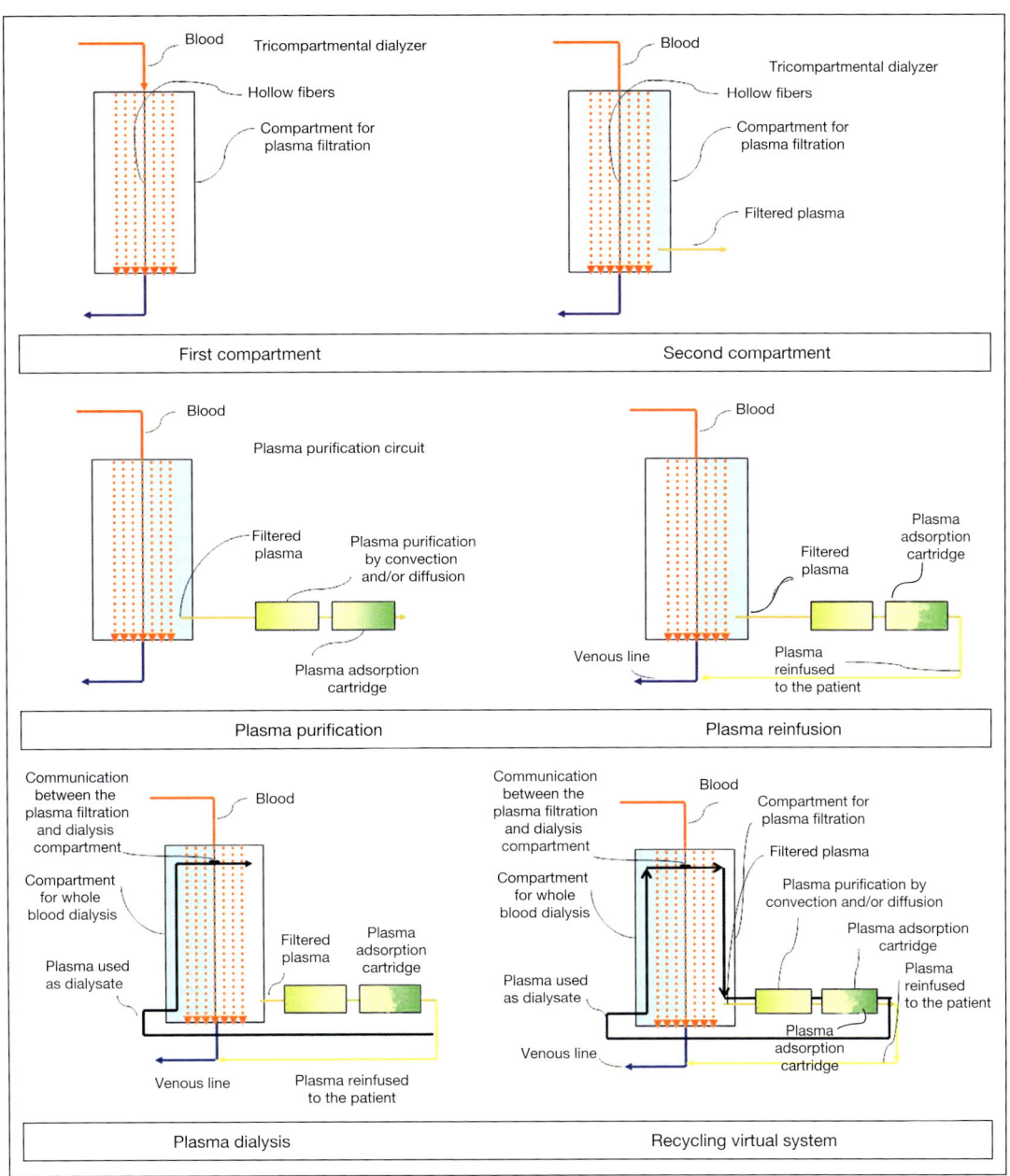

Fig. 2. The TD and the purification processes.

a particular opening in the arterial extremity of dialyzer (fig. 2, not shown in fig. 1). In the first human prototype the second and the third compartment are separated and formed by two different devices, as shown in the figures.

The first step of the process aims to separate the plasma from whole blood (fig. 1). This process determines the plasma filtration from the inner space of hollow fibers to the space of the second compartment. The obtained patient's plasma goes from the second compartment to the plasma purification circuit where it is purified by two different and separate methods: convection and adsorption (fig. 1: 3 = convection process, 4 = adsorption). The plasma flow is obtained by a roller pump (fig. 1, b).

The plasma purification circuit (fig. 1) is composed of two separate processes in order to remove first the water-soluble and dialyzable toxic molecules by convection and then hydrophobic and nondialyzable molecules by adsorption on a specific adsorber (fig. 1, numbers 3 and 4). The convection process is obtained by performing a high volume hemofiltration directly on plasma. It is known that high volume ultrafiltration using a super high flux filter has achieved better cytokine clearances compared to those currently achieved by urea during standard continuous renal replacement therapy [5]. The convective process is able to reestablish hydroelectrolytes, acid-base equilibrium and fluid balance acting directly on plasmatic water.

After the convective purification, the plasma is adsorbed by a specific adsorber to remove hydrophobic or nondialyzable molecules (fig. 1, number 4). The adsorber is specific for the molecules implicated in the patient's disease (sepsis, hepatorenal syndrome, acute and chronic liver failure). The cartridge for adsorption presents a good pressure-flow performance and excellent mechanical and chemical stability in order to perform the best adsorption of plasma.

After these two different processes, the purified plasma can take two separate paths (fig. 1). In the first pathway, the purified plasma returns to the patient through the venous line, in the other, the patient's regenerated plasma is used as dialysate in the suitable compartment of TD, in order to perform the dialysis procedure based on the 'diffusion and binding process'. In this step, the patient's filtrated plasma from the whole blood and the spent dialysate from the third compartment can be purified again in the plasma circuit (fig. 1). In fact there is a connection between the second and the third compartment at the arterial extremity of TD; thus, the spent dialysate can go to the plasma circuit through the second compartment (fig. 2) generating a virtual open loop. In the human prototype the separation between the second and the third compartment assures the same plasma process in the open loop.

The plasma circuit is an open circuit at the venous line; therefore, the dialysis system is not working in closed recycling modality, thus, having new fresh regenerated plasma from the plasma circuit every time it is fed by new plasma from the second compartment of TD (fig. 1). The unique feature of the plasma circuit is that the plasma flow to obtain the dialysate in the third compartment is a virtual flow inside the open plasma loop and can exceed the plasma filtration flow from the first compartment being a recycling flow, as shown in the drawings.

The PFAD is a technique that can be continuously performed for at least 8 h or more a day.

Discussion

The PFAD combines three different processes in order to purify the patient's whole blood using the patient's plasma. With this technology the plasma is both a medium and a substrate of purification.

This technology allows for the removal of molecules from plasmatic water and protein-bound plasma by the sequential combination of convective treatment and adsorption. According to the water solubility and molecular weight, the PFAD can remove both diffusible and nondiffusible molecules, thanks to the combination of specific techniques. In the first step, water-soluble molecules can be removed by convection. This purification process can improve the selection of successive adsorption which is able to remove hydrophobic and large molecular weight molecules. In the plasma circuit the patient's plasma is the substrate of purification but when it is reinfused or used as dialysate it becomes a medium of purification. Therefore, the regenerated albumin and carriers are able to bind toxic ligands according to the 'law of mass action'. This is an important physical process that allows ligands to be shifted from the whole blood and tissues and bound to the regenerated carriers across the TD or capillary membrane in the PFAD system and patient's body, respectively.

In liver failure the accumulation of albumin-bound toxins has been demonstrated; these toxins are responsible, to variable extents, for multiple-organ dysfunction (kidney, cardiovascular instability) [6]. The function of albumin as a transporter and as a possible purification vector have been described in albumin dialysis, in which the removal of these molecules improves the clinical condition of the patient. The best known and most widely used extracorporeal device for liver function support is MARS [7], which uses albumin only to perform purification by adsorption and by classic dialysis. The current literature demonstrates that this approach is capable of improving patient survival [8, 9]. Moreover, this type of approach is useful for intoxications caused by exogenous pathogens that are scarcely water-soluble but are plasma protein-bound.

In several pathologies (sepsis, hepatorenal syndrome, systemic inflammatory response syndrome) there is an involvement of cytokines, and much of the damage that affects the various organs and systems that are not primarily involved in the basic pathological process are determined by molecular factors that circulate in the blood or are present in the plasma. In the PFAD the specificity of the adsorber used for the plasma adsorption is important to remove the toxic molecules retained or produced during liver failure.

References

1 Saliba F: The molecular adsorbent recirculating system (MARS) in the intensive care unit: a rescue therapy for patients with hepatic failure. Crit Care 2006;10:118.
2 Chawla LS, Georgescu F, Abell B, Seneff MG, Kimmel PL: Modification of continuous venovenous hemodiafiltration with single-pass albumin dialysate allows for removal of serum bilirubin. Am J Kidney Dis 2005;45:e51–e56.
3 Rifai K, Manns MP: Clinical experience with Prometheus. Ther Apher Dial 2006;10:132–137.
4 Nalesso F: Patent of PFAD. WO2004091694, 2004.
5 Uchino S, Bellomo R, Goldsmith D, Davenport P, Cole L, Baldwin I, Panagiotopoulos S, Tipping P: Super high flux hemofiltration: a new technique for cytokine removal. Intensive Care Med 2002;28: 651–655.
6 Sen S, Jalan R, Williams R: Liver failure: basis of benefit of therapy with the molecular adsorbents recirculating system. Int J Biochem Cell Biol 2003;35:1306–1311.
7 Sen S, Williams R, Jalan R: Emerging indications for albumin dialysis. Am J Gastroenterol 2005;100: 468–475.
8 Isoniemi H, Koivusalo AM, Repo H, Ilonen I, Hockerstedt K: The effect of albumin dialysis on cytokine levels in acute liver failure and need for liver transplantation. Transplant Proc 2005;37: 1088–1090.
9 Sen S, Jalan R, Williams R: Liver failure: basis of benefit of therapy with the molecular adsorbents recirculating system. Int J Biochem Cell Biol 2003;35:1306–1311.

Prof. Claudio Ronco
Department of Nephrology, Dialysis and Transplantation
St. Bortolo Hospital, Viale Rodolfi 31
IT–36100 Vicenza (Italy)
Tel. +39 0444 753 689, Fax +39 0444 753 949, E-Mail Cronco@goldnet.it

Renal Assist Device and Treatment of Sepsis-Induced Acute Kidney Injury in Intensive Care Units

Naim Issa[a], *Jennifer Messer*[b], *Emil P. Paganini*[b]

[a]Department of Nephrology and Hypertension, and [b]Section of Dialysis and Extracorporeal Therapy, Department of Nephrology and Hypertension, Cleveland Clinic, Cleveland, Ohio, USA

Abstract

Acute kidney injury (AKI) is a frequent and serious complication of sepsis in ICU patients and is associated with a very high mortality. Despite the advent of sophisticated renal replacement therapies (RRT) employing high-dose hemofiltration and high-flux membranes, mortality and morbidity from sepsis-induced AKI remained high. Moreover, these dialytic modalities could not substitute for the important functions of renal tubular cells in decreasing sepsis-induced AKI biological dysregulations. The results from the in vitro and preclinical animal model studies were very intriguing and led to the development of a bioartificial kidney consisting of a renal tubule assist device containing human proximal tubular cells (RAD) added in tandem to a continuous venovenous hemofiltration circuit. The results from the phase I safety trial and the recent phase II clinical trial showed that the RAD not only can replace many of the indispensable biological kidney functions, but also modify the natural history of sepsis-induced AKI by ameliorating patient survival.

Copyright © 2007 S. Karger AG, Basel

Sepsis-Associated Acute Kidney Injury and Available Renal Replacement Therapies

Acute kidney injury (AKI) occurs in approximately 20% of patients with sepsis and 51% of patients with septic shock with positive blood cultures [1]. Sepsis-induced AKI can be associated with 70% mortality as compared with 40% mortality among patients with AKI alone [2]. Sepsis stimulates the induction of nitric oxide synthase leading to nitric oxide-mediated arterial vasodilatation. The arterial vasodilatation induces a decrease in systemic vascular resistance resulting in increased sympathetic tone and the release of vasopressin from the central nervous system along with activation of the renin-angiotensin-aldosterone system. The resulting renal vasoconstriction induces sodium and water retention and predisposes to AKI [3–5]. Sepsis also induces the generation of oxygen radicals that scavenge renal endothelial nitric oxide thereby causing peroxynitrite-related acute tubular injury and necrosis [6]. Following or more likely concomitant to this renal vasoconstrictor phase, a proinflammatory phase involving cytokines and chemokines leads to further acute injury of the renal endothelium and an increased rate of patient death [7]. Patients with AKI especially when associated with sepsis are extremely hypercatabolic and frequently require renal replacement therapy (RRT) in the form of intermittent or continuous RRT. Neither form of RRT has proved to be superior in terms of survival and renal recovery. Despite the advent of RRT, mortality rates from AKI have not changed significantly and the appropriate dose of dialysis in AKI has not been defined to date. A recent study showed that intensive daily hemodialysis (HD) compared with alternate-day HD reduced mortality (28 vs. 46%, $p < 0.01$) without increasing hemodynamically-induced morbidity [8]. Moreover intensive daily HD was associated with less systemic inflammatory response syndrome or sepsis (22 vs. 46%, $p = 0.005$) and a shorter duration of acute renal failure (ARF; mean $\pm$ SD, 9 ± 2 vs. 16 ± 6 days, $p = 0.001$). A randomized study by Ronco et al. [9] showed that hemofiltration rates of 35 or 45 ml/kg/h improve survival in ARF ($p < 0.001$) as compared with 20 ml/kg/h. To date any survival benefit of cytokine removal by convective or diffusive mode in patients with sepsis-induced AKI remains to be proven. In a recent study, the use of a polyflux hemofilter with a high membrane cutoff point (approximately 60 kDa) convection had an advantage over diffusion in the clearance capacity of cytokines, but was associated with greater plasma protein losses [10]. Neither of these dialytic modalities could substitute for the important biological functions of renal tubular cells in decreasing sepsis-induced AKI-associated mortality. Moreover, adding proximal tubular cells to the RRT circuit may offer a more complete and physiological form of RRT.

Proximal Tubular Cell Therapy

Renal tubular epithelial cells are the main site of blood purification by solute and fluid clearance as well as of reclamation of the essential electrolytes and metabolites from the glomerular ultrafiltrate in order to maintain the 'milieu interieur' of the human body. These highly differentiated epithelial cells actively transport electrolytes and water and perform other metabolic and endocrinological activities as well. Tubular cells are very sensitive to ischemic injury and sepsis [5] that can lead to acute tubular necrosis resulting in tubular dysfunction, solute, electrolyte and fluid dysregulation [11] that may ultimately require RRT. Fortunately, tubular cells have the capacity to regenerate and regain their original functionality providing the underlying basement membrane is not deranged. This regenerative capability led researchers to postulate the presence of putative resident or migrating progenitor stem cells that may furnish newly formed tubules as opposed to the same tubular cells undergoing mitosis to form new tubular cells [12–17]. Humes et al. [18, 19] successfully isolated human proximal tubular cells from deceased donor kidneys not suitable for transplantation because of excessive fibrosis. These proximal tubular cells were employed to create the renal assist device (RAD) as described later in the text. In their preclinical animal and subsequent human clinical studies [18, 19], the researchers demonstrated the vitality and functionality of these proximal tubular cells seeded in the RAD cartridge in regulating active transport of electrolytes and glucose, as well as glutathione metabolism, ammonia excretion and 1α-hydroxylation of 25-dihydroxyvitamin D_3, and regulating immune response by decreasing proinflammatory cytokine levels. The procedure of human renal tubular cell isolation is described in detail elsewhere [20] and is beyond the scope of this review.

Bioartificial Kidney: in vitro Studies, Preclinical Animal Studies, Circuit, and Phase I/II Clinical Studies

In vitro Studies
Humes et al. [21] developed the RAD that was tested in vitro for a variety of differentiated tubular functions.

In these in vitro studies, they demonstrated that the RAD seeded with porcine proximal tubular cells allowed vectorial transport of fluid from the intraluminal space to the antiluminal space through the Na^+, K^+-ATPase pumps of the tubular cells. The RAD also facilitates other important metabolic and endocrine activities of the renal proximal tubular cells analogously to a nephron [21, 22]. These important functions included active bicarbonate and glucose transports, intraluminal glutathione breakdown into its constituent amino acids, ammonia production, and conversion of 25-(OH)-vitamin D_3 to the active form 1,25-$(OH)_2$ vitamin D_3.

Preclinical Animal Studies
Humes et al. [18] developed animal models to support their hypothesis that the RAD provides incremental renal replacement support, thus decreasing morbidity and mortality rates observed in patients with AKI. They postulated

that the RAD tubular cells supplemented the standard RRT circuit, conferring benefits by adding the functional properties of the nephron such as clearance and other important metabolic and endocrinological functions. They also postulated that the RAD can play a role in normalizing proinflammatory cytokine imbalance that characterizes systemic inflammatory response syndrome and multiorgan failure. Their initial preclinical animal model consisted of nephrectomized dogs with endotoxin lipopolysaccharide-induced hypotension. In this study, the RAD increased ammonia excretion, glutathione metabolism, and 1,25-dihydroxy-vitamin D_3 production in uremic dogs. Moreover, the dogs had excellent cardiovascular stability during the extracorporeal therapy with the RAD [18].

Further intriguing preclinical animal trials on acutely uremic animals with induced septic shock showed that animals treated with cell RAD demonstrated significantly better cardiovascular performance and survival times than those treated with sham RAD lacking the tubular cells [23, 24]. Interestingly enough, these preclinical trials showed that plasma cytokine levels were also altered during the treatment with RAD. Specifically the IL-10 levels ($p < 0.01$) as well as the mean arterial pressures ($p < 0.04$) were both significantly higher during the treatment interval in the cell RAD animals compared to their sham controls [23].

The same group performed a more realistic sepsis animal model employing nonnephrectomized pigs [25]. The pigs were administered 30×10^{10} bacteria/kg body weight of *Escherichia coli* intraperitoneally to induce septic shock and thereby causing acute tubular necrosis [26]. The pigs were started on continuous venovenous hemofiltration (CVVH) along with a RAD in tandem to the CVVH extracorporeal circuit. They were divided in two groups: the sham RAD (without cells) or a cell RAD with renal proximal tubular cells. Cell RAD therapy resulted in significantly higher cardiac outputs and renal blood flow and was associated with significantly lower plasma circulating proinflammatory cytokine concentrations (IL-6 and interferon-γ), resulting in a nearly twice the average survival time in the cell-RAD-treated group compared with the sham RAD control group.

Human Bioartificial Kidney: Hemofilter and Circuit
The hemofilter of the RAD that was used in human clinical studies consisted of seeding human renal proximal tubular cells obtained from cadaveric kidneys (deemed not suitable for kidney transplantation) into the hollow into the hollow fiber membranes of a standard polysulfone high-flux hemofiltration cartridge pretreated with a synthetic extracellular matrix protein [21]. This hemofilter typically consists of confluent monolayers containing up to 1.5×10^9 cells. These seeded cells along the inner surface of the hollow fibers

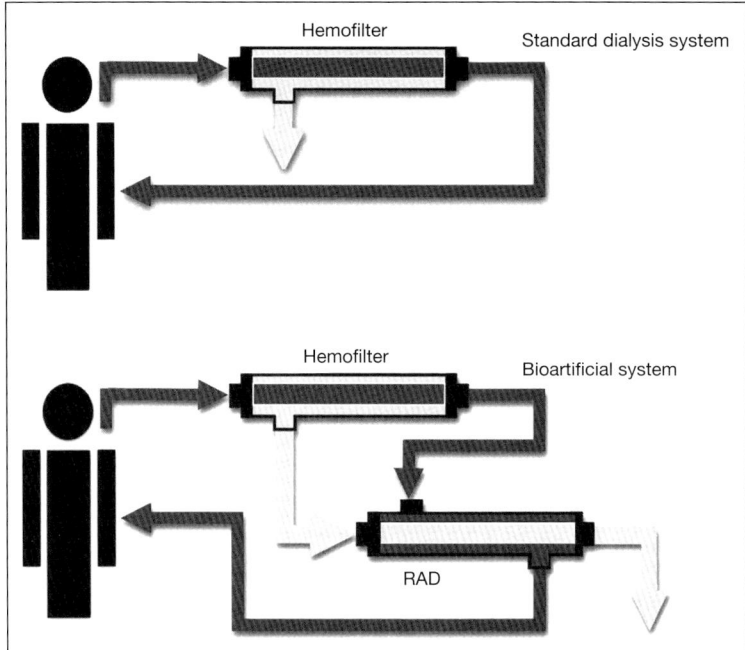

Fig. 1. Schematic diagram of the extracorporeal circuit for the bioartificial kidney; flow rates are detailed in the text.

remain immunoprotected from the patient's blood by the semipermeable membrane. The RAD is kept horizontally oriented and at a temperature of 37°C to assure vitality and maximum functioning of the tubular cells.

The bioartificial kidney was then created by attaching the RAD to a conventional CVVH system as shown in figure 1. The major difference between the RAD and the conventional CVVH circuit is that the blood and ultrafiltrate flow in opposite directions inside the hemofilter. In the RAD circuit, the ultrafiltrate emanating from the CVVH circuit is diverted via a pump to the inside of the hollow fibers and inundates the proximal tubular cells. On the other hand, the blood derived from the CVVH circuit is shunted to the dialysate compartment of the hemofiltration cartridge. This shunting process creates a 'bioartificial nephron' (fig. 2) in which the ultrafiltrate circulates inside the hollow fibers of the dialyzer, bathes the proximal tubular cells allowing their interaction, and thereby replaces the indispensable metabolic functions. Once processed, the ultrafiltrate emanating from the RAD is discarded as urine. The blood exiting the RAD is returned to the patient via a pump (fig. 1).

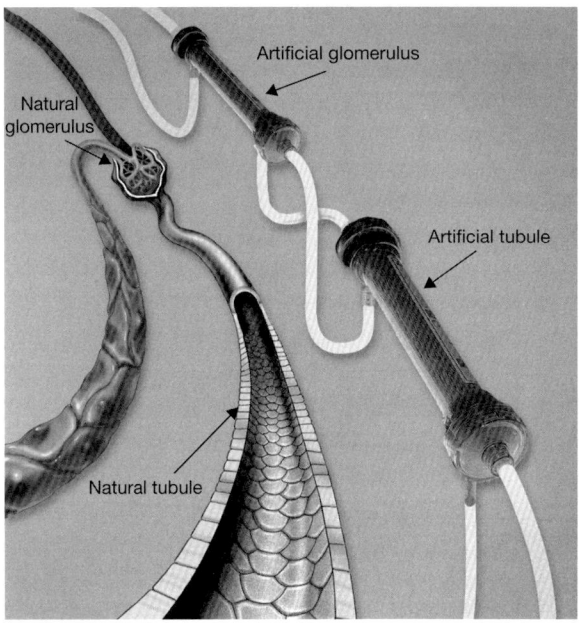

Fig. 2. Schema depicting the analogy of the bioartificial kidney with a human nephron.

Phase I/II Clinical Studies

These promising preclinical animal studies led the US Food and Drug Administration to approve a phase I safety clinical trial on 10 patients with ARF and multiorgan failure on CVVH, with predicted hospital mortality rates averaging above 85% according to the APACHE III score [19, 26]. A RAD, seeded with human renal proximal tubule cells from discarded deceased donor kidneys as described earlier, was placed in tandem with the CVVH system. The effluent ultrafiltrate from the hemofilter was shunted to the inside of the dialyzer fibers at a rate of 10 ml/min; the blood emanating from the hemofilter was pumped to the outside space of the dialyzer of the RAD with the aid of a pump at 150 ml/min. The hydraulic pressure inside the RAD filter was adjusted to reabsorb 5 ml/min to the blood compartment so that the processed 5 ml of ultrafiltrate was subsequently discarded. The blood was then returned to the patient with the aid of a pump as described earlier (fig. 1).

This phase I safety study showed that the RAD therapy can be safely done in combination with CVVH for up to 24 h under the protocol guidelines; it also demonstrated cardiovascular stability as well as increased urine output from the native kidneys and less vasopressor requirements with minimal side effects. No evidence of death directly related to RAD was shown among the study population with overall 30-day survival of 60% (6/10 patients). The study authors also showed that for the subset of patients who had excessive proinflammatory cytokine levels, RAD therapy resulted in significant declines in granulocyte colony-stimulating factor, interleukin-6 (IL-6), IL-10 and IL-6/IL-10 ratios. Moreover, RAD therapy evidenced metabolic and endocrinological activities with increased glutathione degradation and conversion of 25-OH-vitamin D_3 to 1,25-$(OH)_2$-vitamin D_3 analogously to the in vitro and the preclinical animal studies. These auspicious results led the FDA to approve a randomized, open-labeled multicenter phase II trial recently completed in which the RAD was employed in 58 critically ill ICU patients with dialysis-dependent AKI [27]. The major endpoint of the study was 28-day all-cause mortality. Patients were randomized 2 to 1 after 6 h of CVVH initiation to CVVH with or without RAD for 72 h. The impressive results showed lower 28-day mortality in patients receiving any duration of RAD therapy in comparison to patients receiving CVVH without RAD (34.3 vs. 55.6%); these results warrant a larger clinical study.

Conclusion

All the preclinical and clinical trials demonstrate a safety profile of the RAD therapy. RAD therapy not only replaces solute and water clearance but also replaces active reabsorptive transport, metabolic functions, and beneficial systemic effects that decrease the morbidity and mortality associated with AKI in critically ill patients. This research also shows that cell therapy can be a very promising step toward achieving better outcomes in patients with sepsis-induced AKI by using a 'living membrane', thereby crossing the limitations of diffusive and convective therapies by going beyond physics, artificial membranes, and urea clearance. While the use of cell therapy has not yet been tested in end-stage renal disease patients on chronic HD, it has the potential to become complementary to a variety of dialytic therapies because of its ability to replace many indispensable renal physiological functions.

References

1 Rangel-Frausto MS, Pittet D, Costigan M, Hwang T, Davis CS, Wenzel RP: The natural history of the systemic inflammatory response syndrome (SIRS). A prospective study. JAMA 1995;273: 117–123.
2 Neveu H, Kleinknecht D, Brivet F, Loirat P, Landais P: Prognostic factors in acute renal failure due to sepsis. Results of a prospective multicentre study. The French Study Group on Acute Renal Failure. Nephrol Dial Transplant 1996;11:293–299.
3 Benedict CR, Rose JA: Arterial norepinephrine changes in patients with septic shock. Circ Shock 1992;38:165–172.

4 Cumming AD, Driedger AA, McDonald JW, Lindsay RM, Solez K, Linton AL: Vasoactive hormones in the renal response to systemic sepsis. Am J Kidney Dis 1988;11:23–32.
5 Schrier RW, Wang W: Acute renal failure and sepsis. N Engl J Med 2004;351:159–169.
6 Wang W, Jittikanont S, Falk SA, Li P, Feng L, Gengaro PE, Poole BD, Bowler RP, Day BJ, Crapo JD, Schrier RW: Interaction among nitric oxide, reactive oxygen species, and antioxidants during endotoxemia-related acute renal failure. Am J Physiol Renal Physiol 2003;284:F532–F537.
7 Schor N: Acute renal failure and the sepsis syndrome. Kidney Int 2002;61:764–776.
8 Schiffl H, Lang SM, Fischer R: Daily hemodialysis and the outcome of acute renal failure. N Engl J Med 2002;346:305–310.
9 Ronco C, Bellomo R, Homel P, Brendolan A, Dan M, Piccinni P, La Greca G: Effects of different doses in continuous veno-venous haemofiltration on outcomes of acute renal failure: a prospective randomised trial. Lancet 2000;356:26–30.
10 Morgera S, Slowinski T, Melzer C, Sobottke V, Vargas-Hein O, Volk T, Zuckermann-Becker H, Wegner B, Muller JM, Baumann G, Kox WJ, Bellomo R, Neumayer HH: Renal replacement therapy with high-cutoff hemofilters: impact of convection and diffusion on cytokine clearances and protein status. Am J Kidney Dis 2004;43:444–453.
11 Schrier RW, Wang W, Poole B, Mitra A: Acute renal failure: definitions, diagnosis, pathogenesis, and therapy. J Clin Invest 2004;114:5–14.
12 Fujigaki Y, Goto T, Sakakima M, Fukasawa H, Miyaji T, Yamamoto T, Hishida A: Kinetics and characterization of initially regenerating proximal tubules in S3 segment in response to various degrees of acute tubular injury. Nephrol Dial Transplant 2006;21:41–50.
13 Maeshima A, Sakurai H, Nigam SK: Adult kidney tubular cell population showing phenotypic plasticity, tubulogenic capacity, and integration capability into developing kidney. J Am Soc Nephrol 2006;17:188–198.
14 Yamashita S, Maeshima A, Nojima Y: Involvement of renal progenitor tubular cells in epithelial-to-mesenchymal transition in fibrotic rat kidneys. J Am Soc Nephrol 2005;16:2044–2051.
15 Bussolati B, Camussi G: Adult stem cells and renal repair. J Nephrol 2006;19:706–709.
16 Oliver JA, Maarouf O, Cheema FH, Martens TP, Al-Awqati Q: The renal papilla is a niche for adult kidney stem cells. J Clin Invest 2004;114:795–804.
17 Lin F, Moran A, Igarashi P: Intrarenal cells, not bone marrow-derived cells, are the major source for regeneration in postischemic kidney. J Clin Invest 2005;115:1756–1764.
18 Humes HD, Fissell WH, Weitzel WF, Buffington DA, Westover AJ, MacKay SM, Gutierrez JM: Metabolic replacement of kidney function in uremic animals with a bioartificial kidney containing human cells. Am J Kidney Dis 2002;39:1078–1087.
19 Humes HD, Weitzel WF, Bartlett RH, Swaniker FC, Paganini EP, Luderer JR, Sobota J: Initial clinical results of the bioartificial kidney containing human cells in ICU patients with acute renal failure. Kidney Int 2004;66:1578–1588.
20 Smith PL, Buffington DA, Humes HD: Kidney epithelial cells. Methods Enzymol 2006;419:194–207.
21 Humes HD, MacKay SM, Funke AJ, Buffington DA: Tissue engineering of a bioartificial renal tubule assist device: in vitro transport and metabolic characteristics. Kidney Int 1999;55:2502–2514.
22 Nikolovski J, Gulari E, Humes HD: Design engineering of a bioartificial renal tubule cell therapy device. Cell Transplant 1999;8:351–364.
23 Fissell WH, Dyke DB, Weitzel WF, Buffington DA, Westover AJ, MacKay SM, Gutierrez JM, Humes HD: Bioartificial kidney alters cytokine response and hemodynamics in endotoxin-challenged uremic animals. Blood Purif 2002;20:55–60.
24 Fissell WH, Lou L, Abrishami S, Buffington DA, Humes HD: Bioartificial kidney ameliorates gram-negative bacteria-induced septic shock in uremic animals. J Am Soc Nephrol 2003;14:454–461.
25 Humes HD, Buffington DA, Lou L, Abrishami S, Wang M, Xia J, Fissell WH: Cell therapy with a tissue-engineered kidney reduces the multiple-organ consequences of septic shock. Crit Care Med 2003;31:2421–2428.

26 Humes HD, Weitzel WF, Bartlett RH, Swaniker FC, Paganini EP: Renal cell therapy is associated with dynamic and individualized responses in patients with acute renal failure. Blood Purif 2003;21:64–71.
27 Tumlin J, Wali R, Brennan K, Humes HD: Effect of the renal assist device (RAD) on mortality of dialysis-dependent acute renal failure: a randomized, open-labeled, multicenter, phase II trial (abstract). American Society of Nephrology (ASN) 38th Annual Meeting, Philadelphia, 2005.

Emil P. Paganini
Head, Section of Dialysis and Extracorporeal Therapy
Department of Nephrology and Hypertension Cleveland Clinic
9500 Euclid Avenue, M82
Cleveland, OH 44195 (USA)
Tel. +1 216 444 5792, Fax +1 216 444 7577, E-Mail paganie@ccf.org

Renal Replacement Therapy in Neonates with Congenital Heart Disease

Stefano Morelli, Zaccaria Ricci, Luca Di Chiara, Giulia V. Stazi, Angelo Polito, Vincenzo Vitale, Chiara Giorni, Claudia Iacoella, Sergio Picardo

Department of Pediatric Cardiology and Cardiac Surgery, Bambino Gesù Hospital, Rome, Italy

Abstract

Background: The acute renal failure (ARF) incidence in pediatric cardiac surgery intensive care unit (ICU) ranges from 5 to 20% of patients. In particular, clinical features of neonatal ARF are mostly represented by fluid retention, anasarca and only slight creatinine increase; this is the reason why medical strategies to prevent and manage ARF have limited efficacy and early optimization of renal replacement therapy (RRT) plays a key role in the outcome of cardiopathic patients. **Methods:** Data on neonates admitted to our ICU were prospectively collected over a 6-month period and analysis of patients with ARF analyzed. Indications for RRT were oligoanuria (urine output less than 0.5 ml/kg/h for more than 4 h) and/or a need for additional ultrafiltration in edematous patients despite aggressive diuretic therapy. **Results:** Incidence of ARF and need for RRT were equivalent and occurred in 10% of admitted neonates. Eleven patients of 12 were treated by peritoneal dialysis (PD) as only RRT strategy. PD allowed ultrafiltration to range between 5 and 20 ml/h with a negative balance of up to 200 ml over 24 h. Creatinine clearance achieved by PD ranged from 2 to 10 ml/min/1.73 m^2. We reported a 16% mortality in RRT patients. **Conclusion:** PD is a safe and adequate strategy to support ARF in neonates with congenital heart disease. Fluid balance control is easily optimized by this therapy whereas solute control reaches acceptable levels.

Copyright © 2007 S. Karger AG, Basel

The incidence of acute renal failure (ARF) in the neonatal intensive care unit (ICU) ranges between 5 and 20% of patients admitted [1, 2]. It is typically classified as prerenal disease, intrinsic renal disease (including vascular insults), and obstructive uropathy [3]. ARF in the newborn commonly occurs in the postnatal period because of hypoxic ischemic injury and toxic insults. Nephrotoxic ARF in newborns is usually associated with aminoglycoside antibiotics and nonsteroidal anti-inflammatory medications used to close a patent ductus arteriosus. Such alterations are usually reversible. Neonates with congenital heart disease are a particular subset of critically ill patients at high risk of ARF [2]. Low cardiac output and hypoxic/ischemic/nephrotoxic insults can induce prerenal/intrinsic kidney injury in the early phase of ICU admission. Neonates who undergo cardiac surgery for correction or palliation of congenital heart disease are exposed to additional risk factors for ARF. Once intrinsic renal failure has become established, management of the metabolic complications of ARF requires appropriate management of fluid balance, electrolyte status, acid-base balance, nutrition and, when appropriate, the initiation of renal replacement therapy (RRT). Renal replacement may be provided by peritoneal dialysis (PD), intermittent hemodialysis, or continuous hemofiltration. PD is the preferred modality of therapy for ARF in the neonate since it is relatively easy to perform, it does not require heparinization, and it can be safely administered to a hemodynamically unstable patient [1, 3].

The present chapter will discuss the epidemiology of RRT, its indication and prescription in neonates admitted to our cardiologic intensive care unit (CICU) with particular attention to several aspects of PD management and administration.

Patients

Between June and December 2006 we prospectively collected clinical data on neonates admitted to the CICU of our department, who represented 31% of the overall patients (114 over 367). Data are presented as median and interquartile range (IQR). Tests for nonparametric data were performed when necessary. A p value <0.05 was considered significant. The weight of the patients was 2.9 kg (1.8–3.9). Forty-one (36%) of these patients were premature. Eighty-two (72%) of them underwent elective and 22 (19%) urgent surgery. Nine percent of the admitted patents did not require any surgical procedure. Twelve patients (10%) required RRT. Six of these patients had aortic arch obstruction (coarctation/interruption); other children had miscellaneous diagnoses (2 cases of anomalous pulmonary venous return, 1 transposition of the great arteries, 1 double outlet right ventricle and 2 cases of right ventricular obstruction). Indications for renal support were oligoanuria (urine output less than 0.5 ml/kg/h for more than 4 h; 8–66%), need for additional ultrafiltration in edematous patients (2–16%) or both (2–16%), despite aggressive diuretic therapy (continuous furosemide infusion at a dose of 10 mg/kg/day). Four patients (3%) required preoperative RRT, whereas 9 (8%) were treated after surgery; 1 patient received RRT before and after intervention. Eleven patients were treated by PD. One patient was administered predilution hemofiltration during postoperative extracorporeal membrane oxygenation. Indications for stopping RRT included return of sufficient urine output to maintain or achieve negative fluid balance and normalization of serum electrolytes and acid-base status. When PD was prescribed, access to the peritoneal cavity was obtained through a periumbilical Tenckhoff catheter. Commercially available 1.36 and 2.7% glucose solutions were utilized. Solutions were lactate buffered. An exchange volume of 10 ml/kg was infused and left to dwell for 10–15 min, then drained for 10–15 min before the procedure was repeated; this technique allowed a median (IQR) exchange volume of 64 (40–100) ml/h and 1,500–2,000 liters/day (about 25 ml/kg/h and 600 ml/kg/day). Creatinine clearance achieved by PD ranged from 2 to 10 ml/min/1.73 m^2. Median (IQR) net ultrafiltration was 3 (1.5–6) ml/kg/h. This management allowed a median (IQR) negative balance of 15 (10–60) ml/kg/24 h. PD institution never induced hemodynamic instability neither did it trigger inotropic infusion increase. Transient hyperglycemia (serum glucose over 150 mg/dl) was observed on the first treatment day in 6 patients: this condition was generally not observed after the first 24 h and did not require any treatment. Persistent hyperlactatemia due to lactate-buffered PD solutions was present in 2 patients: these neonates were already hyperlactatemic at PD institution. Peritonitis was never observed. Catheter site induration and/or leakage from catheter insertion were observed in 2 patients but did not require any specific intervention.

Median (IQR) duration of RRT was 4 (3–8) days and in 11 (91%) patients full renal function had recovered after 8 days from ARF onset. Median (IQR) CICU length of stay among RRT patients was 24 (15–58) days, which was significantly higher than in other neonates [4 (6–11); $p < 0.05$]. Ten patients were transferred to the cardiology ward and 2 died. Mortality among RRT patients (16%) was significantly higher than among other neonates (8%; $p < 0.05$).

Discussion

Surgical interventions in case of complex congenital cardiac pathology are often performed in the neonatal period. With the use of improved cardiopulmonary bypass systems, new treatment modalities in preoperative, perioperative, and postoperative patient care and follow-up, morbidity and mortality rates have decreased [4]. There is also evidence that the neonatal kidney is more vulnerable to conditions of hemodynamic stress, with loss of autoregulation leading to blood-pressure-dependent renal blood flow and ischemia-induced renal injury. All of these conditions render the neonate more prone to complications of ischemia than the older infant or child [5, 6].

Hence, perioperative care of the neonate with congenital heart disease is challenging: organ failure should be avoided and, when it occurs, failing organs should be effectively replaced. In particular, clinical features of neonatal ARF are essentially represented by fluid retention, anasarca and only slight creatinine increase. In this case, the fluid balance must be optimized and acid-base/electrolyte equilibrium achieved. Medical interventions are often disappointing either for prevention or for the therapy of established ARF [7]. Neonatal RRT is mostly performed by PD [1, 3]. In the adult setting renal replacement is generally achieved by extracorporeal techniques, such as dialysis and hemofiltration, whether continuous or intermittent. These techniques have reached a good standard of care, specific practice patterns, dedicated technology and relatively high levels of consensus to the point that extracorporeal RRT is administered worldwide to 99% of adult ARF patient, PD being limited to underdeveloped countries [8]. Nonetheless, extracorporeal RRT requires an adequate vascular access, the heparinization of the patient and a relative amount of extracorporeal circulating blood volume; all these aspects identify PD as by far the ideal RRT in the neonate patient with ARF [1].

PD is an RRT modality where solutes and water are transported across a membrane that separates two compartments: the blood in the peritoneal capillaries and the dialysis solution in the peritoneal cavity, which is rendered hyperosmolar by a high concentration of glucose. Three transport processes are simultaneously involved during PD: uremic solutes and potassium diffuse from the peritoneal capillary blood into the PD solution whereas glucose, lactate and calcium diffuse in the opposite direction; simultaneously, the relative hyperosmolarity of dialytic solution leads to ultrafiltration of water (and associated solutes) across the membrane; finally, water and solutes are absorbed into the lymphatic system [9].

PD prescriptions during neonatal RRT generally tend to involve short dwell times (10–15 min) with relatively high exchange volumes (10–15 ml/kg/h) [10]. This technique enhances solute diffusion from blood to dialysate solution because a high concentration gradient is constantly maintained between the solutions. Ultrafiltration is optimized for the same reasons. Nonetheless, one of the main disadvantages of PD in the setting of adult renal disease is a relative lack of efficiency especially when the treatment of a highly catabolic patient is required. It must be said that during neonatal kidney dysfunction this is not often the case, and that serum creatinine levels are generally maintained below 1 mg/dl even in oligoanuric children. A long debate has been ongoing in recent years about the beneficial effect of removing inflammatory mediators through RRT [11]. This issue seems of outstanding importance in postoperative patients, and some authors have measured significant levels of proinflammatory molecules and cytokines on peritoneal drainage after cardiopulmonary bypass in neonates [12]. Several studies, however, noted a statistical difference in the percentage of fluid overload of children with severe renal dysfunction requiring RRT: at the time of dialysis initiation, survivors tended to have less fluid overload than nonsurvivors, especially in the setting of the multiorgan dysfunction syndrome [13–15]. Prevention of volume overload prompted some authors to deliver postoperative prophylactic PD in neonates and infants after complex congenital cardiac surgery [16].

We acknowledged this aspect of neonatal kidney dysfunction to the point that our ARF population substantially corresponded to patients who received PD: urine output and fluid balance needs triggered our intervention rather than serum creatinine levels. This led to a relatively high incidence of ARF/PD (10%), but also to early and timely treatments: after 4 h of oligoanuria PD catheter is inserted if not already present and RRT is started without the need for nephrologic counseling or dedicated staff. These PD schedules allowed ultrafiltration to range between 5 and 20 ml/h with a negative balance of up to 200 ml over 24 h. Creatinine levels at the stop of PD were significantly lower than initial values and they never exceeded 1 mg/dl (data not shown); solute control was presumably adequate with this strategy. We reported a 16% mortality in PD patients and, even if significantly higher than in non-RRT patients, it appeared lower than that reported by other authors (20–70%) [3, 17].

Conclusions

PD is a safe and adequate renal replacement technique to support ARF in neonates with congenital heart disease. Fluid balance control is easily optimized by this therapy, whereas solute control only reaches acceptable levels. New technology has recently been made available for diuretic resistant heart failure: this kind of miniaturized, highly accurate, slow efficiency ultrafiltration device that utilizes peripheral vascular access and allows relatively high ultrafiltration rates could be promising and so be adopted by pediatric critical care nephrology and significantly impact future strategies in pediatric RRT [18].

References

1 Andreoli SP: Acute renal failure in the newborn. Semin Perinatol 2004;28:112–123.
2 Gouyon JB, Guignard JP: Management of acute renal failure in newborns. Pediatr Nephrol 2000;14:1037–1044.
3 Moghal NE, Embleton ND: Management of acute renal failure in the newborn. Semin Fetal Neonatal Med 2006;11:207–213.
4 Feltes TF: Postoperative recovery of congenital heart disease; in Garson A, Bricker JT, Fisher DJ, Neish SR (eds): The Science and Practice of Pediatric Cardiology, ed 2. Baltimore, Williams & Wilkins, 1997.

5 Vanpee M, Blennow M, Linne T, et al: Renal function in very low birth weight infants: normal maturity reached during childhood. J Pediatr 1992;121:784–788.
6 Drukker A, Guignard JP: Renal aspects of the term and preterm infant: a selective update. Curr Opin Pediatr 2002;14:175–182.
7 Kellum JA: What can be done about ARF. Minerva Anestesiol 2004;70:181–188.
8 Uchino S, Kellum JA, Bellomo R, Doig GS, Morimatsu H, Morgera S, Schetz M, Tan I, Bouman C, Macedo E, Gibney N, Tolwani A, Ronco C; Beginning and Ending Supportive Therapy for the Kidney (BEST Kidney) Investigators: Acute renal failure in critically ill patients: a multinational, multicenter study. JAMA 2005;294:813–818.
9 Korbet SM, Kronfol NO: Acute peritoneal dialysis prescription; in Daugirdas JT, Blake PG, Ing TS (eds): Handbook of Dialysis, ed 3. Lippincott, Williams & Wilkins, 2001.
10 McNiece KL, Ellis EE, Drummond-Webb JJ, Fontenot EE, O'Grady CM, Blaszak RT: Adequacy of peritoneal dialysis in children following cardiopulmonary bypass surgery. Pediatr Nephrol 2005;20:972–976.
11 Venkataraman R, Subramanian S, Kellum JA: Clinical review: extracorporeal blood purification in severe sepsis. Crit Care 2003;7:139–145.
12 Bokesch PM, Kapural MB, Mossad EB, Cavaglia M, Appachi E, Drummond-Webb JJ, Mee RBB: Do peritoneal catheters remove pro-inflammatory cytokines after cardiopulmonary bypass in neonates? Ann Thorac Surg 2000;70:639–643.
13 Goldstein SL, Currier H, Graf JM, et al: Outcome in children receiving continuous veno-venous hemofiltration. Pediatrics 2001;107:1309–1312.
14 Goldstein SL, Somers MJ, Baum MA, et al: Pediatric patients with multi-organ dysfunction syndrome receiving continuous renal replacement therapy. Kidney Int 2005;67:653–658.
15 Foland JA, Fortenberry JD, Warshaw BL, et al: Fluid overload before continuous hemofiltration and survival in critically ill children: a retrospective analysis. Crit Care Med 2004;32:1771–1776.
16 Alkan T, Akcevin A, Turkoglu H, Paker T, Sasmazel A, Bayer, Ersoy C, Askn D, Aytac A: Postoperative prophylactic peritoneal dialysis in neonates and infants after complex congenital cardiac surgery. ASAIO J 2006;52:693–697.
17 Sorof JM, Stromberg D, Brewer ED, Feltes TF, Fraser CD: Early initiation of peritoneal dialysis after surgical repair of congenital heart disease. Pediatr Nephrol 1999;13:641–645.
18 Liang KV, Hiniker AR, Williams AW, Karon BL, Greene EL, Redfield MM: Use of a novel ultrafiltration device as a treatment strategy for diuretic resistant, refractory heart failure: initial clinical experience in a single center. J Card Fail 2006;9:707–714.

Zaccaria Ricci
Department of Pediatric Cardiology and Cardiac Surgery
Bambino Gesù Hospital
Piazza S. Onofrio
IT–00100 Rome (Italy)
Tel. +39 06 6859 3333, E-Mail z.ricci@libero.it

The DOse REsponse Multicentre International Collaborative Initiative (DO-RE-MI)[1]

G. Monti[a], M. Herrera[g], D. Kindgen-Milles[h], A. Marinho[i], D. Cruz[b], F. Mariano[c], G. Gigliola[d], E. Moretti[e], E. Alessandri[f], R. Robert[j], C. Ronco[b]

[a]Department of Anesthesiology and Intensive Care, Hospital Niguarda, Milan, [b]Department of Nephrology, Hospital San Bortolo, Vicenza, [c]Department of Nephrology and Dialysis Unit, CTO Hospital, Turin, [d]Department of Nephrology, Hospital Santa Croce e Carle, Cuneo, [e]Department of Anesthesiology and Intensive Care, Hospital Riuniti di Bergamo, Bergamo, and [f]Department of Anesthesiology and Intensive Care, Umberto I Hospital, Rome, Italy; [g]Regional Hospital, Malaga, Spain; [h]Anesthesiology Clinic, University of Düsseldorf, Düsseldorf, Germany; [i]Anesthesiology and Intensive Care Unit, Hospital Geral Sant Antonio, Porto, Portugal; [j]I Department of Intensive Care, University of Poitier, Poitier, France

Abstract

Background: Current practices for renal replacement therapy (RRT) in ICU remain poorly defined. The observational DOse REsponse Multicentre International collaborative initiative (DO-RE-MI) survey addresses the issue of how the different modes of RRT are currently chosen and performed. The primary endpoint of DO-RE-MI will be the delivered dose versus in ICU, 28-day, and hospital mortality, and the secondary endpoint, the hemodynamic response to RRT. Here, we report the first preliminary descriptive analysis after 1-year recruitment. **Methods:** Data from 431 patients in need of RRT with or without acute renal failure (mean age 61.2 + 15.9) from 25 centers in 5 countries (Spain, Italy, Germany, Portugal, France) were entered in electronic case report forms (CRFs) available via the website acutevision.net. **Results:** On admission, 51% patients came from surgery, 36% from the

---

[1]*Scientific Committee: Germany:* D. Kindgen-Milles; *France:* D. Journois (Paris), R. Robert (Poitiers); *Italy:* R. Fumagalli (Milan), C. Ronco (Vicenza), S. Vesconi (Milan); *Spain:* J. Maynar (Vitoria); *Portugal:* A. Marinho (Porto); *Steering Committee: Germany:* J.A. Amman (Düsseldorf); *Italy:* A. Brendolan (Vicenza), G. Monti (Milano), M. Formica (Turin), F. Mariano (Turin), M. Marchesi (Bergamo), S. Livigni (Turin), M. Maio (Turin), D. Silengo (Turin).

emergency department, and 16% from internal medicine. On admission, mean SOFA and SAPS II were 13 and 50, respectively. The first criteria to initiate RRT was the RIFLE in 38% (failure: 70%, injury: 25%, risk: 22%), the second the high urea/creatinine, and the third immunomodulation. A total of 3,010 cumulative CRF were reported: continuous venovenous hemodiafiltration (CVVHDF) 60%, continuous venovenous hemofiltration (CVVH) 15%, intermittent hemodialysis (IHD) 15%, high-volume hemofiltration (HVHF) 7%, continuous venovenous hemodialysis (CVVHD) 1%, and coupled plasma filtration adsorption/CVVD 2%. In 15% of cases, the patient was shifted to another modality. Mean blood flow rates (ml/min) in the different modalities were: 145 (CVVHDF), 200 (CVVH), 215 (IHD), 283 (HVHF), and 150 (CVVHD). Downtime ranged from 8 to 28% of the total treatment time. Clotting of the circuit accounted for 74% of treatment interruptions. **Conclusions:** Despite a large variability in the criteria of choice of RRT, CVVHDF remains the most used (49%). Clotting and clinical reasons were the most common causes for RRT downtime. In continuous RRT, a large variability in the delivered dose is observed in the majority of patients and often in the same patient from one day to another. Preliminary analysis suggests that in a large number of cases the delivered dose is far from the 'adequate' 35 ml/h/kg.

Copyright © 2007 S. Karger AG, Basel

Various continuous and intermittent modalities of renal replacement therapy (RRT) are currently used. In recent years remarkable advances in continuous RRT (CRRT) technology have been made, driven by nephrologists dedicated to improving efficiency and function. Today, however, intensivists are

Participating Doctors/Centers: Belgium: P. Honoré (Saint-Pierre Para-University Hospital, Ottignies-Louvain-La-Neuve); *France:* R. Robert (Hôpital Jean Bernard, CHU, Poitiers), D. Journois (Hôpital European Georges Pompidou, Paris), V. Labiotte (Hôpital de Lens), B. Thievenin (Hôpital de Maubeuge), O. Joannes Boyau (University Hospital of Bordeaux; *Germany:* A. Amman, D. Kindgen-Milles (University of Düsseldorf, Düsseldorf); *Portugal:* A. Marinho (Hospital Geral Sant Antonio, Porto), A. Lafuente (Hospital de Penafiel, Penafiel), A. Santos (Hospital Geral Sant Antonio, Porto); *Italy:* E. Moretti (Ospedale Riuniti di Bergamo, Bergamo), M. Cerisara (Ospedale Maggiore, Crema), P. Inguaggiato, G. Gigliola (Ospedale Santa Croce e Carle, Cuneo), W. Morandini (Ospedale Valle Camonica, Esine), R. Fumagalli, R. Rona (University of Milan, Ospedale S. Gerardo, Monza), A. Sicignano (Policlinico di Milano, Milan), G.P. Monti, S. Vesconi (Ospedale Niguarda, Milan), G. Slaviero (IRCCS San Raffaele, Milan), F. Mariano, L. Tedeschi (CTO Ospedale, Turin), S. Livigni, M. Maio (Ospedale G. Bosco, Turin), Z. Ricci, E. Alessandri (Policlinico Umberto 1, Rome), A. Brendolan, D. Cruz (Ospedale San Bortolo, Vicenza), G. Marchesi (Ospedale Bolognini di Seriate, Seriate); *Spain:* J. Maynar (Hopital de Vitoria), Teresa Doñate, A. Leon (Hopital Gral. De Catalunya, Sant Cugat del Vallés), M. Herrera (Hospital Carlos Haya, Malaga), F. Labayen, J. Maynar (Hospital Santiago Apostol, Vitoria), Á. Montero, J. Sánchez-Izquierdo (Hospital 12 De Octubre, Madrid), J. Luño, E. Junco (Hospital Gregorio Marañon, Madrid), J.A. Sánchez Tomero, C. Bernix (Hospital De La Princesa, Madrid), J. Bustos (Hospital Virgen De La Salud, Toledo), J. Cruz, J. Moll (Hospital La Fe, Valencia), R. Cabadas (Policlinico De Vigo, Vigo).

the most familiar with these techniques. Nevertheless, in some countries such as the USA, CRRT is still infrequently employed [1]. Other modalities include intermittent hemodialysis (IHD), slow extended daily dialysis [2], or daily hemodialysis [3]. Some of the reasons for the considerable variability worldwide in extracorporeal treatment of acute renal failure (ARF) include local practice (e.g. whether management is by nephrologists or intensivists), the center's experience with the various techniques, organization and health resources. Various methods of extracorporeal treatment, whether intermittent or continuous, are currently being employed and no guidelines exist. This variability was highlighted in the Beginning and Ending of Supportive Therapy for the Kidney (BEST Kidney) trial, which collected data on ARF management in 1,743 patients in 54 ICU from 23 countries worldwide [4].

The practice of CRRT has apparently not changed, even following the prospective studies conducted by Ronco et al. [5]. Despite the positive findings of that prospective trial, the practice of a higher intensity CRRT has not been widely adopted into routine ICU practice. The most outstanding examples are Australia and New Zealand, where almost 100% of treatments are CRRT. A survey of several units active in the Australian and New Zealand Intensive Care Society Clinical Trials Group (Bellomo, unpubl. data, 2002) found that very few units had adopted the intensive CRRT regimen proposed by Ronco and coworkers [4]. Data from such Australian units show instead that the vast majority ($>$90%) prescribe a 'fixed' standard CRRT dose of 2 l/h, which is not adjusted for body weight. Thus, a 100-kg man would receive 20 ml/kg · h – the dose shown to have the worst outcome in the study by Ronco and coworkers [4]. In another recent study that involved several Australian units (the BEST Kidney study), the median body weight for Australian patients was 80 kg, thus indicating that the vast majority receive a CRRT intensity of approximately 25 ml/kg · h of effluent. Finally, although in the study conducted by Ronco and colleagues [4] the technique of CRRT was uniformly in the form of continuous venovenous hemofiltration (CVVH) with postfilter fluid replacement, current practice includes a variety of techniques in addition to CVVH, such as continuous venovenous hemodiafiltration (CVVHDF). Furthermore, scarce information exists on the practice of CRRT in Europe, particularly regarding the actually delivered dose of therapy in critically ill patients with ARF (i.e. in those who could potentially derive more benefit from high-volume convective therapy).

In a recent preliminary collaborative study [6] we reported that there was no significant difference between prescribed and delivered ultrafiltration rate (both in ml/min and in l/h), which was related to the reduced downtime associated with the technique. However, of greater relevance is that the dose of dialysis was over 40 ml/kg · h.

If we are to understand how dialysis doses are actually delivered in routine clinical practice in ICUs, an observational clinical study is needed to confirm how, to what extent and with what clinical indication the different modalities of RRT are administered. With this in mind we initiated the DOse REsponse Multicentre International collaborative initiative (DO-RE-MI) survey. The protocol was published [7]. The survey was listed as CRG110600093 in the Cochrane Renal Group. The primary endpoint of DO-RE-MI is mortality (ICU mortality, 28-day mortality and hospital mortality), and the secondary endpoint is the hemodynamic response to RRT, expressed as percentage reduction in noradrenaline (norepinephrine) requirement to maintain blood pressure. Here, we present the preliminary descriptive results from the DO-RE-MI survey. The survey started on June 1, 2005 and will be terminated in December 2007.

Materials and Methods

All data from incident patients admitted to the ICU in need of RRT with or without ARF are entered into electronic case report forms (CRFs) and downloaded via the internet onto a server [8, 11]. The following rules are applied without exception:

First, all patient data are entered anonymously. To this aim, each center has a code, and patients are consecutively assigned a progressive number. Under no circumstances is there any written or oral transmission of data that may make the identification of the patient possible. Failure to adhere to this is immediately followed by cancellation of the data from the website by the webmaster.

Second, data for each patient are entered into a separate CRF. These data may be copied from paper CRFs in order to make the reporting of data from bed to computer station easier. All fields may be amended at any time until the patient's CRF is completed and closed. At this point, one may access the patient's CRF but it is no longer possible to amend it. In the case of overt inconsistency, corrections must be detailed in writing (E-Mail) by the person responsible for the data quality of the center. In no cases are corrections permitted in the absence of an express written request. The person responsible for data quality will have access to his or her center's CRF. A registry collects the correspondence between the person responsible for data quality and the center.

Third, completion of some fields in the CRF is mandatory. Failure to complete them prevents progression to the following CRF and closure of the opened CRF. Failure to complete a CRF electronically results in the patient being excluded from the study.

Finally, each center is enabled to open CRFs of its own patients but under no circumstances the CRFs of patients from other centers.

Definitions

'Treatment interruption' is defined as when the treatment is stopped and resumed within 18 h. In the case of treatment interruption, the CRF is continued and the treatment that follows is considered in the context of the preceding one. The only exception is when, after RRT interruption, the modality is changed.

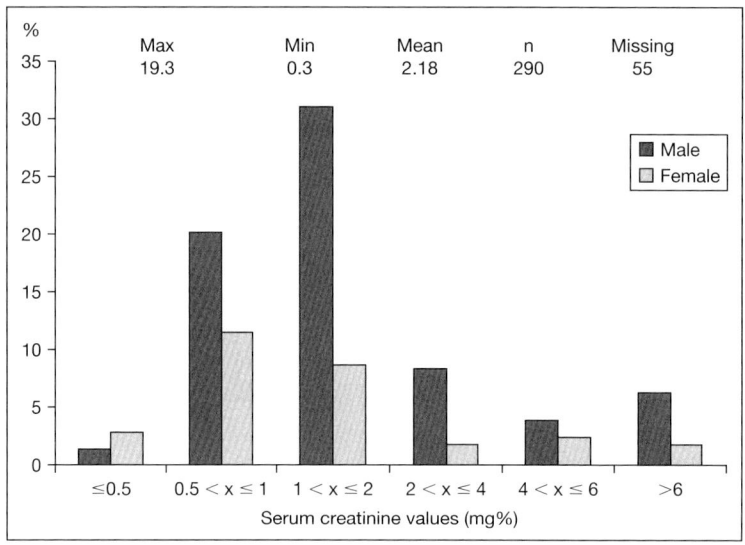

Fig. 1. Percentage of patients on admission categorized according to six ranges of creatinine values.

'Treatment end' is defined as when a given RRT is stopped because of clinical or other factors for more than 12 h or when clinical or other factors have changed since the start of RRT. Should the patient be started on another RRT, then the latter is considered a new one.

In case the modality is changed, a new CRF is to be filled in. This is followed by a new CRF.

Each center is asked to define the clinical/practical reasons for changing a modality. The change of modality may be necessary after treatment is interrupted. In this case, the following treatment is considered a new treatment. This CRF aims to provide information on why the modality was chosen. It is similar to the CRF.

Results

In the first year, 434 patients were recruited from a total of 37 centers (15 in Italy, 14 in Spain, 5 in Portugal, 2 in France, 1 in Germany; mean age: 61.2 ± 15.9 years, mean weight: 68.9 + 18.2 kg). The most common diagnosis for admission was septic shock. However, the assignment to admission diagnosis is still fraught by the highest percent of missing data (56%). On admission, patients had serum creatinine between 1 and 2 mg%, and in 16% creatinine values were over 4 mg% with a clearcut difference between male and female patients (fig. 1). The most common indication for initiating treatment was high

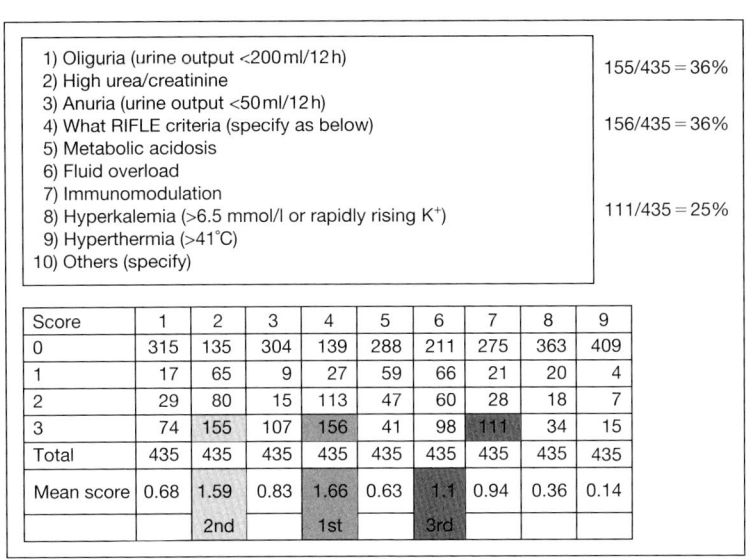

Fig. 2. Criteria to initiate RRT as listed. Score 0 = No priority; score 3 = top priority. Shaded fields indicate the top three criteria either as those with the highest values for score 3 (horizontal order) or as mean score (vertical order).

plasma urea/creatinine levels followed by one of the RIFLE criteria and by hoping to achieve some kind of immunomodulation (fig. 2). The different centers remarkably differed as to the type of RRT of choice. In some centers, all RRT modalities were available. In some cases, patients were started on high-volume hemofiltration (HVHF) to CVVH or CVHDF, in some others pulse HVHF was initiated and finally some other centers were using almost only one type of RRT modality either CVVHDF or IHD (fig. 3). A total of 3,010 cumulative CRF were reported: CVVHDF 60%, CVVH 15%, IHD 15%, HVHF 7%, continuous venovenous hemodialysis (CVVHD) 1%, and coupled plasma filtration adsorption/CVVD 2%. In 15% of cases, the patient was shifted to another modality. Mean blood flow rates (ml/min) in the different modalities were: 145 (CVVHDF), 200 (CVVH), 215 (IHD), 283 (HVHF), and 150 (CVVHD). The parameters to deliver RRT also appeared to differ among the different modalities but evidently in the same modality among different centers (fig. 4). Downtime was precisely calculated as it will allow to calculate the really delivered dose of dialysis. Downtime ranged from 8 to 28% of the total treatment time. The major causes of treatment interruption were clotting of the circuit in 74%, failure of the vascular access in 11%, clinical reasons in 10%, and machine alarms in 2%. In 3% there was no specific reason. As shown in figure 5, the anticoagulation regimen remarkably differed in the different modalities of RRT.

Center No.	CVVHDF	CVVH	IHD	HVHF	CVVHD	CPFA	
1	16						16
2		29	4	8	1	2	44
3	7	10		3			20
4	1	16	1	9			27
6	7	1	11	1			20
7	8		29		5		42
8	5	2				2	9
9	12					1	13
13	4	4					8
14	82	3	2				87
17		1					1
19		1					1
24	12				1		13
25			1	2			3
27		1					1
29		1					1
36		2	1	2			5
44		2		1			3
45		1					1
Total	154	74	49	26	7	5	315
%	49%	23%	16%	8%	2%	2%	

Fig. 3. Each center is listed by its identification number. Each column indicates the number of treatments per each type of RRT modality. CPFA = Coupled plasma filtration adsorption.

Discussion

The practice of CRRT has been the subject of much debate. Only a few prospective randomized studies have been performed and published on the relationship between CRRT and outcome, and so conclusions are difficult to draw [9, 10]. As emphasized in a recent editorial [11], in the field of artificial organs, prospective observational studies, despite their inherent limitations, have been performed because they are more affordable but are also capable of providing useful information from a practical and medical standpoint.

The results of the present study were obtained from the analyses of the CRFs and they are at the present time descriptive in nature. Calculation of the actually delivered dose of dialysis is ongoing for each patient undergoing different modalities, taking into consideration the downtimes and the change of modality if any. It is anticipated that the dose of dialysis is surprisingly much lower than that which is recognized as being the minimally required one to ensure enhanced survival in a randomized clinical trial [4]. The large variability

	Buffer		
	Bicarbonate	Lactate	%Bic
CVVHDF	986	208	83%
CVVH	519	100	83%
IHD	291	0	100%
HVHF	125	53	70%
CVVHD	41	0	100%
CPFA	32	0	100%

Blood flow rates			
Treatment	Q_B	SD	n
HVHF	250	63.4	175
IHD	209	46.8	291
CVVH	207	45.1	609
CPFA	204	51.9	32
CVVHD	175	33.4	41
CVVHDF	146	37.3	1,200

Vascular access

Vascular access	CVVH		CVVHDF		IHD		CVVHD		CPFA		HVHF	
Femoral catheter	60	71%	150	87%	18	35%	1	14%	4	67%	26	74%
Jugular catheter	15	18%	9	5%	15	29%	2	29%		0%	7	20%
Subclavian catheter	10	12%	13	8%	19	37%	4	57%	2	33%	2	6%
	85		172		52		7		6		35	

Fig. 4. Summary of data showing the use of the kind of buffer, the blood flow rate and vascular access in each type of RRT modality. CPFA = Coupled plasma filtration adsorption.

in current practices even in a relatively restricted geographical area allows us already to make the hypothesis that RRT in ICU is still undefined not only conceptually but also in very practical terms.

The present survey has the well-known limitations of any observational study. First, due to the largely geographically dispersed sample of centers, the information will not make it possible for us to sort out any practice more specific for any of the countries represented by the centers. Nevertheless, due to the high number of patients and the design of the CRF [see 7], we feel that the survey may throw lighten on how RRT is currently done in ICUs and should hopefully provide tentative answers to as yet undefined questions such as: what are the criteria for beginning and ending treatment?; what is the currently delivered dose of dialysis?; how is fluid control taken care of?; what schedules are mostly used?; how is technology used (or not used)?, and finally what are the reasons for downtime in RRT? Finally, the survey will probably show the 'real world'

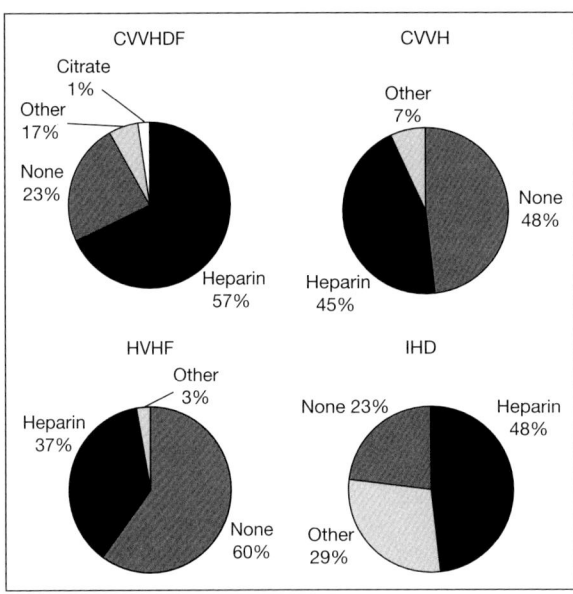

Fig. 5. Anticoagulation regimen in the different RRT modalities.

in terms of the actually delivered dose outside the rigor of a randomized controlled study. It will also imply that comparisons of therapies in observational studies may be fraught by the large variability which might be at the origin of a great background noise that possible disturbs the emergence of as yet undefined advantages in outcome measures.

References

1 Mehta RL, Letteri JM: Current status of renal replacement therapy for acute renal failure. A survey of US nephrologists. The National Kidney Foundation Council on Dialysis. Am J Nephrol 1999;19:377–382.
2 Marshall MR, Golper TA, Shaver MJ, Alam MG, Chatoth DK: Sustained low-efficiency dialysis for critically ill patients requiring renal replacement therapy. Kidney Int 2001;60:777–785.
3 Schiffl H, Lang SM, Fischer R: Daily hemodialysis and the outcome of acute renal failure. N Engl J Med 2002;346:305–310.
4 Uchino S, Kellum JA, Bellomo R, et al: Acute renal failure in critically ill patients. A multinational, multicentre study. JAMA 2005;294:813–818.
5 Ronco C, Bellomo R, Homel P, Brendolan A, Dan M, Piccinni P, La Greca G: Effects of different doses in continuous veno-venous haemofiltration on outcomes of acute renal failure: a prospective randomized trial. Lancet 2000;356:26–30.
6 Brendolan A, D'Intini V, Ricci Z, Bonello M, Ratanarat R, Salvatori G, Bordoni V, DeCal M, Andrikos E, Ronco C: Pulse high volume haemofiltration. Int J Artif Organs 2004;27:398–403.

7   Kindgen-Milles D, Journois D, Fumagalli R, et al: Study protocol: the dose response Multicentre International Collaborative Initiative (DO-RE-MI). Crit Care 2005;9:R396–R406.
8   Acute Vision. http://www.acutevision.it.
9   Metha R, McDonald B, Gabbai FB, Pahl M, Pascual MT, Farkas A, Kaplan RM; Collaborative Group for Treatment of ARF in the ICU: A randomized clinical trial of continuous versus intermittent dialysis for acute renal failure. Kidney Int 2001;60:1154–1163.
10 Tonelli M, Manns B, Feller-Kopman D: Acute renal failure in the intensive care unit: a systematic review of the impact of dialytic modality on mortality and renal recovery. Am J Kidney Dis 2002;100:158–160.
11 Ronco C: Evidence-based medicine: can we afford it? Int J Artif Organs 2004;27:819–820.

Dr. Gianpaola Monti
Department of Anesthesiology and Intensive Care, Hospital Niguarda
IT–20162 Milan (Italy)
E-Mail Gianpaola.monti@ospedaleniguarda.it

Clinical Effects of Polymyxin B-Immobilized Fiber Column in Septic Patients

Dinna N. Cruz[a,b], Rinaldo Bellomo[c], Claudio Ronco[a]

[a]Department of Nephrology, Ospedale San Bortolo, Vicenza, Italy; [b]Section of Nephrology, Department of Medicine, St. Luke's Medical Center, Quezon City, Philippines; [c]Department of Intensive Care, Austin Hospital, Melbourne, Vic., Australia

Abstract

Endotoxin is one of the principal biological substances that cause gram-negative septic shock. Lack of clinical success with antiendotoxin or anticytokine therapy has shifted interest to extracorporeal therapies to reduce circulating levels of the mediators of sepsis. Direct hemoperfusion with polymyxin-B-immobilized fiber (PMX-F) is a promising treatment of gram-negative sepsis in critically ill patients. Because of the high affinity of polymyxin B for endotoxin, the rationale underlying extracorporeal therapy would be to remove circulating endotoxin by adsorption, thus preventing progression of the biological cascade of sepsis. In a systematic review of 28 studies (pooled sample size 1,390 patients), the preliminary results of which are described here, PMX-F therapy appeared to significantly lower endotoxin levels, improve blood pressure, and reduce mortality. However, publication bias and lack of blinding need to be considered. These encouraging results need to be verified with large-scale controlled clinical trials.

Copyright © 2007 S. Karger AG, Basel

Sepsis is a common problem encountered in the hospital population and is responsible for spending a large proportion of hospital resources. It involves a complex interaction between bacterial factors and the host immune system producing a systemic inflammatory response. When uncontrolled, it is associated with multiple organ failure and high mortality rates.

Endotoxin, a lipopolysaccharide released from gram-negative bacteria, has been implicated as a potent, prototypical stimulus of the immune response to bacterial infection [1]. It causes the release of cytokines such as interleukin-1 and TNF-α, activates complements and coagulation factors, and is an ideal potential therapeutic target to treat septic shock. However, antiendotoxin drug therapies, such as monoclonal antibodies or lipopolysaccharide-neutralizing

Table 1. Summary of the results of a systematic review of direct hemoperfusion with PMX-F column

Parameter	Studies n	Patients n	Effect after PMX-F	Statistically significant
Blood pressure	12	275	↑	yes
Dopamine/dobutamine dose	4	96	↓	yes
PaO$_2$/FiO$_2$ ratio	7	151	↑	yes
Endotoxin level	17	435	↓	yes
Mortality	15	885	↓	yes

proteins, failed to demonstrate a clinical benefit in trials [1]. Polymyxin B is a cationic cyclic polypeptide antibiotic which binds with high affinity to endotoxin, neutralizing its effects. Unfortunately, clinical use is limited by significant nephro- and neurotoxicity.

Lack of clinical success with antiendotoxin or anticytokine therapy has shifted interest to extracorporeal therapies to reduce circulating levels of the mediators of sepsis. To overcome the toxicity issues of systemic administration of polymyxin B yet take advantage of its ability to neutralize lipopolysaccharide, polymyxin B bound and immobilized to polystyrene fibers (PMX-F) was developed [2]. The resulting cartridge can be used in direct hemoperfusion, the rationale being the removal of circulating endotoxin by adsorption, thus preventing progression of the biological cascade of sepsis.

PMX-F has been shown to bind and neutralize endotoxin in both in vitro and in vivo studies [1, 2]. An improvement in survival in both murine and canine models of endotoxic shock has also been demonstrated, especially when combined with antibiotic therapy [1]. Clinical trials, both randomized and observational, have reported encouraging results but are limited by small sample size. To assimilate the published clinical experience with PMX-F a systematic review of 28 published studies was conducted [3]. Some preliminary results of this review are presented here (table 1).

Effects on Endotoxin Level

Summary data from 19 studies [9 randomized controlled trials (RCT), 10 observational] confirmed that PMX-F was able to reduce levels of circulating endotoxin [4–22]. Overall, endotoxin levels decreased by 22 pg/ml (95% CI 18.1–25.8 pg/ml), representing a decrease of 33–80% from baseline levels [3]. Studies which enrolled patients with higher baseline endotoxin levels tended to show a greater reduction after PMX-F. Work by Uriu et al. [20] suggests that reduction in blood endotoxin concentration by PMX-F therapy positively correlated with the reduction in cardiac output. In a European multicenter study, significant improvements in cardiac index and left ventricular stroke work were noted after treatment with PMX [21]. This change was not accompanied by a change in heart rate or in pulmonary capillary wedge pressure, and can therefore be interpreted as reflecting an increase in stroke volume. The clinical effects of endotoxin removal, specifically on hemodynamics, were therefore examined in further detail.

Effects on Blood Pressure

Mean arterial pressure (MAP) data were reported in 12 studies (2 RCT, 10 observational) [11, 14, 17–19, 21–27], while systolic blood pressure (SBP) was reported in 2 studies [14, 28]. Baseline MAP ranged from 68 to 87 mm Hg and SBP from 80 to 108 mm Hg. All the individual studies showed an improvement in blood pressure. Overall, the MAP increased by 19 mm Hg (95% CI 15–22 mm Hg, $p < 0.001$) [3] and SBP increased by 24 mm Hg in both studies. This represented an increase of approximately 27% from baseline values. For MAP, effect sizes of RCTs were smaller than of nonrandomized studies. This is consistent with the observation in meta-analyses that trials of lower quality also tend to show larger treatment effects [29, 30]. Similar to the results with endotoxin, studies which enrolled patients with lower baseline blood pressure demonstrated a bigger change after PMX-F therapy.

In critically ill patients, it is often difficult to interpret blood pressure in isolation, as vasoactive agents can be manipulated to alter the blood pressure. In 4 studies (1 RCT, 3 observational), there was a trend toward a decrease in the dose of dopamine or dobutamine after PMX-F [11, 18, 20, 22]. Overall, the dose was decreased by 1.8 µg/kg/min (95% CI 0.4–3.3 µg/kg/min, $p = 0.01$). In these studies, there was also an increase in mean MAP after PMX-F (range 16–28 mm Hg). Only two cohort studies reported noradrenaline doses [18, 27]. The mean noradrenaline dose decreased by 0.2–0.9 µg/kg/min after PMX-F treatment, while the MAP increased by 16–26 mm Hg. It has been reported that left ventricular stroke work index (LVSWI) is less likely to be affected by vasopressors than is blood pressure. Therefore, LVSWI serves as a good indicator of improvement in circulation, without having to take the vasopressor dose into consideration. Unfortunately there was insufficient data on this and other hemodynamic parameters for meta-analysis. Individual studies have demonstrated that PMX-F therapy elevated the stroke volume and improved LVSWI in patients with septic shock [11, 21]. In fact, an improvement in LVSWI within 24 h of PMX therapy appeared to predict survival [11].

Table 2. Selected studies on sepsis and its effect on mortality

Intervention	Ref. No.	Group	APACHE II score	Mortality %
PMX-F	3	Standard	10.6–28	61.8
		PMX-F	16.7–28.5	34.4
Low-dose steroids	33	Placebo	SAPS II 57 ± 19	63.0
		Steroids	SAPS II 60 ± 19	53.0
Protective ventilation	34	Conventional	24.0 ± 7.0	71.0
		Protective	24.0 ± 6.0	38.0
EGDT	35	Standard	20.4 ± 7.4	49.2
		EGDT	21.4 ± 6.9	33.3
APC	36	Placebo	25.0 ± 7.8	30.8
		APC	24.6 ± 7.6	24.7

APC = Activated protein C; EGDT = early goal-directed therapy.

Effects on Mortality

Data on mortality, variably reported as 14-day [17], 28-day [10, 11, 15, 21, 24, 31], 30-day [6, 25, 32], and 60-day mortality [8], were available from 11 studies. Mortality was reported but the length of follow-up was not clearly stated in another 4 studies [4, 7, 9, 23]. Pooled mortality for the 15 studies (8 RCT, 7 non-RCT, 885 patients) was 61.8% in the conventional therapy group and 34.4% in the PMX-F group. Looking specifically at the 9 studies that reported 28- to 30-day mortality (704 patients), PMX-F therapy appeared to significantly reduce mortality compared with conventional medical therapy (RR 0.54; 95% CI 0.43–0.68). However, this must be viewed in the context of the mortality of the patients under standard medical management (table 2). The overall mortality of 61.8% seen in the conventional therapy group was comparable to that seen in a French multicenter study on moderate-dose corticosteroid therapy (63%) [33] and a Brazilian study on protective ventilation (71%) [34], but higher than that reported in a study on early goal-directed therapy (46.5%) [35] and activated protein C (30.8%) [36]. Moreover, mortality in the conventional therapy group within the various studies averaged 58% (range 0–88.6%), which is higher than the predicted mortality based on APACHE II scores (mean 44.8%; range 9.7–63.9%). Whether the apparent beneficial effect of PMX-F is due to a reduction in deaths related to endotoxin removal by the therapy itself or due to a higher than expected mortality in the conventional therapy group is not clear. Nevertheless, these results are provocative, and should pave the way for large-scale RCTs. The presence of publication bias was explored with a funnel plot analysis. Interestingly, contrary to expectation, smaller studies (3 studies, n = 17–35) had point estimates for mortality favoring standard medical therapy (data not shown).

The efficacy of therapies for sepsis may be affected by the baseline severity of illness of the patients. For instance, activated protein C was deemed relatively cost effective when targeted to patients with severe sepsis, greater severity of illness (an APACHE II score of 25 or more), and a reasonable life expectancy if they survive the episode of sepsis [37]. We attempted a supplementary analysis of the pooled mortality data along these lines of thinking. There were 9 studies (n = 678) in which the baseline APACHE scores of enrolled patients were <25, and 6 studies (n = 207) in which the baseline APACHE scores were 25 or higher. There appeared to be a greater effect seen in the sicker patients (RR 0.45, 95% CI 0.30–0.68) as opposed to the less severely ill patients (RR 0.58, 95% CI 0.45–0.75). However, such exploratory analyses are performed post hoc with the use of summary data, rather than patient level data, and should therefore be interpreted with caution. At this point, these findings remain hypothesis-generating rather than conclusive.

As with all systematic reviews, these findings are limited by the quality of the primary studies. One third of the included studies were nonrandomized, and the vast majority of published clinical experience comes from Japan, where PMX-F has been in clinical use for over 10 years. Differences between intrinsic patient characteristics and/or medical practices may limit the generalizability of these results to a more heterogeneous septic population. In addition, because of the small number of controlled studies, meta-analyses on blood pressure and vasopressor dose were performed on data from single cohorts (i.e. pre- and post-PMX-F), regardless of study design. With the use of single-arm studies, there will tend to be a bias towards improvement since the data will tend to overrepresent the survivors, particularly in a high-mortality disease such as sepsis. Nevertheless, this systematic review remains the most comprehensive summary to date of the clinical effects of direct hemoperfusion with PMX-F.

Conclusion

Polymyxin B binds endotoxin, one of the principal biological substances that cause gram-negative septic shock, but has adverse nephro- and neurotoxin effects. Direct hemoperfusion with PMX-F column would theoretically allow removal of circulating endotoxin without systemic side effects. Based on published literature, PMX-F therapy appears to effectively reduce endotoxin levels and have some positive effects on blood pressure, use of vasoactive agents, and mortality. Despite these encouraging results, randomized, controlled clinical trials are necessary to definitively determine its efficacy as a form of therapy in sepsis. Since the publication of this article, this work has been updated.

Acknowledgment

This work has been made possible by the International Society of Nephrology-funded fellowship of Dr. Dinna Cruz, and was presented in part at the annual meeting of the American Society of Nephrology, San Diego, Calif., November 2006.

References

1 Manocha S, Feinstein D, Kumar A, Kumar A: Novel therapies for sepsis: antiendotoxin therapies. Expert Opin Investig Drugs 2002;11:1795–1812.
2 Shoji H: Extracorporeal endotoxin removal for the treatment of sepsis: endotoxin adsorption cartridge (Toraymyxin). Ther Apher Dial 2003;7:108–114.
3 Cruz D, Corradi V, Polanco N, Bellomo R, Ocampo C, de Cal M, Ronco C: Effectiveness of polymyxin B-immobilized fiber column in sepsis: a metaanalysis (abstract). J Am Soc Nephrol 2006;17:768A.
4 Nakamura T, Ushiyama C, Sukuzi Y, Inoue T, Shoji H, Shimada N, Koide H: Combination therapy with polymyxin B-immobilized fibre haemoperfusion and teicoplanin for sepsis due to methicillin-resistant *Staphylococcus aureus*. J Hosp Infect 2003;53:58–63.
5 Nakamura T, Kawagoe Y, Matsuda T, Ueda Y, Koide H: Effects of polymyxin B-immobilized fiber on urinary N-acetyl-B-glucosaminidase in patients with severe sepsis. ASAIO J 2004;50:563–567.
6 Nakamura T, Ebihara I, Shoji H, Ishiyama C, Suzuki S, Koide H: Treatment with polymyxin B-immobilized fiber reduces platelet activation in septic shock patients: decrease in plasma levels of soluble P-selectin, platelet factor-4 and beta-thromboglobulin. Inflamm Res 1999;68:171–175.
7 Nakamura T, Ushiyama C, Suzuki Y, Shoji H, Shimada N, Koide H: Hemoperfusion with polymyxin-B immobilized fiber for urinary albumin excretion in septic patients with trauma. ASAIO J 2002;48:244–248.
8 Nakamura T, Ushiyama C, Suzuki Y, Osada S, Inoue T, Shoji H, Hara M, Shimada N, Koide H: Hemoperfusion with polymyxin-B immobilized fiber in septic patients with methicillin-resistant *Staphylococcus aureus*-associated glomerulonephritis. Nephron Clin Pract 2003;94:C33–C39.
9 Nakamura T, Kawagoe Y, Matsuda T, Koide H: Effect of polymyxin B-immobilized fiber on bone resorption in patients with sepsis. Intensive Care Med 2004;30:1838–1841.
10 Nemoto H, Nakamoto H, Okada H, Sugahara S, Moriwaki K, Arai M, Kanno Y, Suzuki H: Newly developed polymyxin B-immobilized fibers improve the survival of patients with sepsis. Blood Purif 2001;19:361–369.
11 Suzuki H, Nemoto H, Nakamoto H, Okada H, Sugahara S, Kanno Y, Moriwaki K: Continuous hemodiafiltration with polymyxin B-immobilized fiber is effective in patients with sepsis syndrome and acute renal failure. Ther Apher 2002;6:234–240.
12 Nakamura T, Ebihara I, Shimada N, Suzuki S, Ushiyama C, Shoji H, Koide H: Effects of hemoperfusion with polymyxin B-immobilized fibre on serum neopterin and soluble interleukin-2 receptor concentrations in patients with septic shock. J Infect 1998;37:241–247.

13 Nakamura T, Ebihara I, Shimada N, Koide H: Changes in plasma erythropoietin and interleukin-6 concentrations in patients with septic shock after hemoperfusion and polymyxin B-immobilized fiber. Intensive Care Med 1998;24:1271–1276.
14 Nakamura T, Kawagoe Y, Matsuda T, Shoji H, Ueda Y, Tamura N, Ebihara I, Koide H: Effect of polymyxin B-immobilized fiber on blood metalloproteinase-9 and tissue inhibitor of metalloproteinase-1 levels in acute respiratory distress syndrome patients. Blood Purif 2004;22: 256–260.
15 Nakamura T, Kawagoe Y, Sukuzi T, Shoji H, Ueda Y, Kobayashi N, Koide H: Changes in plasma interleukin-18 by direct hemoperfusion with polymyxin B-immobilized fiber in patients with septic shock. Blood Purif 2005;23:417–420.
16 Shimada N, Nakamura T, Takayashi Y, Tanaka A, Shoji H, Sekizuka K, Ebihara I, Koide H: Effects of polymyxin B-immobilized fiber on serum phosphate concentrations in patients with sepsis. Nephron 2000;86:359–360.
17 Tani T, Hanasawa K, Endo Y, Yoshioka T, Kodama M, Kaneko M, Uchiyama Y, Akizawa T, Takahashi K, Sugai T: Therapeutic apheresis for septic patients with organ dysfunction: hemoperfusion using a polymyxin-B immobilized column. Artif Organs 1998;22:1038–1044.
18 Tojimbara T, Sato S, Nakajima I, et al: Polymyxin B-immobilized fiber hemoperfusion after emergency surgery in patients with chronic renal failure. Ther Apher Dial 2004;8:286–292.
19 Ueno T, Sugino M, Nemoto H, Shoji H, Kakita A, Watanabe M: Effect over time of endotoxin adsorption therapy in sepsis. Ther Apher Dial 2005;9:128–136.
20 Uriu K, Osajima A, Kamochi M, Watanabe H, Aibara K, Kaizu K: Endotoxin removal by direct hemoperfusion with an adsorbent column using polymyxin B-immobilized fiber ameliorates systemic circulatory disturbance in patients with septic shock. Am J Kidney Dis 2002;39:937–947.
21 Vincent J, Laterre P, Cohen J, Burchardi H, Bruining H, Lerma F, Wittebole X, de Backer D, Brett S, Marzo D, Nakamura H, John S: A pilot-controlled study of a polymyxin B-immobilized hemoperfusion cartridge in patients with severe sepsis secondary to intra-abdominal infection. Shock 2005;23:400–405.
22 Kojika M, Sato N, Yaegashi Y, Suzuki Y, Suzuki K, Nakae H, Endo S: Endotoxin adsorption therapy for septic shock using polymyxin B-immobilized fibers (PMX): evaluation by high-sensitivity endotoxin assay and measurement of the cytokine production capacity. Ther Apher Dial 2006;10: 12–18.
23 Ono S, Tsujinomoto H, Matsumoto A, Ikuta S, Kinoshita M, Michizuki H: Modulation of human leukocyte antigen-DR on monocytes and CD16 on granulocytes in patients with polymyxin B-immobilized fiber. Am J Surg 2004;188:150–156.
24 Tsujimoto H, Ono S, Hiraki S, Majima T, Kawarabayashi N, Sugasawa H, Kinoshita M, Hiraide H, Mochizuki H: Hemoperfusion with polymyxin B-immobilized fibers reduced the number of CD16+CD14+ monocytes in patients with septic shock. J Endotoxin Res 2004;10:229–237.
25 Tsushima K, Kubo K, Koizumi T, Yamamoto H, Fujimoto K, Hora K, Kan-nou Y: Direct hemoperfusion using a polymyxin B immobilized column improves acute respiratory distress syndrome. J Clin Apheresis 2002;17:97–102.
26 Ikeda T, Ikeda K, Nagura M, et al: Clinical evaluation of PMX-DHP for hypercytokinemia caused by septic multiple organ failure. Ther Apher Dial 2004;8:293–298.
27 Casella G, Monti G, Terzi V, Pulici M, Ravizza A, Vesconi S: Terapie 'non-convenzionali' nello shock settico refrattario: esperienza clinica con Polymyxin B. Minerva Anestesiol 2006;72(suppl 1): 63–67.
28 Ebihara I, Nakamura T, Shimada N, Shoji H, Koide H: Effect of hemoperfusion with polymyxin B-immobilized fiber on plasma endothelin-1 and endothelin-1 mRNA in monocytes from patients with sepsis. Am J Kidney Dis 1998;32:953–961.
29 Moher D, Pham B, Jones A, Cook DJ, Jadad AR, Moher M, Tugwell P, Klassen TP: Does quality of reports of randomised trials affect estimates of intervention efficacy reported in meta-analyses? Lancet 1998;352:609–613.
30 Schulz KF, Chalmers I, Hayes RJ, Altman DG: Empirical evidence of bias. Dimensions of methodological quality associated with estimates of treatment effects in controlled trials. JAMA 1995;273:408–412.

31   Nakamura T, Matsuda T, Suzuki Y, Shoji H, Koide H: Polymyxin B-immobilized fiber in patients with sepsis. Dial Transplant 2003;32:602–607.
32   Tsugawa K, Koyonagi N, Hashizume M, Wada H, Ayukawa K, Akahoshi K, Tomikawa M, Sugimachi K: Results of endotoxin absorption after a subtotal resection of the small intestine and a right hemicolectomy for severe superior mesenteric ischemia. Hepatogastroenterology 2002;49: 1303–1306.
33   Annane D, Sébille V, Charpentier C, Bollaert PE, Francois B, Korach JM, Capellier G, Cohen Y, Azoulay E, Troche G, Chaumet-Riffaut P, Bellissant E: Effect of a treatment with low doses of hydrocortisone and fludrocortisone on mortality in patients with septic shock. JAMA 2002;288: 862–971.
34   Amato MB, Barbas CS, Medeiros DM, Magaldi RB, Schettino GP, Lorenzi-Filho G, Kairalla RA, Deheinzelin D, Munoz C, Oliveira R, Takagaki TY, Carvalho CR: Effect of a protective-ventilation strategy on mortality in the acute respiratory distress syndrome. N Engl J Med 1998;338:347–354.
35   Rivers E, Nguyen B, Havstad S, Ressler J, Muzzin A, Knoblich B, Peterson E, Tomlanovich M; Early Goal-Directed Therapy Collaborative Group: Early goal-directed therapy in the treatment of severe sepsis and septic shock. N Engl J Med 2001;345:1368–1377.
36   Bernard GR, Vincent JL, Laterre PF, LaRosa SP, Dhainaut JF, Lopez-Rodriguez A, Steingrub JS, Garber GE, Helterbrand JD, Ely EW, Fisher CJ Jr; Recombinant Human Protein C Worldwide Evaluation in Severe Sepsis (PROWESS) Study Group: Efficacy and safety of recombinant human activated protein C for severe sepsis. N Engl J Med 2001;344:699–709.
37   Manns BJ, Lee H, Doig CJ, Johnson D, Donaldson C: An economic evaluation of activated protein C treatment for severe sepsis. N Engl J Med 2002;347:993–1000.

Dinna N. Cruz
Department of Nephrology, San Bortolo Hospital
Viale Rodolfi 37
IT–36100 Vicenza (Italy)
Tel. +39 0444 753650, Fax +39 0444 753973, E-Mail dinnacruzmd@yahoo.com

Author Index

Adib-Conquy, M. 101
Alessandri, E. 434
Arroyo, V. 17
Aucella, F. 287

Bagshaw, S.M. 1, 236, 340
Bainotti, S. 405
Baldwin, I. 178, 191
Bellomo, R. 1, 10, 75, 167, 236, 309, 340, 411, 444
Berger, M.M. 267
Bonventre, J.V. 39, 213
Brendolan, A. 411
Bussolati, B. 250

Calzavacca, P. 167
Camussi, G. 250
Cano, N.J.M. 112
Cavaillon, J.-M. 101
Chioléro, R. 267
Crepaldi, C. 411
Cruz, D.N. 309, 411, 434, 444

Davenport, A. 259, 333
De Becker, W. 185
de Cal, M. 411
Delaney, A. 236
Devarajan, P. 203
Di Chiara, L. 428
Di Paolo, S. 287
Dragun, D. 75

Faenza, S. 396
Ferramosca, E. 396
Formica, M. 405

Gesualdo, L. 287
Gigliola, G. 434
Giorni, C. 428
Gressens, B. 354, 371, 387

Haase, M. 75, 340
Haase-Fielitz, A. 75, 340
Herrera, M. 434
Hirasawa, H. 365
Honoré, P.M. 354, 371, 387
Hoste, E.A.J. 32

Iacoella, C. 428
Inguaggiato, P. 405
Issa, N. 419

Joannes-Boyau, O. 354, 371, 387
Joannidis, M. 92
Jones, D. 236

Kellum, J.A. 10, 32, 158
Kindgen-Milles, D. 434

Lameire, N. 325
Langenberg, C. 1
Leverve, X.M. 112
Licari, E. 167

Mancini, E. 396
Mariano, F. 434
Marinho, A. 434
Matsuda, K. 365
Messer, J. 419
Molitoris, B.A. 227
Monti, G. 434
Morelli, S. 428
Moretti, E. 434

Nalesso, F. 411

Oda, S. 365
Opal, S.M. 220

Paganini, E.P. 419
Picardo, S. 297, 428
Pinsky, M.R. 47, 133
Polanco, P.M. 133
Polito, A. 428

Ricci, Z. 197, 297, 428
Robert, R. 434
Ronco, C. XI, 1, 10, 167, 197, 236, 297, 309, 340, 411, 434, 444

Sandoval, R.M. 227
Santoro, A. 396
Schetz, M. 275
Schmit, X. 64
Schusterschitz, N. 92
Sever, M.S. 325

Singer, M. 119
Stazi, G.V. 428

Taccone, F. 64
Tolwani, A.J. 320

Van Biesen, W. 304, 325
Vanholder, R. 304, 325
Veys, N. 304
Vincent, J.-L. 24, 64
Vitale, V. 428

Wheeler, T.S. 320
Wille, K.M. 320
Wratten, M.L. 405

Yassin, J. 119

Subject Index

N-acetyl-β-D-glucosaminidase (NAG), acute kidney injury biomarker studies 216
Acid-base balance, see Anion gap; Strong ion gap
Activated partial thromboplastin time (aPTT), waveform analysis in sepsis-induced renal injury 224, 225
Acute renal failure (ARF)
  definitions 32, 33
  epidemiology 33, 34
  outcomes
    end-stage kidney disease 35
    length of hospital stay 35
    long-term outcome 36
    mortality 35, 36
  recovery 250, 251
Acute tubular necrosis (ATN)
  conventional view 3, 4
  conversion from prerenal azotemia 4–6
  sepsis association evidence 7, 8
Albumin
  blood detoxification approaches
    molecular adsorbent recirculation system 412, 413, 417
    overview 411, 412
    plasma filtration adsorption dialysis circuits 413, 414, 416
      purification process 415–417
      tricompartmental dialyzer 413, 414
    Prometheus system 413
    single-pass albumin dialysis 413
  fluid resuscitation in sepsis-associated acute kidney injury 171

hepatorenal syndrome management in spontaneous bacterial peritonitis 22, 23
toxin transport 412
Alveolar oxygen (P_AO_2), calculation 122–124
Anion gap (AG)
  calculation 158–160
  interpretation 161–163
APACHE scoring
  limitations 97, 98
  mortality prediction 93
  performance 97
Apolipoprotein E, cardiopulmonary-bypass-associated acute kidney injury alleles 84–86
Argatroban, heparin-induced thrombocytopenia anticoagulation for renal replacement therapy 263, 265
Arterial pressure
  catheterization indications 139
  determinants 137, 138
  physiological significance 136, 137
  polymyxin-B-immobilized column effects in sepsis 446
  variations during ventilation 138, 139

Beginning and Ending Supportive Therapy for the Kidney (BEST Kidney) investigators, continuous renal replacement therapy survey 298, 299
Bicarbonate
  cardiopulmonary-bypass-associated acute kidney injury prevention 348, 349, 436

continuous renal replacement therapy
    buffer 290–291
Bioartificial kidney
  animal studies 421, 422
  clinical trials 424, 425
  hemofilter and circuit 422, 423
  in vitro studies 421
  proximal tubular cell therapy 420, 421
  sepsis-induced acute kidney injury
    management rationale 419, 420
Bioartificial liver (BAL)
  approaches 397, 398
  prospects 397
Biomarkers
  acute kidney injury
    N-acetyl-β-D-glucosaminidase 216
    cystatin C 207, 208, 217
    ideal characteristics 204
    interleukin-18 208, 209, 217
    kidney injury molecule-1
      208, 215, 216
    neutrophil gelatinase-associated
      lipocalin 205–207, 216, 217
    prospects 209, 217, 218
    rationale 204, 214
  definition 214, 215
  immunodysregulation in sepsis
    cellular markers 106–109
    plasma biomarkers 102–106
  sepsis-induced renal injury
    activated partial thromboplastin time
      waveform analysis 224, 225
    interleukin-6 223
    lipopolysaccharide assays 222, 223
    monocyte function markers 223, 224
    overview 220, 221
    procalcitonin 225
    triggering receptor expressed on
      myeloid cells (TREM-1) 225
  validation 215

Calcitonin, *see* Procalcitonin
Calcium, replacement fluid composition
    293
Cardiac output (CO)
  measurement 149, 150
  resuscitation end point 151, 152

Cardiopulmonary-bypass-associated acute
    kidney injury
  costs and outcomes 341
  epidemiology 82, 341
  hemolysis
    free hemoglobin and kidney injury
      induction 346, 347
    pigment nephropathy 343, 344
    sources and magnitude during bypass
      344–346
  ischemia-reperfusion injury 342, 343
  oxidative stress 343
  pathophysiology 341, 342
  prevention 342, 347, 349
  single-nucleotide polymorphisms
    82–87
Catechol-O-methyltransferase (COMT),
    cardiopulmonary-bypass-associated acute
    kidney injury single-nucleotide
    polymorphisms 83, 86, 87
CD11b, sepsis pathophysiology 59
CD14
  immunodysregulation marker in sepsis
    104, 105
  inflammation modulation 48
Central venous catheter (CVC)
  complications
    acute malfunction 282
    central vein stenosis and thrombosis
      282, 283
    infection 279–281
  insertion and care 277, 278
Central venous pressure (CVP)
  determinants 140
  measurement 139, 140
  monitoring 140–142
Cerebral perfusion pressure (CPP), renal
    replacement therapy effects
  anticoagulant effects 337
  continuous renal replacement therapy
    336, 337
  intermittent hemodialysis 336
  peritoneal dialysis 335, 336
Chemokines, acute kidney injury
    pathophysiology 43, 44
Citrate, continuous renal replacement
    therapy buffer 290

Congenital heart disease, renal replacement therapy in neonates 428–432
Continuous hemodiafiltration, cytokine-adsorbing hemofilter 365–369
Continuous renal replacement therapy (CRRT)
  alarms 193
  circuit patency
    access catheter 180
    anticoagulant use 179, 180
    blood pump flow speed 181
    clotting sites 179
    membrane factors 181
    substitution fluid administration 181, 182
    venous chamber 182
  computerized monitoring 188
  cytokine-adsorbing hemofilter in continuous hemodiafiltration 365–369
  dialysate and replacement fluid composition
    administration route 294, 295
    buffers 288–292
    calcium 293
    glucose 293, 294
    physical properties 294
    potassium 292, 293
    sodium 292
  emergency team 192
  equipment 188
  heparin-induced thrombocytopenia patients, see Heparin-induced thrombocytopenia
  information technology applications
    dialysis accuracy and safety 201
    dialysis dose calculation 198–201
    prospects 202
  intermittent hemodialysis comparison 305–308, 310–317
  intracranial pressure and cerebral perfusion pressure effects 336, 337
  medical emergency team
    exchange rate 359
    global approach in intensive care unit 355–357
    rationale 360–363
  Saint-Pierre Para-University Hospital experience 357, 358
  vascular access 358, 359
  modes 311
  nursing
    emergency team
      challenges 194, 195
      overview 360–363
      rationale 193, 194
    models 192, 193
    tasks in intensive care unit 188, 189
  nutrition, see Nutrition, acute kidney disease
  program features for intensive care unit 186, 187
  proinflammatory mediator removal 312
  Renal Disaster Relief Task Force 330
  staff training and education 182, 183, 187
  standard settings 189
  surveys
    Beginning and Ending Supportive Therapy for the Kidney 298, 299
    Dose Response Multicenter International 301, 302
    International Course on Critical Care Nephrology 299, 300
    summary of findings 302
  thermal balance 313
  vascular access
    catheter insertion and care 277, 278
    complications
      acute malfunction 282
      central vein stenosis and thrombosis 282, 283
      infection 279–281
    temporary dialysis catheter
      characteristics 276, 277
      insertion sites 277
Coupled plasma filtration adsorption (CPFA)
  sepsis management
    clinical trials 408
    prospects 409
    rationale 406
    technical characteristics 407

Critical care nephrology
   collaboration with nephrologists and
      other specialists 29, 30
   continuous renal replacement therapy, see
      Continuous renal replacement therapy
   general intensivist role 28, 29
   overview 25, 27
Crush syndrome, see also Renal Disaster
   Relief Task Force
   mortality 326
   rhabdomyolysis and acute kidney injury
      pathophysiology 326, 327
Cystatin C, acute kidney injury biomarker
   studies 207, 208, 217

Danaparoid, heparin-induced
   thrombocytopenia anticoagulation
   for renal replacement therapy
   262, 263
Dermatan sulfate, heparin-induced
   thrombocytopenia anticoagulation for
   renal replacement therapy 263
Disaster, see Renal Disaster Relief Task Force
Diuretics, acute kidney injury management
   clinical trials 237
   survey
      additional comments 245
      clinical outcomes 244, 245
      clinical response assessment
         243, 244
      demographics 240
      drug administration
         dosing 241
         drug types 240, 241
         indications and timing 241–243
         route 241
      sampling frame 239
      study design 237–239
      summary of findings 245–248
Dose Response Multicenter International
   (DO-RE-MI)
   participating centers 435
   renal replacement therapy use patterns
      301, 302, 436, 438–442
   study design 437, 438

Endotoxin, see Lipopolysaccharide

Fluid resuscitation, sepsis-associated acute
   kidney injury
   albumin 171
   animal studies 173–175
   clinical trials 169–171
   gelatin 172
   lactate 172, 173
   rationale 168, 169
   saline 172
   starches 171, 172
Fondaparinux, heparin-induced
   thrombocytopenia anticoagulation for
   renal replacement therapy 263
Furosemide, see Diuretics, acute kidney
   injury management

Gelatin, fluid resuscitation in sepsis-
   associated acute kidney injury 172
Glucose
   control in renal replacement therapy 272
   replacement fluid composition 293, 294

Heat shock proteins (HSPs), intracellular
   inflammatory response in sepsis 54–57
Hemodialysis, see Continuous renal
   replacement therapy; Intermittent
   hemodialysis; Slow extended daily
   dialysis
Hemodynamic monitoring
   arterial pressure
      catheterization indications 139
      determinants 137, 138
      physiological significance 136, 137
      variations during ventilation 138, 139
   cardiac output
      measurement 149, 150
      resuscitation end point 151, 152
   central venous pressure
      determinants 140
      measurement 139, 140
      monitoring 140, 141
   mixed venous oxygen saturation
      central venous oxygen saturation 150,
         151
      measurement 150
      resuscitation end point 151, 152
   overview 133, 134

Hemodynamic monitoring (continued)
　physiological basis 134, 135
　pulmonary artery occlusion pressure
　　catheterization controversies 152, 153
　　left ventricular function
　　　afterload 147, 148
　　　performance 148, 149
　　measurement 144, 145
　　pleural pressure effects 145, 146
　　pulmonary edema 146
　　pulmonary vasomotor tone 146
　pulmonary artery pressure
　　catheterization
　　　controversies 152, 153
　　　indications 141
　　determinants 144
　rationale 134
Hemofiltration, see Continuous
　hemodiafiltration; Continuous renal
　replacement therapy; High-volume
　hemofiltration
Heparin-induced thrombocytopenia (HIT)
　anticoagulant alternatives for renal
　　replacement therapy
　　direct thrombin inhibitors 263, 264
　　heparinoids 262, 263
　　monitoring 265
　autoantibody detection 261, 262
　incidence 259
　pathophysiology 261, 262
　scoring 261
Hepatorenal syndrome (HRS)
　liver support approaches, see also
　　Bioartificial liver
　　artificial systems 398
　　conventional systems 399
　　novel treatments 399–403
　　overview 397, 398
　pathogenesis of type 2 disease 18, 19
　spontaneous bacterial peritonitis and
　　type 1 disease
　　pathophysiology 20, 21
　　prevention of syndrome 21
　　treatment
　　　liver transplantation 22
　　　vasoconstrictors and albumin 22, 23
　types 17–19

High-volume hemofiltration (HVHF)
　animal studies 374–376
　clinical trials in acute kidney injury 372,
　　373, 376–380, 382, 392
　dosing 377, 380
　filter porosity 391, 392
　indications 372, 380
　mediator delivery hypothesis 390, 391
　peak concentration hypothesis 388, 389
　prospects for study 380–382, 391–393
　rationale 388–391
　threshold immunomodulation hypothesis
　　389, 390
Hybrid hemodialysis, see Slow extended
　daily dialysis
Hypoxia
　causes 129
　classification 129
　definition 128

Immunodysregulation, sepsis
　cellular markers 106–109
　definition 101, 102
　plasma biomarkers 102–106
Incidence, acute kidney injury 34, 203
Inflammation, acute kidney injury
　pathophysiology
　　innate immune response 42
　　intracellular inflammatory response
　　　heat shock proteins 54–57
　　　mitochondria role 57, 58
　　　nuclear factor-κB 51–54
　　leukocyte-endothelial cell interactions
　　　42, 43
　　overview 40, 41
　　prospects for study 45
　　response modulation by resolvins and
　　　protectins 44
　　systemic and cellular events 48–50
　　tubular contribution to injury 43, 44
Information technology (IT)
　continuous renal replacement therapy
　　applications
　　dialysis accuracy and safety 201
　　dialysis dose calculation 198–201
　　prospects 202
　definition 197

Intensive care unit (ICU), *see* Critical care nephrology
Intercellular adhesion molecule-1 (ICAM-1), acute kidney injury pathophysiology 42, 43
Interleukin-1 (IL-1), inflammatory response 49, 50
Interleukin-6 (IL-6)
  cardiopulmonary-bypass-associated acute kidney injury single-nucleotide polymorphisms 83, 84, 86
  inflammatory response 49, 50
  removal by cytokine-adsorbing hemofilter 368
  sepsis-induced renal injury biomarker 223
Interleukin-10 (IL-10)
  immunodysregulation marker in sepsis 105
  inflammatory response 49, 50
  removal by cytokine-adsorbing hemofilter 368
  sepsis-associated acute kidney injury single-nucleotide polymorphisms 79, 81
Interleukin-18 (IL-18), acute kidney injury biomarker studies 208, 209, 217
Intermittent hemodialysis
  continuous renal replacement therapy comparison 305–308, 310–317
  dialysate and replacement fluid composition
    administration route 294, 295
    buffers 288–292
    calcium 293
    glucose 293, 294
    physical properties 294
    potassium 292, 293
    sodium 292
  intracranial pressure and cerebral perfusion pressure effects 336
  Renal Disaster Relief Task Force 330
  vascular access
    catheter insertion and care 277, 278
    complications
      acute malfunction 282
      central vein stenosis and thrombosis 282, 283
      infection 279–281
    temporary dialysis catheter
      characteristics 276, 277
      insertion sites 277
Intracranial pressure (ICP), renal replacement therapy effects
  anticoagulant effects 337
  continuous renal replacement therapy 336, 337
  intermittent hemodialysis 336
  peritoneal dialysis 335, 336
  sustained intracranial pressure surge management 337
Ischemia-reperfusion injury, cardiopulmonary-bypass-associated acute kidney injury 342, 343

Kidney injury molecule-1 (KIM-1), acute kidney injury biomarker studies 208, 215, 216

Lactate
  anaerobic metabolism 120, 122
  fluid resuscitation in sepsis-associated acute kidney injury 172, 173
Left ventricular function, pulmonary artery occlusion pressure studies
  afterload 147, 148
  performance 148, 149
Lepirudin, heparin-induced thrombocytopenia anticoagulation for renal replacement therapy 264
Lipopolysaccharide (LPS)
  inflammatory response 444
  polymyxin B removal, *see* Polymyxin-B-immobilized column
  sepsis-induced renal injury biomarker assays 222, 223
Lipopolysaccharide-binding protein (LBP), immunodysregulation marker in sepsis 104
Liver transplantation, hepatorenal syndrome management in spontaneous bacterial peritonitis 22

Mesenchymal stem cell (MSC)
  acute kidney injury pathophysiology 44
  plasticity 251
  renal repair role 251, 252, 256
  therapy in acute renal failure 252, 253

Mitochondria, intracellular inflammatory response in sepsis 57, 58
Mixed venous oxygen saturation (SvO2)
  central venous oxygen saturation 150, 151
  measurement 150
  resuscitation end point 151, 152
Molecular adsorbent recirculating system (MARS)
  blood detoxification 412, 413, 417
  liver support 400, 401
Mortality
  acute kidney injury 36, 37, 203
  continuous renal replacement therapy versus intermittent hemodialysis 314, 315
  crush syndrome 326
  multiple organ failure 24, 25
  polymyxin-B-immobilized column outcomes in sepsis 447, 448
  prediction scores
    APACHE score 93
    limitations 97, 98
    mortality prediction model 95, 96
    performance 97
    physician judgement 98
    Sequential Organ Failure Assessment 96
    simplified acute physiology score 94, 95
    therapeutic intervention scoring system 96
  sepsis 68, 167, 405
  slow extended daily dialysis 323
Mortality prediction model (MPN) 95, 96
Multi-photon microscopy
  applications 227–229
  challenges and prospects 233, 234
  kidney dynamic process studies 229, 231–233
Multiple organ failure (MOF), see also Critical care nephrology
  classification 65, 66
  incidence 67, 68
  mortality 24, 25
  predisposing factors 66
  resuscitation and reversal of organ injury 60, 61

Sequential Organ Failure Assessment score 25, 26

NADPH oxidase, sepsis-associated acute kidney injury single-nucleotide polymorphisms 80, 81
Neonates, congenital heart disease renal replacement therapy 428–432
Neutrophil-gelatinase-associated lipocalin (NGAL), acute kidney injury biomarker studies 205–207, 216, 217
Nuclear factor-κB (NF-κB)
  activation 52–54
  intracellular inflammatory response in sepsis 51–54, 60
Nursing, see Continuous renal replacement therapy
Nutrition, acute kidney disease
  amino acids 115, 116
  assessment 267, 268
  energy 115, 268
  lipids 115
  malnutrition prevalence 267
  pathophysiology considerations 113, 114
  renal replacement therapy considerations 114, 268
  glucose control 272
  hyperglycemia 269
  nutritional support 270, 271
  protein catabolism 268, 269
  vitamin and mineral loss 269, 270
  vitamins and minerals 116, 117, 271, 272

Oxygen delivery
  alveolar oxygen 122–124
  anaerobic metabolism and lactate excess 120, 122
  capillary to pulmonary vein 126
  cellular uptake 127, 128
  oxidative metabolism 119, 120
  red blood cells 124–126
  supply dependency 130
  tissue transfer 126, 127

Peritoneal dialysis (PD)

congenital heart disease renal
  replacement therapy in neonates
  429–432
 intracranial pressure and cerebral
  perfusion pressure effects 335, 336
 Renal Disaster Relief Task Force 331
PIRO, sepsis staging 66
Plasma filtration adsorption dialysis
  (PFAD)
 circuits 413, 414, 416
 purification process 415–417
 tricompartmental dialyzer 413, 414
Plasmapheresis, liver support 400
Platelet factor 4 (PF4), heparin-induced
  thrombocytopenia pathophysiology 261
Polymorphisms, see Single-nucleotide
  polymorphisms 7
Polymyxin-B-immobilized column
 blood pressure effects 446
 endotoxin removal efficiency 445, 446
 lipopolysaccharide binding 445
 mortality studies in sepsis 447, 448
Potassium, replacement fluid composition
  292, 293
Prerenal azotemia
 acute tubular necrosis conversion 4–6
 conventional view 3, 4
 therapeutic implications 6, 7
Procalcitonin, sepsis-induced renal injury
  biomarker 225
Prometheus system
 blood detoxification 413
 liver support 400–403
Protectins, acute kidney injury
  pathophysiology 44
Pulmonary artery occlusion pressure
  (Ppao)
 catheterization controversies 152, 153
 left ventricular function
  afterload 147, 148
  performance 148, 149
 measurement 144, 145
 pleural pressure effects 145, 146
 pulmonary edema 146
 pulmonary vasomotor tone 146
Pulmonary artery pressure
 catheterization

 controversies 152, 153
 indications 141
 determinants 144
Pulmonary edema, pulmonary artery
  occlusion pressure measurement 146

Renal assist device, see Bioartificial kidney
Renal blood flow (RBF), sepsis-associated
  acute kidney injury 168
Renal Disaster Relief Task Force
  (RDRTF)
 crush syndrome
  mortality 326
  rhabdomyolysis and acute kidney
   injury pathophysiology 326, 327
 dialysis
  continuous renal replacement therapy
   330
  intermittent hemodialysis 330
  peritoneal dialysis 331
 earthquakes as renal disasters
  326, 327
 interventions 328
 origins 327
 tasks
  field hospitals 330
  fluid administration 328, 329
  severity assessment 328
  transport 330
Renal stem cell
 distribution in kidney 255
 markers 254, 255
 properties 253, 254
 renal repair role 251, 252, 256
Resolvins, acute kidney injury
  pathophysiology 44
Rhabdomyolysis, acute kidney injury
  pathophysiology 326, 327
RIFLE criteria
 acute kidney injury 2, 11, 14, 15, 35, 75,
   167, 168
 cardiac patients 13, 14
 classes 33, 34
 extracorporeal membrane oxygenation
   patients 14
 origins 11
 validation studies 12, 13

Saline, fluid resuscitation in sepsis-
  associated acute kidney injury 172
Sepsis
  acute kidney injury association
    epidemiology 77, 406
    fluid resuscitation
      albumin 171
      animal studies 173–175
      clinical trials 169–171
      gelatin 172
      lactate 172, 173
      rationale 168, 169
      saline 172
      starches 171, 172
    pathophysiology 168
    single-nucleotide polymorphisms
      77–82
  acute tubular necrosis association
    evidence 7, 8
  bioartificial kidney, see Bioartificial
    kidney
  classification 65, 66
  coupled plasma filtration adsorption
    clinical trials 408
    prospects 409
    rationale 406
    technical characteristics 407
  critical care nephrology 25, 27
  endotoxin removal, see Polymyxin-B-
    immobilized column
  hot response 66
  immunodysregulation
    cellular markers 106–109
    definition 101, 102
    plasma biomarkers 102–106
  incidence 67, 68
  infection 66
  intracellular inflammatory response
    heat shock proteins 54–57
    mitochondria role 57, 58
    nuclear factor-κB 51–54
  mortality 68, 167, 405
  multiple organ failure, see Multiple organ
    failure
  organ dysfunction 66, 67
  outcome studies 68–72
  predisposing factors 66

  pro-inflammatory and anti-inflammatory
    activity 58–60, 406
  systemic cellular events 48–50
  systemic inflammatory response
    syndrome 47, 65
Sequential Organ Failure Assessment
  (SOFA)
  mortality prediction 96
  multiple organ failure scoring 25, 26
  outcome studies 72
Simplified Acute Physiology Score (SAP)
  mortality prediction 94, 95
  outcome studies 68–72
Single-nucleotide polymorphisms (SNPs)
  abundance in human genome 76
  cardiopulmonary-bypass-associated acute
    kidney injury 82–87
  clinical significance 76, 77
  sepsis-associated acute kidney injury
    77–82
Single-pass albumin dialysis, blood
  detoxification 413
Slow extended daily dialysis (SLEDD)
  advantages 316, 323
  anticoagulation 322
  costs 323
  dialysate composition 321, 322
  dialysate and ultrafiltrate rates 321
  duration and timing 321
  hemodynamic tolerance 322, 323
  machinery 321
  mortality 323
  principles 320
  solute control 322
Sodium, replacement fluid composition
  292
Sodium/potassium ATPase, acute kidney
  injury pathophysiology 40
Spontaneous bacterial peritonitis (SBP),
  type 1 hepatorenal syndrome features
  pathophysiology 20, 21
  prevention of syndrome 21
  treatment
    liver transplantation 22
    vasoconstrictors and albumin 22, 23
Starches, fluid resuscitation in sepsis-
  associated acute kidney injury 171, 172

Stem cell, *see* Mesenchymal stem cell; Renal stem cell
Strong ion gap (SIG)
  etiology 163, 164
  interpretation 161–163
  strong ion difference calculation 160, 161
Survey, *see* Continuous renal replacement therapy; Diuretics, acute kidney injury management
Sustained low-efficiency dialysis, *see* Slow extended daily dialysis
Systemic inflammatory response syndrome (SIRS)
  high-volume hemofiltration 380, 381
  sepsis 47, 65

Terlipressin, hepatorenal syndrome management in spontaneous bacterial peritonitis 22, 23
Therapeutic intervention scoring system (TISS), mortality prediction 96
Thrombocytopenia
  causes in intensive care unit 260, 261
  heparin-induced thrombocytopenia, *see* Heparin-induced thrombocytopenia
Toll-like receptors (TLRs)
  acute kidney injury pathophysiology 42

TLR4 as immunodysregulation marker in sepsis 102, 104, 105, 107, 109
Transforming growth factor-β (TGF-β), immunodysregulation marker in sepsis 105, 106
Traumatic brain injury (TBI)
  acute kidney injury association 334
  management 334, 335
  renal replacement therapy effects on intracranial pressure and cerebral perfusion pressure
    anticoagulant effects 337
    continuous renal replacement therapy 336, 337
    intermittent hemodialysis 336
    peritoneal dialysis 335, 336
    sustained intracranial pressure surge management 337
Triggering receptor expressed on myeloid cells-1 (TREM-1)
  immunodysregulation marker in sepsis 102, 104, 106
  sepsis-induced renal injury biomarker 225
Tumor necrosis factor-α (TNF-α)
  inflammatory response 49, 50
  sepsis pathophysiology 59, 60
  sepsis-associated acute kidney injury single nucleotide polymorphisms 78, 79, 81